Stress Physiology of Tea in the Face of Climate Change

Wen-Yan Han • Xin Li • Golam Jalal Ahammed
Editors

Stress Physiology of Tea in the Face of Climate Change

Springer

Editors
Wen-Yan Han
Chinese Academy of Agricultural Sciences
Tea Research Institute
Hangzhou, PR China

Xin Li
Chinese Academy of Agricultural Sciences
Tea Research Institute
Hangzhou, PR China

Golam Jalal Ahammed
Department of Horticulture, College of Forestry
Henan University of Science and Technology
Luoyang, PR China

ISBN 978-981-13-2139-9 ISBN 978-981-13-2140-5 (eBook)
https://doi.org/10.1007/978-981-13-2140-5

Library of Congress Control Number: 2018959848

This Springer imprint is published by the registered company Springer Nature Singapore Pte Ltd.
The registered company address is: 152 Beach Road, #21-01/04 Gateway East, Singapore 189721, Singapore

This book is dedicated to the Tea Research Institute, Chinese Academy of Agricultural Sciences, for advancing tea research and industry in China.

Foreword

The world has witnessed a notable upsurge in climate change that is expected to continue at a greater pace in the coming decades with negative impact on the tea plant and the entire tea production and consumption system. Climate change involves not only rising temperatures and elevated CO_2 concentrations; it also involves extreme weather events including drought, heavy precipitation, and late spring frosts that are becoming more frequent and unpredictable. In addition, environmental stressors that accompany climate change including the outbreak of pests and diseases, UV radiation, ozone depletion, and nutrient deficiency are posing threats to the resilience of tea production systems. Therefore, it is critical to understand the effects of climate change on the growth and development of the tea plant and its ecosystem. This evidence is crucial for developing mitigation and adaptation strategies to cope with climate change for the sustainable development of the tea industry.

Tea is one of the most important cash crops worldwide, playing a significant role in rural development, poverty reduction, and food security in many developing countries. It is planted in 58 countries on 5 continents, with the majority of tea production countries being in Asia and Africa. The total area of land under tea cultivation is 4.37 million ha, with an annual production of 5.30 million tons in 2015. Smallholder farmers are the main stakeholders in tea production in many tea-producing areas, particularly in mountainous regions. It has been estimated that China and India have 20 million and 3 million rural laborers, respectively, involved in tea production and processing. Smallholders constitute 73%, 60%, and 47% of the total tea production in Sri Lanka, Kenya, and Indonesia, respectively. Tea also plays a vital role in relation to economic development in many tea-producing countries. For example, in Sri Lanka, tea generates 1.3 billion US dollars in exports, comprising 15% of the total export earnings or 60% of the agricultural export earnings. Tea exports contribute 20% to the total national foreign exchange earnings in Kenya.

Tea is an ideal model perennial plant for studying the effects of climate change on terrestrial plants due to its wide distribution, stable ecosystem, quality parameters, and long economic life. From a quality perspective, tea is rich in secondary metabolites, which account for 35–50% on dry weight basis; these tea constituents are sensitive to environmental variation and could thus serve as suitable indicators of climate change. Professor Wen-Yan Han and his team have focused on the study of climate change effects on the growth and development of tea plants, the soil carbon and nitrogen transformation, and the development of mitigation and adaptation technologies for the past few years. This effort by his team has resulted in the achievement of fruitful results regarding the effects of CO_2 concentration, temperature, and monsoon on the yield and quality of tea. In addition, his group has had success in using plant hormones including brassinosteroid and melatonin to improve the responses of tea plants to heat and cold stress. Professor Han has successfully implemented a National Key Research and Development project "Effects of Climate Change on Tea Agro-Ecosystems and Development of Key Technological Measures" (2017YFE0107500) from the Ministry of Science and Technology, People's Republic of China, and a seed project "Carbon-Neutral Tea Production in China: Developing Technology System" from the Food and Agriculture Organization of the United Nations. He also participated in the US Natural Science Foundation project "Climate Effects on Tea Quality and Social-economic Responses." *Science*, the frontline science magazine, reported their work with the title "Reading the tea leaves for effects of climate change" on May 29, 2015.

I am pleased to know that Springer, one of the leading publishers in the world, will host the publication and distribution of the book *Stress Physiology of Tea in the Face of Climate Change*, edited by Prof. Han, Dr. Li, and Dr. Ahammed. This book has fully reviewed the climate change effect on tea plant physiology, biochemistry, and metabolism as well as economic and societal aspects of the tea industry, with major foci on recent advances in the physiological and molecular mechanisms of the responses of tea plants to various abiotic and biotic stresses. I believe that this book could provide a solid foundation for understanding the stress physiology of tea plant and developing stress management strategies in tea ecosystem in the face of climate change. I sincerely congratulate Prof. Han and his team for the publication of this book, and I wish them a great success in the future regarding their work on climate change and tea toward a sustainable future for the tea industry.

Tea Research Institute
Chinese Academy of Agricultural Sciences,
Hangzhou, Zhejiang, PR China
May 8, 2018

Zongmao Chen

Preface

Climate change notably impacts global crop production as well as food security through influencing carbon sequestration, soil fertility, plant physiology, insect herbivory, and plant-pathogen interactions that eventually alter plant growth and productivity. These changes are linked to increasing frequency and intensity of extreme weather events such as heat waves, drought, heavy precipitation, and frost. As a woody perennial cultivated in a rain-fed mono-cropping system, the sustainability of tea (*Camellia sinensis* (L.) O. Kuntze) largely depends on weather conditions and is thus vulnerable to climate change. Tea is primarily cultivated in Asia and Africa and is the most widely consumed nonalcoholic beverage in the world after water. Most often, a bud and the adjacent two young leaves of tea plants are used to manufacture tea. In recent years, the popularity of green tea is rising across the world due to its pleasant flavor and numerous health benefits such as anticancer, anti-inflammatory, anti-allergic, and anti-obesity effects on human. To meet the increasing demand, tea cultivation has been expanding in many countries in tropical and subtropical regions including China, the world's largest producer of tea. The pleasant taste and health benefits of tea are due to the presence of various bioactive compounds. The *in planta* synthesis of these bioactive compounds, including their presence and concentration, is greatly influenced by multiple environmental and endogenous signals. Unlike annual crops that are planted each year, tea plants grow for multiple years (up to a century in some cases) and thus face seasonal variation and multiple abiotic and biotic stressors year after year that affect both yield and quality of tea. It is speculated that tea plants will respond to climate change through various physiological adaptations due to their long life span. Therefore, it is critical to understand the stress physiology of tea for sustaining tea production in the face of climate change.

Our goal for this book is to provide a comprehensive overview of stress-induced changes in tea physiology, biochemistry, and quality as well as physiological and molecular mechanisms of stress tolerance and associated stress management strategies. In addition, the current status of the global tea industry and future challenges are discussed in the context of climate change. This book is composed of a total of 15 chapters, each of which is dedicated to various aspects of tea production relating to climate change. There are individual chapters dedicated to address specific stressors, including high temperature, low temperature, drought, low light stress, ultraviolet radiation, elevated CO_2, ozone, nutrient deficiency, insect herbivory, and pathogenic agents in tea plants. One chapter is dedicated to the physiological and molecular mechanisms controlling phytohormone-mediated response of tea plants to abiotic stresses and hormone crosstalk. Genetic engineering aspects for improving biotic and abiotic stress tolerance in tea plants along with major challenges and future research directions are suggested in a chapter on genomics approaches in tea. Another chapter highlights the changes in tea antioxidants, a parameter of tea quality, in response to various environmental factors. While the first chapter provides a review on tea ecophysiology and socioeconomic implications of climate change, the last chapter covers important agricultural, physiological, and molecular innovations for tea at the production level toward the development of climate-resilient tea systems. Much attention has been paid to adhere with the focus of each chapter that enabled the authors to avoid repetition of similar topics. However, a few specific topics may reappear in different chapters to better situate the context for each chapter. It is worth mentioning that all authors of this book have recently contributed original research articles to the relevant area of tea research.

Therefore, we believe that this book will provide a comprehensive overview regarding the latest developments in tea research that cover physiological and molecular basis of plant responses to changing climate from the cellular to the systems level. We expect that this book will be of interest to a broad audience, from graduate students to senior researchers in a wide range of disciplines including tea science, plant physiology, plant biochemistry, agriculture, and environmental assessment. It will also be of practical interest to environmentalists, policy-makers, and resource managers working on topics of climate effects on natural resources. Although minor editorial changes were adopted, the author content was kept intact in each chapter. However, some errors may still exist in the book, and thus we would greatly appreciate readers' feedback for potential improvement in future editions. This work is supported by the National Key R&D Program of China (2017YFE0107500) and the Science and Technology Innovation Project of the Chinese Academy of Agricultural Sciences (CAAS-ASTIP-2015-TRICAAS-08). We wish to thank all the authors who joined this book project by contributing their valuable work. Lastly, we extend our sincere thanks to Springer Nature Singapore Pte. Ltd., especially Dr. Zhu Yu, Associate Editor, Life Sciences, Beijing Office; Mr. Rameshbabu Rathinam, Project Coordinator (Books); and all the other staff members of Springer Nature involved in this book project for their generous cooperation.

Hangzhou, PR China — Wen-Yan Han
Hangzhou, PR China — Xin Li
Luoyang, PR China — Golam Jalal Ahammed

Contents

Contributors

Golam Jalal Ahammed Department of Horticulture, College of Forestry, Henan University of Science and Technology, Luoyang, PR China

Selena Ahmed Food and Health Lab, Sustainable Food and Bioenergy Systems Program, Montana State University, Bozeman, MT, USA

Sangeeta Borchetia Department of Biotechnology, Tocklai Tea Research Institute, Tea Research Association, Jorhat, Assam, India

Sean B. Cash Friedman School of Nutrition and Policy, Tufts University, Boston, MA, USA

Lorenzo Cotrozzi Department of Agriculture, Food and Environment, University of Pisa, Pisa, Italy

Xin Cui Tea Science Research Institute, College of Horticulture, Nanjing Agricultural University, Nanjing, China

Murat Dikilitas Department of Plant Protection, Faculty of Agriculture, Harran University, Şanlıurfa, Turkey

ZhaoTang Ding College of Horticulture, Qingdao Agricultural University, Qingdao, Shandong, China

Natthawuddhi Donlao Graduate School of Horticulture, Chiba University, Matsudo, Japan

School of Agro-Industry, Mae Fah Luang University, Chiang Rai, Thailand

Fang-yuan Fan Zhejiang University Tea Research Institute, Hangzhou, China

Tim Griffin Friedman School of Nutrition and Policy, Tufts University, Boston, MA, USA

Roghieh Hajiboland Plant Science Department, University of Tabriz, Tabriz, Iran

Wen-Yan Han Chinese Academy of Agricultural Sciences, Tea Research Institute, Hangzhou, PR China

Gautam Handique Advisory Department, Tocklai Tea Research Institute, Tea Research Association, Jorhat, Assam, India

Xinyuan Hao National Center for Tea Improvement, Tea Research Institute, Chinese Academy of Agricultural Sciences; Key Laboratory of Tea Biology and Resources Utilization, Ministry of Agriculture, Hangzhou, China

Jianhui Hu College of Horticulture, Qingdao Agricultural University, Qingdao, Shandong, China

Aziz Karakaya Department of Plant Protection, Faculty of Agriculture, Ankara University, Ankara, Turkey

Lovedeep Kaur Riddet Institute and Massey Institute of Food Science and Technology, Massey University, Palmerston North, New Zealand

Marco Landi Department of Agriculture, Food and Environment, University of Pisa, Pisa, Italy

Chun-lin Li Zhejiang University Tea Research Institute, Hangzhou, China

Hui Li Tea Science Research Institute, College of Horticulture, Nanjing Agricultural University, Nanjing, China

Xin Li Chinese Academy of Agricultural Sciences, Tea Research Institute, Hangzhou, PR China

Zhi-Wei Liu Tea Science Research Institute, College of Horticulture, Nanjing Agricultural University, Nanjing, China

Chunlin Long College of Life and Environmental Sciences, Minzu University of China, Beijing, China

Giacomo Lorenzini Department of Agriculture, Food and Environment, University of Pisa, Pisa, Italy

Zhou Luo Zhejiang University Tea Research Institute, Hangzhou, China

Corene Matyas Department of Geography, University of Florida, Gainesville, FL, USA

Cristina Nali Department of Agriculture, Food and Environment, University of Pisa, Pisa, Italy

Colin M. Orians Department of Biology, Tufts University, Medford, MA, USA

Elisa Pellegrini Department of Agriculture, Food and Environment, University of Pisa, Pisa, Italy

Wenjun Qian College of Horticulture, Qingdao Agricultural University, Qingdao, Shandong, China

Albert Robbat Department of Chemistry, Tufts University, Medford, MA, USA

Somnath Roy Department of Entomology, Tocklai Tea Research Institute, Tea Research Association, Jorhat, Assam, India

Eric R. Scott Department of Biology, Tufts University, Medford, MA, USA

John Richard Stepp Department of Anthropology, University of Florida, Gainesville, FL, USA

Gui-zhen Tang Zhejiang University Tea Research Institute, Hangzhou, China

Lu Wang National Center for Tea Improvement, Tea Research Institute, Chinese Academy of Agricultural Sciences; Key Laboratory of Tea Biology and Resources Utilization, Ministry of Agriculture, Hangzhou, China

Wen-Li Wang Tea Science Research Institute, College of Horticulture, Nanjing Agricultural University, Nanjing, China

Xinchao Wang National Center for Tea Improvement, Tea Research Institute, Chinese Academy of Agricultural Sciences; Key Laboratory of Tea Biology and Resources Utilization, Ministry of Agriculture, Hangzhou, China

Yong-Xin Wang Tea Science Research Institute, College of Horticulture, Nanjing Agricultural University, Nanjing, China

Yu Wang College of Horticulture, Qingdao Agricultural University, Qingdao, Shandong, China

Shabir Hussain Wani Mountain Research Centre for Field Crops, Sher-e-Kashmir University of Agricultural Sciences and Technology of Kashmir, Jammu, Jammu and Kashmir, India

Department of Plant, Soil and Microbial Sciences, Michigan State University, East Lansing, MI, USA

Dayuan Xue College of Life and Environmental Sciences, Minzu University of China, Beijing, China

Peng Yan Key Laboratory of Tea Quality and Safety Control, Ministry of Agriculture, Tea Research Institute, Chinese Academy of Agricultural Sciences, Hangzhou, People's Republic of China

Yajun Yang National Center for Tea Improvement, Tea Research Institute, Chinese Academy of Agricultural Sciences; Key Laboratory of Tea Biology and Resources Utilization, Ministry of Agriculture, Hangzhou, China

Jianming Zeng National Center for Tea Improvement, Tea Research Institute, Chinese Academy of Agricultural Sciences; Key Laboratory of Tea Biology and Resources Utilization, Ministry of Agriculture, Hangzhou, China

Lan Zhang Key Laboratory of Tea Quality and Safety Control, Ministry of Agriculture, Tea Research Institute, Chinese Academy of Agricultural Sciences, Hangzhou, People's Republic of China

Liping Zhang Key Laboratory of Tea Quality and Safety Control, Ministry of Agriculture, Tea Research Institute, Chinese Academy of Agricultural Sciences, Hangzhou, People's Republic of China

Xinfu Zhang College of Horticulture, Qingdao Agricultural University, Qingdao, Shandong, China

Lei Zhao College of Horticulture, Qingdao Agricultural University, Qingdao, Shandong, China

Jing Zhuang Tea Science Research Institute, College of Horticulture, Nanjing Agricultural University, Nanjing, China

About the Editors

Dr. Wen-Yan Han is a Professor at the Tea Research Institute, Chinese Academy of Agricultural Sciences (TRI CAAS), Hangzhou, China. Dr. Han is the Chief Scientist of the Science and Technology Innovation Team of Tea Cultivation in CAAS. He obtained his Ph.D. in Soil Science from Zhejiang University. He has been actively involved in tea research since 1984. His fields of specialization include soil carbon and nitrogen transformation, tea physiology, plant nutrition, organic tea, abiotic stress, and climate change effect on tea ecosystem. Dr. Han is one of the dedicated tea scientists in China. He has been involved in developing intergovernmental strategies to tackle climate change in tea system as a Co-Chair of the Climate Change Working Group in Food and Agriculture Organization (FAO) Intergovernmental Group on Tea. Dr. Han was awarded a number of grants and honors such as Outstanding Contributor by Zhejiang Provincial Government and Albert Nelson Marquis Lifetime Achievement Award. He published more than 160 research articles in national and international peer-reviewed journals.

Dr. Xin Li is an Associate Professor at the Tea Research Institute, Chinese Academy of Agricultural Sciences, Hangzhou, China. He received his B.Sc. in Horticulture in 2009 from Hebei Agricultural University, Baoding, China. Dr. Li earned his Ph.D. in Olericulture in 2014 from Zhejiang University, China. Then he joined the Tea Research Institute, Chinese Academy of Agricultural Sciences, Hangzhou, and got involved in tea research. His major research interests include tea stress physiology, plant hormones and growth regulators, and climate change effect on tea plants. He authored over 40 research papers in peer-reviewed journals. Dr. Li is a regular reviewer of some prestigious journals. Dr. Li was awarded several grants including one by the National Natural Science Foundation of China. Currently, Dr. Li is actively engaged in unveiling the effect of elevated CO_2 on tea quality.

Dr. Golam Jalal Ahammed is an Associate Professor at the Department of Horticulture, College of Forestry, Henan University of Science and Technology, Luoyang, China. He obtained his B.Sc. in Agriculture in 2004 and M.S. in Horticulture in 2006 from Bangladesh Agricultural University, Mymensingh, Bangladesh. Dr. Ahammed received his Ph.D. in Olericulture with major focus on Plant Stress Physiology and Hormonal Regulation in 2012 from Zhejiang University, China. Afterward he completed two consecutive postdoctoral tenure at the Institute of Pesticide and Environmental Toxicology and Institute of Crop Science of Zhejiang University. Dr. Ahammed has been involved in tea research since 2015 in collaboration with the Key Laboratory of Tea Biology and Resources Utilization, Tea Research Institute of Chinese Academy of Agricultural Sciences, Hangzhou, China. His major research interests include plant stress physiology, phytohormones, climate change effect on plants, and environmental pollution. Dr. Ahammed authored over 60 research papers in peer-reviewed journals. He is an Editorial Board Member of *Scientific Reports*. Dr. Ahammed was awarded an International Young Scientist Fellowship by the National Natural Science Foundation of China. He was also a recipient of several grants from China Postdoctoral Science Foundation. Currently, Dr. Ahammed is actively involved in unraveling the role of melatonin in plant stress response.

Chapter 1
Global Climate Change, Ecological Stress, and Tea Production

Selena Ahmed, Tim Griffin, Sean B. Cash, Wen-Yan Han, Corene Matyas, Chunlin Long, Colin M. Orians, John Richard Stepp, Albert Robbat, and Dayuan Xue

Abstract Tea is the second most consumed beverage worldwide, and the production of tea is of economic importance in over 50 countries. As a woody perennial, tea plants are cultivated in production systems for numerous decades and thus experience the multiple decadal effects of climate change including influences on tea yields and quality. Changes in tea yields and quality can have notable impacts on the livelihoods of laborers and farmers as well as on regional economies more broadly. This chapter provides an overview on the effects of global climate change on tea production. We start with a review on global climate change trends that highlight robust multi-decadal warming and changes in extreme weather events that have increased. This review on climate change trends is followed with a synopsis of the major effects of climate change on agriculture broadly as well as on tea plants more

S. Ahmed (✉)
Food and Health Lab, Sustainable Food and Bioenergy Systems Program,
Montana State University, Bozeman, MT, USA
e-mail: selena.ahmed@montana.edu

T. Griffin · S. B. Cash
Friedman School of Nutrition and Policy, Tufts University, Boston, MA, USA

W.-Y. Han
Chinese Academy of Agricultural Sciences, Tea Research Institute, Hangzhou, PR China
e-mail: hanwy@tricaas.com

C. Matyas
Department of Geography, University of Florida, Gainesville, FL, USA

C. Long · D. Xue
College of Life and Environmental Sciences, Minzu University of China, Beijing, China

C. M. Orians
Department of Biology, Tufts University, Medford, MA, USA

J. R. Stepp
Department of Anthropology, University of Florida, Gainesville, FL, USA

A. Robbat
Department of Chemistry, Tufts University, Medford, MA, USA

W.-Y. Han et al. (eds.), *Stress Physiology of Tea in the Face of Climate Change*,
https://doi.org/10.1007/978-981-13-2140-5_1

specifically. We provide a review on tea ecophysiology and thresholds followed by a synthesis on the key abiotic and biotic stressors associated with climate change that impact tea plants including carbon dioxide concentrations, temperature, rainfall, humidity, solar radiation, wind, soil conditions and microorganisms, pests, and pathogens. This chapter ends with a discussion on socioeconomic implications of climate change in major tea-producing areas globally in order to highlight the need to better understand tea physiology in the context of climate change. Overall, the scientific literature and news reports highlight that climate change is already impacting tea systems and that expected climate changes can increase the likelihood of severe and irreversible impacts for tea production and associated livelihoods. Evidence-based adaptation and mitigation strategies through a community-based approach are called for toward a more sustainable tea sector.

Keywords Abiotic stress · Environmental variation · Crop thresholds · Tea quality · Secondary metabolites

1 Introduction

Tea (*Camellia sinensis* (L.) Kuntze) is among the most important beverage crops globally and is vulnerable to climate change. Successful cultivation of tea is highly dependent on agroclimatic conditions including specific thresholds of temperature, rainfall, humidity, and solar radiation that influence crop yields and quality (Wijeratne 1996). These agroclimatic conditions are experiencing variability in tea-producing areas globally with climate change, thus presenting ecological stress and constraints, as well as opportunities, for tea production. The past six decades have experienced a trend of gradual systematic changes in average climate conditions including warming and increased interannual variability of temperatures and precipitation (IPCC 2014) as well as more extreme weather conditions (Ewert et al. 2005). The aforementioned climate trends are linked to an increase of anthropogenic greenhouse gas emissions (IPCC 2014) and are recognized to influence agricultural systems including tea production (Chang and Bratloff 2015). As woody perennials, tea plants are cultivated in production systems for numerous decades and thus experience the multiple decadal effects of climate change including influences on tea quality and yields.

Climate variables have been shown to influence tea yields due to shifts in rainfall and solar radiation (Boehm et al. 2016) in tea-growing countries throughout Asia and Africa (Chang and Bratloff 2015). Tea yield is greatly influenced by droughts which can cause irreparable yield losses (Wijeratne 1996; Ahmed et al. 2014a, b). Heavy rains on the other hand can cause water logging, erode top soil, and cause extensive leaching of critical soil nutrients (Wijeratne 1996). Multiple dimensions of tea quality, including phytochemical profiles and organoleptic properties (color, visual appeal, aroma, taste, texture), have also been demonstrated to vary with climate change (Ahmed et al. 2014a, b; Kowalsick et al. 2014). As a specialty crop consumed for its flavor, cultural, physiological, and health attributes versus calories, tea quality is particularly important (Ahmed and Stepp 2016). Tea quality is largely determined by nutrient, secondary metabolite, and mineral profiles via their effects on flavor profiles and their physiological effects on human consumers (Ahmed and Stepp 2016). Secondary metabolites are plant defense compounds that vary in concentration depending on genetic (van Dam and Vrieling 1994), environmental (Björkman et al. 2011; Lower and Orians 2003), and management (Ahmed et al. 2013) conditions. In addition, climate change has been shown to shift the geographic range for high-quality tea (Ethical Tea Partnership 2015). Ultimately, climate effects on tea yields, quality, and geographic range have direct and indirect effects on prices, consumer demand, farmer livelihoods, regional economies, and the physiological implications of tea consumption.

Tea is among the most economically valued crops in areas in which it is cultivated, with a noteworthy place in the economy of tea-producing countries (Chang and Bratloff 2015). Over 50 countries on 5 continents grow tea, with over three million hectares of total land under tea cultivation producing almost five million tonnes annually (FAOSTAT 2016). Tea is a major foreign currency earning commodity in most countries where it is produced (Wijeratne 1996; Mukhopadhyay and Mondal 2017). It is the primary income for millions of laborers, both male and female, in many tea-producing countries and provides economic opportunity to laborers in remote and economically depressed rural areas (Chang and Bratloff 2015). In addition, for smallholder tea farmers, tea production can be the primary source of livelihood (Ahmed et al. 2010). Thus, any changes in tea yields and quality have notable impacts on the livelihoods of laborers and farmers and regional economies more broadly. As the most widely consumed beverage in the world after water, changes in tea yields and quality influence buying decisions and consumption practices of its consumers (Boehm et al. 2016), as well as the health implications associated with this consumption (Ahmed et al. 2014a, b).

This chapter provides an overview on the effects of global climate change on tea production. We start with a review on global climate change trends. This is followed with a synopsis of the major effects of climate change on crop physiology broadly as well as on tea plants more specifically. We provide a review on tea ecophysiology, thresholds, and ecological stressors in order to understand the vulnerability of tea plants and tea quality to specific climate factors. This chapter ends with a discussion on socioeconomic implications of climate change in major tea-producing areas globally in order to highlight the need to better understand tea physiology in the context of climate change.

2 Global Climate Change Trends

The climate system is comprised of interrelated variables including temperature, precipitation, and rainfall where a change in one variable through shifts in regulatory mechanisms often triggers changes in other variables (Sombroek and Gommes

1996). Global climate change is influencing the climate system and interactions among climate variables in multiple ways. Here we summarize key global climate change trends as synthesized by the Intergovernmental Panel on Climate Change's (IPCC) "Climate Change IPCC 2014: Synthesis Report," from the fifth and most recent round of IPCC assessment reports.

The past decade has experienced robust multi-decadal warming with an increase of the globally averaged combined land and ocean surface temperature of 0.85 °C over the period 1880–2012 (IPCC 2014). The dominant cause of the observed warming since the mid-twentieth century is most likely an increase in anthropogenic greenhouse gas emissions that have increased since the pre-industrial era due to population growth and increased economic activity coupled with changes in lifestyle, energy use, land use patterns, technology, and climate policy (IPCC 2014). Anthropogenic greenhouse gas emissions since the pre-industrial era are responsible for notably heightened levels of atmospheric concentrations of carbon dioxide (CO_2), methane (CH_4), and nitrous oxide (N_2O) (IPCC 2014). Increases in greenhouse gas emissions have affected the global water cycle since 1960 including driving the retreat of glaciers, causing increased surface melting of the Greenland ice sheet, and contributing to Arctic sea ice loss that have very likely influenced the increases in global upper ocean heat content and the global rise in mean sea levels (IPCC 2014).

Global climate change is associated with changes in extreme weather climate events including heat waves, droughts, floods, cyclones, and wild fires that have increased since the 1950s (IPCC 2014). This includes a decrease in cold temperature extremes, an increase in warm temperature extremes, an increase in extreme high sea levels, and an increase in the number of heavy precipitation events (IPCC 2014). On a global scale, the number of cold days and nights has decreased, while the number of warm days and nights has increased (IPCC 2014). In addition, the frequency of heat waves has likely increased in large parts of Asia including doubling the probability of the occurrence of heat waves in some locations (IPCC 2014). Increased areas of land have experienced heightened number of heavy precipitation events compared to areas where the number of heavy precipitation events has decreased (IPCC 2014).

Changes in climate have influenced natural and human systems on all continents across the globe including agricultural systems and crops. For example, changing precipitation is altering hydrological systems and influencing water resources in terms of both quantity and quality (IPCC 2014). Climate change has been shown to directly influence terrestrial species through shifts in their geographic ranges, seasonal activities, migration patterns, abundances, and species interactions (IPCC 2014). Continued increase in greenhouse gas emissions is likely to cause continued warming including increased surface temperatures and long-lasting changes in all components of the climate system. Changes in precipitation will vary globally depending on latitude (IPCC 2014). High latitude areas are expected to experience an increase in annual mean precipitation based on some climate scenarios. Many mid-latitude and subtropical dry regions are expected to experience decreased mean precipitation, while many mid-latitude wet regions are expected to experience increased mean precipitation (IPCC 2014). Most mid-latitude land masses and wet tropical regions will experience increased, more intense, and more frequent extreme precipitation events (IPCC 2014). These expected climate changes will increase the likelihood of severe and irreversible impacts for ecosystems and humans without the implementation of adaptation and mitigation strategies (IPCC 2014). As tea plants grow in many mid-latitude, tropical, and subtropical regions, these climate effects are likely to impact tea production with notable impacts for livelihoods and associated economies.

3 Climate Change Effects on Agriculture and Tea Physiology

Climate change is impacting crops in agricultural systems through multiple mechanisms. Crop physiology is directly impacted by shifts in the greenhouse gas CO_2, tropospheric O_3 levels, ultraviolet radiation (UV-B), temperature, and the hydrological cycle associated with climate change (Sombroek and Gommes 1996). This section reviews the effects of abiotic and biotic variables on crop physiology broadly and follows with specific effects of key abiotic and biotic environmental factors on tea physiological processes. Our discussion includes the specific thresholds of these variables to which tea is adapted.

The primary physiological processes responsible for tea growth are determined by a range of environmental and genotypic factors including the climate to which a particular tea genotype is adapted (De Costa et al. 2007). Tea plants are woody perennials that are found primarily in rain-fed monocropping terraced landscapes around the world in tropical, subtropical, and temperate regions at altitudes between sea level to 3000 m and latitudes between 30° S and 42° N that fit the specific ecophysiological requirements of tea cultivation including thresholds of temperature, rainfall, humidity, solar radiation (Fig. 1.1; Williges 2004; Ahmed and Stepp 2012; Chang and Bratloff 2015). The most southern range of tea is in South Africa, and the most northern range is in Russia (Caffin et al. 2004). Tea plants can grow profusely in the equatorial regions up to 3000 m; the optimum altitude for tea cultivation drops as the distance from the equator increases (Willson 1999). Injury from frost increases with an increase in altitude (Zhou and Chen 1985). Variation of environmental conditions, including both abiotic factors and biotic factors, above or below tea's thresholds present ecological stress to tea plants with constraints for growth and consequences for tea yields and quality.

The key ecophysiological processes intimately linked to tea yield include photosynthesis, respiration, transpiration, stomatal conductance, shoot initiation, and plant-water relations and responses (De Costa et al. 2007). Tea plants demonstrate the C_3 mechanism of photosynthesis (Roberts 2015) with this physiological process primarily occurring in tea leaves and some CO_2 assimilation occurring in the stems (De Costa et al. 2007). The external abiotic environmental factors that influence tea physiological processes include altitude and latitude, carbon dioxide concentrations, temperature, rainfall, humidity, solar radiation, wind, and soil conditions

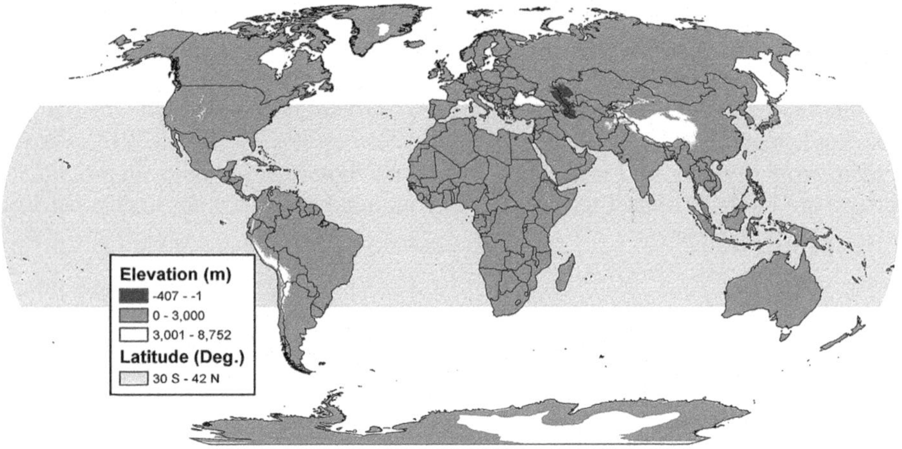

Fig. 1.1 Latitude belt (30°S – 42°N) and elevation (0–3000 m) ideal for tea growth. Global digital elevation model data obtained from ASTER GDEM, which is a product of METI and NASA. (URL: http://www.webgis.com/terr_world.html)

(Ahmed et al. 2015). The biotic environmental factors that influence these tea physiological processes include arthropod pests, disease pests, pollinators, cultivated plants (and associated plant hormones), and soil microorganisms (Ahmed et al. 2015). Internal plant processes at the cellular level including leaf dormancy, photoinhibition, biomass production, biomass partitioning, and transpiration efficiency interact with key tea ecophysiological processes and environmental factors to influence tea leaf growth (De Costa et al. 2007).

The degree of impact of abiotic and biotic factors on tea plants further depends on the variety, stage of growth, plant age, maturity of the branches, maturity of leaves, and plucking intensity. Young tea plants are generally more vulnerable to environmental factors compared to older tea plants. As a consequence, young tea plants are often raised in controlled and protected nursery conditions for their first 2–4 years while they are immature (Ahmed and Stepp 2012). Once mature, tea plants are transplanted to the field where abiotic and biotic factors impact tea production.

Carbon Dioxide The photosynthetic assimilation of carbon dioxide (CO_2) is central to the metabolism of plants with the rise of atmospheric CO_2 concentrations impacting plant growth, physiology, and chemistry (Taub 2010). Plants uptake atmospheric CO_2 during the process of photosynthesis and chemically reduce the carbon as an acquisition of stored chemical energy (Taub 2010). This stored chemical energy provides the carbon skeletons for the organic molecules that make up a plants' structure along with hydrogen and oxygen assimilated into organic molecules by photosynthesis (Taub 2010).

Rising atmospheric CO_2 concentrations that are associated with climate change have been shown to notably influence plant growth and responses to biotic and abiotic stresses (Unsworth and Hogsett 1996) including stimulating photosynthesis and enhancing respiration in plants and consequently resulting in increased yield (Li et al. 2017). Changes in CO_2 levels act on plants at the leaf level to impact the rate of net photosynthesis and stomatal conductance (Unsworth and Hogsett 1996). Stomatal conductance is the regulation of the degree of stomatal opening to allow CO_2 to diffuse into leaves for photosynthesis while preventing water to diffuse out of leaves in order to maintain high rates of photosynthesis and low rates of water loss (Taub 2010). Elevated CO_2 results in an increase in the rate of net photosynthesis, in a phenomena known as the CO_2 fertilization effect, with decreased stomatal conductance if other conditions remain optimal (Unsworth and Hogsett 1996; Taub 2010). Increased concentrations of CO_2 due to increased carbon emissions worldwide can thus have a positive influence on plant photosynthesis under optimal growing conditions of light, temperature, nutrients, and moisture (Sombroek and Gommes 1996). Between 10% and 20% of the approximate doubling of crop productivity in the past century is attributed to the CO_2 fertilization effect (Tans et al. 1990). However, the beneficial aspects of the CO_2 fertilization effect are only experienced if all other growth conditions are optimal such as adequate water; with climate change, these optimal conditions are experiencing variability and interact with elevated CO_2 levels in complex ways (Unsworth and Hogsett 1996). Overall, the effects of elevated CO_2 levels on crops depend on other environmental factors (Taub 2010).

A variety of free-air carbon dioxide enrichment (FACE) experiments have shown that elevated CO_2 levels decrease stomatal conductance of water by an average of 22% (Ainsworth and Rogers 2007) and decreases whole plant water use by 5–20% (Taub 2010). These changes in water use by plants can have consequences for the hydrological cycle of entire ecosystems including increased soil moisture levels and runoff under elevated CO_2 (Leakey et al. 2009). The CO_2 fertilization effect depresses photorespiration more for C_3 plants (such as the tea plant) compared to C_4 plants and leads to higher biomass given all other growth conditions being optimal.

In addition to impacting plant growth, elevated CO_2 levels also influence the chemical composition of plant tissues. FACE experiments of elevated CO_2 levels have shown increased photosynthetic activity to be associated with a 30–40% increase in leaf nonstructural carbohydrates including sugars and starches per unit leaf area, while leaf nitrogen concentrations typically decrease on average by 13% (Ainsworth and Long 2005). This decrease in nitrogen concentration in plant tissue is likely driven by multiple factors including (1) dilution of nitrogen from increased carbohydrate concentrations (Taub 2010), (2) decreased uptake of minerals from the soil due to decreased stomatal conductance (Taub and Wang 2008), and (3) decreased rate of assimilation of nitrate into organic compounds (Bloom et al. 2010). Changes in the nitrogen status of plants are linked to protein concentrations in plant tissues (Taub 2010) and can lead to increased consumption of plant tissues as herbivores compensate for decreased food quality (Stiling and Cornelissen 2007; Gliessman 2015). Concentrations of minerals important for human nutrition and food security including calcium, magnesium, and phosphorus may also be decreased under elevated CO_2 (Gilani and Nasim 2007; Taub and Wang 2008; Myers et. al 2014).

Atmospheric ozone (O_3) is another greenhouse gas that is increasing in concentrations with climate change and pollution (Vingarzan 2004). High concentrations of ozone have been shown to be detrimental for crops by causing plants to close their stomata and thus slowing down photosynthesis and plant growth (Feng et al. 2008; Morgan et al. 2003). However, increased ozone concentrations are primarily having detrimental effects near major cities.

Carbon Dioxide Effects on Tea Tea productivity directly responds to increasing carbon dioxide concentrations (Smith et al. 1993). For example, a linear correlation was found between net photosynthetic rate and carbon dioxide concentrations in a chamber experiment (Smith et al. 1993). Likewise, previous research found that the maximum net photosynthetic rate of tea plants can be raised temporarily by artificial carbon dioxide enrichment up to a level (Anandacoomaraswamy et al. 2000). The specific growth response of tea plants to carbon dioxide concentrations varies with agricultural management factors including shade levels and nitrogen levels (De Costa et al. 2007). In addition, carbon dioxide levels can mitigate the effects of water stress on tea plants (Anandacoomaraswamy et al. 2000). The net photosynthetic rate of water-stressed tea leaves was raised by artificial carbon dioxide enrichment in a chamber experiment (Anandacoomaraswamy et al. 2000).

Exposure of tea seedlings to elevated CO_2 (800 μmol mol-1 for 24 day) in controlled environment growth chambers was found to improve both photosynthesis and respiration in tea leaves and result in enhanced plant growth and biomass accumulation (Li et al. 2017). In addition, elevated CO_2 was found to increase the concentrations of key constituents in tea leaves including soluble sugar, starch, and total carbon while resulting in a decrease in total nitrogen concentration and an increased carbon to nitrogen ratio (Li et al. 2017). Further, tea quality was altered with elevated CO_2; specifically, there was an increase in concentrations of total catechins and other polyphenols, free amino acids, and theanine, while caffeine decreased. These chemical findings are in sync with real-time qPCR analysis that revealed that the expression levels of biosynthetic genes for catechins and theanine were upregulated in tea leaves under elevated CO_2 conditions. Concurrently, the expression levels of biosynthetic genes for caffeine were downregulated. Another study found that elevated CO_2 levels can notably decrease the negative effects of high ozone on photosynthesis because of the decreased openness of leaf stomata under elevated CO_2 and thus decrease ozone exposure of sensitive leaf tissues (Taub 2010).

Temperature Temperature is a primary factor that affects the rate of plant photosynthesis, metabolism, and growth (Hatfield and Prueger 2015). Each species has a specific temperature range represented by a minimum, maximum, and optimum (Hatfield and Prueger 2015). Temperatures that fall below or above these specific temperature thresholds can notably impact plant productivity (Hatfield and Prueger 2015). The effect of temperature changes on plant responses depends on the specific stage of development of plants during their life cycle (Hatfield and Prueger 2015). Typically, plant vegetative development increases with a rise of temperatures rise to the optimum level for a specific species (Hatfield and Prueger 2015). The stage of vegetative development typically has a higher optimum temperature compared to the reproductive development stage (Hatfield and Prueger 2015). At the same time, extreme high temperatures during the reproductive stage of plants influence pollen viability, fertilization, and grain or fruit formation; acute exposure of plants to extreme events may be most detrimental during the reproductive stages of development (Hatfield and Prueger 2015). Plants are further vulnerable to chronic exposures to extreme temperatures during their pollination stage with notable consequences for yield (Hatfield and Prueger 2015).

Higher temperatures are associated with higher solar radiation and water use by crops that effect both ecological processes within agricultural systems as well as physiological processes at the plant and plant organ levels. Specifically, changes in surface temperature directly impact evaporation, atmospheric moisture, and precipitation (Sombroek and Gommes 1996). Rising temperatures are diminishing the yields of many crops (Lobell and Burke 2010), and the expectation of increased occurrence of extreme temperature events will have increasing negative impacts on plant production (Hatfield and Prueger 2015). On the other hand, rising temperatures could have a positive effect on the growth of other plants such as CAM plants and those at high latitudes and altitudes. Increased temperatures during the cold season can lead to earlier bloom and ripening of crops as well as a lengthening of the growing season. At the same time, cool season species can have a constrained growing season with increased temperatures because of average temperatures exceeding their temperature thresholds (Hatfield and Prueger 2015).

Higher temperatures can strengthen the effect of the CO_2 fertilization effect unless temperatures reach levels above plant thresholds and plants get overheated (Sombroek and Gommes 1996). Winter kill of pests and pathogens is also being compromised with higher temperatures and calling for heightened pest control. The effects of temperature changes on crops are exacerbated by water interactions including water deficits and excess soil water (Hatfield and Prueger 2015). These interactions point to the need to understanding the interaction of temperature and water in order to develop more effective adaptation strategies to mitigate climate risk.

Temperature Effects on Tea Growth of tea is highly temperature dependent with ecophysiological constraints when temperatures are either too low or too high, regardless of other climatic factors (Lemmesa 1996). Temperature directly influences the photosynthetic rate, shoot growth, and rate of shoot initiation of tea plants (De Costa et al. 2007). Many tea-growing regions have a climate similar to the cool tropical conditions under which tea originated (Kingdon-Ward 1950). Newer tea-growing regions range from a Mediterranean climate to colder climates in the Northern and Southern latitudes including Russia and Georgia, respectively (Carr 1972).

Cultivation of tea is possible in areas that have annual mean temperatures between 12 and 30°C (Fig. 1.2; Williges 2004). Atmospheric temperatures below and above this range have been found to reduce tea shoot growth (Carr 1972). The

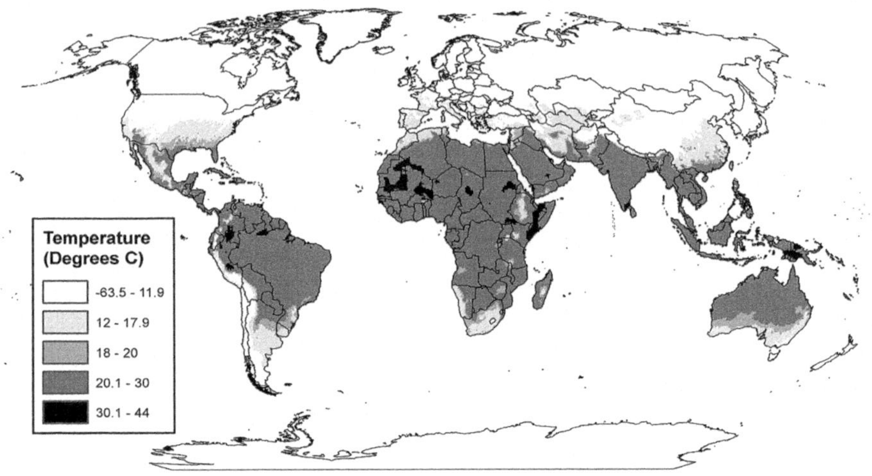

Fig. 1.2 Average annual surface air temperature 1981–2010. Optimum temperature for tea growth is 18–20 ° C, acceptable range is 12–30 °C. (Global Historical Climatology Network Gridded V2 data provided by the NOAA/OAR/ESRL PSD, Boulder, Colorado, USA, from their Web site at http://www.esrl.noaa.gov/psd/. *Citation*: Fan, Y., and H. van den Dool (2008), A global monthly land surface air temperature analysis for 1948-present, J. Geophys. Res., 113, D01103, doi:10.1029/2007JD008470)

maximum mean daily air temperature that tea can tolerate is between 30 °C and 35 °C (Shoubo 1989). The minimum mean daily air temperature that tea can tolerate varies based on tea variety and cultivar. For the broad-leaf assamica variety of tea, the minimum mean daily air temperature is approximately −5 °C, while it is −10 °C for the small-leaf variety *sinensis* (Shoubo 1989). Optimum average ambient temperatures for tea based on photosynthetic rate vary between 18 and 20° C (Fig. 1.2; Williges 2004). In general, the *assamica* variety can tolerate higher temperatures and grows in areas with annual mean temperature above 16.5 °C, while the *sinensis* variety grows in areas with annual mean temperature above 14 °C (Shoubo 1989).

In addition to variety, optimum temperatures vary based on cultivar. A study found that four out of six cultivars tested in tea plantations in South India had an optimum maximum temperature based on photosynthetic rate of 25 °C (Joshi and Palni 1998). Temperatures beyond this resulted in a drastic decline of the net photosynthetic rate (Joshi and Palni 1998). The other two cultivars had an optimum maximum temperature of 30 °C and did not experience as drastic a decline in the net photosynthetic rate with temperatures beyond this (Joshi and Palni 1998). Furthermore, optimum temperatures for tea plants vary based on plant management practices. For example, temperatures interact with water levels such as soil moisture based on irrigation and soil fertility such as nitrogen levels to influence photosynthetic rates and optimum temperatures for tea plants (Smith et al. 1994).

Annual mean temperatures notably vary in different tea growing geographies. For example, annual mean temperatures for tea range from 15 to 19 °C in China's southwest tea zone, 17–22 °C in China's southern tea zone, and 12–18 °C in China's eastern tea zone (Yang 2005). In addition to annual mean temperatures and daily mean temperatures, tea plant growth can be viewed as being dependent on accumulated thermal units (Zhao and Wu 1979). For example, 110–124 degree-days of accumulated temperature above 10 °C have been found to be beneficial for tea yields (Zhao and Wu 1979). Tea can grow in a hardiness zone of eight and above as classified by the United States Department of Agriculture (Mukhopadhyay and Mondal 2017).

The tea harvest, bud burst of new leaves, and rate of shoot initiation are highly dependent on temperature (Mukhopadhyay and Mondal 2017). The timing of bud burst of tea plants in the spring in areas where plants have a winter dormancy is a direct response to an increase in temperature (Shoubo 1989). Bud burst occurs when mean daily air temperature is over 10 °C, and the specific temperature varies depending on the variety and cultivar. Variation in mean air temperature below approximately 20 °C in the late spring has been shown to decrease spring tea yield, while mean air temperature below 24 °C in June has been found to increase summer tea yield (Shoubo 1989). Atmospheric temperatures variably influence tea leaf temperature based on leaf size and structure as well as wind velocity, relative humidity, and rate of transpiration (Rahman 1988).

Excessive heat and frost are harmful for the growth of tea plants and influence photosynthetic rates. Increased temperatures can cause soils of tea agricultural systems to dry, particularly if mulching is not applied, and consequently decrease yields (Chang and Bratloff 2015). A more recent study suggested that tea yields in Assam has varied inversely with warmer monthly average temperatures in recent years (Duncan et al. 2016). If temperatures become too high for tea cultivation in a given area, new suitable land areas need to be found (Chang and Bratloff 2015). This tea expansion can have detrimental environmental impacts including deforestation if tea plantations are created in forest areas. In addition, such forest clearing would release carbon dioxide into the atmosphere and thereby exacerbate the effects of climate change on tea in a detrimental feedback loop.

Previous studies have documented an inverse relationship of temperature and tea quality based on the levels of catechins, phenolic secondary metabolites, and antioxidants (Wang et al. 2011; Lee et al. 2010; Wei et al. 2011). Overall, the literature suggests an inverse relationship between temperature and tea quality on the basis of catechins, phenolic secondary metabolites, and antioxidant properties (Wang et al. 2011; Lee et al. 2010; Wei et al. 2011). However, one study showed that increased temperature resulted in increased catechin compounds in tea (Yao et al. 2005). In addition, higher altitudes that are associated with cooler temperatures are also associated with higher quality tea (Abeywickrama et al. 2010, 2011; Ohno et al. 2011; Akhlas 2003). For example, levels of catechins and other polyphenols were found to be higher at higher elevations (Abeywickrama et al. 2010, 2011; Ohno et al. 2011; Akhlas 2003). Caffeine levels have also been found to be higher in tea samples from higher elevation (Ohno et al. 2011; Abeywickrama 2010; Akhlas 2003).

Precipitation The hydrological cycle is directly impacted by global climate change with an increase in surface temperatures (Sombroek and Gommes 1996) and is predicted to become more intense in future climates (Zeppel et al. 2011). This includes increased rainfall intensities, rainfall variability, changes in the timing of precipitation, and heightened extreme events including drought and flooding. Similar to temperature, water is a primary factor that affects the rate of plant photosynthesis, metabolism, and growth (Galmes et al. 2005). Changes of precipitation within and across seasons are notably impacting plants through shifts in soil water content (Zeppel et al. 2011). In addition, the size and timing of rain events are strong drivers of ecological processes (Zeppel et al. 2011) that impact various dynamics within agricultural systems.

Shifts in precipitation including extreme precipitation and changing seasonal precipitation may influence plant-water relations, hydraulic architecture, and phenology including fruit, leaf, and wood development (Zeppel et al. 2011). The impacts of shifts in precipitation vary greatly across regions. Increased precipitation in subhumid and semiarid areas can increase plant growth in these areas and led to improved land surface protection (Sombroek and Gommes 1996). However, increased rainfall in humid area may impair adequate crop drying and result in increased runoff and erosion with a higher frequency of floods (Sombroek and Gommes 1996).

Numerous studies have shown that water stress is the major environmental stressor that influences agricultural production through shifts in crop yields and quality (Boutraa et al. 2010). Drought is a chemical-physical phenomena that impacts the metabolic processes of photosynthesis and respiration as well as the organization of biomolecules including nucleic acids, proteins, carbohydrates, fatty acids, hormones, ions, and nutrients (Boutraa et al. 2010). Leaf respiration has been shown to decrease during the initial stages of water stress (relative water content (RWC) > 60%) and increase as RWC decreases below 50% (Galmes et al. 2005). Plants respond to water stress by producing reactive oxygen species (ROS) that ultimately damage cells that suffer from water deficit (Boutraa et al. 2010).

Precipitation Effects on Tea Cultivation of tea is possible in areas that receive at least 120–150 cm of rain annually with optimum annual rainfall for tea production being 250–300 cm (Fig. 1.3; Williges 2004). In general, the *assamica* variety grows in areas with higher annual mean rainfall compared to the *sinensis* variety (Shoubo 1989). The *assamica* variety optimally grows in areas with higher annual mean rainfall of 120 cm, and the *sinensis* variety optimally grows in areas with annual mean rainfall 100 cm (Shoubo 1989). Tea plants are under tremendous ecological stress and growth constraint in situations where they receive less than 75 cm of mean annual rainfall (Shoubo 1989). Annual mean rainfall notably varies in different tea growing geographies. For example, annual mean rainfall for tea ranges from 100 to170 cm in China's southwest tea zone, 150–260 cm in China's southern tea zone, and 70–170 cm in China's eastern tea zone (Yang 2005).

Excessive drought and water logging are harmful for the growth of tea plants. A lack of soil moisture in tea fields is directly related to growth with decreased growth of branches, hardening of leaves, and increased number of sterile buds which ultimately reduce the yield and quality (Lemmesa 1996). The use of irrigation sys-

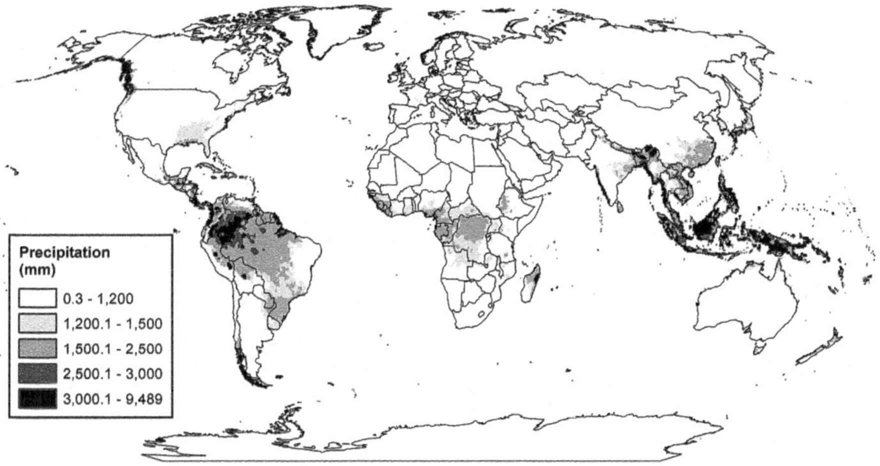

Fig. 1.3 Average annual precipitation 1981–2010. Tea cultivation is possible with 1200–1500 mm of rainfall, while optimum rainfall is 2500–3000 mm. Global Precipitation Climatology Centre data provided by the NOAA/OAR/ESRL PSD, Boulder, Colorado, USA, from their Web site at http://www.esrl.noaa.gov/psd/. *Citation:* Schneider, Udo; Becker, Andreas; Finger, Peter; Meyer-Christoffer, Anja; Rudolf, Bruno; Ziese, Markus (2011): GPCC Full Data Reanalysis Version 6.0 at 0.5°: Monthly Land-Surface Precipitation from Rain-Gauges built on GTS-based and Historic Data. https://doi.org/10.5676/DWD_GPCC/FD_M_V7_050)

tems has consequently increased in some tea production areas. Heavy rainfall patterns can result in erosion of top soil and agricultural input runoff, calling for increased usage and production costs for fertilizers and pesticides (Chang and Bratloff 2015). In a study assessing historical tea yields across China, it was found that increases in rainfall during periods of high rain were inversely related to total tea yields, with a 1% increase in average daily precipitation occurring during the monsoon period associated with a 0.184–0.262% reduction in tea yields (Boehm et al. 2016). In addition to rainfall patterns, limited ground and river water is a constraining factor for tea productivity (Chang and Bratloff 2015).

Tea plants require well-drained, sandy, thoroughly aired, deep, and nutritious soils with a healthy layer of humus and low pH (Lemmesa 1996). Soils in which tea grows soils should have a high capacity of water retention but should be well drained so that water does not remain stagnant (Willson 1999). Decreased soil water potential is linked to a decrease in net photosynthetic productivity (Anandacoomaraswamy et al. 2000). Roots of tea plants grow poorly and accumulate mildew and die when soil moisture content is above the threshold of 93% soil moisture content (Shoubo 1989). Soil moisture levels between 73.4% and 83.2% in 0–30 cm deep soil are optimal for tea growth (Shoubo 1989). Fertile soils with a minimum depth of 2 m are beneficial for tea growth (Mukhopadhyay and Mondal 2017). Roots of tea plants have been found to grow up to 15 m deep under ideal soil conditions (Willson 1999). Soil temperature is also related to tea growth with the most favorable range being between 19 °C and 22 °C (Mukhopadhyay and Mondal 2017). Photosynthesis and growth of tea are highly related to nutrient levels including availability of nitrogen (Aoki 1987). Increased application of nitrogen and potassium is related to increased levels of chlorophyll content in immature and mature leaves.

Tea is notably responsive to humidity with optimum relative humidity of ambient conditions of tea gardens above 70–75% (Mukhopadhyay and Mondal 2017) and up to 80–90% during the tea growth period (Shoubo 1989). Lower relative humidity of the atmosphere has been found to have a depressing influence on tea shoot development (Lemmesa 1996). Specifically, relative mean monthly humidity below 60% is stressful to tea plants with levels below 50% inhibiting shoot growth and humidity below 40% adversely impacting shoot growth (Shoubo 1989). Excessive wind negatively influences the growth of tea plants (Mukhopadhyay and Mondal 2017). Tea plants growing in areas without windbreaks that experience excessive wind are vulnerable to high evapotranspiration and water stress (Lemmesa 1996).

Excessive drought and water logging are harmful for the growth of tea plants and may result in a lower quality product in terms of chemistry, taste, aroma, and bioactivity (Ahmed and Stepp 2012). While numerous studies demonstrate that tea quality is significantly impacted by changes in precipitation levels, this varies based on geographic factors and type of cultivar (Hernandez et al. 2006; Bhattacharya 2014; Ahmed et al. 2013). In some cases, tea quality based on concentrations of catechins, phenolics, and/or antioxidant activity has been shown to increase with drought stress (Hernandez et al. 2006; Bhattacharya 2014; Ahmed et al. 2013), while in other cases, tea quality has been shown to decrease with drought stress (Cheruiyot et al. 2007, 2008).

Soil moisture content has been shown to influence the aromatic constituents in tea leaves with the highest levels under a soil relative water content of 53.90% and lowest under a soil relative water content of 99.75% (Cao et al. 2007). Nutrient and mineral content of soils including levels of Al, Zn, Fe, N, P, Mg, Sel, Mn, K, and B and micronutrients notably impact secondary metabolite compounds (Duan et al. 2012; Sae-Lee 2012; Chen et al. 2011; Ruan et al. 2010; Ruan et al. 2013; Ruan et al. 2007; Hu 2003; Jayanganesh et al. 2011; Lin 2011; Yang et al. 2014).

Ultraviolet Radiation Depletion of the stratospheric ozone layer driven by increased greenhouse gas emissions is resulting in increased ultraviolet radiation (UV-B) reaching the earth's surface. Increased levels of UV-B have been shown to be harmful to living organisms by damaging DNA, proteins, lipids, and membranes and can thus be detrimental to plants by impacting photosynthesis and growth (Hollósy 2002). Plants have evolved protective structures and mechanisms in response to increased UV-B including the upregulations of phytochemicals (Hollósy 2002). Increased UV-B in experimental conditions have shown to decrease plant yields by 10% while increasing plant phytochemicals that serve as defense compounds (Sombroek and Gommes 1996).

Ultraviolet Radiation Effects on Tea As tea originally evolved as an understorey plant, its photosynthetic apparatus is likely adapted to function with maximum capacity under shade (De Costa et al. 2007). Increasing light intensity displays an asymptotic effect on the net photosynthetic rate of fully expanded tea leaves with variability occurring with interactions of specific tea genotypes and environmental conditions (De Costa et al. 2007). Previous studies have shown that increased shade levels resulted in increased net photosynthetic rate of tea young tea plants under controlled environmental conditions (Mohotti et al. 2000). The net photosynthetic rate of tea has shown to decrease due to photoinhibition with an increase in light intensity beyond 1400–1500 μmol m^{-2} s^{1} (Mohotti and Lawlor 2002). Shade can indirectly influence net photosynthetic rates by regulating leaf temperatures and canopy temperatures (Mohotti and Lawlor 2002). In addition to solar radiation, light quality impacts tea growth with shorter wavelength of ultraviolet light inhibiting the growth of tea plants and longer wavelengths stimulating growth.

A minimum of 5 h of direct daily sunlight or 11 h of indirect daily sunlight are required for tea successful tea cultivation and productivity (Ahmed and Stepp 2012). Total days of sunshine further impact tea productivity (TRI ADD Year). A 1% decrease in solar radiation in the previous growing season in China was associated with 0.554–0.864% decrease in tea yields (Boehm et al. 2016). The *assamica* tea variety has a higher threshold for shade and lower solar radiation compared to the *sinensis* tea variety due to their differences in leaf angle (Willson 1999).

Interacting Effects While understanding the effects of single variables associated with climate change on crops is beneficial, it is essential to understand the interaction of the multiple mechanisms impacted by climate change. For example, carbon dioxide levels and temperatures interact in ways that alter the temperature optimum of plants (Unsworth and Hogsett 1996). Likewise, carbon dioxide levels and water interact in ways that alter water use of plants (Unsworth and Hogsett 1996). In addition, shifts in temperature and precipitation that are becoming more variable with climate change are associated with the increased incidence of new pests and diseases that attack tea plants (Chang and Bratloff 2015) including arthropod pests and disease pests. Tea production environments are thus becoming more vulnerable to pests and diseases that can result in crop loss (Mukhopadhyay and Mondal 2017) and impact quality (Ahmed et al. 2014b).

Effects of Pests on Tea Changes in temperature and precipitation driven by climate change may influence the prevalence of pests and disease in tea fields. Arthropod pests can be classified as chewing pests, sucking pests, and mite pests. Major chewing pests of the tea plant include bunch caterpillar (*Andraca bipunctata*), looper caterpillar (*Buzura suppressaria*), red slug caterpillar (*Eterusia magnifica*), and flush worm (*Lespeyrasia leucostoma*). Prominent sucking pests of the tea plant include tea mosquito bug (*Helopeltis theivora*), jassid (*Empoasca flavescens*), and thrips (*Scirtothrips dorsalis*). Prevalent mite pests of the tea plant include purple mite (*Calacarus carinatus*), pink mite (*Acaphylla theae*), scarlet mite (*Brevipalpus phoenicis*), and red spider mite (*Oligonychus coffeae*). The presence of herbivory has found to impact tea quality in multiple ways including a decrease in catechins (Chakraborty et al. 2005), increase in catechins (Dong et al. 2011), increase in volatiles (Dong et al. 2011), and increase in some volatiles and decrease in other volatiles (Han and Chen 2002).

In addition to pests, tea plants are vulnerable to multiple diseases including blister blight caused by the *Exobasidium vexans* pathogen, red rust caused by the *Cephaleuros parasiticus* algae, black rot caused by *Corticiumtheae* and *Corticium invisum* fungi, and anthracnose caused by *Colletotrichum theae-sinensis*. For example, blister blight that typically occurs during the rainy season may become more prevalent in areas with a longer duration of the monsoon season. These increases in pests are resulting in increased usage of pesticides associated with increased costs of production, environmental concerns, and toxicology concerns for human consumers. Microorganisms have further been shown to influence tea quality with a decrease in certain secondary metabolites with pathogen infection (Ponmurugan and Baby 2007) or an increase in the presence and levels of secondary metabolites with deliberate microorganism inoculation.

4 Socioeconomic Implications of Climate Change in Major Tea-Producing Areas Globally

Tea plays a notable role in rural development, poverty reduction, and food security in tea-producing countries (Chang and Bratloff 2015). Climate change effects on tea production thus have notable socioeconomic implications through pathways including crop quality, farmer well-being, consumer satisfaction, and human nutrition. Here we review key trends reported in news outlets, policy reports, and scientific literature on the socioeconomic implications of climate effects on tea with the objective to highlight the need to better understand tea physiology in the context of climate change in order to mitigate climate risk on tea agriculture.

China In China, harsh weather conditions including drought have been reported to decrease tea production between 11% and 35% and result in fluctuating tea prices with implications for multiple sectors within the tea industry. Analysis of historical weather tea production data in all tea-producing counties in China found a decrease in tea yields with longer monsoon seasons that are accompanied with increased daily rainfall associated with climate change (Boehm et al. 2016). The study revealed that a 1% increase in the date of the monsoon retreat in China could be linked to a reduction in yields of between 0.48% and 0.535%, while an increase of 1% in average daily rainfall could be associated with a drop in yield of 0.18–0.26%. In Yunnan Province of China, the monsoon season is also associated with lower tea quality as well as a decrease in farmer livelihoods (Ahmed et al. 2014a, b).

India In India, extreme rainfall and temperature conditions have been reported during multiple recent years to impact tea production including drought, high-intensity rainfall, long rainless spells, changes in rainfall patterns, increases of maximum temperatures, and decreases of minimum temperatures (TNT Desk 2016,

"Assam tea industry bears the brunt of climate change"). Tea plantations in Assam are experiencing heightened climate risk due to their location at the upper end of the range suitable for tea cultivation coupled with their proximity to the flood-prone Brahmaputra River (AVP 2016, "Changing Climate's Challenges to TATA tea"). Slight changes in temperature pose risks for areas in Assam becoming unsuitable for tea cultivation; tea productivity is forecast to decline up to 40% by 2050 in Assam due to climate change without the adoption of climate mitigation and adaptation measures (AVP 2016 "Changing Climate's Challenges to TATA tea"). These changes in tea productivity present risks to the approximately two million laborers in Assam that are dependent on tea production for their livelihoods including both tea yields and quality and have contributed to political and ethnic unrest (Das 2015). Adaptation measures include implementing irrigation systems and a major change of operations to relocate tea plantations to higher altitudes with cooler temperatures. These adaptation measures are associated with major costs which are resulting in less employment opportunities on many estates (Chang and Bratloff 2015).

Kenya In Kenya, unpredictable and extreme weather events including unpredictable rainfall patterns, drought, and rising temperatures have been notably shown to impact multiple aspects of the tea industry (Kiarii 2017; Roberts 2015; Ethical Tea Partnership 2015). Changes in weather events are associated with increased incidence of pests and diseases in Kenya. The abiotic and biotic stressors have been reported to reduce tea quantities and affect how and where tea can be grown. The detrimental effects of climate change have already been witnessed for Kenya's tea industry with the 2009 drought in the Rift Valley that led to a 30% drop in production (Ethical Tea Partnership 2015). Climate change is forecast to reduce land suitable for tea production by 40% by 2050 (Ethical Tea Partnership 2015).

These changes in tea productivity amid unchanged tea prices are resulting in notable decrease in farmer income as well as employment opportunities linked to the tea sector (Kiarii 2017, "Adverse weather conditions hurting tea farmers"). The tea sector contributes to approximately 11% of total agricultural sector output to Kenya's gross domestic product with over three million people either directly or indirectly deriving income from the tea sector. It is further anticipated that foreign exchange earnings from tea sales will plummet with the negative effects of climate change on tea productivity with ramifications across the economy at large. Smallholder farmers, which account for approximately 60% of Kenyan tea, are likely to bear the greatest burden of climate effects on the tea sector due to their lack of financial stability and ability for adaptation measures such as purchasing input materials to manage increased incidence of pests.

Sri Lanka In Sri Lanka, extreme weather including drought, extreme rainfall intensities, and warmer temperatures as a result of climate change have been reported to influence tea systems with multiple socioeconomic and environmental implications (Wijeratne 1996). Heavy rains have been reported to erode top soil in tea plantations and wash away fertilizers and chemicals (Wijeratne 1996). Both drought and heavy rains have been reported to decrease tea yields (Wijeratne 1996).

Droughts have already shown to have irreparable losses to Sri Lanka's tea industry since the 1980s including in 1983 and 1992 (Central Bank 1983, 1992). The drought of 1992 in Sri Lanka was reported to decrease tea yields 26% from the previous year and result in increased production costs of 19% (Central Bank 1992). It is expected that there will be an increase in the length of both the dry and wet seasons in Sri Lanka's main plantation areas which will exacerbate the effects of weather extremes. In particular, low-elevation areas are most vulnerable to the detrimental effects of climate change with increases in temperature, soil moisture deficit, and saturation vapor pressure deficit adversely affecting tea yields (Wijeratne 1996).

5 Conclusion

Global climate change is associated with robust multi-decadal warming and shifts in extreme weather events including heat waves, droughts, floods, cyclones, and wild fires that have increased since the 1950s. The overall increased unpredictability of weather and climate is making planning of farm operations more difficult (Sombroek and Gommes 1996). Agroclimatic zones suitable for specific crops are shifting spatially including altitudinally, while specific crop species may not be able to adapt in time. Higher temperatures are resulting in seasonally longer plant growth in some areas, allowing for increased production in cooler areas and reduced productivity in already warm areas. The incidence of pests and disease is increasing, particularly the prevalence of introduced species. Rises in sea level threaten to submerge coastal agricultural land.

Tea plants are experiencing the effects of multiple decadal climate change with implications for livelihoods of laborers and farmers and regional economies more broadly. The key environmental factors and processes that influence tea physiological processes that are shifting with climate change include carbon dioxide concentrations, temperature, rainfall, humidity, solar radiation, wind, soil conditions and microorganisms, pests, and pathogens. Reports in the scientific literature and news highlight that climate change is already impacting tea systems and that the expected impacts of climate change may result in severe and irreversible impacts for tea production and associated livelihoods. Adaptation and mitigation strategies are thus called for that are based on scientific evidence coupled with community-based approaches toward a more sustainable tea sector.

Acknowledgments This work was supported by funding from (1) the United States National Science Foundation NSF (Award Numbers CNH BCS-1313775 and NSF RII Track-2 FEC 1632810) and (2) the National Institute of General Medical Sciences of the National Institutes of Health (Award Numbers P20GM103474 and 5P20GM1044-17). The content is solely the responsibility of the authors and does not necessarily represent the official views of the US National Science Foundation and the National Institutes of Health.

References

Abeywickrama K, Ratnasooriya W, Amarakoon A (2010) Oral diuretic activity of hot water infusion of Sri Lankan black tea (*Camellia sinensis* L.) in rats. Pharmacogn Mag 6(24):271

Abeywickrama K, Ratnasooriya W, Amarakoon A (2011) Oral hypoglycaemic, antihyperglycaemic and antidiabetic activities of Sri Lankan broken orange pekoe fannings (BOPF) grade black tea (*Camellia sinensis* L.) in rats. J Ethnopharmacol 135(2):278–286. https://doi.org/10.1016/j.jep.2011.02.035

Ainsworth EA, Long SP (2005 Feb) What have we learned from 15 years of free-air CO2 enrichment (FACE)? A meta-analytic review of the responses of photosynthesis, canopy properties and plant production to rising CO2. New Phytol 165(2):351–371

Ainsworth EA, Rogers A (2007) The response of photosynthesis and stomatal conductance to rising [CO2]: mechanisms and environmental interactions. Plant Cell Environ 30(3):258–270. https://doi.org/10.1111/j.1365-3040.2007.01641.x

Ahmed S, Stepp JR, Toleno R, Peters CM (2010) Increased market integration, value, and ecological knowledge of tea agro-forests in the akha highlands of Southwest China. Ecol Soc 15(4):27

Ahmed S, Stepp JR (2012) Green tea: the plants, processing, manufacturing and production. In: Preedy V (ed) Tea in health and disease prevention. Academic Press/Elsevier Science and Technology, Missouri

Ahmed S, Stepp JR (2016) Beyond yields: climate effects on specialty crop quality and agroecological management. Elementa 4:000092

Ahmed S, Peters CM, Chunlin L, Meyer R, Unachukwu U, Litt A, Kennelly E, Stepp JR (2013) Biodiversity and phytochemical quality in indigenous and state-supported tea management systems of Yunnan, China. Conserv Lett 5(6):28–23

Ahmed S, Stepp JR, Orians C, Griffin T, Matyas C, Robbat A, Cash S, Xue D, Chunlin L, Unachuckwu U, Buckley S, Small D, Kennelly E (2014a) Effects of extreme climate events on tea (*Camellia sinensis*) functional quality validate indigenous farmer knowledge and sensory preferences in Tropical China. PLoS One 9(10):e109126

Ahmed S, Orians C, Griffin T, Buckley S, Unachukwu U, Stratton AE, Stepp JR, Robbat A, Cash S, Kennelly E (2014b) Effects of water availability and pest pressures on tea (*Camellia sinensis*) growth and functional quality. AoB Plants 6

Ahmed S, Stepp JR, Xue D (2015) Cultivating botanicals for sensory quality: from good agricultural practices (GAPs) to taste discernment by smallholder tea farmers. In: Reynertson K (ed) Botanicals: methods for quality and authenticity. CRC Press/Taylor & Francis Group, LLC, Boca Raton

Akhlas M (2003) Qualitative assessment of fresh tea produced in Pakistan growing under different agro-ecological conditions and fertilizer treatments. Pak J Bot 35(5):779–790

Anandacoomaraswamy A, De Costa WAJM, Shyamalie HW, Campbell GS (2000) Factors controlling transpiration of mature field-grown tea and its relationship with yield. Agric For Meteorol 103:375–386

Aoki S (1987) Varietal differences and effects of nitrogen fertilization on decline of photosynthetic rate in overwintered tea leaves. Jpn J Crop Sci 56(2):252–256

AVP (2016) Changing climate's challenges to TATA tea. https://rctom.hbs.org/submission/changing-climates-challenges-to-tata-tea/

Bhattacharya A, Saini U, Joshi R, Kaur D, Pal AK, Kumar N et al (2014) Osmotin-expressing transgenic tea plants have improved stress tolerance and are of higher quality. Transgenic Res 23(2):211–223

Björkman M, Klingen I, Birch AN, Bones AM, Bruce TJ, Johansen TJ et al (2011) Phytochemicals of Brassicaceae in plant protection and human health–Influences of climate, environment, and agronomic practice. Phytochemistry 72(7):538–556

Bloom AJ, Burger M, Asensio JSR, Cousins AB (2010) Carbon dioxide enrichment inhibits nitrate assimilation in wheat and Arabidopsis. Science 328(5980):899–903

Boehm R, Cash S, Anderson B, Ahmed S, Griffin T, Robbat A Jr, Stepp JR, Han W, Hazel M, Orians C (2016) Association between empirically estimated monsoon dynamics and other weather factors and historical tea yields in China: results from a yield response model. Climate 4(2)

Boutraa T, Akhkha A, Al-Shoaibi AA, Alhejeli AAM (2010) Effect of water stress on growth and water use efficiency (WUE) of some wheat cultivars (*Triticum durum*) grown in Saudi Arabia. J Taibah Univ Sci 3:39–48

Caffin N, D'Arcy B, Yao L, Rintoul G (2004) Developing an index of quality for Australian tea. University of Queensland: RIRDC publication no. 04/033. RIRDC project no. UQ-88A

Cao P, Liu C, Liu K (2007) Aromatic constituents in fresh leaves of lingtou dancong tea induced by drought stress. Front Agric China 1(1):81–84. https://doi.org/10.1007/s11703-007-0015-x

Carr MKV (1972) The climatic requirements of the tea plant: a review. Exp Agric 8:1–14

Central Bank (1983) Annual Report. Central Bank of Sri Lanka, Colombo

Central Bank (1992\1983), Annual Report. Central Bank of Sri Lanka, Colombo

Chakraborty U, Chakraborty N (2005) Impact of environmental factors on infestation of tea leaves by helopeltis theivora, and associated changes in flavonoid flavor components and enzyme activities. Phytoparasitica 33(1):88–96. https://doi.org/10.1007/bf02980930

Chang K, Brattlof M (2015) Socio-economic implications of climate change for tea producing countries. Rome, Italy: Food and Agriculture Organization of the United Nations (FAO). Retrieved from http://www.fao.org/3/a-i4482e.pdf

Chen YM, Tsao TM, Liu CC, Lin KC, Wang MK (2011) Aluminum and nutrients induce changes in the profiles of phenolic substances in tea plants (*Camellia sinensis* CV TTES, no. 12 (TTE)). J Sci Food Agric 91(6):1111–1117

Cheruiyot EK, Mumera LM, Ng'etich WK, Hassanali A, Wachira F (2007) Polyphenols as potential indicators for drought tolerance in tea (*Camellia sinensis* L.). Biosci Biotechnol Biochem 71(9):2190–2197

Cheruiyot EK, Mumera LM, Ng'etich WK, Hassanali A, Wachira F, Wanyoko JK (2008) Shoot epicatechin and epigallocatechin contents respond to water stress in tea (*Camellia sinensis* (l) o. kuntze). Biosci Biotechnol Biochem 72(5):1219–1226. https://doi.org/10.1271/bbb.70698

De Costa WAJM, Mohotti AJ, Wijeratne MA (2007) Ecophysiology of Tea. Braz J Plant Physiol 19(4):299–332

Dong F, Yang Z, Baldermann S, Sato Y, Asai T, Watanabe N (2011) Herbivore-induced volatiles from tea (*Camellia sinensis*) plants and their involvement in intraplant communication and changes in endogenous nonvolatile metabolites. J Agric Food Chem 59(24):13131–13135

Duan X, Hu X, Chen F, Deng Z (2012) Bioactive ingredient levels of tea leaves are associated with leaf Al level interactively influenced by acid rain intensity and soil Al supply. J Food Agric Environ 10:1197–1204

Duncan JMA, Saikia SD, Gupta N, Biggs EM (2016) Observing climate impacts on tea yield in Assam, India. Appl Geogr 77:64–71

Ethical Tea Partnership (2015) Addressing the Effects of Climate Change, Kenya

Ewert F, Rounsevell MDA, Reginster I, Metzger MJ, Leemans R (2005) Future scenarios of European agricultural land use: I. Estimating changes in crop productivity. Agric Ecosyst Environ 107(2–3):101–116

FAOSTAT (2016) http://www.fao.org/statistics/en/

Feng ZZ, Kobayashi K, Ainsworth EA (2008) Impact of elevated ozone concentration on growth, physiology, and yield of wheat (L.): a metaanalysis. Glob Chang Biol 14(11):2696–2708

Galmes J, Flexas J, Keys AJ, Cifre J, Mitchell RAC, Madgwick PJ, Haslam RP, Medrano H, Parry MAJ (2005) Rubisco specificity factor tends to be larger in plant species from drier habitats and in species with persistent leaves. Plant Cell Environ 28(5):571–579

Gilani GS, Nasim A (2007) Impact of foods nutritionally enhanced through biotechnology in alleviating malnutrition in developing countries. J AOAC Int 90(5):1440–1444

Gliessman S (2015) Agroecology: the ecology of sustainable food systems. CRC Press, Boca Raton

Han B, Chen Z (2002) Composition of the volatiles from intact and mechanically pierced tea aphid– tea shoot complexes and their attraction to natural enemies of the tea aphid. J Agric Food Chem 50(9):2571–2575

Hatfield JL, Prueger JH (2015) Agroecology: implications for plant response to climate change. In: Yadav SS, Redden RJ, Hatfield JL, Lotze-Campen H, Hall AE (eds). Wiley-Blackwell, West Sussex, pp 27–43

Hernández I, Alegre L, Munné-Bosch S (2006) Enhanced oxidation of flavan-3-ols and proanthocyanidin accumulation in water-stressed tea plants. Phytochemistry 67(11):1120–1126

Hollósy F (2002) Effects of ultraviolet radiation on plant cells. Micron 33(2):179–197

Hu Q, Xu J, Pang G (2003) Effect of selenium on the yield and quality of green tea leaves harvested in early spring. J Agric Food Chem 51(11):3379–3381

IPCC (2014) Climate Change 2014: Synthesis Report. Contribution of Working Groups I, II and III to the Fifth Assessment Report of the Intergovernmental Panel on Climate Change Core Writing Team, R.K. Pachauri and L.A. Meyer (eds.). IPCC, Geneva, p 151

Jayanganesh S, Venkatesan S, Senthurpandian V (2011) Impact of different sources and doses of magnesium fertilizer on biochemical constituents and quality parameters of black tea. Asian J Biochem 6:273–281

Joshi SC, Palni LMS (1998) Clonal variation in temperature response of photosynthesis in tea. Plant Sci 137:225–232

Kiarii (2017) https://www.standardmedia.co.ke/article/2001230909/adverse-weather-conditions-hurting-tea-farmers

Kingdon-Ward F (1950) Does wild tea exist? Nature 165:297–299

Krishna N. Das (2015) "Climate change sparks tension in India's tea gardens." Reuters, 4 May

Kowalsick A, Kfoury N, Robbat A, Ahmed S, Orians C, Griffin T, Cash S, Stepp JR (2014) Metabolite profiling of *Camellia sinensis* by automated sequential, multidimensional gas chromatography/mass spectrometry (GC-GC/MS) reveals strong monsoon effects on tea constituents. J Chromatogr A 1370(28):230–239

Leakey ADB, Ainsworth EA, Bernacchi CJ, Rogers A, Long SP, Ort DR (2009 1 July) Elevated CO2 effects on plant carbon, nitrogen, and water relations: six important lessons from FACE. J Exp Bot 60(10):2859–2876. https://doi.org/10.1093/jxb/erp096

Lee JE, Lee BJ, Chung JO, Hwang JA, Lee SJ, Lee CH et al (2010) Geographical and climatic dependencies of green tea (*Camellia sinensis*) metabolites: a 1H NMR-based metabolomics study. J Agric Food Chem 58(19):10582–10589

Lemmesa F (1996) Tea production and management. Teaching handout, Department of Plant Sciences, Jimma College of Agriculture, Jimma, pp 1–14

Li X, Zhang L, Ahammed GJ, Li ZX, Wei JP, Shen C et al (2017) Stimulation in primary and secondary metabolism by elevated carbon dioxide alters green tea quality in *Camellia sinensis* L. Sci Rep 7(1):7937. https://doi.org/10.1038/s41598-017-08465-1

Lin BB (2011) Resilience in agriculture through crop diversi cation: adaptive management for environmental change. Bioscience 61:183–193

Lobell DB, Burke MB (2010) On the use of statistical models to predict crop yield responses to climate change. Agric For Meteorol 150(11):1443–1452

Lower SS, Orians CM (2003) Soil nutrients and water availability interact to influence willow growth and chemistry but not leaf beetle performance. Entomol Exp Appl 107(1):69–79

Mohotti AJ, Lawlor DW (2002) Diurnal variation of photosynthesis and photoinhibition in tea: effects of irradiance and nitrogen supply during growth in the field. J Exp Bot 53:313–322

Mohotti AJ, Dennett MD, Lawlor DW (2000) Electron transport as a limitation to photosynthesis of tea (*Camellia sinensis* (L.) O. Kuntze): a comparison with sunflower (*Helianthus annus* L.) with special reference to irradiance. Trop Agric Res 12:1–10

Morgan PB, Ainsworth EA, Long SP (2003) How does elevated ozone impact soybean? A meta-analysis of photosynthesis, growth and yield. Plant Cell Environ 26(8):1317–1328

Mukhopadhyay M, Mondal T (2017) Cultivation, improvement, and environmental impacts of tea. doi: https://doi.org/10.1093/acrefore/9780199389414.013.373

Myers SS, Zanobetti A, Kloog I, Huybers P, Leakey ADB, Bloom AJ, Carlisle E et al (2014) Increasing CO2 threatens human nutrition. Nature 510(7503):139–142 and the literature summarized by them

Ohno A, Oka K, Sakuma C, Okuda H, Fukuhara K (2011) Characterization of tea cultivated at four different altitudes using 1H NMR analysis coupled with multivariate statistics. J Agric Food Chem 59(10):5181–5187

Ponmurugan P, Baby U (2007) Morphological, physiological and biochemical changes in resistant and susceptible cultivars of tea in relation to phomopsis disease. Plant Pathol J 6(1):91–94

Rahman F (1988) Physiology of the tea bush. Two Bud 35:1–14

Roberts S (2015) Addressing climate change in one of the world's poorest countries: how a sector programme on living wage and better livelihoods can help http://www.ethicalteapartnership.org/addressing-climate-change-in-one-of-the-worlds-poorest-countries-how-a-sector-programme-on-living-wage-and-better-livelihoods-can-help/

Ruan J, Gerendás J, Haerdter R, Sattelmacher B (2007) Effect of alternative anions (Cl^- vs $SO42^-$) on concentrations of free amino acids in young tea plants. J Plant Nutr Soil Sci 170(1):49–58

Ruan J, Haerdter R, Gerendás J (2010) Impact of nitrogen supply on carbon/nitrogen allocation: a case study on amino acids and catechins in green tea [*Camellia sinensis* (l) o. kuntze] plants. Biol Plant 12(5):724–734

Ruan J, Ma L, Shi Y (2013) Potassium management in tea plantations: its uptake by field plants, status in soils, and efficacy on yields and quality of teas in China. J Plant Nutr Soil Sci 176(3):450–459

Sae-Lee N, Kerdchoechuen O, Laohakunjit N (2012) Chemical qualities and phenolic compounds of Assam tea after soil drench application of selenium and aluminum. Plant Soil 356(1–2):381–393

Shoubo H (1989) Meteorology of the tea plant in China: a review. Agric For Meteorol 47(1):19–30

Smith BG, Stephens W, Burges PJ, Carr MKV (1993) Effects of light, temperature, irrigation and fertilizer on photosynthetic rate in tea (*Camellia sinensis* L.). Exp Agric 29:291–206

Smith BG, Burgess PJ, Carr MKV (1994) Effects of clone and irrigation on the stomatal conductance and photosynthetic rate of tea (*Camellia sinensis*). Exp Agric 30:1–16

Sombroek WG, Gommes R (1996) The climate change - agricultural conundrum. In: Bazzaz F, Sombroek WG (eds) Global climate change and agricultural production. Wiley and Sons, Chichester

Stiling P, Cornelissen T (2007) How does elevated carbon dioxide (CO2) affect plant–herbivore interactions? A fieldexperiment and meta-analysis of CO2-mediated changes on plant chemistry and herbivore performance. Glob Change Biol 13(9):1823–1842

Tans PP, Fung IY, Takahashi T (1990) Observational contrains on the global atmospheric CO2 budget. Science 247(4949):1431–1438

Taub D (2010) Effects of rising atmospheric concentrations of carbon dioxide on plants. Nat Educ Knowl 3(10):21

Taub D, Wang X (2008) Why are nitrogen concentrations in plant tissues lower under elevated CO2? A critical examination of the hypotheses. J Integr Plant Biol 50(11):1365–1374

TNT Desk (2016) Assam tea industry bears the brunt of climate change. https://thenortheasttoday.com/assam-tea-industry-suffer-due-to-climate-change/

Unsworth MH, Hogsett WE (1996) Combined effects of changing CO2, temperature, UVB radiation and O3 on crop growth. In: Bazzaz F, Sombroek W (eds) Global climate change and agricultural production. FAO and John Wiley & Sons, New York, pp 171–198

van-Dam NM, Vrieling K (1994) Genetic variation in constitutive and inducible pyrrolizidine alkaloid levels in *Cynoglossum officinale* L. Oecologia 99(3–4):374–378

Vingarzan R (2004) A review of surface ozone background levels and trends. Atmos Environ 38(21):3431–3442

Wang L, Wei K, Jiang Y, Cheng H, Zhou J, He W et al (2011) Seasonal climate effects on flavanols and purine alkaloids of tea (*Camellia sinensis* L.). Eur Food Res Technol 233(6):1049–1055

Wei K, Wang L, Zhou J, He W, Zeng J, Jiang Y et al (2011) Catechin contents in tea (*Camellia sinensis*) as affected by cultivar and environment and their relation to chlorophyll contents. Food Chem 125(1):44–48

Wijeratne MA (1996) Vulnerability of Sri Lanka tea production to global climate change. Water Air Soil Pollut 92:87–94

Williges U (2004) Status of organic agriculture in Sri Lanka with special emphasis on tea production systems (*Camellia sinensis* (L.) O. Kuntze). Fachbereich Pflanzenbau der Justus-Liebig-Universitat Gießen

Willson KC (1999) Coffee, cocoa, and tea. Crop production science in horticulture series 8. CABI Publishing, Cambridge

Yang Y (2005) Chinese tea cultivation. Shanghai Scientific & Technical Publishers, Shanghai

Yang T, Li H, Hu X, Li J, Hu J, Liu R et al (2014) Effects of fertilizing with N, P, Se, and Zn on regulating the element and functional component contents and antioxidant activity of tea leaves planted in red soil. J Agric Food Chem 62(17):3823–3830

Yao L, Caffin N, D'arcy B, Jiang Y, Shi J, Singanusong R et al (2005) Seasonal variations of phenolic compounds in Australia-grown tea (camellia sinensis). J Agric Food Chem 53(16):6477–6483

Zeppel MJB, Adams HD, Anderegg WRL (2011) Mechanistic causes of tree drought mortality: recent results, unresolved questions and future research needs. New Phytol 192:800–803

Zhao J, Wu X (1979) A preliminary analysis of effects of tea sprinkler irrigation. Tea 2:1–6 in Chinese

Zhou Z, Chen X (1985) The relationship between the ecoclimate for tea plants and tea production in the hilly land of Zhejiang province. J Hangzhou Univ 12(2):258–267 in Chinese

Chapter 2
Understanding Response of Tea Plants to Heat Stress and the Mechanisms of Adaptation

Yong-Xin Wang, Zhi-Wei Liu, Hui Li, Wen-Li Wang, Xin Cui, and Jing Zhuang

Abstract Tea plants are mainly distributed in the tropic and subtropic regions. Tea leaves provide the resources of healthy nonalcoholic beverages and have great economical value. Global warming appears to be a trend of climate change in the future. Heat stress is one of the disastrous factors that considerably restrict the growth and development of tea plant and affect quality and yield of tea products. A series of complex physiological and biochemical adaptations have evolved to adapt and cope with the diverse environmental stresses. Therefore, a comprehensive understanding of the mechanism in response to high temperature might be a great strategy for improving the heat resistance of tea plants. Previous studies have reported functional and regulatory genes involved in heat tolerance of tea plants. Transgenic verification has been implemented mainly in heat shock protein (HSP). The integration of new technologies, such as transcriptome analysis, small RNAs, proteomics, metabolomics, etc., are still relatively poor in tea plants, which need more attention for studying heat resistance in tea plants.

Keywords *Camellia sinensis* · Heat stress · Heat shock proteins · MicroRNAs · Reference genes · Transcription factors

1 Introduction

Plants are continuously exposed to all kinds of abiotic and biotic stresses under natural conditions, which lead to a disorder of physiological metabolism and cause considerable loss to agricultural production worldwide (Heil and Bostock 2002; Swarbrick et al. 2006; Massad et al. 2012; Rejeb et al. 2014). Abiotic stress is the

Y.-X. Wang · Z.-W. Liu · H. Li · W.-L. Wang · X. Cui · J. Zhuang (✉)
Tea Science Research Institute, College of Horticulture, Nanjing Agricultural University, Nanjing, China
e-mail: zhuangjing@njau.edu.cn

W.-Y. Han et al. (eds.), *Stress Physiology of Tea in the Face of Climate Change*, https://doi.org/10.1007/978-981-13-2140-5_2

primary cause of crop loss worldwide, resulting in growth reductions over 50% in most plant species (Wang et al. 2003). Abiotic stress causes losses worth hundreds of millions of dollars each year due to reduction in crop productivity and crop failure. Biotic stress, such as pathogens and insect attack, is another challenge inducing a strong pressure on plants (Hovmøller 2002; Strauss and Zangerl 2002; Mordecai 2011). In order to adapt and cope with the diverse environmental stresses, a series of complex physiological and biochemical adaptations have evolved, which are critical for the survival of all plants.

Tea plant [*Camellia sinensis* (L.) O. Kuntze], an important perennial evergreen woody crop, provides the resources for producing natural nonalcoholic beverages "tea," rich in bioactive components, such as tea polyphenols, theanine, and polysaccharides (Sharangi 2009; Kumar et al. 2016; Wang et al. 2017c). Tea plants are grown in a wide range of latitudes in the world ranging from 45°N (Russia) to 30°S (South Africa) and longitude from 150°E (New Guinea) to 60°W(Argentina) (Panda and Upadhyaya 2013). Tea plants frequently encounter various stresses during their life cycle, such as salinity (Li et al. 2010), drought (Das et al. 2012), heavy metal (Basak et al. 2001), radiation, temperature stress (Liu et al. 2016), microbic infection, and soil nutrient deficiency (Panda and Upadhyaya 2013).

Among these stresses, heat (high-temperature) stress is one of the disastrous factors that considerably affect the quality and yield of tea products (Wang et al. 2017a). Given the changes in climate, especially global warming, breeding elite cultivars with high heat resistance and recoverability is an important target for crop breeders. After being treated with high temperature, proline content, peroxidase (POD) activity, catalase (CAT) activity, contents of electrolytic leakage (EL), and malondialdehyde (MDA) could increase, which could seriously affect the normal metabolism of tea plant (Han et al. 2016). Under prolonged exposure to heat stress, tea plants presented different levels of damage, such as leaf discoloration, scorched or detached, dieback of branches, or the death of the entire plant. Tea plants that grow in a specific environment are adapted to these conditions and can also cope with changes that might be adverse for their growth and development (Panda and Upadhyaya 2013). Until now, little is still known about the way that tea plants modulate responses to heat stress.

In this chapter, the primary objective is to report recent advances in adaptation mechanisms of the responses to heat stress and their biotechnological applications in tea plants.

2 Identification of Heat Stress-Related Functional Genes from Tea Plant

Over the course of evolution, plants have to cope with complex and diverse interplay of many environmental factors. Therefore, plants evolved specific mechanisms and allowed them to adapt and survive with adverse environmental conditions (Rejeb et al. 2014). Exposure with heat stress, the critical antioxidant system could be activated to eliminate reactive oxygen species (ROS) and protect the plant from high-temperature damage (Choi et al. 2006). Functional genes directly protected cells from heat stress by the production of important metabolic proteins. Some enzymatic systems including catalases, superoxide dismutases, peroxiredoxins ascorbate peroxidases, and nonenzymatic mechanisms were established to protect plants against oxidative damage (Agrawal et al. 2002; Navrot et al. 2006; Anjum et al. 2012).

ROS was generated under various stresses, including heat stress. High concentration of ROS is harmful to plants. The glutathione peroxidase gene has positive function to damage induced by ROS (Milla et al. 2003; Navrot et al. 2006; Ramos et al. 2009). A glutathione peroxidase gene, *CsGPX2*, was cloned and showed strong response to abiotic stress and hormone treatments, especially heat stress, indicating the potential functions of *CsGPX2* genes in protecting tea plant from peroxidative damage (Fu 2014). Heat shock proteins (HSPs) and ROS-scavenging enzymes are two major functional proteins that are induced by heat stress and are well-known target genes of heat stress-responsive transcription factors (TFs) (Kotak et al. 2007; Ohama et al. 2017). Heat shock proteins (HSPs) and chaperones play crucial roles in protecting plants against stress by reestablishing normal protein conformation and thus cellular homeostasis (Wang et al. 2004). Recently, a heat shock protein genes, *CsHSP17.2* (GenBank accession number: KU244518), was isolated from tea plant and demonstrated highly induced by heat stress (Wang et al. 2015a).

3 Identification of Heat Stress-Related Transcription Factor Genes of Tea Plant

Transcription factors (TFs) are proteins that operated as molecular switches for gene expression by interacting with their corresponding *cis*-regulatory sequences (Zhuang et al. 2011a, b; Puranik et al. 2012). Transcription factors are one of the most important regulatory proteins that modulate the expression of downstream target genes in response to cold, heat, drought, salt, hormones, and pathogen signals (Nakashima and Yamaguchishinozaki 2009). Thus, transcription factors played essential roles in the regulation of gene networks for different kinds of biological processes and defense responses. As for the timing of the induction of stress-inducible genes, transcription factor genes encoding proteins are normally induced rapidly and transiently in response to various environmental stresses.

Multiple transcription factors are involved in heat stress, including AP2/ERF, MYB, WRKY, NAC, and HSF (Kotak et al. 2007; Wahid et al. 2007). Heat shock transcription factors (HSFs) played important roles in regulating heat stress response in plants (Wang et al. 2004; Ohama et al. 2017). HSFA1a/b (heat shock transcription factor A1a/b) were the vital factors in the signal transduction pathway of plant heat stress response (Ohama et al. 2017).

Some transcription factor genes involved in the regulation of heat stress signaling pathways have been identified by RT-PCR in tea plant. Several AP2/ERF transcription factor genes (*CsERF-B1*, *CsERF-B4*, *CsERF-B3*, *CsDREB-A4b*, *CsDREB-A1*, *CsDREB-A4*, and *CsRAV2*) were cloned and validated could be induced by high-temperature treatments by real-time quantitative PCR (Liu et al. 2014a, b; Wu et al. 2014a, b; Liu et al. 2015). Some genes encoding heat shock factor (HSF) (*CsHsfA5* and *CsHsfA4*), NAC (*CsNAC1* and *CsNAC2*), and DOF (*CsDof1* and *CsDof2*) transcription factor genes were also identified and exhibited as different expression profiles under high-temperature treatments among three tea plant cultivars, "Yingshuang," "Anjibaicha," and "Yunnanshilixiang" (Wang et al. 2015b; Xiong et al. 2015; Li et al. 2016b).

Some families of transcription factor genes were identified and retrieved from the transcriptome database of tea plant (Wu et al. 2015, 2016a; Liu et al. 2016). Liu and her colleagues identified 16 CsHsfs based on the transcriptome database of tea plant cultivars and validated that heat stress positively or negatively regulated all of the selected six *CsHsf* genes (*CsHsfA1a*, *CsHsfA1b*, *CsHsfA6*, *CsHsfB1*, *CsHsfB2b*, and *CsHsfC1*). The gene of *CsHsfA6* showed the largest increment among the tested genes under heat treatment, in three tea plant cultivars, namely, 'Yunnanshilixiang,' 'Chawansanhao,' and 'Ruchengmaoyecha' (Liu et al. 2016). Wu et al. identified 89 putative AP2/ERF transcription factors with full length AP2 domain and 50 putative WRKY with complete WRKY domain in tea plant and validated those *CsAP2/ERF* and *CsWRKY* genes involved in responses to heat stress among four tea plant cultivars, 'Yunnanshilixiang,' 'Chawansanhao,' 'Ruchengmaoyecha,' and 'Anjibaicha'(Wu et al. 2015, 2016a). Li et al. identified 29 *CsDOF* genes in tea plant and validated some selected genes encoding CsDOF transcription factor related to different abiotic stresses including heat stress (Li et al. 2016c). Wang et al. identified 45 *CsNAC* genes, and 9 predicted stress-related genes were validated involved in diverse environmental stresses including heat stress, in two tea plant cultivars, 'Huangjinya' and 'Yingshuang' (Wang et al. 2016e).

4 Reliable Reference Genes for Normalization of Expression Profiles of Genes Related Heat Stress in Tea Plant

Quantitative real-time polymerase chain reaction (qRT-PCR) has become a reliable and rapid method to evaluate the expression level of target gene because of its simplicity, high sensitivity, and specificity (Li et al. 2016d, e). To ensure the accuracy of results, the selection and validation of appropriate internal reference genes or housekeeping genes for qRT-PCR are necessary (Wang et al. 2017c).

Some suitable reference genes have been investigated and selected under different experimental conditions or different organs and tissues in tea plant. Five genes (*CsPTB1*, *CsEF1*, *CsSAND1*, *CsCLATHRIN1*, and *CsUBC1*) were identified as suitable reference genes for analyses under complex experimental conditions (Hao et al. 2014). Ma et al. commended *TUA1* (alpha-1 tubulin) as the most stably expressed reference gene under damage stresses (Ma et al. 2016). Wu et al. reported that *CsTBP* and *CsTIP41* displayed most stably expressed reference genes in tea leaf development, and *CsTBP* is also most suitable reference gene in response to hormones (Wu et al. 2016b). *PP2AA3* and *18S rRNA* genes displayed the maximum stability in tea plants exposed to metal stress (i.e., Mn, Al, Cu, Fe, and Zn) (Wang et al. 2017c). In addition, several miRNAs were also evaluated and selected as stably expressed genes among different tissues and under different experimental conditions in the tea plant (Song et al. 2016).

Till now, there were no reports of reference genes in the conditions of high-temperature stress, which needs to be studied in the future. Now, generally, the *β-actin* gene (GenBank accession number: HQ420251) of tea plant was selected as an internal reference to normalize the expression levels under heat stress and showed satisfactory results (Wang et al. 2015a, 2017a, b).

5 Verification of Heat Stress-Related Genes of Tea Plant in Transgenic Plants

The in vitro culture system has not been successfully established in transgenic tea plant, up to now. In order to validate the function of the genes from tea plants, overexpression of these target genes in transgenic plants mainly using the model plant, such as tobacco and *Arabidopsis*, have become effective strategies.

The *CsHis* (H1 histone) gene, a tea plant stress-induced gene, overexpression in transgenic tobacco plants exhibited higher tolerance to various abiotic stresses than the wild-type tobacco plants. The transgenic tobacco plants displayed reduced wilting and senescence and exhibited greater net photosynthetic rate (Pn), stomatal conductance (Gs), and maximal photochemical efficiency (Fv/Fm) values (Wang et al. 2017b). Wang et al. reported that overexpression of the stress-responsive gene *CsHSP17.2* in transgenic *Arabidopsis* plants improved the thermo-tolerance than wild type plants and enhanced the scavenging of ROS and induced the expression of heat-responsive genes (Wang et al. 2017a). Wang and his colleagues also isolated three tea plant *CsHSP* genes (i.e., *CsHSP17.7*, *CsHSP18.1*, and *CsHSP21.8*) using suppression subtractive hybridization (SSH) technology. With heat and cold treatments, transgenic *A. thaliana* plants hosting these tea plant *CsHSP* genes exhibited lower malondialdehyde contents, ion leakage, higher proline contents, and transcript levels of stress-related genes (Wang et al. 2017b).

6 Transcriptome Analysis of Tea Plant

In recent years, transcriptome analysis was widely used in tea plant. Based on transcriptome database, some genes involved in metabolism of specific secondary metabolites closely related with tea quality were revealed (Shi et al. 2011; Wu et al.

2013; Li et al. 2015; Wei et al. 2015). The different expression genes in different tea plant cultivars (Wang et al. 2014, 2016a; Wu et al. 2014c; Zhang et al. 2017) and development stages (Paul et al. 2014; Wang et al. 2014; Wei et al. 2014) were also identified. Furthermore, the genes response to abiotic stresses, such as cold (Wang et al. 2013; Zheng et al. 2015), drought (Maritim et al. 2015; Tony et al. 2016; Wang et al. 2016c), and common biotic stresses (Jayaswall et al. 2016; Wang et al. 2016b, 2016d) were discovered. The influences at molecular level on tea plant treated with different colored lights (Hao et al. 2016), nitrogen sources (Liu et al. 2017), and aluminum levels (Li et al. 2017b) were also analyzed. The integrated analysis of transcriptomic and proteomic (Wang et al. 2016a), metabolic (Li et al. 2016a; Liu et al. 2017; Zhang et al. 2017), or miRNA (Zheng et al. 2015) further revealed the metabolic mechanism in tea plant.

However, the specialized transcriptome analysis of tea plants under the heat stress was not reported at present. The transcription factors involved in regulating response mechanism of heat stress, such as heat shock factors (Liu et al. 2016), AP2/ERF (Wu et al. 2015), and WRKY (Wu et al. 2016a) family transcription factors, were identified based on transcriptome and analyzed by bioinformatics and qRT-PCR. Except the regulated genes, the structural genes involved in responding to heat stress were also expected to be identified and analyzed by transcriptome sequencing. Combined with the transcript levels of responding genes under heat stress, the comprehensive analyses of corresponding miRNAs, proteins, and physiological and biochemical indicators will improve the understanding for heat resistance mechanism of tea plant.

7 MiRNA Analysis of Tea Plant

MicroRNAs are a class of endogenously, transcribed single stranded, with a length of about 18–22 nucleotides noncoding RNAs. In higher plants, microRNAs play important roles in gene regulation at the posttranscriptional level (Jia et al. 2015; Hou et al. 2017). MicroRNAs also play critical regulatory roles involved in diverse aspects of plant, such as plant development, signal transduction, protein degradation, metabolism, and response to environmental stress, including heat stress (Yang et al. 2010; Gao et al. 2011; Mohanpuria and Yadav 2012).

Higher temperature resulted in the disruption and disassembly of chloroplasts and usually causes great losses in yields and quality of crops. High temperature suppresses the production of small RNAs from the chloroplast genome of Chinese cabbage, indicating the potential roles in transcriptional or posttranscriptional regulation of these chloroplast small RNAs (csRNAs) (Wang et al. 2011). In poplar, 52 miRNAs were identified as heat-responsive miRNAs at the genome-wide level (Chen et al. 2012). Several heat-inhibitive miRNAs (bra-miR398a and bra-miR398b) and heat-induced miRNAs (bra-miR156h and bra-miR156g) were identified in Chinese cabbage (Xiang et al. 2012).

A great many of conserved and novel miRNAs have been identified using computational and experimental approaches in tea plant (Das and Mondal 2010; Prabu and Mandal 2010; Mohanpuria and Yadav 2012). Zhang et al. identified and characterized numerous microRNA respond to cold stress through degradome analysis (Zhang et al. 2014). Sun et al. combined small RNA, and degradome sequencing reveals complex microRNA regulation of catechin biosynthesis in tea plant (Sun et al. 2017). Jeyaraj et al. discovered miRNAs were closely related with the dormancy of bud in tea plant (Jeyaraj et al. 2014).

There is no report of miRNAs in response to heat stress in tea plant till now. That is certainly true some miRNAs function in positive modulation of heat tolerance and should be clarified later.

8 Proteomic, Metabolomic, and Genomic Approaches of Heat Stress in Tea Plant

With the rapid development of molecular biology and the emergence of a suite of new technologies, including degradome sequencing, metabolism sequencing, and genome sequencing, a large amount of data resources and efficient research methods have been applied in tea plant.

The first whole-genome sequencing of tea plant has already completed, which provided insights into of tea flavor and independent evolution of caffeine biosynthesis and laid a foundation for the further research of metabolomic and functional genomic refinement of characteristic biosynthesis pathways (Xia et al. 2017). Ji et al. applied proton nuclear magnetic resonance (1H NMR)-based metabolomics to characterize the metabotype and to understood and the metabolic mechanism of several tea plant cultivars, which created a useful method of comprehensive metabotyping to assess and develop a new tea plant cultivar (Ji et al. 2017). Li et al. combined the transcriptomes and related metabolites helped clarify the molecular regulatory mechanisms of albino phenomenon in ‘Anji Baicha’ underlying the secondary metabolic pathways in different stages (Li et al. 2016a).

The ever-evolving sequencing technology has been widely used and greatly facilitated the development of “Omics.” Proteomic approach provides an effective way for better understanding of the molecular mechanisms involved in heat stress in tea plant. The sequencing of the tea plant genomes has been completed, which will lead to a positive effect to faster and more accurate understanding of the tea plant genome. The implementation of proteomic, metabolomic, and genomic will no doubt make it more effective and productive for improvement of heat stress tolerance of tea plant.

9 Future Perspectives

Tea plants are mainly distributed in the tropics and subtropics regions. In the course of life cycle, tea plants inevitably undergo various adverse stresses. The distribution of tea plants determined they are more likely to suffer from heat stress. High temperature seriously affects the development and growth of tea plant, spoiling the flavor of tea products and declining of tea productions. Global warming appears to be a trend of climate change in the future. Greenhouse planting technology to grow tea plant is not an economically viable option. Alter of genetic machinery may be a more realistic and effective approach for tea plants to grow and reproduce under high-temperature environments. Tea plants have evolved complex physiological and biochemical adaptations to adjust and adapt to a variety of environmental stresses. Enhancing the ability of heat tolerance could enable tea plants to adapt and survive high-temperature conditions. The cultivation of the heat-tolerant breeds has recently received much attention.

Functional proteins and regulatory proteins are two groups of the products of stress-inducible genes. HSPs are a kind of primary heat stress proteins and have been researched in the tea plants. There are also other heat-induced proteins including ubiquitin (Sun and Callis 1997), cytosolic Cu/Zn-SOD (Herouart et al. 1994), and Mn-POD (Brown et al. 1993), which need to be discovered late in tea plants. Transcription factors acted as one of the most important regulatory proteins which could induce rapidly and transiently in response to various environmental stresses. Multiple transcription factors are involved in heat stress, including AP2/ERF, MYB, WRKY, NAC, and HSF (Kotak et al. 2007; Wahid et al. 2007). Among them, HSFs are one of the most important family transcription factors that facilitate the heat shock proteins (HSPs) accumulate and enhance plant resistance to heat (Pirkkala et al. 2001; Liu et al. 2011, 2016).

The integration of new technologies, such as transcriptome analysis, small RNAs, proteomics, metabolomics, etc., has been used in development or environmental stress, which need more attention for studying heat resistance in tea plants. It would be helpful to the improvement of tea plant heat resistance. However, new technology has been conducted with heat stress, and the mechanisms of adaptation in tea plants are still behind in model plant, even other horticulture plants, such as fruit, vegetable, etc. (Zhuang et al. 2014; Li et al. 2017a). In the near future, we believe that these emerging data resources and efficient research methods will further be applied to the study of heat stress and adaptation mechanisms of tea plant.

Acknowledgments The research was supported by the National Natural Science Foundation of China (31200520; 31570691) and Jiangsu Natural Science Foundation (BK2012774).

References

Agrawal GK, Rakwal R, Jwa NS, Agrawal VP (2002) Effects of signaling molecules, protein phosphatase inhibitors and blast pathogen (Magnaporthe grisea) on the mRNA level of a rice (*Oryza sativa* L.) phospholipid hydroperoxide glutathione peroxidase (OsPHGPX) gene in seedling leaves. Gene 283:227–236

Anjum NA, Ahmad I, Mohmood I, Pacheco M, Duarte AC, Pereira E, Umar S, Ahmad A, Khan NA, Iqbal M (2012) Modulation of glutathione and its related enzymes in plants' responses to toxic metals and metalloids—a review. Environ Exp Bot 75:307–324

Basak M, Sharma M, Chakraborty U (2001) Biochemical responses of *Camellia sinensis* (L.) O. Kuntze to heavy metal stress. J Environ Biol 22:37–41

Brown JA, Li D, Alic M, Gold MH (1993) Heat-shock induction of manganese peroxidase gene-transcription in phanerochaete-chrysosporium. Appl Environ Microbiol 59:4295–4299

Chen L, Ren YY, Zhang YY, Xu JC, Sun FS, Zhang ZY, Wang YW (2012) Genome-wide identification and expression analysis of heat-responsive and novel microRNAs in *Populus tomentosa*. Gene 504:160–165

Choi Y, Lee SM, Chun J, Lee HB, Lee J (2006) Influence of heat treatment on the antioxidant activities and polyphenolic compounds of Shiitake (*Lentinus edodes*) mushroom. Food Chem 99:381–387

Das A, Mondal TK (2010) Computational identification of conserved microRNAs and their targets in tea (*Camellia sinensis*). Am J Plant Sci 1:77–86

Das A, Das S, Mondal TK (2012) Identification of differentially expressed gene profiles in young roots of tea [*Camellia sinensis* (L.) O. Kuntze] subjected to drought stress using suppression subtractive hybridization. Plant Mol Biol Report 30:1088–1101

Fu JY (2014) Cloning of a new glutathione peroxidase gene from tea plant (*Camellia sinensis*) and expression analysis under biotic and abiotic stresses. Bot Stud 55:7

Gao P, Bai X, Yang LA, Lv DK, Pan X, Li Y, Cai H, Ji W, Chen Q, Zhu YM (2011) osa-MIR393: a salinity- and alkaline stress-related microRNA gene. Mol Biol Rep 38:237–242

Han D, Yang F, Yang ZQ, Jin ZF (2016) Effects of high temperature stress and recovery on photosynthesis and stress tolerance of tea leaves. Chin J Agrometeorol 37:297–306

Hao X, Horvath DP, Chao WS, Yang Y, Wang X, Xiao B (2014) Identification and evaluation of reliable reference genes for quantitative real-time PCR analysis in tea plant (*Camellia sinensis* (L.) O. Kuntze). Int J Mol Sci 15:22155–22172

Hao XY, Li LT, Hu YR, Zhou C, Wang XC, Wang L, Zeng JM, Yang YJ (2016) Transcriptomic analysis of the effects of three different light treatments on the biosynthesis of characteristic compounds in the tea plant by RNA-Seq. Tree Genet Genomes 12:118

Heil M, Bostock RM (2002) Induced systemic resistance (ISR) against pathogens in the context of induced plant defences. Ann Bot 89:503–512

Herouart D, Van Montagu M, Inze D (1994) Developmental and environmental regulation of the Nicotiana plumbaginifolia cytosolic Cu/Zn-superoxide dismutase promoter in transgenic tobacco. Plant Physiol 104:873–880

Hou YH, Jeyaraj A, Zhang X, Wei CL (2017) Absolute quantification of microRNAs in green tea (*Camellia sinensis*) by stem-loop quantitative real-time PCR. J Sci Food Agric 97:2975–2981

Hovmøller MS (2002) Aerial dispersal of pathogens on the global and continental scales and its impact on plant disease. Science 297:537

Jayaswall K, Mahajan P, Singh G, Parmar R, Seth R, Raina A, Swarnkar MK, Singh AK, Shankar R, Sharma RK (2016) Transcriptome analysis reveals candidate genes involved in blister blight defense in tea (*Camellia sinensis* (L) Kuntze). Sci Rep 6:30412

Jeyaraj A, Chandran V, Gajjeraman P (2014) Differential expression of microRNAs in dormant bud of tea [*Camellia sinensis* (L.) O. Kuntze]. Plant Cell Rep 33:1053–1069

Ji HG, Lee YR, Lee MS, Hwang KH, Kim EH, Park JS, Hong YS (2017) Metabolic phenotyping of various tea (*Camellia sinensis* L.) cultivars and understanding of their intrinsic metabolism. Food Chem 233:321–330

Jia XL, Li MY, Jiang Q, Xu ZS, Wang F, Xiong AS (2015) High-throughput sequencing of small RNAs and anatomical characteristics associated with leaf development in celery. Sci Rep 5:11093

Kotak S, Larkindale J, Lee U, von Koskull-Doring P, Vierling E, Scharf KD (2007) Complexity of the heat stress response in plants. Curr Opin Plant Biol 10:310–316

Kumar D, Gulati A, Sharma U (2016) Determination of theanine and catechin in Camellia sinensis (Kangra Tea) leaves by HPTLC and NMR techniques. Food Anal Methods 9:1666–1674

Li XW, Feng ZG, Yang HM, Zhu XP, Liu J, Yuan HY (2010) A novel cold-regulated gene from *Camellia sinensis*, CsCOR1, enhances salt- and dehydration-tolerance in tobacco. Biochem Biophys Res Commun 394:354–359

Li CF, Zhu Y, Yu Y, Zhao QY, Wang SJ, Wang XC, Yao MZ, Luo D, Li X, Chen L (2015) Global transcriptome and gene regulation network for secondary metabolite biosynthesis of tea plant (*Camellia sinensis*). BMC Genomics 16:560

Li CF, Xu YX, Ma JQ, Jin JQ, Huang DJ, Yao MZ, Ma CL, Chen L (2016a) Biochemical and transcriptomic analyses reveal different metabolite biosynthesis profiles among three color and developmental stages in 'Anji Baicha' (*Camellia sinensis*). BMC Plant Biol 16:195

Li H, Huang W, Liu Z, Wang Y, Wu Z, Zhuang J (2016b) Isolation and expression analysis of two temperature responsive dof genes from *Camellia sinensis*. J Tea Sci 36:312–322

Li H, Huang W, Liu ZW, Wang YX, Zhuang J (2016c) Transcriptome-based analysis of Dof family transcription factors and their responses to abiotic stress in tea plant (*Camellia sinensis*). Int J Genom 2016:5614142

Li MY, Song X, Wang F, Xiong AS (2016d) Suitable reference genes for accurate gene expression analysis in parsley (*Petroselinum crispum*) for abiotic stresses and hormone stimuli. Front Plant Sci 7:1481

Li MY, Wang F, Jiang Q, Wang GL, Tian C, Xiong AS (2016e) Validation and comparison of reference genes for qPCR normalization of celery (*Apium graveolens*) at different development stages. Front Plant Sci 7:313

Li MY, Hou XL, Wang F, Tan GF, Xu ZS, Xiong AS (2017a) Advances in the research of celery, an important Apiaceae vegetable crop. Crit Rev Biotechnol:1–12

Li Y, Huang J, Song XW, Zhang ZW, Jiang Y, Zhu YL, Zhao H, Ni DJ (2017b) An RNA-Seq transcriptome analysis revealing novel insights into aluminum tolerance and accumulation in tea plant. Planta 246:91–103

Liu HC, Liao HT, Charng YY (2011) The role of class A1 heat shock factors (HSFA1s) in response to heat and other stresses in Arabidopsis. Plant Cell Environ 34:738–751

Liu Z, Zhijun WU, Xinghui LI, Tong LI, Zhuang J, University, N.A. (2014a) Gene cloning of Cs DREB-A1 transcription factor from *Camellia sinensis* and its characteristic analysis. J Plant Resour Environ

Liu ZW, Xiong YY, Li T, Yan YJ, Han HR, Wu ZJ, Zhuang J (2014b) Isolation and expression profiles analysis of Two ERF subfamily transcription factor genes under temperature stresses in *Camellia sinensis*. Zhiwu Shengli Xuebao 50:1821–1832

Liu Z, Xiong Y, Li T, Yan Y, Han H, Wu Z, Zhuang J (2015) The cloning of transcription factor gene CsDREB-A4 and the response to temperature stress in *Camellia sinensis*. J Tea Sci 35:24–26

Liu ZW, Wu ZJ, Li XH, Huang Y, Li H, Wang YX, Zhuang J (2016) Identification, classification, and expression profiles of heat shock transcription factors in tea plant (*Camellia sinensis*) under temperature stress. Gene 576:52–59

Liu MY, Burgos A, Zhang Q, Tang D, Shi Y, Ma L, Yi X, Ruan J (2017) Analyses of transcriptome profiles and selected metabolites unravel the metabolic response to NH 4 + and NO 3 − as signaling molecules in tea plant (*Camellia sinensis* L.). Sci Hortic 218:293–303

Ma QP, Hao S, Chen X, Li XH (2016) Validation of reliability for reference genes under various abiotic stresses in tea plant. Russ J Plant Physiol 63:423–432

Maritim TK, Wachira FN, Kamunya SM, Mireji P, Mwendia C, Muoki RC, Wamalwa M, Stomeo F, Martina KP (2015) Transcriptome-based identification of drought responsive genes in the tea plant. J Plant Biotechnol 43:302–310

Massad TJ, Dyer LA, Vega CG (2012) Costs of defense and a test of the carbon-nutrient balance and growth-differentiation balance hypotheses for two co-occurring classes of plant defense. PLoS One 7:e47554

Milla MAR, Maurer A, Huete AR, Gustafson JP (2003) Glutathione peroxidase genes in *Arabidopsis* are ubiquitous and regulated by abiotic stresses through diverse signaling pathways. Plant J 36:602–615

Mohanpuria P, Yadav SK (2012) Characterization of novel small RNAs from tea (*Camellia sinensis* L.). Mol Biol Rep 39:3977–3986

Mordecai EA (2011) Pathogen impacts on plant communities: unifying theory, concepts, and empirical work. Ecol Monogr 81:429–441

Nakashima K, Yamaguchishinozaki K (2009) Promoters and transcription factors in abiotic stress-responsive gene expression. Springer, Dordrecht

Navrot N, Collin V, Gualberto J, Gelhaye E, Hirasawa M, Rey P, Knaff DB, Issakidis E, Jacquot JP, Rouhier N (2006) Plant glutathione peroxidases are functional peroxiredoxins distributed in several subcellular compartments and regulated during biotic and abiotic stresses. Plant Physiol 142:1364–1379

Ohama N, Sato H, Shinozaki K, Yamaguchi-Shinozaki K (2017) Transcriptional regulatory network of plant heat stress response. Trends Plant Sci 22:53–65

Panda SK, Upadhyaya H (2013) Abiotic stress responses in tea [*Camellia sinensis* L (O) Kuntze]: an overview. Rev Agric Sci 1

Paul A, Jha A, Bhardwaj S, Singh S, Shankar R, Kumar S (2014) RNA-seq-mediated transcriptome analysis of actively growing and winter dormant shoots identifies non-deciduous habit of evergreen tree tea during winters. Sci Rep 4:5932

Pirkkala L, Nykanen P, Sistonen L (2001) Roles of the heat shock transcription factors in regulation of the heat shock response and beyond. FASEB J 15:1118–1131

Prabu GR, Mandal AKA (2010) Computational identification of miRNAs and their target genes from expressed sequence tags of tea (*Camellia sinensis*). Genom Proteom Bioinform 8:113–121

Puranik S, Sahu PP, Srivastava PS, Prasad M (2012) NAC proteins: regulation and role in stress tolerance. Trends Plant Sci 17:369–381

Ramos J, Matamoros MA, Naya L, James EK, Rouhier N, Sato S, Tabata S, Becana M (2009) The glutathione peroxidase gene family of Lotus japonicus: characterization of genomic clones, expression analyses and immunolocalization in legumes. New Phytol 181:103–114

Rejeb IB, Pastor V, Mauch-Mani B (2014) Plant responses to simultaneous biotic and abiotic stress: molecular mechanisms. Plants 3:458

Sharangi AB (2009) Medicinal and therapeutic potentialities of tea (*Camellia sinensis* L.) - A review. Food Res Int 42:529–535

Shi CY, Yang H, Wei CL, Yu O, Zhang ZZ, Jiang CJ, Sun J, Li YY, Chen Q, Xia T, Wan XC (2011) Deep sequencing of the Camellia sinensis transcriptome revealed candidate genes for major metabolic pathways of tea-specific compounds. BMC Genomics 12:131

Song H, Zhang X, Shi C, Wang S, Wu A, Wei C (2016) Selection and verification of candidate reference genes for mature MicroRNA expression by quantitative RT-PCR in the tea plant (*Camellia sinensis*). Genes (Basel) 7:25

Strauss, S.Y. and Zangerl, A.R., 2002. Plant-insect interactions in terrestrial ecosystems. In Plant-Animal Interactions. An Evolutionary Approach; Herrera, C.M., Pellmyr, O., Eds.; Blackwell Science: Oxford, UK 77–106

Sun CW, Callis J (1997) Independent modulation of Arabidopsis thaliana polyubiquitin mRNAs in different organs and in response to environmental changes. Plant J 11:1017–1027

Sun P, Cheng C, Lin Y, Zhu Q, Lin J, Lai Z (2017) Combined small RNA and degradome sequencing reveals complex microRNA regulation of catechin biosynthesis in tea (*Camellia sinensis*). PLoS One 12:e0171173

Swarbrick PJ, Schulze-Lefert P, Scholes JD (2006) Metabolic consequences of susceptibility and resistance (race-specific and broad-spectrum) in barley leaves challenged with powdery mildew. Plant Cell Environ 29:1061–1076

Tony M, Samson K, Charles M, Paul M, Richard M, Mark W, Stomeo F, Sarah S, Martina K, Francis W (2016) Transcriptome-based identification of water-deficit stress responsive genes in the tea plant, *Camellia sinensis*. J Plant Biotechnol 43:302–310

Wahid A, Gelani S, Ashraf M, Foolad MR (2007) Heat tolerance in plants: an overview. Environ Exp Bot 61:199–223

Wang WX, Vinocur B, Altman A (2003) Plant responses to drought, salinity and extreme temperatures: towards genetic engineering for stress tolerance. Planta 218:1–14

Wang WX, Vinocur B, Shoseyov O, Altman A (2004) Role of plant heat-shock proteins and molecular chaperones in the abiotic stress response. Trends Plant Sci 9:244–252

Wang L, Yu X, Wang H, Lu YZ, de Ruiter M, Prins M, He YK (2011) A novel class of heat-responsive small RNAs derived from the chloroplast genome of Chinese cabbage (*Brassica rapa*). BMC Genomics 12:289

Wang XC, Zhao QY, Ma CL, Zhang ZH, Cao HL, Kong YM, Yue C, Hao XY, Chen L, Ma JQ, Jin JQ, Li X, Yang YJ (2013) Global transcriptome profiles of *Camellia sinensis* during cold acclimation. BMC Genomics 14:415

Wang L, Yue C, Cao HL, Zhou YH, Zeng JM, Yang YJ, Wang XC (2014) Biochemical and transcriptome analyses of a novel chlorophyll-deficient chlorina tea plant cultivar. BMC Plant Biol 14:1–13

Wang ML, Zhu XJ, Wang WD, Wang XQ, Lin ML, Li XH (2015a) Molecular cloning and expression analysis of low molecular weight heat shock protein gene CsHSP17.2 from *Camellia sinensis*. J Nanjing Agric Univ 38(3):389–394

Wang Y, Liu Z, Zhijun WU, Hui LI, Xinghui LI, Zhuang J (2015b) Isolation and expression profiles of NAC transcription factor genes under temperature stress in *Camellia sinensis*. Acta Botan Boreali-Occiden Sin 35(11):2148–2156

Wang L, Cao HL, Chen CS, Yue C, Hao XY, Yang YJ, Wang XC (2016a) Complementary transcriptomic and proteomic analyses of a chlorophyll-deficient tea plant cultivar reveal multiple metabolic pathway changes. J Proteome 130:160–169

Wang L, Wang YC, Cao HL, Hao XY, Zeng JM, Yang YJ, Wang XC (2016b) Transcriptome analysis of an anthracnose-resistant tea plant cultivar reveals genes associated with resistance to Colletotrichum camelliae. PLoS One 11:e0148535

Wang W, Xin H, Wang M, Ma Q, Wang L, Kaleri NA, Wang Y, Li X (2016c) Transcriptomic analysis reveals the molecular mechanisms of drought-stress-induced decreases in *Camellia sinensis* leaf quality. Front Plant Sci 7:385

Wang YN, Tang L, Hou Y, Wang P, Yang H, Wei CL (2016d) Differential transcriptome analysis of leaves of tea plant (*Camellia sinensis*) provides comprehensive insights into the defense responses to Ectropis oblique attack using RNA-Seq. Funct Integr Genomics 16:383–398

Wang YX, Liu ZW, Wu ZJ, Li H, Zhuang J (2016e) Transcriptome-wide identification and expression analysis of the NAC gene family in tea plant [*Camellia sinensis* (L.) O. Kuntze]. PLoS One 11:e0166727

Wang M, Zou Z, Li Q, Sun K, Chen X, Li X (2017a) The CsHSP17.2 molecular chaperone is essential for thermotolerance in Camellia sinensis. Sci Rep 7:1237

Wang M, Zou Z, Li Q, Xin H, Zhu X, Chen X, Li X (2017b) Heterologous expression of three *Camellia sinensis* small heat shock protein genes confers temperature stress tolerance in yeast and *Arabidopsis thaliana*. Plant Cell Rep 36:1125–1135

Wang ML, Li QH, Xin HH, Chen X, Zhu XJ, Li XH (2017c) Reliable reference genes for normalization of gene expression data in tea plants (*Camellia sinensis*) exposed to metal stresses. PLoS One 12:e0175863

Wei K, Wang LY, Wu LY, Zhang CC, Li HL, Tan LQ, Cao HL, Cheng H (2014) Transcriptome analysis of indole-3-butyric acid-induced adventitious root formation in nodal cuttings of *Camellia sinensis* (L.). PLoS One 9:e107201

Wei K, Wang L, Zhang C, Wu L, Li H, Zhang F, Cheng H (2015) Transcriptome analysis reveals key flavonoid 3′-hydroxylase and flavonoid 3′,5′-hydroxylase genes in affecting the ratio of dihydroxylated to trihydroxylated catechins in *Camellia sinensis*. PLoS One 10:e0137925

Wu HL, Chen D, Li JX, Yu B, Qiao XY, Huang HL, He YM (2013) De Novo characterization of leaf transcriptome using 454 sequencing and development of EST-SSR markers in tea (*Camellia sinensis*). Plant Mol Biol Rep 31:524–538

Wu Z, Li X, Fang W, Zhou L, Zhao Z, Zhuang J (2014a) Isolation of Cs RAV2 transcription factor gene of tea plant and its expression analysis. J Tea Sci:297–306

Wu Z, Lu L, Li X, Fang W, Zhou L, Tan G, Zhuang J (2014b) Isolation and expression profiles analysis of AP2/ERF-B3 group transcription factor from *Camellia sinensis*. J Nanjing Agric Univ 37:67–75

Wu ZJ, Li XH, Liu ZW, Xu ZS, Zhuang J (2014c) *De novo* assembly and transcriptome characterization: novel insights into catechins biosynthesis in *Camellia sinensis*. BMC Plant Biol 14:277

Wu ZJ, Li XH, Liu ZW, Li H, Wang YX, Zhuang J (2015) Transcriptome-based discovery of AP2/ERF transcription factors related to temperature stress in tea plant (*Camellia sinensis*). Funct Integr Genomics 15:741–752

Wu ZJ, Li XH, Liu ZW, Li H, Wang YX, Zhuang J (2016a) Transcriptome-wide identification of *Camellia sinensis* WRKY transcription factors in response to temperature stress. Mol Gen Genomics 291:255–269

Wu ZJ, Tian C, Jiang Q, Li XH, Zhuang J (2016b) Selection of suitable reference genes for qRT-PCR normalization during leaf development and hormonal stimuli in tea plant (*Camellia sinensis*). Sci Rep 6:19748

Xia EH, Zhang HB, Sheng J, Li K, Zhang QJ, Kim C, Zhang Y, Liu Y, Zhu T, Li W, Huang H, Tong Y, Nan H, Shi C, Shi C, Jiang JJ, Mao SY, Jiao JY, Zhang D, Zhao Y, Zhao YJ, Zhang LP, Liu YL, Liu BY, Yu Y, Shao SF, Ni DJ, Eichler EE, Gao LZ (2017) The tea tree genome provides insights into tea flavor and independent evolution of caffeine biosynthesis. Mol Plant 10:866–877

Xiang Y, Han W, Lu Y, Ruiter MD, Cariaso M, Prins M, Tunen AV, He Y (2012) Identification of conserved and novel microRNAs that are responsive to heat stress in *Brassica rapa*. J Exp Bot 63:1025–1038

Xiong YY, Ya-Qi GU, Liu ZW, Tong LI, Zhi-Jun WU, Jing Z (2015) Isolation and expression profiles analysis of two hsf transcription factor genes under temperature stress in *Camellia sinensis*. Plant Physiol J 51:2143–2151

Yang WZ, Liu X, Zhang JG, Feng JL, Li C, Chen JS (2010) Prediction and validation of conservative microRNAs of *Solanum tuberosum* L. Mol Biol Rep 37:3081–3087

Zhang Y, Zhu XJ, Chen X, Song CNA, Zou ZW, Wang YH, Wang ML, Fang WP, Li XH (2014) Identification and characterization of cold-responsive microRNAs in tea plant (*Camellia sinensis*) and their targets using high-throughput sequencing and degradome analysis. BMC Plant Biol 14:271

Zhang Q, Liu M, Ruan J (2017) Integrated transcriptome and metabolic analyses reveals novel insights into free amino acid metabolism in huangjinya tea cultivar. Front Plant Sci 8:291

Zheng C, Zhao L, Wang Y, Shen J, Zhang Y, Jia S, Li Y, Ding Z (2015) Integrated RNA-Seq and sRNA-Seq analysis identifies chilling and freezing responsive key molecular players and pathways in tea plant (*Camellia sinensis*). PLoS One 10:e0125031

Zhuang J, Anyia A, Vidmar J, Xiong AS, Zhang J (2011a) Discovery and expression assessment of the AP2-like genes in *Hordeum vulgare*. Acta Physiol Plant 33:1639–1649

Zhuang J, Chen JM, Yao QH, Xiong F, Sun CC, Zhou XR, Zhang JA, Xiong AS (2011b) Discovery and expression profile analysis of AP2/ERF family genes from Triticum aestivum. Mol Biol Rep 38:745–753

Zhuang J, Zhang J, Hou XL, Wang F, Xiong AS (2014) Transcriptomic, proteomic, metabolomic and functional genomic approaches for the study of abiotic stress in vegetable crops. Crit Rev Plant Sci 33:225–237

Chapter 3
Response and Adaptation Mechanisms of Tea Plant to Low-Temperature Stress

Xinyuan Hao, Lu Wang, Jianming Zeng, Yajun Yang, and Xinchao Wang

Abstract As global warming continues to intensify, abnormal changes in microclimates have become severe environmental problems threatening tea production. Tea plant is a thermophilic evergreen woody species, and low temperature is the most severe environmental stress that affects the geographical distribution and productivity of tea plant. Cold acclimation and winter bud dormancy are two key tea plant adaptive mechanisms for coping with low temperature. Once tea plant perceives low temperature, a series of changes, including those in gene expression, protein abundance, and metabolite accumulation, occur in tea leaves. Under long-term cold stress during winter, the overwintering buds of tea plant enter a dormant state to survive possible damaging temperature fluctuations. Complex physiological and biochemical modifications can be detected during the formation and release of bud dormancy. Both cold acclimation and bud dormancy in tea plant have been extensively studied: many genes have been characterized, and the major signaling pathways have been investigated. In this chapter, recent advances in the responsive and adaptive mechanisms of tea plant to low-temperature stress are highlighted.

Keywords Tea plant · Low temperature · Cold acclimation · Bud dormancy · Freezing tolerance

1 Introduction

Tea plant (*Camellia sinensis* (L.) O. Kuntze) was first discovered and used for the production of drinks and medicine in China approximately 3000 years ago. Since then, due to its prominent health effects, tea has become the most consumed nonalcoholic beverage worldwide following water, and tea plant is cultivated in more than

X. Hao · L. Wang · J. Zeng · Y. Yang (✉) · X. Wang (✉)
National Center for Tea Improvement, Tea Research Institute, Chinese Academy of Agricultural Sciences; Key Laboratory of Tea Biology and Resources Utilization, Ministry of Agriculture, Hangzhou, China
e-mail: haoxy@tricaas.com; wanglu317@tricaas.com; yjyang@tricaas.com; wangxinchao@caas.cn

W.-Y. Han et al. (eds.), *Stress Physiology of Tea in the Face of Climate Change*,
https://doi.org/10.1007/978-981-13-2140-5_3

60 countries on 5 continents, from 49° N in Ukraine to 33° S in South Africa (Turkozu and Sanlier 2017). Temperature is recognized as a limiting factor affecting the geographical distribution of tea plant. In China, although the regional distribution of tea plant is widespread (from 122° E to 95° E and from 18° N to 37° N), the latitude of 37° N in Rongcheng County, Shandong Province, is likely the northern border for tea plant cultivation without protection from agricultural facilities to help the plants survive the long, freezing winter (Yang 2005). As a thermophilous evergreen woody species that originated from tropical areas in Southwest China, tea plant is mainly planted in the Yangtze valley as well as in southeastern coastal areas and southwestern areas in China and is susceptible to low temperature (Yang 2005; Yu 1986). With the exception of a small proportion of regions from southeastern coastal areas and southwestern areas, most tea plant cultivation regions in China have relatively low temperatures in winter, which in turn causes tea plant to enter winter dormancy. In India, tea plant is also grown across a wide range of climatic variables (latitudes from 8° N to 32° N and altitudes ranging from near sea level to 2414 m above mean sea level) (Hazarika and Muraleedharan 2011). In southern India, tea is harvested year round, whereas tea plants grown in northeastern India exhibit winter dormancy and are affected by winter cold. When tea plant enters dormancy during winter, no fresh leaves are produced during that period. Therefore, dormancy caused by winter cold is a crucial factor affecting annual tea yields. Moreover, low temperatures during the growing season usually negatively affect the accumulation of metabolites in shoots or even damage tea shoots (Zheng et al. 2016) (Fig. 3.1).

The natural distribution, growth and development, yield, and ability to resist winter or late-spring cold as well as the quality of tea are associated with cold tolerance. As plants are unable to escape unfavorable environmental conditions, they have evolved adaptive mechanisms. In response to cold stress, plants are known to induce corresponding protection mechanisms, such as (i) changes in the composition, structure, and function of the cell membrane; (ii) synthesis of the antifreezing compounds and proteins; (iii) reductions in cell water content; etc. Understanding the mechanisms of cold response and adaptation in tea plant is pivotal not only for improving the understanding of cold tolerance regulation in woody plants but also for improving breeding.

In this chapter, recent advances in responsive and adaptive mechanisms of tea plant to low-temperature stress are reviewed.

2 Cold Acclimation: A Necessary Process to Increase Low-Temperature Tolerance

Plant cold acclimation (CA) is a common phenomenon in which freezing tolerance is increased during exposure to low, nonfreezing temperatures; at the basic phenotypic level, this tolerance reflects the increase in the survival of plants, tissues, or cells after being subjected to freeze-thaw cycles that present damaging temperature ranges (Hincha and Zuther 2014). CA is genetically complex and is governed by

Fig. 3.1 Typical effects of low-temperature stress and damage on tea plant. (**a**) and (**b**), frosted and snow-covered tea plants, respectively; (**c**) and (**d**), phenomenon of natural cold-damaged leaves and young shoots, respectively

quantitative traits. Extensive gene expression reprogramming and metabolism modification are triggered when plants sense low environmental temperature (Chinnusamy et al. 2010a). The rapid induction of microdomain membrane rigidity by low temperature is recognized as the primary event during cold stress sensing by plants, followed by a transient increase in cytosolic Ca^{2+} levels within seconds (Knight et al. 1991; Orvar et al. 2000). The induced Ca^{2+} signature can be decoded by different pathways that promote the production of potential protective proteins such as cold-responsive (COR)/late embryogenesis abundant (LEA) proteins. Among several cold signaling pathways, the inducer of C-repeat binding factor (CBF) expression (ICE)-CBF/dehydration-responsive element binding factor 1 (DREB1) pathway is key to the differential regulation of cold-responsive genes (Chinnusamy et al. 2007). Moreover, posttranscriptional regulation, posttranslational regulation, and hormone regulation play important roles in the regulation of cold signaling. In general, the processes by which reactive oxygen species (ROS) are cleared and metabolites such as sugars, proline, and flavonoids accumulate are altered during CA and de-acclimation processes (Hincha and Zuther 2014; Miura and Furumoto 2013). Under natural conditions, CA is an important plant response for coping with seasonal low temperatures. CA has been the focus of researchers since the nineteenth century and has become an enduring subject of intense scientific study since then. In China, tea-producing areas are distributed mainly in subtropical and mid-temperate zones where the minimum air temperature is less than 0 °C in most parts or as low as −15 °C in certain regions. After CA, the cold resistance of tea plant can improve. When the 10-day average air temperature decreases to approximately 7° C, tea plant undergoes the CA process, and once the 10-day average air temperature increases past 9° C, tea plant undergoes the de-acclimation process (Yang et al. 2004). However, despite undergoing CA, different tea plant cultivars present distinct cold resistance abilities. Improving cold resistance or breeding cold-resistant tea plant cultivars is crucial for tea production. Thus, understanding the mechanism of CA or the cold response of tea plant is very important. In recent years, the CA response of tea plant, especially its underlying genetic regulatory mechanism, has been the focus of research and intense studies involving molecular biological technologies and next-generation sequencing (NGS).

2.1 Freezing Tolerance Measurements and Physiological Changes During CA

CA is a quantitative trait and is accompanied by complex changes in gene expression, protein abundance, metabolite accumulation, and lipid composition. After CA, plants generally display stronger cold tolerance than do control or de-acclimatized plants. Therefore, the use of one or multiple approaches to precisely quantify plant freezing tolerance is necessary to intensively study the genetic and molecular mechanisms underlying and determining this complex trait. Electrolyte leakage is a reliable physiological parameter for assessing the integrity of the plasma membrane during freezing, which reflects the cold tolerance of a plant (Thalhammer et al. 2014). Electrolyte leakage assays have been widely applied to tea plant cold tolerance studies in which detached leaves with minimal modifications have been used (Ban et al. 2017; Wang et al. 2013; Yue et al. 2015). Both in natural and artificial CA, the temperature leading to 50% tissue damage due to electrolyte leakage (EL50) and the relative electrical conductivity at freezing temperatures decrease, indicating an increase in cold tolerance of tea leaves. With respect to changes in electrolyte leakage, chlorophyll fluorescence imaging, which is based on variable fluorescence/maximal fluorescence (Fv/Fm) measurements, represents a rapid visual approach for determining the temperature at which 50% of damage occurs (LT50) in whole plants and detached or undetached leaves (Thalhammer et al. 2014). For electrolyte leakage assays, because of large leaf sizes, small discs cut from tea leaves via a hole punch are needed; by contrast, chlorophyll fluorescence imaging causes little or no damage to the leaves. Fv/Fm values are reduced after CA treatment, and more reduction occurs in cold-susceptible cultivars of tea plant than in cold-tolerant cultivars (Ban et al. 2017). With the exception of the above parameters, malondialdehyde (MDA), proline, and soluble carbohydrate (including total soluble sugars, fructose, sucrose, glucose, raffinose, trehalose, etc.) contents are also used as important metabolites to reflect the status of cold-acclimating plants or cold-acclimated plants. MDA is a product of the breakdown of peroxidized polyunsaturated fatty acids in plant membranes and is used as an indicator of lipid peroxidation levels (Weber et al. 2004). A relatively high MDA content in leaves indicates a relatively weak freezing tolerance. A marked decrease in MDA content was detected following tea plant CA during winter, and this decrease was followed by an increase in MDA content during the de-acclimation stage in the spring (Yue et al. 2015). In actuality, the cellular MDA content fluctuates as the levels of ROS induced by freezing and cold accumulate; therefore, the MDA content would be relatively high if plant leaves are damaged by extreme cold during winter, even if the plant had undergone the CA process (Vyas and Kumar 2005). As what happens in other plants, the levels of sugars and proline are high during plant CA but gradually decrease to normal levels during de-acclimation in tea plant, and their contents are highly correlated with freezing tolerance across various genotypes (Ban et al. 2017; Korn et al. 2008; Lee et al. 2012). Though multiple parameters have been applied in cold tolerance measurements to investigate the complex physiological changes during CA, the choice of proper parameter(s) and the careful performance of assays are necessary due to the special physical structure and physiological properties of tea plant. Furthermore, additional measurements at both the physiological and biochemical levels, including the freezing response of plant tissues and cells, increases in superoxide and hydrogen peroxide in leaves, changes in the fatty acid composition of membranes, elevation of external NADH dehydrogenases, increases in the ratio of dry mass to leaf area, decreases in the relative phosphate content of chloroplasts, and the relative nitrogen content in leaves, among others, should be considered in future studies.

2.2 *Identification and Characterization of Low-Temperature-Responsive Genes*

Alpha-tubulin (*CaTUA*) is a low-temperature-inducible gene that has been cloned from tea plant (Paul et al. 2012). Two *dehydrin* genes (*CsDHN1* and *CsDHN2*) have also been identified (Paul and Kumar 2013). In *Arabidopsis* and other plants, ICE1-CBF-COR has been identified as a dominant low-temperature-triggered transcriptional cascade reprogramming gene whose expression increases tolerance to low temperature (Zhan et al. 2015). Recently, the development of stable loss-of-function *Arabidopsis* mutants further supported the recognized roles of three CBF transcription factors (CBF1, CBF2, and CBF3) in CA (Shi et al. 2017; Zhao et al. 2016). RNA-Seq was used to identify more than 3000 *COR* genes whose expression levels significantly change in *cbf* mutants; in particular, in the *cbf* triple mutants, 346 and 68 were CBF-activated and CBF-repressed genes, respectively. More downstream *COR* genes were regulated by CBF2 and CBF3 rather than by CBF1. In addition, more than 2/3 of the *COR* genes were co-regulated by two or three CBFs, indicating their synergistic effect in CA-dependent freezing tolerance. The full-length cDNAs of *CsICE1* and *CsCBF1* were isolated from tea plant by Wang et al. (2012). Expression analysis revealed that *CsCBF1* could be significantly upregulated by low (4 °C) temperature. Nevertheless, *CsICE1* expression could not be induced by low temperature (4 °C) but could be induced by freezing temperature (−5 °C) (Ding et al. 2015). Though the binding character of CsCBF1 to the CRT/DRE core sequence (CCGAC) has been validated by electrophoretic mobility shift assays, the function and downstream regulatory genes of CsCBF1 are mostly unknown. The direct regulatory relationship between CsICE1 and CsCBF1 also needs to be further confirmed in tea plant under cold stress. It is therefore clear that, because of its advantages, whole-genome sequencing should be performed to identify *ICE*, *CBF*, and *COR* gene members at the genomic level in tea plant (Xia et al. 2017).

In plants, gene expression, cell wall modification, and metabolism (mainly that of carbohydrates and lipids) are strongly affected by chilling conditions (Zhao et al. 2016); low-temperature-responsive proteins such as heat shock proteins, regulatory proteins, detoxification-related proteins, and LEA/dehydrin proteins accumulate (Janmohammadi et al. 2015). However, limited numbers of *COR* genes have been intensively studied in tea plant so far. In 2010, a heterologous plant transgenic system was first used to clone and functionally identify a *COR* gene (*CsCOR1*) from tea plant (Li et al. 2010). In recent years, more *COR* genes have been studied. In particular, 16 tea plant heat shock transcription factors (CsHsfs) have been identified; the expression of 6 of these *CsHsf*s was sensitive to high (38 °C) or low (4 °C) temperature in different cultivars (Liu et al. 2016). Ectopic expression of *CsHSP17.7*, *CsHSP18.1*, and *CsHSP21.8* in *Arabidopsis* decreased the MDA content and ion leakage in overexpression plants but increased the proline content and transcription levels of stress-related genes (Wang et al. 2017b). Moreover, three galactinol synthase (*GolS*) genes involved in the synthesis of raffinose family oligosaccharides in tea plant were cloned and characterized as cold-inducible genes (Zhou et al. 2017). The spermine synthase (*CsSPMS*) coding gene, fatty acid desaturase (*CsSAD*) gene, and H1 histone (*CsHis*) gene in tea plant were also identified to be upregulated by low temperature (Ding et al. 2016; Wang et al. 2014a; Zhu et al. 2015). Transient expression of *CsSPMS* in tobacco leaves upregulated the expression of *NbDREB2a*, and compared with wild-type control plants, tobacco plants overexpressing *CsHis* exhibited higher tolerance to cold stress (Wang et al. 2014a, Zhu et al. 2015). Family gene identification and expression profile analyses revealed that the expression patterns of multiple members of the aquaporin (AQP) and basic leucine zipper (bZIP) gene families significantly varied under cold and other stress conditions, indicating their putative importance in the regulation of cold tolerance in tea plant (Cao et al. 2015; Yue et al. 2014). By using overexpression techniques in *Arabidopsis*, Wang et al. (2017a) further identified the function of a cold temperature-induced bZIP transcription factor (CsbZIP6); interestingly, the results showed a negative role in the regulation of freezing tolerance (Wang et al. 2017a). As the application of NGS to the study of tea plant increases, a large number of differentially expressed genes (DEGs) related to cold stimulation have been continually identified. However, the functional identification of these genes is still limited so far. In addition, due to the lack of a stable transgenic system in tea plant, how to investigate the precise roles of target genes in cold-responding rather than overexpressing heterologous plants remains a bottleneck for tea researchers.

2.3 Major Signals and Pathways Involved in the Regulation of Low-Temperature Tolerance

Low-temperature tolerance is an essential adaptive mechanism of plants for coping with various abiotic stresses in nature. Low temperature generally includes chilling temperatures (0–15 °C) and freezing temperatures (< 0 °C), and different mechanisms will be triggered in plants subjected to these low-temperature stresses (Hincha and Zuther 2014). A series of physiological and biochemical changes, such as changes in gene transcription as well as the accumulation of or reduction in proteins and metabolites, can be observed in stressed plants as time progresses (Theocharis et al. 2012). Based on global transcriptome analyses, DEGs in tea leaves under low-temperature stress for short- and long-time courses were recently comprehensively studied; the results provided an overall understanding of the major signals and pathways involved in the regulation of low-temperature tolerance in tea plant (Wang et al. 2013; Zheng et al. 2015) (Fig. 3.2). During the early stages of chilling temperature (4 °C) and freezing temperature (−5 °C) treatments, more DEGs were detected in the leaves subjected to chilling stress than in the leaves subjected to freezing stress; these results corresponded to those in which more changes in the expression of cold-responsive transcription factor genes occurred in chilling temperature-treated samples than in freezing temperature-treated samples (Zheng et al. 2015). Gene enrichment analysis further revealed that the response pathways to both cold and karrikin and the starch biosynthetic process overlapped in both chilling temperature and freezing temperature treatments. Crucial biological processes related to signal transduction, responses to stimuli, hormone-mediated signaling pathways, carbohydrate metabolic processes, and ROS metabolic processes as well as photosynthesis distinctly differed between the two treatments. In particular, plant hormone signal transduction pathways, photosynthesis, the mitogen-activated protein kinase (MAPK) signaling pathway, metabolism of xenobiotics by cytochrome P450, glutathione metabolism, and starch and sucrose metabolism were involved and function in early responses to chilling stress.

Tea plant needs a long period of time to complete the CA and de-acclimation processes under winter cold stress. Among these processes, the comparison of the gene expression profiles of non-acclimated, cold-acclimated, and de-acclimated tea leaves has identified up to 1770 DEGs (Wang et al. 2013). Members of at least nine gene families (AP2/ERF, bHLH, WRKY, MYB, NAC, bZIP, heat shock, GARS, and zinc finger protein families) that are related to cold stress resistance in plants are involved (Chinnusamy et al. 2010b; Theocharis et al. 2012; Zhao et al. 2015). Signal transduction genes related to low-temperature perception, including those encoding Ca^{2+}-dependent protein kinases (CDPKs), MAPKs, calcineurin B-like protein

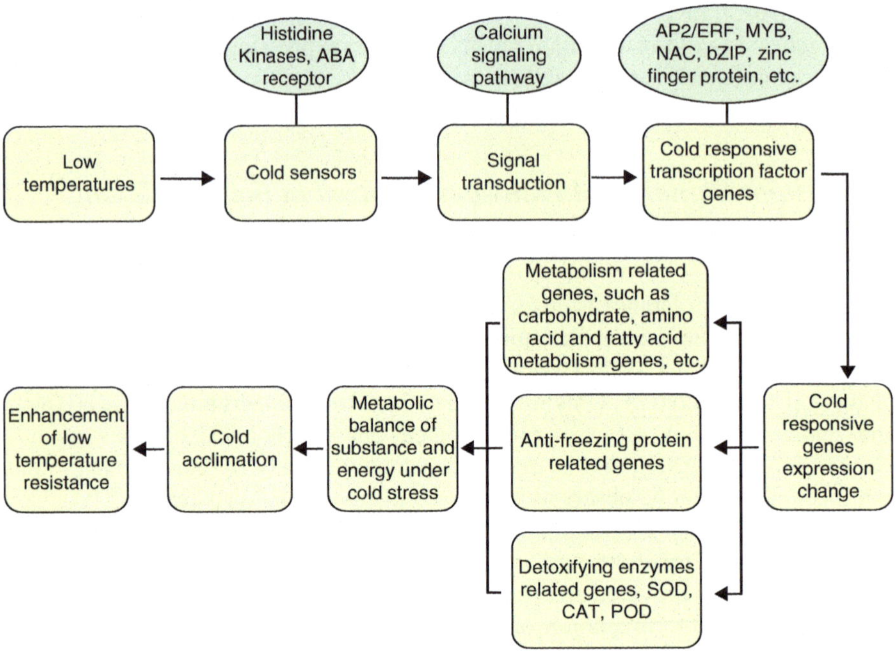

Fig. 3.2 Diagram of the responses of tea plants to low temperature during the CA process. (Wang et al. 2013)

(CBL), calmodulin, calmodulin-binding transcription activator (CAMTA), and the calcium/calmodulin-regulated receptor kinase (CRLK), are differentially regulated and play crucial roles in cold tolerance. Studies have also reported that the expression levels of calmodulin genes in tea plant (*CsCaMs*) are also stimulated by low temperature and $CaCl_2$ treatment, and application of exogenous $CaCl_2$ increased tea plant cold tolerance; electrolyte leakage and the contents of MDA, superoxide anion radical, and proline decreased, whereas superoxide dismutase activity increased (Huang et al. 2015; Huang et al. 2016). Moreover, significant alterations to expression are also observed for genes related to lipid metabolism or stabilization, anti-freezing protein (AFP) accumulation, and osmotic equilibrium maintenance. Importantly, the role of carbohydrate metabolism in freezing tolerance enhancement during CA is highlighted in tea plant. Similarly, in oil tea (*Camellia oleifera*), a sibling species of tea plant, transmembrane transporter genes have been reported to predominate among the DEGs identified from cold-acclimated plants that have different geographical distributions; most of these DEGs encode transmembrane sugar transporters (Chen et al. 2017). By comprehensively determining the content of total sugars and several specific sugars and investigating the expression of sugar-related genes during the winter season, Yue et al. (2015) confirmed the dominant role of starch degradation and sugar accumulation in the enhancement of cold tolerance of tea plant during early periods of CA (Yue et al. 2015). Several sugar-related genes, mainly the beta-amylase (*BAM*) gene, invertase (*INV*) gene, raffinose synthase (*RS*) gene, hexokinase (*HXK*) gene, and Suc non-fermenting 1-related protein kinase 1 (*SnRK1*) gene, and sugars that are ultimately exported by the gene product of the *SWEET* transporter gene are highly important in the regulation of carbohydrate metabolism during CA, and some of these genes have been studied latterly at gene family level (Li et al. 2017; Qian et al. 2016; Yue et al. 2015).

Posttranscriptional regulation based on alternative splicing, pre-mRNA processing, RNA stability, and RNA silencing and export from the nucleus is an important mechanism in CA and cold tolerance (Miura and Furumoto 2013). Increasing numbers of studies have established that microRNA (miRNA)-mediated reprogramming of gene expression is a major defense mechanism in plants and enables them to respond to cold stress (Megha et al. 2017). Zhang et al. identified 106 conserved miRNAs and 98 novel miRNAs out of 215 potential candidate miRNAs from the young leaves of tea plant subjected to 1–48 h of 4 °C temperature (Zhang et al. 2014). A customized miRNA microarray chip loaded with 3228 known miRNA probes and 283 novel probes from tea plant was used to investigate the different expression profiles of cold-responsive miRNAs between the cold-tolerant tea plant cultivar "Yingshuang" and the cold-sensitive cultivar "Baiye 1." The results showed that the cold stress pathways of miR168, miR529, and miR2936 exhibited a certain degree of overlap between both cultivars, while the abundances of miR164, miR408, miR1511, miR5368, miR172, miR482, miR529, and miR1160 significantly differed. Furthermore, several members from the cold stress-specific family, including miR156, miR159, and miR396, present varying regulatory patterns for cold stress responses, suggesting that specificity among species and family members occurs (Karimi et al. 2016; Zhang et al. 2013). Using degradome sequencing, Zhang et al. further confirmed the target genes of partial miRNAs identified in tea plant, including a LEA protein-targeted miRNA (csn-miR164) and six cold-responsive gene-targeted miRNAs (csn-smR35, csn-smR3146, csn-smR5749, csn-smR7277, csn-smR8111, and csn-smR9722) (Zhang et al. 2014). Correlation analysis between differentially expressed miRNAs and differentially expressed mRNAs in response to cold stress revealed complex regulatory relationships, and multiple cold stress-associated signaling pathways or metabolism processes were targeted (Zheng et al. 2015). In actuality, with the exception of posttranscriptional regulation, posttranslational regulation, DNA methylation, histone modification, and the regulation of plant hormone-related gene expression are also important mechanisms involved in CA and cold tolerance and deserve increased attention (Eremina et al. 2016; Miura and Furumoto 2013; Zhou et al. 2015).

3 Bud Dormancy: An Important Strategy for Surviving Freezing Cold During Winter

In perennial woody species native to temperate and boreal regions, the activity-dormancy cycle is an important adaptive trait for winter cold survival (Cooke et al. 2012). Tea plant originates in the tropics, and it produces nearly the same yield every month at or near the equator. Since the tea plant cultivation area has gradually expanded from lower latitudes to higher latitudes, tea plant has developed dormancy traits to cope with winter cold. At latitudes beyond approximately 16°, particularly when the photoperiod is less than 11 h and 15 min and the minimum temperature falls below 13 °C for at least 6 weeks, nearly complete winter dormancy occurs, and this dormancy persists for longer periods of time as the latitudes increase (Barua 1969; Barua 1989). Like for most tree species, tea plant winter dormancy is manifested as overwintering bud formation in conjunction with a long period of suspended growth (termed bud dormancy). Bud dormancy influences plant production not only by promoting survival during inclement climatic conditions but also by affecting biomass accumulation during the growing season (Cooke et al. 2012). Since the cultivation of tea plant represents an important cash crop, the growth and dormancy of buds are related not only to the safe overwintering of tea plant but also to its economic value (Hao et al. 2017). In recent years, as global climate change has intensified, the harmonization of the activity-dormancy cycle of plants under local climatic conditions has been shown to be more significant for tree species than for non-tree species with respect to adaptations to seasonal temperature changes (Hänninen and Tanino 2011).

3.1 Physiological or Biochemical Changes in Overwintering Buds

Bud dormancy is largely described as the inability of a meristem to resume growth under favorable conditions (Rohde and Bhalerao 2007). The formation and release of bud dormancy are correlated with a series of events such as the cessation of apical growth; bud development; the induction, maintenance, and release of dormancy; and bud burst (Cooke et al. 2012). Among the changes that occur during tea plant winter dormancy, changes in plant hormone contents in overwintering buds were studied initially. The levels of conjugated auxin (indole-3-acetic acid [IAA]) and gibberellin (GA) gradually increase as dormancy begins, and the free auxin and GA levels in buds subsequently decrease and remain at their lowest levels during the deep dormancy stage. In contrast, the levels of free auxin and GA increase extensively prior to dormancy release (Nagar and Kumar 2000; Nagar and Sood 2006). However, abscisic acid (ABA) shows the opposite patterns of change during winter dormancy (Nagar 1996). Pan et al. further measured the contents of multiple plant hormones during different seasons and noted that the occurrence of high ratios of both ABA/GA3 and ABA/zeatin is the dominant factor that promotes winter dormancy in tea plant (Pan et al. 2000). In addition, changes in endogenous polyamine levels in tea shoots indicated that high levels of putrescine and low levels of both spermidine and spermine were associated with the induction of dormancy, while high levels of spermidine and spermine were related to the release of dormancy (Kakkar and Nagar 1997). Comparisons of different tea clones that exhibit varying durations of winter dormancy revealed that the accumulation of ROS was strongly positively correlated with the duration of the dormancy period. Compared with those that have longer dormancy periods, clones that have shorter dormancy periods exhibited a greater induction of antioxidative enzymes and less cellular damage caused by oxidative stress in response to freezing cold during winter (Vyas and Kumar 2005; Vyas et al. 2007).

The accurate identification of the dormancy status of buds during different stages of winter is important for intensive study. From a physiological viewpoint, Lang (1987) pioneered a dormancy classification, namely, paradormancy (growth-promoting environmental or endogenous signals from other plant structures are perceived by the dormant structure, which resumes growth when the signal-producing structure is removed), endodormancy (a meristem-containing dormant structure perceives growth-promoting environmental or endogenous signals from itself and does not resume growth even if the external adverse factors are removed), and ecodormancy (a growth-competent structure is dormant due to unfavorable external factors such as poor nutrition, insufficient water, and low temperature and quickly resumes growth in the absence of these unfavorable factors) (Lang et al. 1987; Lang 1987). Later, Rinne et al. (2001) proposed a new model for dormancy cycling. This model depicts the meristem as passing through three sequential states of cellular communication; each state has a characteristic sensitivity to distinct environmental cues depending on the states of cellular communication affected by both the accumulation and hydrolysis of 1,3-beta-d-glucan at the plasmodesmata (Rinne et al. 2001). On the basis of growth ability identification and a comprehensive expression profile clustering analysis of overwintering buds from the onset of winter dormancy to dormancy release, Hao et al. confirmed the dormancy status of overwintering buds and listed a reliable timetable for dormancy status transitions (Hao et al. 2017). Using two tea plant cultivars that present different sprouting phenophases as plant materials, Tang et al. measured the fluorescence signals in calcein-treated overwintering buds to detect the levels of substance exchange (Tang et al. 2017). From the initial formation to dormancy release, the substance exchange in the overwintering buds exhibited a strong-weak-strong variation pattern in both cultivars; however, the duration in the weak exchange stage was much shorter in the early-sprouting cultivar than in the late-sprouting cultivar. Moreover, the pattern of substance exchange was strongly correlated with the expression pattern of beta-1,3-glucanase gene 1 (*CsGLU1*), during which callose hydrolysis activity in tea plant was positive. The results of the dormancy status identification are consistent with the determination of substance exchange. Briefly the apical buds of tea plants located near the region at 30° N gradually stop growing and form overwintering buds in late September, and the overwintering buds stay in a paradormancy stage until November. As the temperature and day length continually decrease, the overwintering buds enter into endodormancy in early December and then transition into ecodormancy in late January. The ecodormant buds rapidly sprout in March, when the environmental temperature remains at relatively high levels.

3.2 *Major Mechanisms Involved in the Regulation of Bud Dormancy*

Bud dormancy has been a hot topic of research for more than a hundred years, but the understanding of its regulatory mechanism is still poorly understood. Photoperiod and temperature are two crucial environmental cues that mediate plant bud dormancy (Singh et al. 2016). Multiple mechanisms that include mainly plant hormones and growth regulation, epigenetic regulation, metabolic regulation, and changes in cell-to-cell communication may control bud dormancy and bud burst regulation, particularly in deciduous woody species (Cooke et al. 2012). The regulatory mechanism of bud dormancy in tea plant, which is an evergreen woody species, has been studied only recently. By using a suppression subtractive hybridization analysis, Paul et al. primarily compared transcriptomic differences in shoots between the winter dormancy stage and the active growth stage and reported that the genes involved in the cell cycle/cell division were significantly downregulated and that the stress-inducible genes were significantly upregulated (Paul and Kumar 2011). Subsequently, using the same method, Wang et al. further identified the different gene expression profiles between winter dormant buds and sprouting axillary buds in both an early-sprouting cultivar and a late-sprouting cultivar (Wang et al. 2014b). In addition to cell division regulation and stress responses, the pathways involved in water metabolism, energy metabolism, and hormone regulation were also significantly enriched in the identified DEGs. According to the precise dormancy status identification, Hao et al. used RNA-Seq data to comprehensively analyze the global gene expression profiles of axillary buds at the paradormancy, endodormancy, ecodormancy, and bud flush stages (Hao et al. 2017). The results of the gene set enrichment analysis of 16,125 DEGs that were identified from each dormancy transition showed that epigenetic regulation, phytohormone regulation, and callose-related cellular communication regulation were the three dominant mechanisms, and GI-, CAL-, SVP-, PHYB-, SFR6-, LHY-, ZTL-, PIF4/6-, ABI4-, EIN3-, ETR1-, CCA1-, PIN3-, CDK-, and CO-related gene sets were enriched.

During the activity-dormancy cycle, high expression levels of histone H3.1 (CsH3) were detected in growing tea plant buds and leaves, while severely low expression levels were observed in all the measured leaves during the dormancy season (Singh et al. 2009). During posttranslational modification, the acetylation, methylation, and phosphorylation of histones can cause structural and functional rearrangements in chromatin, which in turn affect the expression of specific genes or series of nearby genes (Iglesias and Cerdan 2016). Histone modification, DNA methylation, and miRNA synthesis are the three most important epigenetic mechanisms involved in the regulation of dormancy (Ríos et al. 2014). At the global genomic level, relatively high amounts of DNA methylation and low histone H4 acetylation were observed in dormant chestnut (Santamaría et al. 2009). In particular, similar to the epigenetic regulation of FLOWERING LOCUS C, which is considered a key integrator of vernalization effectors for flowering initiation in *Arabidopsis*, the expression of dormancy-associated MADS-box genes (DAMs) is regulated by histone H3 acetylation and H3K4/H3K27 tri-methylation, both of which occur in the promoter region (Bastow et al. 2004; Horvath et al. 2010; Leida et al. 2012; Saito et al. 2015). By comparatively analyzing the proteome of a novel ever-growing tea cultivar during winter, Liu et al. reported that, compared with that in spring tender shoots, the abundance of histone-like 5, histone H4, and histone H2A.1 in winter tender shoots significantly increased (Liu et al. 2017). The inverse expression patterns of histone genes between dormant tea cultivars and evergreen tea cultivars during the winter season indicate the crucial role of epigenetic regulation in tea plant. The results of a global gene expression profile analysis revealed a total of 127 chromatin-associated genes whose expression significantly differed during the dormancy transition; these genes are involved in processes related to the following gene ontology (GO) terms: "histone lysine methylation," "regulation of DNA replication," "DNA methylation," and "DNA replication." In particular, the gene sets "binding partners of HD1 (histone deacetylase 1)," "neighbors of maintenance of DNA methylation," "neighbors of DNA methyltransferases," and "neighbors of MET1 (methyltransferase 1)" were downregulated in the comparison between paradormancy and endodormancy, while "expression targets of histone H3," "binding partners of helicase," and "neighbors of N-acetyltransferase" were enriched in the comparison between endodormancy and ecodormancy when the direction of DEGs was not considered (Hao et al. 2017). Notably, a complex epigenetic mechanism controlling tea plant bud development and dormancy regulation exists.

Seasonal changes in tea plant hormone contents have been correlated with the activity-dormancy cycle (Nagar and Sood 2006; Nagar 1996; Nagar and Kumar 2000). The putative roles of ABA and GA in modulating the expression of common genes responding to winter dormancy and low temperature were highlighted by the identification of DEGs between actively growing shoots and dormant buds (Paul and Kumar 2011). According to the results of the comprehensive gene expression profile analysis by Hao et al., Yue et al. selected 30 genes involved in GA and ABA metabolism and signaling and investigated their variation in expression during the bud activity-dormancy transition or under exogenous GA3 and ABA treatments (Hao et al. 2017; Yue et al. 2017). The results confirmed the close association between the varying expression patterns of GA- and ABA-related genes and the bud activity-dormancy cycle in tea plant. During the initial stage of dormancy induction in poplar, GA signaling repression-related modulators are rapidly upregulated, and ABA signal pathways are strongly induced following GA signaling (Ruttink et al. 2007). Similar patterns of change have also been reported elsewhere (Hoffman 2011; Howe et al. 2015; Olsen 2010). With respect to the regulation of bud dormancy release, by directly regulating the beta-1,3-glucanase gene in dormant buds, GA promotes the hydrolysis of callose near the plasmodesmata (Rinne et al. 2011; Rinne et al. 2016; van der Schoot and Rinne 2011). In addition to GA and ABA, auxin also plays a dominant role in the regulation of tea plant dormancy. The pattern of dynamic changes in free IAA contents during tea plant activity-dormancy alterations was similar to that of GA and tightly matched the changes in substance exchange in overwintering buds with those in other organs (Nagar and Sood 2006;

Nagar and Kumar 2000; Tang et al. 2017). With respect to the plant hormone-related DEGs, auxin-related DEGs were most abundant in the gene expression profile comparisons of overwintering buds under different conditions, and the auxin signaling pathway was strongly upregulated during the transition from endodormancy to ecodormancy (Hao et al. 2017). Dormancy callose that accumulates in phloem can be substantially removed by exogenous applications of synthetic auxin in *Magnolia kobus* and *Vitis vinifera* (Aloni and Peterson 1997; Aloni et al. 1991). The important role of auxin in cell cycle regulation and cross talk with other hormone signals suggested that the changes in both auxin content and auxin-related genes are necessary for dormancy transitions in tea plant and in other species (Faivre-Rampant et al. 2004; Ferguson and Beveridge 2009; Horvath et al. 2003; Shimizu-Sato et al. 2009; Woodward and Bartel 2005). Considering the changes in hormone contents, substance exchange, and the analysis of different gene expression profiles (Hao et al. 2017; Nagar and Sood 2006; Nagar and Kumar 2000; Tang et al. 2017), the establishment of correlations between auxin and GA signals and the regulation of callose-related cellular communication in tea plant will be a meaningful subject for the study of tea plant bud dormancy.

3.3 Key Genes and Transcription Factors Involved in the Regulation of Bud Dormancy

In winter, transcription and metabolism in dormant buds are modulated both by endogenous signals from the buds themselves and by external environmental factors such as freezing temperature. The regulation of dormancy and the regulation of cold tolerance are tightly integrated and even share many signaling pathways, particularly the low-temperature perception signaling pathway in early winter (Foley et al. 2009; Heide and Prestrud 2005; Horvath et al. 2003; Paul and Kumar 2011; Rohde and Bhalerao 2007). Therefore, a large proportion of DEGs identified by comparing the transcriptomic changes between active-growing shoots and dormant buds are cold-responsive genes, including those coding for LEAs, dehydrins, CORs, glutathione S-transferase (GST), BAM, sucrose synthase (SuSy), calmodulin-binding protein (CBP), and AQPs (Paul and Kumar 2011, Wang et al. 2014b). The expression targets or neighbors of MADS, LHY, and CO, whose expression in tea plant differs at different stages of dormancy, might potentially play crucial roles in the regulation of dormancy (Hao et al. 2017). As members of the MIKC-type linage of the MADS-box gene family, DAM genes have been identified as primary candidates for the regulation of growth cessation and formation of terminal buds in peach (Jiménez et al. 2009). DAM deficiency in peach results in an evergreen phenotype during winter dormancy induction (Bielenberg et al. 2008), while DAM overexpression in leafy spurge results in delayed flowering (Horvath et al. 2010). At the genomic level, Hao et al. identified two DAM-like genes (*CsDAM1* and *CsDAM2*) in tea plant. Unlike for other DAMs, overexpressing *CsDAM1* in poplar inhibited growth cessation and bud set under dormancy-induction conditions (Hao 2014). Though DAMs exhibit varying seasonal and photoperiodic expression patterns in multiple perennial species, the expression of *DAM* genes is stimulated at the endodormancy stage and has been directly linked to vegetative endodormancy (Horvath et al. 2010; Li et al. 2009; Yamane et al. 2011). However, the expression levels of both *CsDAM*s are maintained at low levels during the initial dormancy induction and endodormancy stages but are markedly upregulated during the ecodormancy stage (data unpublished). Extensive gene expression profile analyses have also revealed distinct expression patterns of *DAM*s in poplar (Howe et al. 2015). In addition to the important role of epigenetic mechanisms in DAM regulation, feedback regulation between PpDAM1 and the ABA metabolism and signaling pathway during endodormancy in pear was recently discovered (Tuan et al. 2017). In tea plant, additional studies are needed to reveal the action mechanism of CsDAMs in bud set inhibition and whether they are also regulated by methylation, acetylation modification, or plant hormone signals. In *Arabidopsis* and other flowering species, both CO and FT are crucial nodes in the flowering signal network and are involved in the perception of light and day length (Golembeski and Imaizumi 2015; Hajdu et al. 2015; Lazaro et al. 2015). In perennial species, the dual function of flowering-related genes involved in the regulation of both flowering and dormancy is widespread (Horvath 2009). In particular, the multiple roles of the CO/FT module are well understood depending on the functional verification of FT in the regulation of both flowering and seasonal growth (Bohlenius et al. 2006; Hsu et al. 2011; Pin and Nilsson 2012). Direct interactions between CO and DELLA proteins and histone mark readers MRG1/2 further highlight the importance of plant hormones and epigenetic mechanisms involved in the regulation of vegetative growth (Bu et al. 2014; Xu et al. 2016). Two FT transcripts that differ by one amino acid were cloned from tea plant; after the independent overexpression of both transcripts in poplar, one resulted in a strong early flowering phenotype, and the other promoted growth and suppressed bud set under dormancy-induction conditions (Hao 2014). The significant upregulation of CO and FT gene sets during the dormancy transition from endodormancy to ecodormancy indicates the important roles of the CO/FT module in seasonal growth regulation in tea plant. In addition, phytohormone-associated transcription factors/genes such as the expression targets or neighbors of GAI, ABI4, MYC2, BRI1, EIN2/3, ETR1, and PIN3 as well as light signal transduction- and circadian clock-associated gene sets such as those of ZTL, PHYB, GI, CCA1, LHY, and ELF3 were enriched during dormancy transition (Hao et al. 2017). Unfortunately, only a few of them have been characterized or subjected to function identification; therefore, more studies are needed to reveal the molecular mechanisms of bud dormancy in tea plant.

4 Secondary Metabolic Changes that Are Affected by Low Temperature

During tea production, secondary metabolism in tea leaves can be easily affected by ambient environmental conditions, including low temperature, drought and others (Cheng et al. 2014; Wang et al. 2016). Under either cold temperature or freezing temperature stress, the genes associated with terms related to secondary metabolite biosynthesis, transport, and catabolism were significantly differentially regulated in tea leaves (Chen et al. 2017; Wang et al. 2013). In particular, the flavonoid biosynthesis pathway was the top most differentially regulated metabolic pathway. Catechins are the major products of the flavonoid biosynthesis pathway in tea leaves and represent important characteristic compounds, as these molecules play crucial roles in flavor development of tea and cause many benefits to the health of tea drinkers (Jowko 2015; Kahathuduwa et al. 2017; Khan and Mukhtar 2007). Interestingly, the results of a recent study showed that the flavonoid biosynthesis pathway was also greatly differentially regulated in young shoots under cold spell conditions (Hao et al. 2018). However, the roles and regulatory mechanisms of secondary metabolism in tea leaves under cold stress are mostly unknown. Catechins are the major components of flavonoids, and up to 301 isoforms that constitute 90 candidate genes involved in flavonoid biosynthesis have been identified (Xu et al. 2017). The results of a comparative transcriptome analysis revealed that 51 unigenes were differentially expressed during tea plant leaf development and that MYB, bHLH, and MADS were the major transcription factors involved in the regulation of catechin biosynthesis (Guo et al. 2017). In grape, the expression levels of several genes involved in anthocyanin biosynthesis greatly increased under low-temperature stress (Zhang et al. 2015). Flavonoid biosynthesis and anthocyanin biosynthesis shared the majority of these pathways. By increasing the flavonoid content in plants, the overexpression of a tomato gene coding for the flavanone 3-hydroxylase-like protein improved the seed germination, survival, and growth of tobacco plants under chilling stress (Meng et al. 2015). The results of a mutant analysis revealed that, by inhibiting flavonoid accumulation under low-temperature growth conditions, BRASSINOSTEROID-ENHANCED EXPRESSION 1 (*BEE1*) and G2-LIKE FLAVONOID REGULATOR (*GFR*) were important regulators involved in the control of scopolin accumulation (Petridis et al. 2016). Therefore, metabolomic analysis together with the precise determination of the changes in the contents of key metabolites involved in the catechin or anthocyanin biosynthesis pathways as well as comprehensive expression analysis of the related genes involved in these pathways is important for better understanding the roles of secondary metabolism in cold resistance.

5 Challenges and Future Perspectives

As global warming continues to intensify, abnormal changes in microclimate have become severe environmental problems that threaten green tea production. The responses and adaptive mechanisms of tea plant to low-temperature stress have recently become hot research topics and have received much more attention than ever before. However, due to the limitations of genome sequence information and successful genetic transformation systems, current achievements in the cold resistance of tea plant mainly depend on the determination of basic physiological parameters, analyses of gene expression profiles, and analyses of individual omics. The comprehensive and reliable functional identification of candidate genes and the intense study of novel genes or mechanisms involved in tea plant in response to low temperature are challenging subjects for tea researchers. In addition to low temperature in the fall and winter, cold spells, which are a natural type of rapid temperature reduction (also called cold shocks), in the spring frequently destroy the sprouting shoots of tea plant; these phenomena cause immeasurable losses in green tea production and have become the most severe problem for tea growers in China and perhaps other regions. Young sprouting shoots are important raw materials for the manufacture of high-quality green tea; therefore, studies of the cold-responsive mechanism governing the response to low temperature (especially rapid temperature reductions) in young shoots are urgently needed and highly important. Moreover, in future studies, combinations of multi-omic techniques and tea cultivars or genetic populations with various cold resistance or winter dormancy characteristics should be used to intensively study the genetic differences among them and the easily identifiable reliable parameters or gene markers used to rapidly distinguish early-sprouting and cold-resistant tea cultivars/breeding lines, which would promote the breeding of resistant tea lines. Genomic sequencing and the full-scale operation of resequencing in tea plant have recently provided unprecedented opportunities to tea researchers. The extensive use of omics analyses in tea plant studies provides researchers a global view of changes at the transcriptomic, proteomic, and metabolic levels under stress conditions. The lack of a successful genetic transformation system may be overcome in the near future because of abundant investments in and the rapid import of new biotechnology. Together, these will create exciting prospective opportunities for revealing the cold resistance mechanism in tea plant.

Acknowledgments This work was supported by Zhejiang Provincial Natural Science Foundation (LY16C160001), the National Natural Science Foundation of China (31770735; 31600563), the Earmarked Fund for China Agriculture Research System (CARS-19), and the Chinese Academy of Agricultural Sciences through an Innovation Project for Agricultural Sciences and Technology (CAAS-ASTIP-2017-TRICAAS).

References

Aloni R, Peterson CA (1997) Auxin promotes dormancy callose removal from the phloem of *Magnolia kobus* and callose accumulation and earlywood vessel differentiation in *Quercus robur*. J Plant Res 110:37–44

Aloni R, Raviv A, Peterson CA (1991) The role of auxin in the removal of dormancy callose and resumption of phloem activity in *Vitis vinifera*. Can J Bot 69:1825–1832

Ban Q, Wang X, Pan C, Wang Y, Kong L, Jiang H et al (2017) Comparative analysis of the response and gene regulation in cold resistant and susceptible tea plants. PLoS One 12:e0188514

Barua DN (1969) Seasonal dormancy in tea (*Camellia sinensis* L.). Nature 224:514

Barua DN (1989) Science and practice in tea culture. Tea Research Association, Calcutta, p 509

Bastow R, Mylne JS, Lister C, Lippman Z, Martienssen RA, Dean C (2004) Vernalization requires epigenetic silencing of *FLC* by histone methylation. Nature 427:164–167

Bielenberg DG, Wang YE, Li Z, Zhebentyayeva T, Fan S, Reighard GL et al (2008) Sequencing and annotation of the evergrowing locus in peach [*Prunus persica* (L.) Batsch] reveals a cluster of six MADS-box transcription factors as candidate genes for regulation of terminal bud formation. Tree Genet Genomes 4:495–507

Bohlenius H, Huang T, Charbonnel-Campaa L, Brunner AM, Jansson S, Strauss SH et al (2006) CO/FT regulatory module controls timing of flowering and seasonal growth cessation in trees. Science 312:1040

Bu Z, Yu Y, Li Z, Liu Y, Jiang W, Huang Y et al (2014) Regulation of arabidopsis flowering by the histone mark readers MRG1/2 via interaction with CONSTANS to modulate *FT* expression. PLoS Genet 10:821–833

Cao H, Wang L, Yue C, Hao X, Wang X, Yang Y (2015) Isolation and expression analysis of 18 *CsbZIP* genes implicated in abiotic stress responses in the tea plant (*Camellia sinensis*). Plant Physiol Biochem 97:432–442

Chen J, Yang X, Huang X, Duan S, Long C, Chen J et al (2017) Leaf transcriptome analysis of a subtropical evergreen broadleaf plant, wild oil-tea camellia (*Camellia oleifera*), revealing candidate genes for cold acclimation. BMC Genomics 18:211

Cheng H, Wei K, Wang L (2014) The impact of variety, environment and agricultural practices on catechins and caffeine in plucked tea leaves. In: Processing and impact on active components in food. Elsevier Inc., London, pp 597–603

Chinnusamy V, Zhu J, Zhu JK (2007) Cold stress regulation of gene expression in plants. Trends Plant Sci 12:444–451

Chinnusamy V, Zhu J-K, Sunkar R (2010a) Gene regulation during cold stress acclimation in plants. In: Sunkar R (ed) Plant stress tolerance: methods and protocols. Humana Press, Totowa, pp 39–55

Chinnusamy V, Zhu JK, Sunkar R (2010b) Gene regulation during cold stress acclimation in plants. Methods Mol Biol 639:39–55

Cooke JEK, Eriksson ME, Junttila O (2012) The dynamic nature of bud dormancy in trees: environmental control and molecular mechanisms. Plant Cell Environ 35:1707–1728

Ding ZT, Li C, Shi H, Wang H, Wang Y (2015) Pattern of *CsICE1* expression under cold or drought treatment and functional verification through analysis of transgenic *Arabidopsis*. Genet Mol Res 14:11259–11270

Ding ZT, Shen JZ, Pan LL, Wang YU, Li YS, Wang Y et al (2016) CsSAD: a fatty acid desaturase gene involved in abiotic resistance in *Camellia sinensis* (L.). Genet Mol Res 15:15017512

Eremina M, Rozhon W, Poppenberger B (2016) Hormonal control of cold stress responses in plants. Cell Mol Life Sci 73:797–810

Faivre-Rampant O, Cardle L, Marshall D, Viola R, Taylor MA (2004) Changes in gene expression during meristem activation processes in *Solanum tuberosum* with a focus on the regulation of an auxin response factor gene. J Exp Bot 55:613–622

Ferguson BJ, Beveridge CA (2009) Roles for auxin, cytokinin, and strigolactone in regulating shoot branching. Plant Physiol 149:1929–1944

Foley M, Anderson J, Horvath D (2009) The effects of temperature, photoperiod, and vernalization on regrowth and flowering competence in *Euphorbia esula* (Euphorbiaceae) crown buds. Botany 87:986–992

Golembeski GS, Imaizumi T (2015) Photoperiodic regulation of florigen function in *Arabidopsis thaliana*. Arabidopsis Book 13:e0178

Guo F, Guo Y, Wang P, Wang Y, Ni D (2017) Transcriptional profiling of catechins biosynthesis genes during tea plant leaf development. Planta 1:1–14

Hajdu A, Adam E, Sheerin DJ, Dobos O, Bernula P, Hiltbrunner A et al (2015) High-level expression and phosphorylation of phytochrome B modulates flowering time in *Arabidopsis*. Plant J 83:794–805

Hänninen H, Tanino K (2011) Tree seasonality in a warming climate. Trends Plant Sci 16:412

Hao X (2014) Studies on the molecular mechanism of winter bud dormancy in tea plant (*Camellia sinensis* (L.) O. Kuntze). Doctoral Dissertation Northwest A&F University, Shaanxi

Hao X, Yang Y, Yue C, Wang L, Horvath DP, Wang X (2017) Comprehensive transcriptome analyses reveal differential gene expression profiles of *camellia sinensis* axillary buds at para-, endo-, ecodormancy, and bud flush stages. Front Plant Sci 8:553

Hao X, Tang H, Wang B, Yue C, Wang L, Zeng J et al (2018) Integrative transcriptional and metabolic analyses provide insights into cold spell response mechanisms in young shoots of the tea plant. Tree Physiol. https://doi.org/10.1093/treephys/tpy038

Hazarika M, Muraleedharan N (2011) Tea in India: an overview. Two Bud 58:3–9

Heide O, Prestrud A (2005) Low temperature, but not photoperiod, controls growth cessation and dormancy induction and release in apple and pear. Tree Physiol 25:109–114

Hincha DK, Zuther E (2014) Plant cold acclimation and freezing tolerance. Methods Mol Biol 1166:1–6

Hoffman DE (2011) Changes in the transcriptome and metabolome during the initiation of growth cessation in *Hybrid Aspens*. Acta Universitatis Agriculturae Sueciae, Umeå

Horvath D (2009) Common mechanisms regulate flowering and dormancy. Plant Sci 177:523–531

Horvath DP, Anderson JV, Chao WS, Foley ME (2003) Knowing when to grow: signals regulating bud dormancy. Trends Plant Sci 8:534–540

Horvath DP, Sung S, Kim D, Chao W, Anderson J (2010) Characterization, expression and function of DORMANCY ASSOCIATED MADS-BOX genes from leafy spurge. Plant Mol Biol 73:169–179

Howe GT, Horvath DP, Dharmawardhana P, Priest HD, Mockler TC, Strauss SH (2015) Extensive transcriptome changes during natural onset and release of vegetative bud dormancy in *populus*. Front Plant Sci 6:989

Hsu CY, Adams JP, Kim H, No K, Ma C, Strauss SH et al (2011) FLOWERING LOCUS T duplication coordinates reproductive and vegetative growth in perennial poplar. Proc Natl Acad Sci U S A 108:10756–10761

Huang Y, Qian W, Wang B, Cao H, Wang L, Hao X et al (2015) Effects of exogenous calcium and inhibitors of calcium signaling transduction pathway on cold resistance of tea plant. J Tea Sci 35:295–300

Huang Y, Qian W, Wang Y, Cao H, Wang L, Hao X et al (2016) Molecular cloning and expression of CaM family gene *CsCaMs* under cold stress in tea plant [*Camellia sinensis* (L.) O. Kuntze]. J Plant Genet Resour 17:906–913

Iglesias FM, Cerdan PD (2016) Maintaining epigenetic inheritance during dna replication in plants. Front Plant Sci 7:38

Janmohammadi M, Zolla L, Rinalducci S (2015) Low temperature tolerance in plants: changes at the protein level. Phytochemistry 117:76–89

Jiménez S, Lawton-Rauh AL, Reighard GL, Abbott AG, Bielenberg DG (2009) Phylogenetic analysis and molecular evolution of the dormancy associated MADS-box genes from peach. BMC Plant Biol 9:81

Jowko E (2015) Green tea catechins and sport performance. In: Lamprecht M (ed) Antioxidants in sport nutrition. Taylor & Francis Group, LLC, Boca Raton

Kahathuduwa CN, Dassanayake TL, Amarakoon AMT, Weerasinghe VS (2017) Acute effects of theanine, caffeine and theanine-caffeine combination on attention. Nutr Neurosci 20:369–377

Kakkar RK, Nagar PK (1997) Distribution and changes in endogenous polyamines during winter dormancy in tea [*Camellia sinensis* L.(O) Kuntze]. J Plant Physiol 151:63–67

Karimi M, Ghazanfari F, Fadaei A, Ahmadi L, Shiran B, Rabei M et al (2016) The small-RNA profiles of almond (*Prunus dulcis* Mill.) reproductive tissues in response to cold stress. PLoS One 11:e0156519

Khan N, Mukhtar H (2007) Tea polyphenols for health promotion. Life Sci 81:519–533

Knight MR, Campbell AK, Smith SM, Trewavas AJ (1991) Transgenic plant aequorin reports the effects of touch and cold-shock and elicitors on cytoplasmic calcium. Nature 352:524–526

Korn M, Peterek S, Mock HP, Heyer AG, Hincha DK (2008) Heterosis in the freezing tolerance, and sugar and flavonoid contents of crosses between *Arabidopsis thaliana* accessions of widely varying freezing tolerance. Plant Cell Environ 31:813–827

Lang GA (1987) Dormancy: a new universal terminology. Hortscience 22:817–820

Lang G, Early J, Martin G, Darnell R (1987) Endo-, para-, and ecodormancy: physiological terminology and classification for dormancy research. Hortscience 22:371–377

Lazaro A, Mouriz A, Pineiro M, Jarillo JA (2015) Red light-mediated degradation of CONSTANS by the E3 ubiquitin ligase HOS1 regulates photoperiodic flowering in arabidopsis. Plant Cell 27:2437–2454

Lee YP, Babakov A, de Boer B, Zuther E, Hincha DK (2012) Comparison of freezing tolerance, compatible solutes and polyamines in geographically diverse collections of *Thellungiella* sp. and *Arabidopsis thaliana* accessions. BMC Plant Biol 12:131

Leida C, Conesa A, Llácer G, Badenes ML, Ríos G (2012) Histone modifications and expression of *DAM6* gene in peach are modulated during bud dormancy release in a cultivar-dependent manner. New Phytol 193:67–80

Li Z, Reighard GL, Abbott AG, Bielenberg DG (2009) Dormancy-associated MADS genes from the *EVG* locus of peach [*Prunus persica* (L.) Batsch] have distinct seasonal and photoperiodic expression patterns. J Exp Bot 60:3521–3530. https://doi.org/10.1093/jxb/erp195

Li XW, Feng ZG, Yang HM, Zhu XP, Liu J, Yuan HY (2010) A novel cold-regulated gene from *Camellia sinensis*, CsCOR1, enhances salt- and dehydration-tolerance in tobacco. Biochem Biophys Res Commun 394:354–359

Li NN, Qian WJ, Wang L, Cao HL, Hao XY, Yang YJ et al (2017) Isolation and expression features of hexose kinase genes under various abiotic stresses in the tea plant (*Camellia sinensis*). J Plant Physiol 209:95–104

Liu ZW, Wu ZJ, Li XH, Huang Y, Li H, Wang YX et al (2016) Identification, classification, and expression profiles of heat shock transcription factors in tea plant (*Camellia sinensis*) under temperature stress. Gene 576:52–59

Liu S, Gao J, Chen Z, Qiao X, Huang H, Cui B et al (2017) Comparative proteomics reveals the physiological differences between winter tender shoots and spring tender shoots of a novel tea (*Camellia sinensis* L.) cultivar evergrowing in winter. BMC Plant Biol 17:206

Megha S, Basu U, Kav NNV (2017) Regulation of low temperature stress in plants by microRNAs. Plant Cell Environ 41:1–15

Meng C, Zhang S, Deng YS, Wang GD, Kong FY (2015) Overexpression of a tomato flavanone 3-hydroxylase-like protein gene improves chilling tolerance in tobacco. Plant Physiol Biochem 96:388–400

Miura K, Furumoto T (2013) Cold signaling and cold response in plants. Int J Mol Sci 14:5312–5337

Nagar PK (1996) Changes in endogenous abscisic acid and phenols during winter dormancy in tea (*Camellia sinesis* L.(O) Kunze). Acta Physiol Plant 18:33–38

Nagar PK, Kumar A (2000) Changes in endogenous gibberellin activity during winter dormancy in tea (*Camellia sinensis* (L.) O. Kuntze). Acta Physiol Plant 22:439–443

Nagar P, Sood S (2006) Changes in endogenous auxins during winter dormancy in tea (*Camellia sinensis* L.) O. Kuntze. Acta Physiol Plant 28:165–169

Olsen JE (2010) Light and temperature sensing and signaling in induction of bud dormancy in woody plants. Plant Mol Biol 73:37–47

Orvar BL, Sangwan V, Omann F, Dhindsa RS (2000) Early steps in cold sensing by plant cells: the role of actin cytoskeleton and membrane fluidity. Plant J 23:785–794

Pan G-S, Qian L-S, Shen S-R, Wu B-Q (2000) Endogenous hormone level and its regulation mechanism in the growth of tea plant shoost (II) the relationship between endogenous hormone and summer and winter dormancy. J Tea 26:200–204

Paul A, Kumar S (2011) Responses to winter dormancy, temperature, and plant hormones share gene networks. Funct Integr Genomics 11:659–664

Paul A, Kumar S (2013) Dehydrin2 is a stress-inducible, whereas Dehydrin1 is constitutively expressed but up-regulated gene under varied cues in tea [*Camellia sinensis* (L.) O. Kuntze]. Mol Biol Rep 40:3859–3863

Paul A, Lal L, Ahuja PS, Kumar S (2012) Alpha-tubulin (*CsTUA*) up-regulated during winter dormancy is a low temperature inducible gene in tea [*Camellia sinensis* (L.) O. Kuntze]. Mol Biol Rep 39:3485–3490

Petridis A, Doll S, Nichelmann L, Bilger W, Mock HP (2016) *Arabidopsis thaliana* G2-LIKE FLAVONOID REGULATOR AND BRASSINOSTEROID ENHANCED EXPRESSION1 are low-temperature regulators of flavonoid accumulation. New Phytol 211:912–925

Pin PA, Nilsson O (2012) The multifaceted roles of FLOWERING LOCUS T in plant development. Plant Cell Environ 35:1742–1755

Qian W, Yue C, Wang Y, Cao H, Li N, Wang L et al (2016) Identification of the invertase gene family (INVs) in tea plant and their expression analysis under abiotic stress. Mol Biol Rep 35:2269–2283

Rinne PL, Kaikuranta PM, Van der Christiaan SC (2001) The shoot apical meristem restores its symplasmic organization during chilling-induced release from dormancy. Plant J Cell Mol Biol 26:249–264

Rinne PL, Welling A, Vahala J, Ripel L, Ruonala R, Kangasjarvi J et al (2011) Chilling of dormant buds hyperinduces FLOWERING LOCUS T and recruits GA-inducible 1,3-beta-glucanases to reopen signal conduits and release dormancy in *Populus*. Plant Cell 23:130–146

Rinne PLH, Paul LK, Vahala J, Kangasjärvi J, Schoot CVD (2016) Axillary buds are dwarfed shoots that tightly regulate GA pathway and GA-inducible 1,3-β-glucanase genes during branching in *hybrid aspen*. J Exp Bot 67:5975–5991

Ríos G, Leida C, Conejero A, Badenes ML (2014) Epigenetic regulation of bud dormancy events in perennial plants. Front Plant Sci 5:247

Rohde A, Bhalerao RP (2007) Plant dormancy in the perennial context. Trends Plant Sci 12:217–223

Ruttink T, Arend M, Morreel K, Storme V, Rombauts S, Fromm J et al (2007) A molecular timetable for apical bud formation and dormancy induction in poplar. Plant Cell 19:2370–2390

Saito T, Bai S, Imai T, Ito A, Nakajima I, Moriguchi T (2015) Histone modification and signalling cascade of the dormancy-associated MADS-box gene, *PpMADS13-1*, in Japanese pear (*Pyrus pyrifolia*) during endodormancy. Plant Cell Environ 38:1157–1166

Santamaría ME, Hasbún R, Valera MJ, Meijón M, Valledor L, Rodríguez JL et al (2009) Acetylated H4 histone and genomic DNA methylation patterns during bud set and bud burst in *Castanea sativa*. J Plant Physiol 166:1360–1369

Shi Y, Huang J, Sun T, Wang X, Zhu C, Ai Y et al (2017) The precise regulation of different COR genes by individual CBF transcription factors in *Arabidopsis thaliana*. J Integr Plant Biol 59:118–133

Shimizu-Sato S, Tanaka M, Mori H (2009) Auxin-cytokinin interactions in the control of shoot branching. Plant Mol Biol 69:429–435

Singh K, Kumar S, Ahuja PS (2009) Differential expression of Histone H3 gene in tea (*Camellia sinensis* (L.) O. Kuntze) suggests its role in growing tissue. Mol Biol Rep 36:537–542

Singh RK, Svystun T, AlDahmash B, Jönsson AM, Bhalero RP (2016) Photoperiod- and temperature-mediated control of phenology in trees – a molecular perspective. New Phytol 213:511–524

Tang H, Hao X, Wang L, Xiao B, Wang X, Yang Y (2017) Molecular regulation and substance exchange dynamics at dormancy and budbreak stages in overwintering buds of tea plant. Acta Agron Sin 43:669–677

Thalhammer A, Hincha DK, Zuther E (2014) Measuring freezing tolerance: electrolyte leakage and chlorophyll fluorescence assays. Methods Mol Biol 1166:15–24

Theocharis A, Clement C, Barka EA (2012) Physiological and molecular changes in plants grown at low temperatures. Planta 235:1091–1105

Tuan PA, Bai S, Saito T, Ito A, Moriguchi T (2017) Dormancy-associated mads-box (DAM) and the abscisic acid pathway regulate pear endodormancy through a feedback mechanism. Plant Cell Physiol 58:1378–1390

Turkozu D, Sanlier N (2017) L-theanine, unique amino acid of tea, and its metabolism, health effects, and safety. Crit Rev Food Sci Nutr 57:1681–1687

van der Schoot C, Rinne PL (2011) Dormancy cycling at the shoot apical meristem: transitioning between self-organization and self-arrest. Plant Sci 180:120–131

Vyas D, Kumar S (2005) Tea (*Camellia sinensis* (L.) O. Kuntze) clone with lower period of winter dormancy exhibits lesser cellular damage in response to low temperature. Plant Physiol Biochem 43:383–388

Vyas D, Kumar S, Ahuja PS (2007) Tea (*Camellia sinensis*) clones with shorter periods of winter dormancy exhibit lower accumulation of reactive oxygen species. Tree Physiol 27:1253–1259

Wang Y, Jiang CJ, Li YY, Wei CL, Deng WW (2012) CsICE1 and CsCBF1: two transcription factors involved in cold responses in *Camellia sinensis*. Plant Cell Rep 31:27–34

Wang X-C, Zhao Q-Y, Ma C-L, Zhang Z-H, Cao H-L, Kong Y-M et al (2013) Global transcriptome profiles of *Camellia sinensis* during cold acclimation. BMC Genomics 14:415

Wang W, Wang Y, Du Y, Zhao Z, Zhu X, Jiang X et al (2014a) Overexpression of *Camellia sinensis* H1 histone gene confers abiotic stress tolerance in transgenic tobacco. Plant Cell Rep 33:1829–1841

Wang X, Hao X, Ma C, Cao H, Yue C, Wang L et al (2014b) Identification of differential gene expression profiles between winter dormant and sprouting axillary buds in tea plant (*Camellia sinensis*) by suppression subtractive hybridization. Tree Genet Genomes 10:1149–1159

Wang W, Xin H, Wang M, Ma Q, Wang L, Kaleri NA et al (2016) Transcriptomic analysis reveals the molecular mechanisms of drought-stress-induced decreases in *Camellia sinensis* leaf quality. Front Plant Sci 7:385

Wang L, Cao H, Qian W, Yao L, Hao X, Li N et al (2017a) Identification of a novel bZIP transcription factor in *Camellia sinensis* as a negative regulator of freezing tolerance in transgenic arabidopsis. Ann Bot 119:1195–1209

Wang M, Zou Z, Li Q, Xin H, Zhu X, Chen X et al (2017b) Heterologous expression of three *Camellia sinensis* small heat shock protein genes confers temperature stress tolerance in yeast and *Arabidopsis thaliana*. Plant Cell Rep 36:1125–1135

Weber H, Chetelat A, Reymond P, Farmer EE (2004) Selective and powerful stress gene expression in *Arabidopsis* in response to malondialdehyde. Plant J 37:877–888

Woodward AW, Bartel B (2005) Auxin: regulation, action, and interaction. Ann Bot 95:707–735

Xia EH, Zhang HB, Sheng J, Li K, Zhang QJ, Kim C et al (2017) The tea tree genome provides insights into tea flavor and independent evolution of caffeine biosynthesis. Mol Plant 10:866–877

Xu F, Li T, Xu PB, Li L, Du SS, Lian HL et al (2016) DELLA proteins physically interact with CONSTANS to regulate flowering under long days in Arabidopsis. FEBS Lett 590:541–549

Xu Q, Zhu J, Zhao S, Hou Y, Li F, Tai Y et al (2017) Transcriptome profiling using single-molecule direct RNA sequencing approach for in-depth understanding of genes in secondary metabolism pathways of *Camellia sinensis*. Front Plant Sci 8:1205

Yamane H, Ooka T, Jotatsu H, Hosaka Y, Sasaki R, Tao R (2011) Expressional regulation of PpDAM5 and PpDAM6, peach (*Prunus persica*) dormancy-associated MADS-box genes, by low temperature and dormancy-breaking reagent treatment. J Exp Bot 62:3481–3488

Yang Y (2005) Tea plant cultivation in China (Book in Chinese). Shanghai Scientific & Technical Publishers, Shanghai, p 9
Yang Y, Zhen L, Wang X (2004) Effect of cold acclimation and ABA on cold hardiness, contents of proline of tea plants [*Camellia sinensis*(L.) O. Kuntze]. J Tea Sci 24:177–182
Yu F (1986) Discussion on the originating place and the originating centre of tea plant. J Tea Sci 6:1–8
Yue C, Cao H, Wang L, Zhou Y, Hao X, Zeng J et al (2014) Molecular cloning and expression analysis of tea plant aquaporin (AQP) gene family. Plant Physiol Biochem 83:65–76
Yue C, Cao HL, Wang L, Zhou YH, Huang YT, Hao XY et al (2015) Effects of cold acclimation on sugar metabolism and sugar-related gene expression in tea plant during the winter season. Plant Mol Biol 88:591–608
Yue C, Cao H, Hao X, Zeng J, Qian W, Guo Y et al (2017) Differential expression of gibberellin- and abscisic acid-related genes implies their roles in the bud activity-dormancy transition of tea plants. Plant Cell Rep 37:425–441
Zhan X, Zhu JK, Lang Z (2015) Increasing freezing tolerance: kinase regulation of ICE1. Dev Cell 32:257–258
Zhang XN, Li X, Liu JH (2013) Identification of conserved and novel cold-responsive micrornas in trifoliate orange (*Poncirus trifoliata* (L.) Raf.) using high-throughput sequencing. Plant Mol Biol Report 32:1–14
Zhang Y, Zhu X, Chen X, Song C, Zou Z, Wang Y et al (2014) Identification and characterization of cold-responsive microRNAs in tea plant (*Camellia sinensis*) and their targets using high-throughput sequencing and degradome analysis. BMC Plant Biol 14:271
Zhang C, Jia H, Wu W, Wang X, Fang J, Wang C (2015) Functional conservation analysis and expression modes of grape anthocyanin synthesis genes responsive to low temperature stress. Gene 574:168–177
Zhao C, Lang Z, Zhu JK (2015) Cold responsive gene transcription becomes more complex. Trends Plant Sci 20:466–468
Zhao C, Zhang Z, Xie S, Si T, Li Y, Zhu JK (2016) Mutational evidence for the critical role of CBF transcription factors in cold acclimation in *Arabidopsis*. Plant Physiol 171:2744–2759
Zheng C, Zhao L, Wang Y, Shen J, Zhang Y, Jia S et al (2015) Integrated RNA-Seq and sRNA-Seq analysis identifies chilling and freezing responsive key molecular players and pathways in tea plant (*Camellia sinensis*). PLoS One 10:e0125031
Zheng C, Wang Y, Ding Z, Zhao L (2016) Global transcriptional analysis reveals the complex relationship between tea quality, leaf senescence and the responses to cold-drought combined stress in *Camellia sinensis*. Front Plant Sci 7:1858
Zhou Y, Cao H, Yue C, Wang L, Hao X, Wang X et al (2015) Cloning and expression analysis on DNA methyltransferase gene, *CsDRM2*, of *Camellia sinensis*. Acta Tea Sinica 56:1–7
Zhou Y, Liu Y, Wang S, Shi C, Zhang R, Rao J et al (2017) Molecular cloning and characterization of galactinol synthases in *Camellia sinensis* with different responses to biotic and abiotic stressors. J Agric Food Chem 65:2751–2759
Zhu X, Li Q, Hu J, Wang M, Li X (2015) Molecular cloning and characterization of spermine synthesis gene associated with cold tolerance in tea plant (*Camellia sinensis*). Appl Biochem Biotechnol 177:1055–1068

Chapter 4
Response of Tea Plants to Drought Stress

Wenjun Qian, Jianhui Hu, Xinfu Zhang, Lei Zhao, Yu Wang, and ZhaoTang Ding

Abstract The tea plant (*Camellia sinensis* L.) is an important economic crop that is widely cultivated in the tropics and subtropics of Asia. As tea plants are a rain-fed perennial crop, both excessive soil moisture and water deficit can cause water stress in the plants. The most common water stress encountered is moisture deficit, known as drought stress, one of the most adverse factors that severely impairs tea plant growth and development, and limits its distribution, performance, yield, and quality. However, few studies have systematically reported the effects of water stress, especially drought stress, on the tea plant. In this chapter, therefore, we review the current state of drought stress and its progress in tea plants, with the aim being to explore the various morphological, physiological, biochemical, and molecular responses to such stress found in this species. Previous studies have demonstrated some key features in cultivated tea plants challenged by drought stress, such as (1) deep root systems, smaller and succulent leaves, and thickening of cuticle and palisade tissue; (2) water deficit, impairment in photosynthesis and respiration, stomatal closure, reduction of carbohydrate synthesis, acceleration of proteolysis, changes in lipid components in the normal cell wall structure; (3) protein denaturation of the plant tissues constituents, protoplast condensation, and the loss of cell wall semi-permeability; (4) enhanced free radical content, antioxidative systems, and osmoprotectant contents; (5) changes in the contents of minerals required for nutrition, hormones, polyphenols, and amino acids; (6) and changes in the transcription levels of many regulatory and functional genes. Understanding these response mechanisms of tea plants to drought stress is essential for improving the drought resistance of these plants by carrying out appropriate strategies such as mass screening and breeding; marker-assisted breeding; and exogenous application of osmoprotectants and hormones to seeds, seedlings, or growing tea plants, as well as undertaking genetic engineering for drought resistance.

Keywords *Camellia sinensis* · Dought stress · Morphological response · Physiological response · Biochemical responses · Molecular response

---

W. Qian · J. Hu · X. Zhang · L. Zhao · Y. Wang · Z. Ding (✉)
College of Horticulture, Qingdao Agricultural University, Qingdao, Shandong, China
e-mail: qau-WenjunQian@qau.edu.cn

W.-Y. Han et al. (eds.), *Stress Physiology of Tea in the Face of Climate Change*,
https://doi.org/10.1007/978-981-13-2140-5_4

1 Introduction

Water is a core environmental factor for plant growth and development, but both excessive water and water deficit can result in water stress. In regard to excessive water, a key factor is flooding, which mainly refers to soil moisture exceeding field capacity. Flooding primarily results in reduced oxygen supply to the roots, and critical root functions, including respiration and nutrient uptake, are subsequently limited. However, in many cases, the water stress encountered is water deficit, known as drought stress (Mahajan and Tuteja 2005). Drought stress is an important limiting factor for plant growth and development, crop yields, and quality (Sharma and Kumar 2005; Shao et al. 2009), and it is becoming an acute problem all over the world, especially in arid and semi-arid areas. In plants under drought stress conditions, the root systems, photosynthesis and respiration, relative water content, membrane integrity, pigment content, osmotic potential, carbohydrate status, antioxidant metabolism, nutrition content, and phytohormones are affected, as are the transcription levels of both regulatory and functional genes (Benjamin and Nielsen 2006; Farooq et al. 2009; Li et al. 2009; Praba et al. 2009). Therefore, exploring the response mechanisms of plants to drought stress is essential for developing and breeding new drought-resistant varieties, for molecular genetic breeding, and for improving regulation and control measures to enhance plant drought resistance.

The tea plant, an evergreen perennial crop that is mainly grown in the tropical and subtropical regions of Asia, provides one of the world's favorite non-alcoholic beverages. However, both the growth and the distribution of tea plants are affected by many environment factors, such as temperature, light, soil, and water, with water deficit (drought stress) being the major constraint on the growth, yield, and quality of the tea plant. As previous studies have reported, drought caused tea plant crop losses of around 14–40% in different cultivation areas (Wijeratne et al. 1998; Jain 1999; Ng'etich and Stephens 2001; Nixon et al. 2001; Panda et al. 2003; Das et al. 2012; Gupta et al. 2013). In addition, extreme drought frequently leads to tea plant death; Burgess and Carr (1993) reported that drought caused 6–19% plant mortality in varieties of clonal tea cultivars. As mentioned above, in order to withstand drought stress, tea plants show a series of complex morphological, physiological, biochemical, and molecular changes, similar to the changes shown in other plant species. Recently, many studies have found that the contents of some characteristic tea plant components, together with the expressions of many related genes, are also affected by drought stress, such as the contents of caffeine, polyphenols, and theanine, the components of tea aroma, and the corresponding synthetic genes (Cheruiyot et al. 2007; Upadhyaya et al. 2008; Lin et al. 2014; Wang et al. 2016, 2017; Zheng et al. 2016). Although there have been many reports about the drought stress responses of tea plants, current research results have not been summarized. This chapter, therefore, focuses on providing an overview of the morphological, physiological, biochemical, and molecular mechanisms of drought stress responses and tolerance in the tea plant. This review will provide more theoretical knowledge for the drought-resistant breeding of tea plants, and some parameters listed can be employed as drought resistance indexes for the screening and breeding of drought-resistant cultivated varieties of tea plants.

2 Morphological Responses

Under drought-stress conditions, leaf morphology and anatomical structure can simply and intuitively be used to estimate the extent of drought damage in tea plants and the drought resistance of different varieties. The phenotypes of tea plant leaves vary with the processing time of drought stress and the different tea cultivars. As Liu et al. (2015) demonstrated, some leaves of a drought-susceptible tea plant cultivar ('Zhuyeqi') turned brown and curled up, and some leaves showed brown spots when the plant was exposed to mild stress conditions, but no significant changes were observed in a drought-resistant tea plant cultivar ('Ningzhou2') ; under moderate drought stress conditions, most young leaves of the 'Zhuyeqi' cultivar (from the first to the fourth leaves) turned reddish-brown, wilted, curled and withered, while in the 'Ningzhou2' cultivar only needle-like black spots and yellow plaques appeared. Similarly, Chakraborty et al. (2002) reported that, with prolonged drought stress the leaves of two drought-resistant tea cultivars (UPASI-3 and UPASI-26) did not exhibit any morphological wilting, but the leaves of two drought-susceptible tea cultivars (Tocklai and Darjeeling) exhibited severe wilting symptoms.

The leaf is the core organ for plant photosynthesis and is vital for movement; it is also the most sensitive component for responding to various environmental stressors. Under drought-stress conditions, the stem and/or root reflect the effects of drought stress less clearly than the leaf structures (Ennajeh et al. 2010). In particular, many morpho-anatomical features of leaves, such as the leaf area, cuticle, trichome and stomatal densities, and the thicknesses of the spongy parenchyma and palisade parenchyma, will change significantly but differently in different tea cultivars (Bosabalidis and Kofidis 2002; Bacelar et al. 2004; Ennajeh et al. 2010). At present, it is known that the leaf-level morphology and structure of high drought-resistance tea cultivars show greater thickness of the cuticle, total lamina, palisade parenchyma and spongy parenchyma, and smaller but more numerous stomata, and also increased trichome densities. The genotypes of high drought-resistance tea cultivars have smaller leaves than drought-susceptible tea cultivars; this allows the leaves to transpire less water and thus contributes to reducing water loss (Abrams et al.1990; Nevo et al. 2000; Pita and Pardos 2001). Furthermore, in drought-resistant tea cultivars, the thickness of the upper palisade parenchyma is increased, which can promote an increased number of CO_2 fixation sites, contributing to the maintenance of high photosynthetic assimilation rates in spite of the low stomatal conductance values caused by drought stress; the thickness of the spongy parenchyma is also increased in drought-resistant tea cultivars, and this can improve CO_2 diffusion owing to the abundant intercellular spaces; moreover, the increase of stomatal and trichome density, and the thicker upper and lower epidermis allow better control of transpiration and provide more protection against desiccation for the inner leaf tissues (Bolhar 1987; Baldini et al. 1997; Chartzoulakis et al. 1999; Mediavilla et al. 2001; Bosabalidis and Kofidis 2002; Bacelar et al. 2004; Ennajeh et al. 2008, 2010).

3 Physiological and Biochemical Responses

3.1 Water Relations

Relative water content (RWC), leaf water potential, osmotic potential, pressure potential, stomatal resistance, transpiration rate, leaf temperature, and canopy temperature are the major limiting factors of plant water relations when plants are exposed to drought-stress conditions (Anjum et al. 2011). The RWC can reflect the degree of plant water deficit and the metabolic activity in plant tissues, and is a meaningful index of drought tolerance (Anjum et al. 2011). Under drought-stress conditions, the RWC of leaves decreased in a wide variety of plants, and there were great reductions in RWC and water potential in the leaves (Siddique et al. 2001; Nayyar and Gupta 2006). However, drought-resistant plants showed stronger water retaining and regulation capacity and maintained higher RWC than drought-susceptible plants (Farooqui et al. 2000). Numerous studies reported that the RWC and water potential in tea plant leaves decreased in all examined tea cultivars exposed to drought stress, but the values varied in the different cultivars. In particular, the RWC and water potential were both higher in drought-resistant tea cultivars than in drought-susceptible tea cultivars (Cheruiyot et al. 2008; Upadhyaya and Panda 2013; Liu et al. 2015). For example, the exposure of six different tea cultivars (TV-18, TV-26, UPASI-3, UPASI-26, T-78, and HV-39) to drought stress substantially reduced the RWC, but higher RWCs were measured in the two drought-resistant cultivars (UPASI-3 and UPASI-26) than in the others (Chakraborty et al. 2002). Similarly, three drought-resistant tea cultivars (TTL-1, TTL-6, and UPASI-2) showed lower reduction of RWC as compared with four other tea cultivars (TTL-2, TTL-4, TTL-5, and UPASI-3) (Netto et al. 2010). In addition, Handique (1992) discovered that the young shoots of drought-resistant tea clones possessed higher water potential than the drought-susceptible clones.

3.2 Photosynthesis

Photosynthesis is the essential metabolic activity that produces energy and carbohydrates for the growth and development of tea plants. The foliar photosynthetic rate is decreased with the decrease of RWC and water potential under drought stress (Lawlor and Cornic 2002). Many studies have shown that drought stress limits photosynthesis mainly through stomatal limitations or non-stomatal mechanisms (Ahmadi 1998; Tezara et al. 1999; Del Blanco et al. 2000; Lawson et al. 2003; Samarah et al. 2009). However, stomatal limitation was generally considered to be the major determinant of reduced photosynthesis under drought stress (Lawlor and

Cornic 2002); that is, the decrease of leaf water potential will reduce stomatal conductance under drought stress, inhibiting the intracellular entry of CO_2 into the leaf, thereby resulting in a decreased photosynthetic rate owing to CO_2 deficiency. On the other hand, the limitation of photosynthesis through metabolic impairment under drought stress is a more complex phenomenon than stomatal limitation (Reddy et al. 2004). These metabolic pathways mainly refer to the decrease of the photosynthetic rate caused by damage to the chloroplast thylakoid membrane, photosystem I, and photosystem system II; expansion of the grana thylakoids; the increase of the interstitial lamellar space, the decrease of chlorophyll (Chl-a, Chl-b, and total Chl) contents, and the decrease of the synthesis and activity of the ribulose-1,5-bisphosphate enzyme; thereby, photosynthesis is weakened or inhibited (Allen and Ort 2001; Reddy et al. 2004; Anjum et al. 2011).

Under drought-stress conditions, the leaf chloroplasts of the tea plant are deformed, the lamellar structures are damaged, the activities of photosynthetic enzymes are decreased, and the stomatal conductance of the tea leaves is limited, eventually resulting in reductions of the photosynthetic and respiration rates (Barora 1994). However, the net photosynthesis rate (Pn), respiration rate, and stomatal conductance varied significantly among different tea cultivars under drought stress. The Pn of tea plant leaves gradually declined as the drought stress was prolonged, and the trends of Pn were consistent with the degrees of drought resistance in different cultivars, and these trends were also positively correlated with soil moisture content (SMC). At the same time, the transpiration rate of the drought-resistant tea cultivars was higher than that of the drought-sensitive tea cultivars (Netto et al. 2010). A similar changing rhythm, in terms of different tea cultivars, was also observed in the stomatal conductance of tea plant leaves. For example, Lin et al. (2014) found the Pn and photochemical quenching values in full-irrigation and deficit-irrigation tea plants were significantly higher than those in severe deficit-irrigation and non-irrigation tea plants, but the values of non-photochemical quenching were adverse to the Pn and photochemical quenching values in plants with different degrees of drought-stress conditions. Maritim et al. (2015) obtained a result similar to that of Lin et al. (2014), where they demonstrated that the Pn decreased following the reduction of SMC, and the Pn differed significantly between different degrees of drought treatments and different tea cultivars; a significant difference was also observed on the interaction between SMC and tea cultivar.

3.3 Photosynthetic Pigments

The photosynthetic pigments of plant leaves participate in light absorption, transfer, and transformation during photosynthesis. The contents of photosynthetic pigments directly influence the photosynthetic capacity of plants, and these pigment contents are significantly influenced by drought stress. Many studies have reported decreased Chl and carotenoid (Car) contents caused by drought stress in numerous plant species (Loggini et al. 1999; Agastian et al. 2000; Parida et al. 2007; Arivalagan and Somasundaram 2017; Masoumi et al. 2017). In the plant leaves, as Chl is degraded, biosynthesis declines, and the contents of Chl and Car will be reduced. The decrease of photosynthetic pigments under drought-stress conditions may result from photo-oxidation (Ahmedi et al. 2009), which allows reduced light absorption by chloroplasts in the plants (Pastenes et al. 2005). The contents of chloroplast pigments are also important indicators of plant photosynthesis, and these contents may be reduced by the changing ratio of lipid protein in pigment-protein complexes or increasing chlorophyllase activity under drought stress (Iyengar and Reddy 1996; Parida et al. 2004). In addition, the ratios of Chl-a/Chl-b and (Chl-a + Chl-b)/Car are correlated with the ability of plants to endure advere conditions (Parida et al. 2007). Numerous previous studies have observed decreased contents of photosynthetic pigments such as Chl and Car in tea plant leaves, and the contents varied in different tea cultivars (Upadhyaya et al. 2008; Upadhyaya and Panda 2013; Upadhyaya et al. 2013; Zhou et al. 2014). For example, the contents of Chl-a, Chl-b, and Car were decreased both in the drought-tolerant tea cultivar (TV-23) and the drought-sensitive tea cultivar (S.3/A3) by the imposition of drought stress. However, the decrease of these pigments was more rapid in S.3/A3, and the ratio of Chl/Car was higher in TV-23 than that in S.3/A3 in comparison to control plants under drought stress (Das et al. 2015).

3.4 Nutrient Elements

Nutrients functioning in plant growth and biomass production generally grow from the internal cycling of reserve materials, which require water for their solubilization and translocation (Singh and Singh 2004). Nutrient absorption is controlled by the interactions of the soil-root interface, including (1) the root morphology and growth rate, (2) the nutrient absorption kinetics of the root systems, and (3) the soil nutrient supply (Gutierrez-Boem and and Thomas 1999). Therefore, reduced soil water availability slows the diffusion rate of many plant nutrients, and ultimately affects the composition and concentration of the soil solution (Baligar et al. 2001; Singh and Singh 2004; Farooq et al. 2009). Further, the root nutrient absorption of many plants is reduced by the lower transpiration rate and impaired active transport, and by the damage of enzyme activity related to nutrient assimilation under drought-stress conditions (Kramer and Boyer 1995; Baligar et al. 2001; Ashraf and Iram 2005; Gunes et al. 2008).

Many studies found that leaf nutritional status had a close relationship with drought stress in tea plants. Drought stress changed the antioxidant response, with significant decreases in the contents of mineral nutrients, such as potassium (K), sodium (Na), calcium (Ca), magnesium (Mg), zinc (Zn), and iron (Fe) in tea plant leaves, suggesting that mineral nutrient deficiency mediated by drought stress could induce oxidative damage in tea plants (Upadhyaya and Panda 2013). Under drought-stress conditions, the contents of Na^+ and K^+ in plant leaves regulate the osmotic potential, and hence the Na^+/K^+ ratio should be adequate for stress acclimatization by plants through osmotic adjustment (Upadhyaya and Panda 2013). Therefore, the decrease of Na^+ and K^+ contents in tea plant leaves when exposed to drought stress may destroy the osmotic balance and induce oxidative stress. Besides, the accumulation of proline (Pro) content is apparently correlated with the concentration of leaf K^+; therefore, RWC and Pro may be inadequate to maintain leaf water status when the K^+ content is decreased under drought stress in tea plants (Upadhyaya and Panda 2013). In addition, the content of leaf Ca^{2+} is also decreased in growing tea seedlings under drought stress (Upadhyaya and Panda 2013), and this may disturb the cellular signaling responding process (Sanders et al. 2002) and the reactive oxygen species (ROS) scavenging process (Evans et al. 2005). Moreover, as Zn serves as a cofactor for many enzymes and as the key structural motif in transcriptional regulatory proteins, reducing the level of Zn nutrition would reduce stomatal conductance and transpiration rates and modulate the antioxidant function (Hu and Sparks 1991; Sharma et al. 1994; Upadhyaya et al. 2013). Under drought-stress conditions, exogenous Zn treatment could minimize the drought-induced decrease of RWC, Car, dry mass, and antioxidants (ascorbate and glutathione) and increase phenolic content, the activities of antioxidant enzymes, such as catalase (CAT), peroxidase (POX), and glutathione reductase (GR), but decrease the contents of hydrogen peroxide (H_2O_2) and malondialdehyde (MDA), suggesting that Zn modulates the biochemical damage caused by drought stress in tea plants (Upadhyaya et al. 2013).

3.5 Plant Growth Substances

Plant growth substances called phytohormones – such as the well-known phytohormones abscisic acid (ABA), auxins, cytokinins (Cks), gibberellin (GA3), salicylic acid (SA), and ethylene (ET) – or, if applied externally, plant growth regulators (PGRs), participate in plant growth and development in minute concentrations (Farooq et al. 2009; Farooq et al. 2012). In addition, these growth substances are also associated with environmental stress responses, mainly by alterations of the endogenous synthesis and balance of these growth substances. For instance, the contents of ABA, SA, and polyamine (PA) accumulated, but the contents of Cks, indole-3-acetic acid (IAA), GA3, and zeatin declined under drought-stress conditions (Bano et al. 1994; Rivero et al. 2007; Abreu and Munne-Bosch 2008; Figueiredo et al. 2008; Farooq et al. 2009).

Under drought- stress conditions, ABA and SA contents rapidly accumulated in the cells of tea plants, and drought-resistant tea cultivars showed higher ABA and SA contents than drought-susceptible tea cultivars (Liu et al. 2015). ABA is generally emphasized and regarded as a key drought-sense signal from root to shoot (Chaves et al. 2003; Wang et al. 2009), as it is primarily produced in the root and then transported to the shoot for inducing stomatal closure and avoiding water loss via transpiration (Davies and Zhang 1991; Steudle 2000; Holbrook et al. 2002). Exogenous ABA treatment significantly minimized the drought-induced decrease in Chl content, but promoted the increase of MDA content and the accumulation of Pro, and also affected the contents of many functional proteins in tea plants exposed to drought stress (Zhou et al. 2014). In addition, SA, as a secondary metabolite, regulates several physiological processes through signal transduction pathways, thereby enhancing plant drought resistance (Farooq et al. 2012). Furthermore, Cks regulate plant responses through root-to-shoot signals, and may mediate the increase in catalase inside peroxisomes and increases the CO_2 compensation point in response to drought stress (Rivero et al. 2007; Rivero et al. 2009). In brief, the contents of many growth substances in tea plant leaves are changed during drought-stress conditions, but the functions and the regulatory mechanisms of the majority of these substances involved in drought resistance remain unknown..

3.6 Osmotic Adjustment

Plants tend to respond to drought stress through a process known as osmotic adjustment (Mahajan and Tuteja 2005). In this process, plants accumulate various types of organic and inorganic solutes to decrease the osmotic potential, hence maintaining the osmotic equilibrium (Rhodes and Samaras 1994; Serraj and Sinclair 2002). These compatible solutes include amino acids, soluble sugars, sugar alcohols, polyols, amines, organic acids, and inorganic ions like K^+; all of these compatible solutes not only contribute to maintaining turgor pressure but also to protecting the enzymes and macromolecules of cells against the oxidative damage exerted by ROS. The plant membranes, enzymes, and other macromolecules are not damaged by these compatible solutes, even at high concentrations (Mahajan and Tuteja 2005; Cechin et al. 2006; Kiani et al. 2007; Farooq et al. 2008, 2009; Anjum et al. 2011).

Under drought-stress conditions in tea plants, the free amino acid content, such as that of Pro and glycine betaine (GB), may be greatly changed. For instance, the content of arginine (Arg), glycine (Gly), leucine (Leu), lysine (Lys), methionine (Met), glutamic acid (Glu), and phenylalanine (Phe) was significantly decreased in tea plant leaves after drought stress compared with findings in the control. The content of a specific amino acid, theanine, was also significantly, but gradually, decreased in tea plant leaves during a stress period. In contrast, the contents of aspartic acid (Asp), serine (Ser), and Pro were significantly increased after drought-stress treatment, especially that of Pro (Wang et al. 2016). A similar result was also reported by Zhu et al. (2010), where they found the content of total free amino acids was decreased as the drought stress was prolonged. Of these amino acids, Pro has been shown to function in protein solvation, the quarternary structure preservation of complex proteins, the maintenance of membrane integrity and in reducing the oxidation of lipid membranes or photoinhibition under drought stress (Demiral and Turkan 2004). Furthermore, Pro was also shown to participate in stabilizing subcellular structures, scavenging free radicals, and buffering cellular redox potential under stress conditions (Ashraf and Foolad 2007). Many studies found that the content of Pro in tea plants was increased in response to drought stress even though the total amino acid content was reduced, and the Pro concentration was generally higher in drought-resistant tea cultivars than in drought-susceptible tea cultivars (Handique and Manivel 1990; Chakraborty et al. 2002; Upadhyaya et al. 2008; Zhou et al. 2014; Maritim et al. 2015). However, some studies found that the correlation between the Pro concentration and the drought resistance of tea plants was not obvious. Therefore, whether the accumulation of Pro could be supported as a drought-tolerance index in tea plants depends on the tested tea cultivars. In addition to Pro, the content of GB was also increased in tea plants under drought-stress conditions (Maritim et al. 2015). The accumulation of GB is considered to protect the plant by maintaining the osmotic equilibrium between the plant cell and the stresses and by stabilizing macromolecules (Rontein et al. 2002).

3.7 Carbohydrates

Under drought-stress conditions, the contents of sugars, such as sucrose (Suc), glucose (Glc), fructose (Fru), raffinose (Rfo), trehalose (Tre), and galactose (Gal) would be affected in tea plants. Current studies suggest that sugars might function as osmolytes and/or protect specific macromolecules and contribute to stabilizing membrane structures, and that they may also protect cells by forming vitrification during drought stress (Kaplan and Guy 2004; Bartels and Sunkar 2005; Iturriaga et al. 2009; Krasensky and Jonak 2012). In addition, sugars are also considered to prevent membrane fusion by interacting with polar groups of phospholipids in membranes (Bartels and Sunkar 2005). It is clear that soluble sugars, such as Suc, Glc, and Fru, not only serve as metabolic substances and structural constituents of cells, but also function as signal molecules to regulate the expression of genes related to plant metabolism, stress responses, and growth and development by activating specific signal pathways or cascades with hormone signal transduction pathways (Pego and Smeekens 2000; Couée et al. 2006; Rolland et al. 2006; Ramon et al. 2008; Li and Koornneef 2011). Many studies have agreed that drought stress generally reduces the depletion of starch content but induces the accumulation of soluble sugars in tea plant leaves (Chakraborty et al. 2002; Lin et al. 2014; Zhou et al. 2014; Liu et al. 2015, 2016), suggesting that soluble sugars play a great role in the resistance to drought stress through energy supply, signaling, and osmotic protection; however, the corresponding regulatory mechanisms need to be further explored.

In addition to sugars, polyols, such as sorbitol, mannitol, galactinol, and myo-inositol, have also been reported to function in protecting macromolecules and scavenging ROS, thereby preventing membranes and enzymes from undergoing oxidative damage (Stoop et al. 1996; Shen et al. 1997; Sengupta et al. 2008; Patra et al. 2010). However, there are few reports on the functional analysis of polyols and it is necessary to strengthen studies in this area.

3.8 Oxidative Damage

As a natural byproduct of normal oxidative metabolism, ROS, such as O_2^-, $^1O_2$, H_2O_2, and OH^-, play a central role in cell signaling and homeostasis (Devasagayam et al. 2004), and the generation of these ROS is one of the earliest biochemical responses of eukaryotic cells to biotic and abiotic stresses (Anjum et al. 2011).

Drought stress induced the production of ROS and disturbed the balance between generation and quenching; these ROS are highly reactive and, in the absence of any protective mechanism, destroy the normal metabolism of the plant by lipid peroxidation, protein degradation, DNA fragmentation, and ultimately cell death (Hendry 1993; Tambussi et al. 2000; Rout and Shaw 2001; Mittler 2002). In addition, the increase of ROS can induce the accumulation of MDA. MDA is thought to be a suitable marker for reflecting the extent of membrane lipid peroxidation (Moller et al. 2007), which suggests that the higher the MDA content, the more serious is the oxidative damage. In tea plants, an evergreen perennial crop, drought stress can promote the accumulation of both ROS and MDA, which then induces oxidative damage and disturbs antioxidant systems, resulting in different physiological and biochemical processes (Upadhyaya and Panda 2004b; Jeyaramraja et al. 2005; Upadhyaya et al. 2008; Zhou et al. 2014; Liu et al. 2015; Wang et al. 2016).

3.9 Antioxidants

To cope with increased ROS and MDA levels, plants, including tea plants, possess effective antioxidant defense systems, which consist of non-enzymatic components such as ascorbate, glutathione, polyphenol, flavone, tocopherol, salicylates, Pro, and GB, and enzymatic scavengers such as superoxide dismutase (SOD), ascorbate peroxidase (APX), POX, GR, and CAT (Asada 1994; Jebara et al. 2005; Hussain et al. 2008; Ozkur et al. 2009). Non-enzymatic antioxidants may work synergistically to maintain the integrity of the photosynthetic membranes, and the enzymatic scavengers may directly scavenge ROS or may play a role by producing a non-enzymatic antioxidant under oxidative stress (Anjum et al. 2011).

It is generally recognized that SOD primarily catalyzes O_2^- converting it to H_2O_2, which is then metabolized to water by CAT and POX or APX and GR (Foyer and Fletcher 2001). However, during the process of scavenging ROS, the changing antioxidant enzyme activities differ according to the plant variety, type of stress, stress intensity, and treatment time. Similarly, antioxidant efficiency, in terms of the process of ROS scavenging, varied in different clonal varieties of tea (Upadhyaya and Panda 2004a) and the responses to drought stress also varied in different clones of tea (Chakraborty et al. 2002). For example, 20 days of drought stress in four selected 1.5-year-old tea cultivars (TV-1, TV-20, TV-29, and TV-30) decreased the phenolic, ascorbate, and glutathione contents of non-enzymatic antioxidants and the activity of CAT, and increased ROS, MDA, and lipid peroxidation and the activities of GR, POX, and polyphenoloxidase (PPO), but the activity of SOD varied in the four cultivars (Upadhyaya et al. 2008). By comparison, except for a similar increase in the production of O_2^-, H_2O_2, and MDA, and a decrease of phenolic, ascorbate, and glutathione contents, Upadhyaya et al. (2013) found that 7 days of drought stress increased SOD and PPO activities, but reduced the activities of POX, GR, CAT, and APX in all three tested 1.5-year-old tea cultivars (TV-1, TV-17, and TV-29), suggesting that the antioxidant enzyme activities among different tea plant varieties varied with prolonged drought stress.

4 Molecular Responses

The tolerance of plants to environmental stresses such as drought is triggered by the activation of cascades of molecular networks related to stress signal perception, signal transduction, the expression of specific stress-responsive genes, and the synthesis of metabolites (Huang et al. 2012). Drought stress mainly affects the production of two types of proteins: one type is termed regulatory proteins, which participate in drought stress signal perception, transduction, and the regulation of the expression of drought-resistance responsive genes, such as protein kinases, phosphatases, cis-regulatory elements, transcription factors (TFs), and calmodulin. The second type is termed functional proteins, such as osmoregulation protein, membrane proteins, metabolic enzymes, channel proteins, and molecular chaperones and transporters, which are directly involved in drought stress tolerance (Chaves et al. 2009; Hirayama and Shinozaki 2010; Lindemose et al. 2013).

Compared with research on drought resistance mechanisms in model plants such as *Arabidopsis*, tobacco, and tomato, such research in tea plants started relatively later, and the studies on molecular mechanisms are still in the phase of gene isolation and identification. The isolation and identification of drought-responsive genes in tea plants mainly depend upon the techniques of differential display of mRNA (DDRT), cDNA-amplified fragment length polymorphism (cDNA-AFLP), and suppression subtractive hybridization (SSH) (Chakraborty et al. 2002; Sharma and Kumar 2005; Krishnaraj et al. 2011; Das et al. 2012). Based on these molecular biology techniques, many regulatory genes and functional genes have been identified in tea plants under drought-stress conditions. For instance, in tea plants under prolonged drought stress, Muoki et al. (2012), using an SSH technique, found that the expressions of many genes related to chaperones, cell rescue/defense, metabolism, and cellular transport categories (late embryogenesis abundant proteins gene (*CsLEA3*), heat-shock protein (*CsHSP17.6*, *CsHSP1*), glutathione S-transferase gene (*CsGST*), sucrose synthase (*CsSuS*), chitinase (*CsCHT*), and thaumatin like protein (*CsTLP*)) were upregulated earlier in a drought-tolerant tea cultivar (UPASI-9) than in a drought-sensitive tea cultivar (TV2). Using a cDNA-AFLP technique, Gupta et al. (2013) identified 89 differentially expressed and up-regulated sequences in a tea cultivar (TV23) under drought stress, and further functional ontology analysis showed that these genes were mainly involved in carbohydrate metabolism, stress responsiveness, protein modification processes, and translation. Using a DDRT technique, Sharma and Kumar (2005) cloned and identified three drought-modulated expressed sequence tags, *dr1*, *dr2*, and *dr3*. Of these, *dr1* was induced only by drought but not by ABA, and *dr2* and *dr3* were repressed by drought but not by ABA.

With the rapid development and the wide application of Omics, including transcriptomics, proteinomics, metabolomics, and RNomics, studies of the molecular mechanisms of drought resistance in tea plants have been significantly promoted in recent years.

Firstly, the transcriptome has been widely used to identify the differentially expressed drought-responsive genes in tea plants at the transcription level (Liu et al. 2016; Tony et al. 2016; Wang et al. 2016; Zheng et al. 2016). For example, Liu et al. (2016) compared the gene expressions between drought-stress and control conditions, where they found that most of the genes related to phytohormone metabolism and signaling, such as ABA biosynthesis genes and signal transduction-related genes, jasmonic acid (JA) intermediate genes, ET synthesis-related genes and signal transduction-related genes, and Ck synthesis-related genes were all up-regulated, but JA precursor genes and IAA signal transduction-related genes were all reduced under drought-stress conditions. Besides, many protein kinase genes and protein phosphatase genes were also up-regulated. Furthermore, most members of the 12 TF gene families, such as basic leucine zipper (bZIP), squamosa promoter binding protein-like (SPL), homeobox-leucine zipper protein (HD-ZIP), NAC domain-containing proteins (NACs), myeloblastosis oncoprotein (MYB), and APETALA2/ET-responsive element-binding protein (AP2/EREBP) , were also up-regulated. In addition, Suc, Tre, and mannitol synthesis-related genes, starch-degrading genes, and the pyrroline-5-carboxylate synthetase gene (*P5CS*), pyrroline-5-carboxylate reductase gene (*P5CR*), and γ-glutamyl phosphate reductase gene (*GRR*) were all up-regulated, but two key starch synthesis genes were inhibited.

Secondly, the application of proteomics further verified the reliability of the transcriptome results at the translation level. For instance, Wang et al. (2017) analyzed the differentially expressed proteins in tea plants during drought stress, where they found changes in many differentially expressed proteins involved in glycolysis/gluconeogenesis, starch and sucrose synthesis, or degradation metabolism, and secondary metabolism showed a correlation with the changes of the differentially expressed genes during drought as reported by Liu et al. (2016), except for a few differences.

Thirdly, recent studies have shown that drought stress induces the aberrant expression of a large number of microRNAs (miRNAs) in tea plants (Mohanpuria and Yadav 2012; Liu et al. 2016). Liu et al. (2016) demonstrated that a total of 268 conserved miRNAs belonging to 46 miRNA families were identified in tea plants, and about 199 of these miRNAs might be related to the response to drought stress and recovery through inhibiting the expression of miRNA effectors, drought signaling-related receptors and enzymes, TFs, cytoskeleton proteins, cell wall-related enzymes and oxidases, stomatal movement-related proteins, osmoprotectants, phytohormones, and antioxidant synthesis-related proteins.

To summarize, many drought-responsive genes have been identified in tea plants during drought stress by various molecular biology techniques, but the corresponding specific functions of these genes or miRNAs are little known. Also, the activities of proteins are also regulated at the post-translation level, but related studies have not yet been performed. Therefore, it is necessary to strengthen the study of these areas.

5 Conclusion

In conclusion, research on the drought-resistance mechanisms of tea plants has made great progress recently. On the one hand, a series of morphological, physiological, and biochemical changes of tea plants under drought stress have been clearly defined. On the other hand, with the application of different molecular biological techniques, many drought-responsive genes have been identified in tea plants, including *HSPs*, late embryogenesis abundant, *LEAs*, antioxidant metabolism response genes, *TFs*, hormone metabolism-related genes, and sugar metabolism-related genes. In this review, we have summarized the results of recent research on the physiological and molecular mechanisms of tea plants in responding to drought stress, which will provide a theoretical basis for future research on the breeding of drought-resistant tea plants. However, because the research on the drought-resistance mechanisms of tea plants started relatively late, there are still many deficiencies in the current research, namely:

(1) The research lacks a set of systematic and complete morphological identification indexes, including stomatal density, leaf area, waxy layer thickness, and epidermis thickness, to identify the capacity of drought resistance in different tea plant cultivars;
(2) The numbers of physiological and biochemical indexes used for identification of the capacity of drought resistance in different tea plant cultivars are far too few, and studies on the physiological and biochemical changes of tea plants during drought stress still need to be strengthened, including the identification of plant growth substances, osmotic adjustment solutes, sugars, and antioxidants related to drought response; studies also need to target the functional analysis of each of these substances;
(3) Closely linked molecular markers of yield and quality traits under drought stress are lacking, and these should be developed to select drought-resistant tea plant cultivars by using molecular-assisted breeding techniques.
(4) Research on the molecular mechanisms of drought resistance in tea plants is still at the level of the isolation and identification of functional genes. Because the genetic background of the tea plant is complex and the establishment of tea regeneration systems is inadequate, verification of the function of drought-responsive genes can only be made by means of using results in some model plants, such as *Arabidopsis*, tobacco, and tomato. The drawback is whether the functions of these genes in the model plants are similar to those in tea plants. In future research, therefore, we should keep on analyzing and identifying the drought-responsive genes of tea plants, and we should also focus on constructing tea plant regeneration systems and exploring the transgenic technology of the tea plant.

References

Abrams MD, Kubiske ME, Steiner KC (1990) Drought adaptations and responses in five genotypes of *Fraxinus Pennsylvanica* Marsh: photosynthesis, water relations and leaf morphology. Tree Physiol 6:305–315

Abreu ME, Munne-Bosch S (2008) Salicylic acid may be involved in the regulation of drought-induced leaf senescence in perennials: a case study in field-grown *Salvia officinalis* L. plants. Environ Exp Bot 64:105–112

Agastian P, Kingsley SJ, Vivekanandan M (2000) Effect of salinity on photosynthesis and biochemical characteristics in mulberry genotypes. Photosynthetica 38:287–290

Ahmadi AA (1998) Effect of post-anthesis water stress on yield regulating processes in wheat (*Triticum aestivum* L.). Ph. D. Thesis. University of London, Wye College, Wye, Ashford

Ahmedi CB, Rouina BB, Sensoy S, Boukhris M, Abdallah FB (2009) Changes in gas exchange, proline accumulation and antioxidative enzyme activities in three olive cultivars under contrasting water availability regimes. Environ Exp Bot 67:345–352

Allen DJ, Ort DR (2001) Impacts of chilling temperatures on photosynthesis in warm-climate plants. Trends Plant Sci 6:36–42

Anjum SA, Wang LC, Farooq M, Hussain M, Xue LL, Zou CM (2011) Brassinolide application improves the drought tolerance in maize through modulation of enzymatic antioxidants and leaf gas exchange. J Agron Crop Sci 197:177–185

Arivalagan M, Somasundaram R (2017) Alteration of photosynthetic pigments and antioxidant systems in tomato under drought with Tebuconazole and Hexaconazole applications. J Sci Agric 1:146

Asada K (1994) Mechanisms for scavenging reactive molecules generated in chloroplasts under light stress. In: Baker NR, Bower JR (eds) Photoinhibition of photosynthesis. Bios Scientific Publishers, Oxford, pp 131–145

Ashraf M, Foolad MR (2007) Roles of glycine, betaine and proline in improving plant abiotic stress resistance. Environ Exp Bot 59:206–216

Ashraf M, Iram A (2005) Drought stress induced changes in some organic substances in nodules and other plant parts of two potential legumes differing in salt tolerance. Flora 200:535–546

Bacelar EA, Colrreia CM, Mouninho-Pereira JM, Goncalves BC, Lopes JI, Torres-Pereira JM (2004) Sclerophylly and leaf anatomical traits of five field-grown olive cultivars growing under drought conditions. Tree Physiol 24:233–239

Baldini E, Facini O, Nerozzi F, Rossi F, Rotondi A (1997) Leaf characteristics and optical properties of different woody species. Trees 12:73–81

Baligar VC, Fageria NK, He ZL (2001) Nutrient use efficiency in plants. Commun Soil Sci Plant Anal 32:921–950

Bano A, Hansen H, Dorffling K, Hahn H (1994) Changes in the contents of free and conjugated abscisic acid, phaseic acid and cytokinins in xylem sap of drought stressed sunflower plants. Phytochemistry 37:345–347

Barora AC (1994) Gas exchange in tea (*Camellia sinensis* L.) under water stress. Two Bud 41:19–24

Bartels D, Sunkar R (2005) Drought and salt tolerance in plants. Crit Rev Plant Sci 24:23–58

Benjamin JG, Nielsen DC (2006) Water deficit effects on root distribution of soybean, field pea and chickpea. Field Crop Res 97:248–253

Bolhar-Nordenkampf HR (1987) Shoot morphology and leaf anatomy in relation to photosynthesis. In: Techniques in Bioproductivity and Photosynthesis. Pergamon Press, Oxford, pp 107–117

Bosabalidis AM, Kofidis G (2002) Comparative effects of drought stress on leaf anatomy of two olive cultivars. Plant Sci 163:375–379

Burgess PJ, Carr MKV (1993) Responses of tea (*Camellia sinensis*) clones to drought. I. Yield, dry matter production and partitioning. Asp Appl Biol 34:249–258

Cechin I, Rossi SC, Oliveira VC, Fumis TF (2006) Photosynthetic responses and proline content of mature and young leaves of sunflower plants under water deficit. Photosynthetica 44:143–146

Chakraborty V, Dutta S, Chakraborty BN (2002) Responses of tea plant to water stress. Biol Plant 45:557–562

Chartzoulakis K, Patakas A, Bosabalidis A (1999) Changes in water relations, photosynthesis and leaf anatomy induced by intermittent drought in two olive cultivars. Environ Exp Bot 42:113–120

Chaves MM, Maroco JP, Periera S (2003) Understanding plant responses to drought from genes to the whole plant. Funct Plant Biol 30:239–264

Chaves MM, Flexas J, Pinheiro C (2009) Photosynthesis under drought and salt stress: regulation mechanisms from whole plant to cell. Ann Bot 103:446–473

Cheruiyot EK, Mumera LM, Ng'etich WK, Hassanali A, Wachira F (2007) Polyphenols as potential indicators for drought tolerance in tea (*Camellia sinensis* L.). Biosci Biotechnol Biochem 71:2190–2197

Cheruiyot EK, Mumera LM, Ng'etich WK, Hassanali A, Wachira F, Wanyoko JK (2008) Shoot epicatechin and epigallocatechin contents respond to water stress in tea [*Camellia sinensis* (L.) O. Kuntze]. Biosci Biotechnol Biochem 72:1219–1226

Couée I, Sulmon C, Gouesbet G, El AA (2006) Involvement of soluble sugars in reactive oxygen species balance and responses to oxidative stress in plants. J Exp Bot 57:449–459

Das A, Das S, Mondal TK (2012) Identification of differentially expressed gene profiles in young roots of tea [*Camellia sinensis* (L.) O. Kuntze] subjected to drought stress using suppression subtractive hybridization. Plant Mol Biol 30:1088–1101

Das A, Mukhopadhyay M, Sarkar B, Saha D, Mondal TK (2015) Influence of drought stress on cellular ultrastructure and antioxidant system in tea cultivars with different drought sensitivities. J Environ Biol 36:875–882

Davies W, Zhang J (1991) Root signals and the regulation of growth and the development of plants in drying soil. Annu Rev Plant Physiol Plant Mol Biol 42:55–76

Del Blanco IA, Rajaram S, Kronstad WE, Reynolds MP (2000) Physiological performance of synthetic hexaploid wheat-derived populations. Crop Sci 40:1257–1263

Demiral T, Turkan I (2004) Does exogenous glycinebetaine affect antioxidative system of rice seedlings under NaCl treatment? J Plant Physiol 161:1089–1110

Devasagayam TP, Tilak JC, Boloor KK, Sane KS, Ghaskadbi SS, Lele RD (2004) Free radicals and antioxidants in human health: current status and future prospects. J Assoc Physicians India 52:794–804

Ennajeh M, Tounekti T, Vadel AM, Khemira H, Cochard H (2008) Water relations and drought-induced embolism in two olive (*Olea europaea* L.) varieties 'Meski' and 'Chemlali' under severe drought conditions. Tree Physiol 28:971–976

Ennajeh M, Vadel AM, Cochard H, Khemira H (2010) Comparative impacts of water stress on the leaf anatomy of a drought-resistant and a drought-sensitive olive cultivar. J Pomol Hortic Sci 85:289–294

Evans NH, McAinsh MR, Hetherington AM, Knight MR (2005) ROS perception in *Arabidopsis thaliana*: the ozone induced calcium response. Plant J 41:615–626

Farooq M, Basra SMA, Wahid A, Cheema ZA, Cheema MA, Khaliq A (2008) Physiological role of exogenously applied glycinebetaine in improving drought tolerance of fine grain aromatic rice (*Oryza sativa* L.). J Agron Crop Sci 194:325–333

Farooq M, Basra SMA, Wahid A, Ahmad N, Saleem BA (2009) Improving the drought tolerance in rice (*Oryza sativa* L.) by exogenous application of salicylic acid. J Agron Crop Sci 195:237–246

Farooq M, Hussain M, Wahid A, Siddique KHM (2012) Drought stress in plants: an overview. In: Aroca R (ed) Plant responses to drought stress. Springer-Verlag, Berlin/Heidelberg, pp 1–33

Farooqui AHA, Kumar R, Fatima S, Sharma S (2000) Responses of different genotypes of lemon grass (*Cymbopogon flexuosus* and *C. pendulus*) to water stress. J Plant Biol 27:277–282

Figueiredo MVB, Buritya AH, Martınez CR, Chanway CP (2008) Alleviation of drought stress in the common bean (*Phaseolus vulgaris* L.) by co-inoculation with *Paenibacillus polymyxa* and *Rhizobium tropici*. Appl Soil Ecol 40:182–188

Foyer CH, Fletcher JM (2001) Plant antioxidants: colour me healthy. Biologist 48:115–120
Gunes A, Kadioglu YK, Pilbeam DJ, Inal A, Coban S, Aksu A (2008) Influence of silicon on sunflower cultivars under drought stress, II: Essential and non-essential element uptake determined by polarized energy dispersive X-ray fluorescence. Commun Soil Sci Plant Anal 39:1904–1927
Gupta S, Bharalee R, Bhorali P, Das SK, Bhagawati P, Bandyopadhyay T, Gohain B, Agarwal N, Ahmed P, Borchetia S, Kalita MC, Handique AK, Das S (2013) Molecular analysis of drought tolerance in tea by cDNA-AFLP based transcript profiling. Mol Biotechnol 53:237–248
Gutierrez-Boem FH, Thomas GW (1999) Phosphorus nutrition and water deficits in field grown soybeans. Plant Soil 207:87–96
Handique AC (1992) Some silent features in the study of drought resistance in tea. Two Bud 39:16–18
Handique AC, Manivel L (1990) Selection criteria for drought tolerance in tea. Assam Rev. Tea News 79:18–21
Hendry GAF (1993) Evolutionary origins and natural functions of fructans: a climatological, biogeography and mechanistic appraisal. New Phytol 123:3–14
Hirayama T, Shinozaki K (2010) Research on plant abiotic stress responses in the post-genome era: past, present and future. Plant J 61:1041–1052
Holbrook NM, Shashidhar VR, James RA, Munns R (2002) Stomatal control in tomato with ABA-deficient roots: response of grafted plants to soil drying. J Exp Bot 53:1503–1514
Hu H, Sparks D (1991) Zinc deficiency inhibits chlorophyll synthesis and gas exchange in 'Stuart' pecan. Hortic Sci 26:267–268
Huang GT, Ma SL, Bai LP, Zhang L, Ma H, Jia P, Liu J, Zhong M, Guo ZF (2012) Signal transduction during cold, salt, and drought stresses in plants. Mol Biol Rep 39:969–987
Hussain M, Malik MA, Farooq M, Ashraf MY, Cheema MA (2008) Improving drought tolerance by exogenous application of glycinebetaine and salicylic acid in sunflower. J Agron Crop Sci 194:193–199
Iturriaga G, Suãrez R, Nova-Franco B (2009) Trehalose metabolism: from osmoprotection to signaling. Int J Mol Sci 10:3793–3810
Iyengar ERR, Reddy MP (1996) Photosynthesis in high salt-tolerant plants. In: Pesserkali M (ed) Handbook of photosynthesis. Marshal Dekar, Baton Rouge, pp 56–65
Jain NK (1999) Global advances in tea science. Aravali Books International, NewDelhi
Jebara S, Jebara M, Limam F, Aouani ME (2005) Changes in ascorbate peroxidase, catalase, guaiacol peroxidase and superoxide dismutase activities in common bean (*Phaseolus vulgaris*) nodules under salt stress. J Plant Physiol 162:929–936
Jeyaramraja PR, Meenakshi SN, Kumar RS, Joshi SD, Ramasubramanian B (2005) Water deficit induced oxidative damage in tea (*Camellia sinensis*) plants. J Plant Physiol 162:413–419
Kaplan F, Guy CL (2004) Beta-amylase induction and the protective role of maltose during temperature shock. Plant Physiol 135:1674–1684
Kiani SP, Talia P, Maury P, Grieu P, Heinz R, Perrault A, Nishinakamasu V, Hopp E, Gentzbittel L, Paniego N, Sarrafi A (2007) Genetic analysis of plant water status and osmotic adjustment in recombinant inbred lines of sunflower under two water treatments. Plant Sci 172:773–787
Kramer PJ, Boyer JS (1995) Water relations of plants and soils. Academic, San Diego
Krasensky J, Jonak C (2012) Drought, salt, and temperature stress-induced metabolic rearrangements and regulatory networks. J Exp Bot 63:1593–1608
Krishnaraj T, Gajjeraman P, Palanisamy S, Subhas Chandrabose SR, Azad Mandal AK (2011) Identification of differentially expressed genes in dormant (banjhi) bud of tea [*Camellia sinensis* (L.) O. Kuntze] using subtractive hybridization approach. Plant Physiol Bioch 49:565–571
Lawlor DW, Cornic G (2002) Photosynthetic carbon assimilation and associated metabolism in relation to water deficits in higher plants. Plant Cell Environ 25:275–294
Lawson T, Oxborough K, Morison JI, Baker NR (2003) The responses of guard and mesophyll cell photosynthesis to CO_2, O_2, light, and water stress in a range of species are similar. J Exp Bot 54:1743–1752

Li P, Koornneef M (2011) Fructose sensitivity is suppressed in *Arabidopsis* by the transcription factor ANAC089 lacking the membrane-bound domain. Proc Natl Acad Sci U S A 108:3436–3441
Li YP, Ye W, Wang M, Yan XD (2009) Climate change and drought: a risk assessment of crop yield impacts. Clim Res 39:31–46
Lin SK, Lin J, Liu QL, Ai YF, Ke YQ, Chen C, Zhang ZY, He H (2014) Time-course of photosynthesis and non-structural carbon compounds in the leaves of tea plants (*Camellia sinensis* L.) in response to deficit irrigation. Agr Water Manage 144:98–106
Lindemose S, Oshea C, Jensen MK, Skriver K (2013) Structure, function and networks of transcription factors involved in abiotic stress responses. Int J Mol Sci 14:5842–5878
Liu SC, Yao MZ, Ma CL, Jin JQ, Ma JQ, Li CF, Chen L (2015) Physiological changes and differential gene expression of tea plant under dehydration and rehydration conditions. Sci Hortic 184:129–141
Liu SC, Xu YX, Ma JQ, Wang WW, Chen W, Huang DJ, Fang J, Li XJ, Chen L (2016) Small RNA and degradome profiling reveals important roles for microRNAs and their targets in tea plant response to drought stress. Physiol Plant 158:435–451
Loggini B, Scartazza A, Brugnoli E, Navari-Izzo F (1999) Antioxidative defense system, pigment composition, and photosynthetic efficiency in two wheat cultivars subjected to drought. Plant Physiol 119:1091–1099
Mahajan S, Tuteja N (2005) Cold, salinity and drought stresses: an overview. Arch Biochem Biophys 444:139–158
Maritim TK, Kamunya SM, Mireji P, Mwendia V, Muoki RC, Cheruiyot EK, Wachira FN (2015) Physiological and biochemical responses of tea (*Camellia sinensis* L. O Kuntze) to water-deficit stress. J Hortic Sci Biotechnol 90:395–400
Masoumi A, Nabati J, Mohammad K (2017) The influence of drought stress on photosynthetic pigments and some metabolic contents of two *Kochia scoparia* ecotypes in saline condition. Adv Biores 8:55–61
Mediavilla S, Escudero A, Heilmeier H (2001) Internal leaf anatomy and photosynthetic resource-use efficiency: interspecific and intraspecific comparisons. Tree Physiol 21:251–259
Mittler R (2002) Oxidative stress, antioxidants and stress tolerance. Trends Plant Sci 7:405–410
Mohanpuria P, Yadav SK (2012) Characterization of novel small RNAs from tea (*Camellia sinensis* L.). Mol Biol Rep 39:3977–3986
Moller IM, Jensen PE, Hansson A (2007) Oxidative modifications to cellular components in plants. Annu Rev Plant Biol 58:459–481
Muoki RC, Paul A, Kumar S (2012) A shared response of thaumatin like protein, chitinase, and late embryogenesis abundant protein 3 to environmental stresses in tea [*Camellia sinensis* (L.) O. Kuntze]. Funct Integr. Genomics 12:565–571
Nayyar H, Gupta D (2006) Differential sensitivity of C_3 and C_4 plants to water deficit stress: association with oxidative stress and antioxidants. Environ Exp Bot 58:106–113
Netto LA, Jayaram KM, Puthur JT (2010) Clonal variation of tea [*Camellia sinensis* (L.) O. Kuntze] in countering water deficiency. Physiol Mol Biol Plants 16:359–367
Nevo E, Bolshakova MA, Martyn GI, Musatenko LI, Sytnik K, Pavléek T, Behatav A (2000) Drought and light anatomical adaptive leaf strategies in three woody species caused by microclimatic selection at "Evolution Canyon" Israel. Israel J Plant Sci 48:33–46
Ng'etich WK, Stephens W (2001) Responses of tea to environment in Kenya. 1. Genotype x environment interactions for total dry matter production and yield. Exp Agric 37:333–342
Nixon DJ, Burgess PJ, Sanga BNK, Carr MKV (2001) A comparison of the responses of mature and young clonal tea to drought. Exp Agric 37:391–402
Ozkur O, Ozdemir F, Bor M, Turkan I (2009) Physiochemical and antioxidant responses of the perennial xerophyte *Capparis ovata* Desf. to drought. Environ Exp Bot 66:487–492
Panda RK, Stephens W, Matthews R (2003) Modelling the influence of irrigation on the potential yield of tea (*Camellia sinensis*) in North-East India. Exp Agric 39:181–198
Parida AK, Das AB, Mittra B (2004) Effects of salt on growth, ion accumulation, photosynthesis and leaf anatomy of the mangrove *Bruguiera parviflora*. Trees Struct Funct 18:167–174

Parida AK, Dagaonkar VS, Phalak MS, Umalkar GV, Aurangabadkar LP (2007) Alterations in photosynthetic pigments, protein and osmotic components in cotton genotypes subjected to short-term drought stress followed by recovery. Plant Biotech Rep 1:37–48
Pastenes C, Pimentel P, Lillo J (2005) Leaf movements and photoinhibition in relation to water stress in field-grown beans. J Exp Bot 56:425–433
Patra B, Ray S, Richter A, Majumder AL (2010) Enhanced salt tolerance of transgenic tobacco plants by co-expression of *PcINO1* and *McIMT1* is accompanied by increased level of myo-inositol and methylated inositol. Protoplasma 245:143–152
Pego JV, Smeekens SCM (2000) Plant fructokinases: a sweet family get-together. Trends Plant Sci 5:531–536
Pita P, Pardos JA (2001) Growth, leaf morphology, water use and tissue water relations of Eucalyptus globulus clones in response to water deficit. Tree Physiol 21:599–607
Praba ML, Cairns JE, Babu RC, Lafitte HR (2009) Identification of physiological traits underlying cultivar differences in drought tolerance in rice and wheat. J Agron Crop Sci 195:30–46
Ramon M, Rolland F, Sheen J (2008) Sugar sensing and signaling. The Arabidopsis Book 6:e0117
Reddy AR, Chaitanya KV, Vivekanandan M (2004) Drought-induced responses of photosynthesis and antioxidant metabolism in higher plants. J Plant Physiol 161:1189–1202
Rhodes D, Samaras T (1994) Genetic control of osmoregulation in plants. In: Strange SK (ed) Cellular and molecular physiology of cell volume regulation. CRC Press, Boca Raton, pp 347–361
Rivero RM, Kojima M, Gepstein A, Sakakibara H, Mittler R, Gepstein S, Blumwald E (2007) Delayed leaf senescence induces extreme drought tolerance in a flowering plant. Proc Natl Acad Sci U S A 104:19631–19636
Rivero RM, Shulaev V, Blumwald E (2009) Cytokinin-dependent photorespiration and the protection of photosynthesis during water deficit. Plant Physiol 150:1530–1540
Rolland F, Baena-Gonzalez E, Sheen J (2006) Sugar sensing and signaling in plants: conserved and novel mechanisms. Annu Rev Plant Biol 57:675–709
Rontein D, Bassat G, Hanson HD (2002) Metabolic engineering of osmoprotectant accumulation in plants. Metab Eng 4:49–56
Rout NP, Shaw BP (2001) Salt tolerance in aquatic macrophytes: possible involvement of the antioxidative enzymes. Plant Sci 160:415–423
Samarah NH, Alqudah AM, Amayreh JA, McAndrews GM (2009) The effect of late-terminal drought stress on yield components of four barley cultivars. J Agron Crop Sci 195:427–441
Sanders D, Pelloux J, Brownlee C, Harper JF (2002) Calcium at the crossroads of signaling. Plant Cell 14:S401–S417
Sengupta S, Patra B, Ray S, Majumder AL (2008) Inositol methyl transferase from a halophytic wild rice, *Porteresia coarctata* Roxb. (Tateoka): regulation of pinitol synthesis under abiotic stress. Plant Cell Environ 31:1442–1459
Serraj R, Sinclair TR (2002) Osmolyte accumulation: can it really help increase crop yield under drought conditions? Plant Cell Environ 25:333–341
Shao HB, Chu LY, Jaleel CA, Manivannan P, Panneerselvam R, Shao MA (2009) Understanding water deficit stress-induced changes in the basic metabolism of higher plants–biotechnologically and sustainably improving agriculture and the ecoenvironment in arid regions of the globe. Crit Rev Biotechnol 29:131–151
Sharma PN, Kumar S (2005) Differential display-mediated identification of three drought-responsive expressed sequence tags in tea [*Camellia sinensis* (L.) O. Kuntze]. J Bios 30:231–235
Sharma PN, Kumar N, Bisht SS (1994) Effect of zinc deficiency on chlorophyll contents, photosynthesis, and water relations of cauliflower plants. Photosynthetica 30:353–359
Shen B, Jensen RG, Bohnert HJ (1997) Increased resistance to oxidative stress in transgenic plants by targeting mannitol biosynthesis to chloroplasts. Plant Physiol 113:1177–1183
Siddique KHM, Regan KL, Tennant D, Thomson BD (2001) Water use and water use efficiency of cool season grain legumes in low rainfall Mediterranean-type environments. Eur J Agron 15:267–280

Singh B, Singh G (2004) Influence of soil water regime on nutrient mobility and uptake by *Dalbergia sissoo* seedlings. Trop Ecol 45:337–340

Steudle E (2000) Water uptake by plant roots: an integration of views. Plant Soil 226:45–56

Stoop JHM, Williamson JD, Pharr DM (1996) Mannitol metabolism in plants: a method for coping with stress. Trends Plant Sci 1:139–144

Tambussi EA, Bartoli CG, Beltrano J, Guiamet JJ, Arans JL (2000) Oxidative damage to thylakoid proteins in water stressed leaves of wheat (*Triticum aestivum*). Physiol Plant 108:398–404

Tezara W, Mitchell VJ, Driscoll SD, Lawlor DW (1999) Water stress inhibits plant photosynthesis by decreasing coupling factor and ATP. Nature 401:914–917

Tony M, Samson K, Charles M, Paul M, Richard M, Mark W, Stomeo F, Sarah S, Martina K, Francis W (2016) Transcriptome-based identification of water-deficit stress responsive genes in the tea plant, *Camellia sinensis*. J Plant Biotechnol 43:302–310

Upadhyaya H, Panda SK (2004a) Responses of *Camellia sinensis* to drought and rehydration. Biol Plant 48:597–600

Upadhyaya H, Panda SK (2004b) Antioxidant efficiency and biochemical variations in five clones of *Camellia sinensis* L. Physiol Mol Biol Plants 10:115–120

Upadhyaya H, Panda SK (2013) Abiotic stress responses in tea [*Camellia sinensis* L (O) Kuntze]: an overview. Rev Agr Sci 1:1–10

Upadhyaya H, Panda SK, Dutta BK (2008) Variation of physiological and antioxidative responses in tea cultivars subjected to elevated water stress followed by rehydration recovery. Acta Physiol Plant 30:457–468

Upadhyaya H, Dutta BK, Panda SK (2013) Zinc modulates drought-induced biochemical damages in tea [*Camellia sinensis* (L) O Kuntze]. J Agric Food Chem 61:6660–6670

Wang Z, Zhu Y, Wang L, Liu X, Liu Y, Phillips J, Deng X (2009) A WRKY transcription factor participates in dehydration tolerance in *Boea hygrometrica* by binding to the W-box elements of the galactinol synthase (BhGolS1) promoter. Planta 230:1155–1166

Wang W, Xin H, Wang M, Ma Q, Wang L, Kaleri NA, Wang Y, Li X (2016) Transcriptomic analysis reveals the molecular mechanisms of drought-stress-induced decreases in *Camellia sinensis* leaf quality. Front Plant Sci 7:385

Wang Y, Fan K, Wang J, Ding ZT, Wang H, Bi CH, Zhang YW, Sun HW (2017) Proteomic analysis of *Camellia sinensis* (L.) reveals a synergistic network in the response to drought stress and recovery. J Plant Physiol 219:91–99

Wijeratne MA, Fordhan R, Anandacumaraswamy A (1998) Water relations of clonal tea (*Camellia sinensis* L.) with reference to drought resistance: II. Effect of water stress. Trop Agri Res Exten 1:74–80

Zheng C, Wang Y, Ding Z, Zhao L (2016) Global transcriptional analysis reveals the complex relationship between tea quality, leaf senescence and the responses to cold-drought combined stress in *Camellia sinensis*. Front Plant Sci 7:1858

Zhou L, Xu H, Mischke S, Meinhardt LW, Zhang D, Zhu X, Li X, Fang W (2014) Exogenous abscisic acid significantly affects proteome in tea plant (*Camellia sinensis*) exposed to drought stress. Hortic Res 1:14029

Zhu M, Simons B, Zhu N, Oppenheimer DG, Chen S (2010) Analysis of abscisic acid responsive proteins in Brassica napus guard cells by multiplexed isobaric tagging. J Proteome 73:790–805

Chapter 5
Molecular and Physiological Adaptations of Tea Plant in Response to Low Light and UV Stress

Lorenzo Cotrozzi and Marco Landi

Abstract The sensory quality, economic value, and health functions of tea liquor depend on several metabolites in leaves of tea (*Camellia sinensis* L.), such as polyphenols (the most representative quality- and function-related compounds, especially catechins), caffeine, amino acids, aroma compounds, vitamins, and carbohydrates. The biosynthesis of these molecules is significantly affected by both management practices in the field and environmental factors, such as shading and ultraviolet (UV) stress. In this chapter, therefore, we review existing literature reporting the effects of shade and UV irradiances on tea plants with the aim to reappraise morphological, biochemical, physiological, and molecular responses found in this species. Cultivation of tea plants under shade can (i) induce anatomical and ultrastructural adaptations; (ii) influence leaf color, providing characteristic vivid greenness due to incremented chlorophyll content; (iii) decrease the risk of photoinhibition; and finally (iv) improve the quality of tea beverages by causing the reduction of the concentration of phenylpropanoids and caffeine (which contribute to astringency) while increasing the level of amino acids (the compounds that contribute to sweetness). However, the shade practice in tea is still subjected to controversy, and the correct managing of the shade practice is essential. Elevated UV irradiances, especially UV-B, can alter the plant morphology and affect several biochemical and physiological processes; most plant species respond to UV in a dose dependent manner. Protracted exposition to low UV-B doses (similar to UV peaks actually perceive by plants in the ambient) leads to an enhancement of phenylpropanoid metabolism and in particular stimulates the biosynthesis of UV-protecting flavonoids (mainly catechins). Conversely, UV-B doses much higher than those usually experienced in natural environment lead to the downregulation of the phenylpropanoid metabolism likely due to irreversible (to some extent) UV-triggered damages. The understandings of the reactions of tea plants to different light conditions, including situations of low light and high UV, are of pivotal importance to enhance both plant yield and tea quality by "driving" the accumulation of desired compounds in tea leaves.

L. Cotrozzi · M. Landi (✉)
Department of Agriculture, Food and Environment, University of Pisa, Pisa, Italy
e-mail: marco.landi@agr.unipi.it

W.-Y. Han et al. (eds.), *Stress Physiology of Tea in the Face of Climate Change*,
https://doi.org/10.1007/978-981-13-2140-5_5

Keywords Amino acids · Caffeine · Catechins · Flavonoids · Low light · Tea plant · UV stress

1 Codifying the Sunlight: "Bright" Information for Tea Plant Metabolism

Light plays a key role in the development of higher plants which use sunlight not simply as a source of energy but also as a source of environmental information (Robson et al. 2015). Photomorphogenesis, photoperiodicity, and phototropism are just few examples of the complex roles of light quality, intensity, direction, and timing. In higher plants, natural radiation simultaneously activates more than one photoreceptor intercepting visible wavebands, namely, phytochromes, cryptochromes, and phototropin (Casal 2000), as well as UV wavebands (i.e., UV Resistance Locus 8; UVR8) (Rizzini et al. 2011). Experiments carried out in natural light conditions are beginning to show that the interaction and cross talk between photoreceptors create a photo-transduction network with emergent properties, which provide a more robust system for light perception in plants. However, it is believed that much is yet to be discovered about light perception in higher plants (Costa Galvão and Fankhause 2015).

Although light is an unavoidable resource for photosynthesis, both low and high sunlight exposure can seriously affect plant performances. Shortages of pivotal resources, such as light, impair plant survival and growth, whereas high sunlight leads to heat stress, desiccation, excessive irradiance, and ultraviolet (UV) radiation stresses in plants, which require high protective investments to cope with both these harmful conditions (Valladares and Niinemets 2008). Plants are typically genetically predisposed (adapted) to grow within a specific range of photon flux density (PFD). Furthermore, sun or shade plants may also possess the capacity to respond to different PFDs within the range which they are adapted to grow (acclimation). Both adaptation and acclimation to different PFDs are associated with several changes in plant morphology, physiology, and biochemistry (Seemann 1989).

The presence of nearby individuals, as well as self-shading within the canopy, could reduce the availability of photosynthetic active radiation (PAR) and impair light quality plants. So, the shade condition is ubiquitous in nature, and most plants (or their parts) are shaded to some extent throughout their life (Valladares and Niinemets 2008). As a consequence, there is a decrease in the red/far-red ratio in shaded vegetation, which is sensed by plants as a cue for competitors by the phytochromes. In combination with a curtailment of blue light and total light intensity, these light cues can result in two alternative strategies to cope with shade and, namely, avoidance and tolerance (Gommers et al. 2013). All shaded plants optimize photosynthesis to adapt themselves to the decrease quality/quantity of light. However, most species in open habitats express the shade-avoidance syndrome when shaded. Shade avoidance includes a set of traits to escape shade and reach the light, such as elongation of stem and petioles, hyponasty (upward bending of leaves), and reduced branching; accelerated flowering may be another opportunity (Seemann 1989; Valladares and Niinemets 2008; Ruberti et al. 2012; Casal 2013; Ballaré and Pierik 2017). Conversely, shade-tolerant species from forest understories cannot outgrow surrounding trees and adopt a tolerance response (Gommers et al. 2013). Shade tolerance is a set of traits that typically optimize the carbon gain under low light conditions, such as an increased chlorophyll (Chl) content and a reduction of the Chl *a*/*b* ratio, specific leaf area, and the photosystem II (PSII)/photosystem I (PSI) ratio (Valladares and Niinemets 2008; Gommers et al. 2013; Gong et al. 2015). Unlike shade avoidance, little is still known about the mechanisms involved in shade tolerance (Gommers et al. 2013).

Among various environmental stressors, UV radiation is one of the most harmful due to its interaction with UV-absorbing biological molecules such as nucleic acids, proteins, lipids, and phytohormones and the consequent damage that it causes at whole-plant level (Kataria et al. 2014). The doses of UV and the proportion of UV into the full sunlight spectrum reaching the plant are of crucial importance for plant responses. In particular, the spectral balance between PAR and UV-B has been shown to be important in determining plant sensitivity in field studies (Sharma et al. 2017). For example, PAR can alleviate some of the negative effects of enhanced UV-B radiation (Nithia et al. 2005). This seems mainly attributable to blue light which stimulates the production of photolyases which are involved in the repair of UV-induced pyrimidine dimers of DNA (Ballaré et al. 2011). In addition, sessile nature of plants forces them to adapt to challenging environmental conditions, and plants have developed protective mechanisms that, among others, include increased level of UV-absorbing compounds (Caldwell et al. 1983). Some of the UV-screening molecules accumulated in plant leaves due to UV radiation are also responsive to PAR intensity (e.g., flavonoids) (Jansen et al. 2008).

Tea plant (*Camellia sinensis* (L.) O. Kuntze) is the most important nonalcoholic beverage crop in the world, which possesses typical taste and flavor and has many antioxidants and beneficial health effects such as reduced risk of cardiovascular disease, certain types of cancer, inflammatory bowel, liver, and neurodegenerative diseases, and diabetes, and even weight loss (Kaur et al. 2014). The productivity and the sensory quality, economic value, and health functions of tea depend on physiological behavior and several metabolites in tea plant, such as polyphenols (the most representative quality- and function-related compounds in the tea leaves), caffeine, amino acids, aroma compounds, vitamins, and carbohydrates (Chen et al. 2017). All these compounds are significantly affected by management practices and environmental factors, such as shading and UV stress (Zhang et al. 2014; Zheng et al. 2008). Thus, given the different growing environments and the economic interest of tea plant, substantial research has focused on the effects of different light conditions on these quality-related compounds with the aim of producing high-value tea. At the same time, a further added value of that research topic is that tea is a good candidate for understanding (i) the morphological and metabolic adjustments which occur in plants under condition of shade and UV irradiances, (ii) the effect of shade UV radiation on different biochemical compounds that represent the economic prerogative of this widely cultivated species, and (iii) the adaptation mechanism of plants to shade and UV stress conditions by an ecological and evolutionary point of view. In this chapter, therefore, we review existing literature reporting the effects of shade and UV irradiances on tea plants with the aim to reappraise morphological, biochemical, physiological, and molecular responses found in this species. Particular attention will be devoted to the phenylpropanoid metabolism of leaves under the aforementioned light conditions.

2 Shade in *Camellia sinensis*: Is It Beneficial?

In its native habitats, tea plant is a shade-adapted species, which inhabits the understory of tropical rainforests. Today, shade is considered as an important management practice in tea (which is relatively easy to apply by planting plants under the shade of trees or by sunscreens) but also subjected to controversy (Mohotti 2004). The history of the establishment of shade in tea plantations dates back to 1800s, when the practice of using shade trees spread from Northeast India to South India, Sri Lanka, Indonesia, and Africa (Murthy 1996). Since then, shade has been highly debated with contrasting conclusions. For instance, increased, unchanged, or even decreased yields have been reported for tea grown under shade (Mohotti 2004). Certainly, directly or indirectly, shade interferes with all the three major sets of processes on which ecophysiological research in tea has focused, which are strictly linked to tea yield: (i) photosynthesis, (ii) shoot growth, and (iii) plant-water relations, including the responses to drought (De Costa et al. 2007). On the other hand, extended recent research has concerned the effects of shading culture on chemical composition, mostly focusing on phenylpropanoids (especially catechins), caffeine, volatile organic compounds, amino acids, and pigments which are strongly related with tea quality (Ku et al. 2010). Shade adaptations in tea plants are reviewed in this section.

2.1 *Ecophysiological Adaptations*

Net photosynthetic rate (*A*) of fully expanded mature leaves of tea shows the typical asymptotic response to increasing light level. However, a wide range of saturating light intensities has been reported, extending from a PAR of 200 μmol m^{-2} s^{-1} (Mohotti et al. 2000), through 600–800 (Squire 1977; Gee et al. 1982; Mohotti 2004) and 1000 μmol m^{-2} s^{-1} (Smith et al. 1993, 1994), up to 1200–1500 μmol m^{-2} s^{-1} (Okano et al. 1995). Similarly, the reported values of light-saturated maximum rate of photosynthesis (A_{max}) largely range from 2 to 14 μmol CO_2 m^{-2} s^{-1} (Squire 1977; Roberts and Keys 1978; Smith et al. 1993, 1994;

Mohotti and Lawlor 2002; Karunaratne et al. 2003). This is likely ascribable to different genotypes and environmental conditions under which tea is grown, such as shade (De Costa et al. 2007).

Examinations of photosynthetic partial processes (i.e., light capture, electron transport, photochemical and non-photochemical quenching of energy and carboxylation) have shown that the whole photosynthetic apparatus of tea is likely adapted to operate with the maximum capacity under shade (Mohotti et al. 2000; Mohotti and Lawlor 2002; Mohotti 2004). Mohotti et al. (2000) observed that A_{max}, the apparent quantum yield (mol CO_2 assimilated per mol of incident PAR) and the efficiency of PSII (i.e., maximum quantum yield of PSII, F_v/F_m; photochemical efficiency of PSII in light conditions, Φ_{PSII}) of 5-month-old tea seedlings of clone TRI 2025 grown in controlled environment facilities were significantly lower under unshaded (650 μmol m^{-2} s^{-1} of PAR) as compared to shaded (150 μmol m^{-2} s^{-1} of PAR) conditions. This is in agreement with several works which report the occurrence of photoinhibition in mature tea leaves at light intensities over 1400–1500 μmol m^{-2} s^{-1} (Smith et al. 1993; Mohotti and Lawlor 2002; Mohotti 2004). Photoinhibition is caused principally by reversible/irreversible damages to the photosynthetic apparatus under high irradiances as a consequence of an excess of excitation energy and an over-reduced state of electron acceptors at the electron transport chain (Lawlor 2001).

Decreased stomatal conductance (g_s) under increasing irradiance can also play a key role in inducing photoinhibition of tea leaves. In a field study on mature tea bushes (clone TRI 2025) grown under three shade treatments (100%, 65%, and 30% of full sunlight) combined with three levels of N fertilization (0, 360 and 720 kg ha^{-1} $year^{-1}$), Mohotti and Lawlor (2002) showed that increasing irradiance, leaf temperature (T_L), and air vapor pressure deficit (VPD) throughout the day induced reductions of g_s as well as of intercellular CO_2 concentration (C_i), reaching the minima values at midday. This led to the reduction of *A*, which was not restored in the afternoon when PAR, T_L, and VPD decreased, despite both g_s and C_i recovered. A possible explanation of this phenomenon was given as follows. Early in the morning, with low PAR, T_L, and VPD (compared to the afternoon), g_s was large and *A* low, so that C_i was high. With increasing irradiance, A rose to a maximum. Throughout the morning, temperature and VPD increased, and g_s decreased but relatively less than *A*, so C_i also decreased. During this period, the greater percentage of excitation energy absorbed was not used in photochemistry (photochemical quenching, qP, decreased), but it was dissipated by non-photochemical quenching (qNP, increased). However, this did not preserve the leaf from photoinhibition, which persisted during the afternoon, so impairing the net carbon gain even when PAR and C_i returned to be favorable. A complete recovery of the photosynthetic rate, *A*, was however observed overnight. Although this sensibility of plants to radiation changes throughout the day, the ecophysiological responses were similar among the shade treatments; and there were no differences in g_s and C_i between treatments. However, shade maintained larger A and minimized the rise in qNP. Photoinhibition was exacerbated in unshaded leaves at large PAR flux, as indicated by the lower values of F_v/F_m ratio of unshaded plants at approximately midday when the decrease in *A* was already apparent. This confirms that photoinhibition may be decreased by shading.

Mohotti and Lawlor (2002) also observed an interaction between photoinhibition and N nutrition as the photoinhibition of tea was minimized (but not completely eliminated) by abundant N supply (720 kg N ha^{-1} $year^{-1}$). This is in agreement with a previous work by Smith et al. (1993) where only plants treated with low N fertilization rates (<225 kg N ha^{-1} $year^{-1}$) were photoinhibited at high light intensities (1400–2000 μmol m^{-2} s^{-1}), whereas those treated with higher N concentrations (375 kg N ha^{-1} $year^{-1}$) did not show photoinhibition.

The low values of *A* observed for tea leaves (2–14 μmol CO_2 m^{-2} s^{-1}) as compared with several tropical species (20–40 μmol CO_2 m^{-2} s^{-1}) have been associated to a low efficiency of PSII in capture of excitation energy, as well as low rates of linear electron transport that contributes to the great tendency for photoinhibition in tea (Mohotti et al. 2000). Furthermore, it was observed that a small proportion of the light energy absorbed by chloroplasts supported the activity of ribulose bis phosphate carboxylase-oxygenase (Rubisco) whose activity was generally lower than in what observed in several other tropical species. The high stomatal and mesophyllic resistance of tea leaves make also CO_2 poorly available for the carboxylation sites.

All these features establish a source limitation on tea yield (De Costa et al. 2007). However, there have been some conflicting opinions on how important the photosynthetic rate is in determining the productivity of tea plants: some authors concluded that the current rate of A is not directly linked to the foliar biomass production of tea as it is controlled mainly by the rates of shoot initiation and extension than the supply of assimilates from current *A* (sink-limited yield); other authors claimed that the photoinhibition could reduce the leaf source capacity and thereby could impose a source limitation on tea yield as well (De Costa et al. 2007). Nevertheless, it has been observed that the shade conditions increase the efficiency of PSII and the rate of linear electron transport, as well as g_s (allowing greater CO_2 influx) and qNP, thus lowering the possibility of leaf photoinhibition. These adaptations to shade have been reported in both controlled conditions and field studies on young and mature tea (Mohotti et al. 2000; Mohotti and Lawlor 2002; Karunaratne et al. 2003). Furthermore, it seems that tea has an enhanced xanthophyll pool, which may have a key role in photoprotection. For example, Wei et al. (2004) found a large amount of violaxanthin, antheraxanthin, and zeaxanthin in fresh tea leaves under high sunlight at 15, 22, and 30 °C though little is known on how these pigments could act in tea plants exposed to different irradiances (De Costa et al. 2007).

Even though the direct effect of the shade is to lower the light burden to the leaves, shade can also influence the plant physiology indirectly by modifying leaf and canopy temperatures (Gee et al. 1982; Mohotti and Lawlor 2002). Maximum *A* of tea occurs at T_L between 25 and 30 °C at saturating light and atmospheric CO_2, whereas it decreases at T_L above 35 °C and stops completely above 40 °C (Mohotti and Lawlor 2002). Under the warm, tropical conditions of India and Sri Lanka, shading allows the leaves to maintain a reduced T_L, which otherwise would exceed the optimum for photosynthesis on sunny days. For example, a reduction of T_L by

10–12 °C at midday was observed in shaded plants of a drought-susceptible clone (Gee et al. 1982). On the other hand, in other areas of the world, for example, in the cool highlands of Kenya, where sub-optimal temperatures prevail, shade is not considered necessary (Smith et al. 1993). That the leaf/canopy temperature modifies the balance between shoot and root growth by influencing the physiology of shoot growth, which is a process intimately linked to tea yield, has also worth to be considered (De Costa et al. 2007). Generally, the optimum air temperature for tea plants ranges between 18 and 25 °C, while air temperatures below 13 °C and above 30 °C have been found to reduce the plant growth (Carr and Stephens 1992).

Shade also influences tea-water relations as radiation intensity reaching the leaves is the main determinant of transpiration when the soil is at or near the saturation, providing the latent heat energy necessary for water evaporation (De Costa et al. 2007). This was observed, for example, in tea growing under the shade of *Grevillea robusta* which had substantially lower transpiration rates (0.42–1.07 L $plant^{-1}$ d^{-1}) than unshaded one (3.51 L $plant^{-1}$ d^{-1}). Anandacoomaraswamy et al. (2000) observed that the transpiration rate of field-grown mature tea decreased linearly with decreasing irradiance throughout all the radiation levels tested (from 100% to 15% of full sunlight) at the rate of 0.03 L $plant^{-1}$ d^{-1} for each 1% reduction of solar irradiance.

As shade has helpful effects on the physiology and productivity of tea plants, a correct management of the shade practices is essential. In a field study on tea bushes subjected to three shade treatments (30% artificial shade supplied by nylon netting; natural shade tree provided by *G. robusta*; unshaded), Karunaratne et al. (2003) observed significant photoinhibition on clear, sunny days, but not on cloudy days following the severity of light exposure. Therefore, in environments which only have a small number of clear sunny days per year, a condition of excessive shade could cause yield drop due to a reduced radiation budget by the canopy (De Costa et al. 2007). In a study across a range of locations with different levels of natural shading, Gamage et al. (2007) reported that the optimum level of shade for tea growing in the humid zone of Sri Lanka at altitudes up to 600 m a.s.l. is around 30–40%. Furthermore, Karunaratne et al. (2003) observed that shade tree plots had significantly higher dry weight compared to unshaded and artificial treatments. There was a 42% and 51% reduction of dry matter yield in unshaded and artificially shaded plants, respectively, indicating the loss of yield due to excess shade. This is in agreement with Mohotti (2004), who reported that the best microclimate to tea was provided under the shade of *G. robusta* also in terms of physiological responses. In commercial plantations, partial shading is provided by growing tea plants under taller shade trees, among the tea bushes at densities which are not too competitive with tea. Canopies of tall tree species (10–15 m) such as *G. robusta* and *Albizia moluccana* and shrubs (about 3–5 m tall) such as *Gliricidia sepium*, *Erythrina lithosperma*, and *Acacia* spp. provide the partial shade at different times of the day depending on the direction of incident radiation (De Costa et al. 2007). Identification of alternative species of shade trees with favorable microclimate and low competition with tea plantations will have to be carried out in order to sustain the long-term stability (Xue and Tang 1998; Mohotti 2004).

Furthermore, due to the nature in which the bush is managed, the leaves at the bottom of the canopy receive a very low intensity of light, which can also create unwanted effects on plant physiology. Therefore, the changes in management of tea can cause significant changes in the photosynthetic features at both leaf and canopy level. Wijeratne et al. (2008) examined the responses of mature tea when subject to long-term exposure to three different levels of shade (NS, 100% PAR; MS, 65% PAR; and HS, 35% PAR). The responses of leaves were also examined at different depths of the canopy. On bright, clear days, NS leaves at the top of the canopy received excessive light incurring, consequently, to photoinhibition. MS leaves received light intensities adequate to photosystems requirement and therefore had the highest rates of A (12.2% higher than NS). HS leaves received less light than that potentially usable for photochemistry; hence they had the lowest rates of A. The MS plants also operated under favorable leaf temperature and had the highest g_s. The leaves inside the canopy received much lower radiation than saturation intensities and resultantly had lower rates of photosynthesis. This is in agreement with Okano et al. (1995) who reported that leaf layers located more than 10 cm below the surface contributed only less than 3% of canopy photosynthesis and that 85% of canopy photosynthesis was performed by the leaf layers as far as 5 cm below the surface, in mature tea bush at late autumn when the growth of shoots was in rest. However, the photosynthetic characterization of tea leaves in different canopy layers is still poorly understood.

2.2 Anatomical and Ultrastructural Changes and Color Variation

Tea leaves grown under different light intensities also show significant anatomical changes. Usually, shade leaves are thinner than non-shaded ones. Moreover, they lack a well-organized palisade layer, and the mesophyll cells are primarily spongy, with much more air spaces compared with sun-developed leaves. These adaptations were confirmed by Wijeratne et al. (2008) in tea plants: the specific leaf weight decreased with increased shade, and it was significantly higher in the leaves on top of the canopy than in those located inside the canopy. A similar trend was reported in the thickness of palisade layer which decreased with the increased shade and depth inside canopy. Similar anatomical changes have been shown by Zhang et al. (2004) in Oolong tea covered in the field with black shading net. These changes in tea leaves under shade conditions favor light absorption in such environments. The palisade function in light piping properties for direct beam sunlight is not necessary in diffused light in shade environments, while spongy mesophyll is quite effective in scattering light within the leaf, making light absorption more effective (Wijeratne et al. 2008).

At the ultrastructural level, Chen et al. (2017) reported that transmission electron microscopy showed significantly reduced numbers of chloroplasts due to complete dark treatment protracted for 2 weeks in leaves of green tea (var. Jinxuan), although the structure of chloroplast was not affected by dark conditions. Differently, a study focused on the shade-induced (18 μmol m^{-2} s^{-1} for 4 consecutive days) conversion from pale/yellow phenotype to a green-leafed phenotype of the albino tea cultivar Yu-Jin-Xiang (YJX, 6 years old). Liu et al. (2017) revealed that non-shaded pale leaves contained fewer and underdeveloped chloroplasts per transectioned cell than the shade-induced green leaves of YJX. Chloroplasts of shaded leaves had a more marked thylakoid membrane with more intensively stained chloroplast components (grana stacks) and starch granules as compared to chloroplasts of pale leaves. The ultrastructure of chloroplasts from shaded leaves of YJX plants was similar to those isolated from the leaves of Shu-Cha-Zao (SCZ, a common green leaf cultivar used as control) collected at the same developmental stage (whether grown under shade or full sunlight). These features suggested that pale green leaves in YJX had anomalous chloroplast development, but that shade was able to partially recover normal chloroplast development (Liu et al. 2017). Similar results have been also reported for cv. Bai Jiguan, which similarly exhibits a yellow phenotype under high light intensity with disordered grana stacking. After 6 days of dark, the chloroplasts of Bai Jiguan exhibited stacked thylakoids that were distributed around the cell wall similar to common green cultivars, except that fewer starch grains were observed (Wu et al. 2016). These observations suggest that both the albino cultivars were able to re-establish (at least partially) both the normal chloroplast development once grown under reduced light intensity.

The color variation of tea leaves has attracted much attention because this is strictly related to tea quality (Wu et al. 2016). Shade and dark treatments have notable influence on leaf color of tea, with different effects. The color of tea leaves grown under a shade net for 21 days was dark green, whereas in full sunlight, the leaves were yellowish green (Wang et al. 2012). Pale or yellow leaves of albino cultivars (i.e., as YJX and Bai Jiguan) also recovered a green color under reduced light irradiation after shading (Wu et al. 2016; Liu et al. 2017). Furthermore, Lee et al. (2013) and Topuz et al. (2014) observed that color parameters of the green tea powders also were affected by the shading level. The shading treatments provided the characteristic vivid green color that is the mark of high-quality green tea powders. Differently, tea plants cultivated in complete dark for 10 days showed the symptoms of etiolation, such as absence of greening, reduced leaf size, failure of leaves to unroll, and elongation of the stem as compared to those grown under low intensity white light (Lu et al. 2014). The leaf etiolation due to dark treatment is in agreement with the findings of Yang et al. (2012) and Chen et al. (2017). However, Lu et al. (2014) observed that the etiolated seedlings turned green after being kept under low light intensity for a few days, confirming that light acts as a signal or a developmental trigger rather than a direct energy source in inducing the photomorphogenesis.

2.3 Biochemical Responses

As stated above, shade influences the biosynthesis of the main quality-related tea molecules, such as phenylpropanoids, caffeine, volatile organic compounds, amino acids, and pigments (Ku et al. 2010). For example, "matcha," the highest-quality green tea from Japan, is a fine powder made from tencha which is grown under shade and is shown to have high amino acid together with a low catechin content. Sweetness is associated to amino acids, particularly theanine, which has a taste that is described as "umami" or "brothy," while catechins and caffeine contribute to the astringency of the liquor (Liu et al. 2017). Matcha has not only a characteristic taste but also a high content of chlorophylls. Different to other green teas, tencha is usually grown under 60–90% shade for about 1–3 weeks before harvest. This treatment leads green teas to be greener, sweeter, and less astringent in taste (Ku et al. 2010). Many attempts have been made to improve or modify the quality-related metabolites of teas or tea beverages, by treating tea leaves during either the plant growth or the manufacturing process. To date, shading is the most common agronomic practice adopted to enhance the flavor and reducing astringency in manufactured teas (Ku et al. 2010).

Phenylpropanoids, powerful antioxidant compounds, are the major category of secondary metabolites in tea plants, which account approximately for 18–36% of leaf dry weight (Lu et al. 2014). Even though the main health benefits of tea are mainly associated with these secondary metabolites, unfortunately, the accumulation/buildup of these compounds is inversely related to the taste and the sensor experience of tea (Lee et al. 2013). Flavonoids (including anthocyanins, flavan 3 ols, flavonols, and proanthocyanins) and phenolic acids are a wide class of secondary metabolites synthetized by multiple branches of the phenylpropanoid pathway (Guidi et al. 2015) (Fig. 5.1), one of the most studied secondary metabolic pathway in plants (Grotewold 2006). Major phenylpropanoid components in tea plants are catechins (a group of flavan-3-ols) which account alone for more than 60–80% of total phenylpropanoids in green tea (Liang et al. 2001) and comprise as high as 20% of the dry weight of the green tea (Haslam 2003). Differently, in black tea, catechins form key precursors in enzymatic oxidation (polyphenols oxidase) products forming the tea pigments (theaflavins and thearubigins) comprising 10–20% of the black teas (Wang et al. 2000). The tea plant is unique in the ability to synthesize and accumulate large amounts of catechins, and, in addition, it produces a wide range of these compounds (Arts et al. 2000). Catechins including (−)-epigallocatechin (EGC), (−)-epicatechin (EC), (+)-gallocatechin (GC), and (+)-catechin (C), as well as their respective gallate esters (EGCG, ECG, GCG, and CG), are abundant in the young leaves and buds of tea plants (Hong et al. 2014). More than 60% of the total catechins (Haslam, 2003; Liang et al., 2001) and over 10% of the dry weight of green tea in fresh green leaf is represented by EGCG (Friedman et al. 2007). Other potent chemical components in leaves of tea are the flavonols, mainly quercetin, kaempferol, and myricetin (Mamati 2005) which are the major precursors of catechin biosynthesis (Wang et al. 2000).

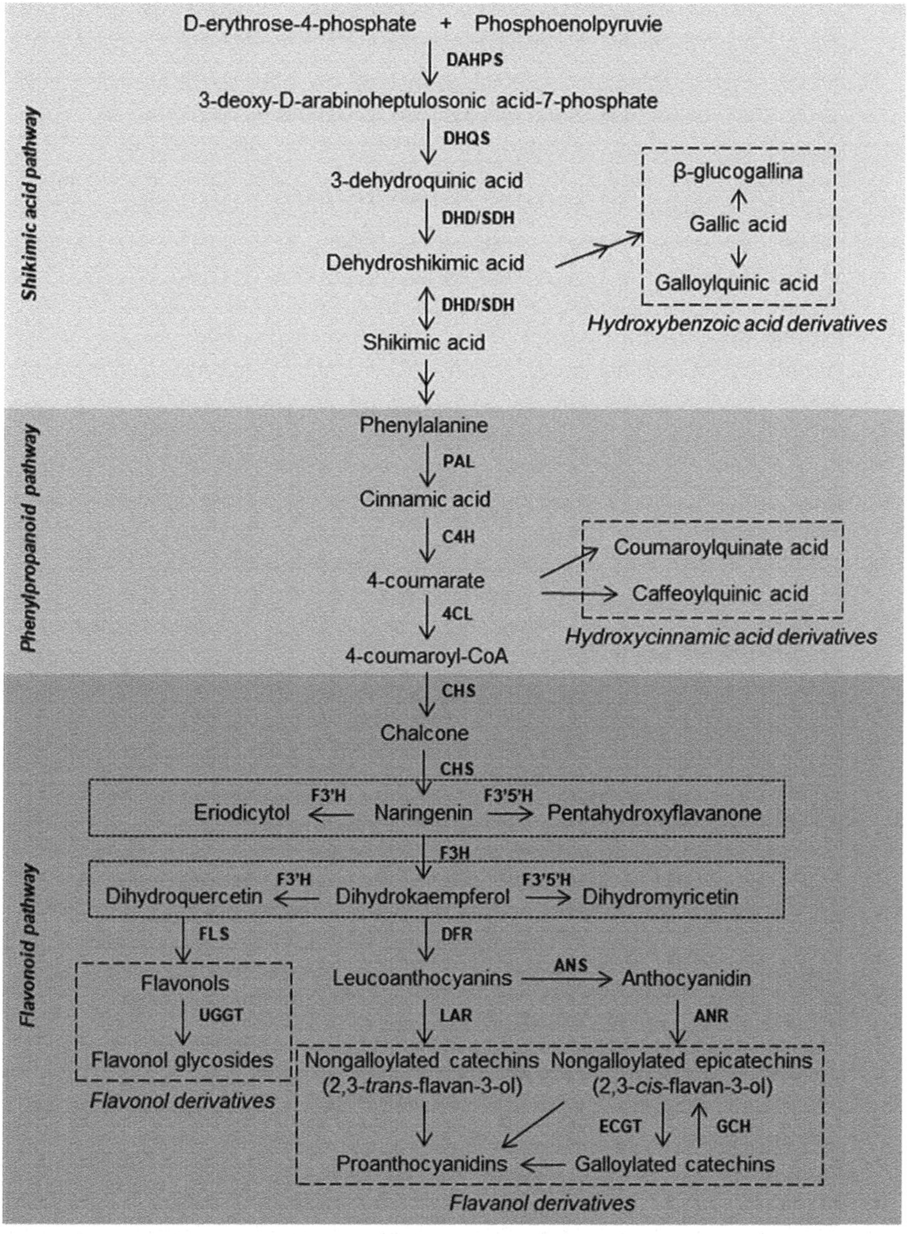

Fig. 5.1 Simplified biosynthesis pathway of phenylpropanoid compounds in tea plants modified from Lu et al. (2014). Abbreviations: 4CL, 4-hydroxy-cinnamoyl CoA ligase; ANR, anthocyanidin reductase; ANS, anthocyanidin synthase; C4H, cinnamate 4-hydroxylase; CHI, chalcone isomerase; CHS, chalcone synthase; DAHPS, 3-deoxy-D-arabino-heptulosonate 7-phosphate synthase; DFR, dihydroflavonol 4-reductase; DHD/SDH, 3-dehydroquinate dehydratase/shikimate 5 dehydrogenase; DHQS, 3-dehydroquinate synthase; ECGT, epicatechin:1-*O*-galloyl-*β*-D glucose *O*-galloyltransferase; F3H, flavanone-3-hydroxylase; F3′5′H, flavonoid-3′,5′-hydroxylase; F3'H, flavonoid-3′-hydroxylase; FLS, flavonol synthase; GCH, galloylated catechins hydrolase; LAR, leucoanthocyanidin 4-reductase; PAL, phenylalanine ammonia-lyase; UGGT, UDP-glucose galloyl-1-*O*- *β* -D-glucosyltransferase

The experience of tea cultivation suggests that low light conditions may negatively affect the biosynthesis of phenylpropanoids, also based on the fact that opportune shading lowers the astringency of tea (Wang et al. 2012; Liu et al. 2017). However, due to the complexity of the pathway and the regulation mechanism, the molecular and biochemical regulation of low light on phenylpropanoid biosynthesis remains unclear and sometimes controversial (Zhang et al. 2014). For example, Song et al. (2012) observed an increase of catechins (EGCG, EGC, EC, ECG) due to shade in tea leaves collected in a tea plantation of Hawaii. Wang et al. (2012) reported that shade treatment (20% light transmitted) had notably negative effects on accumulation of both flavonoids (including catechins, *O*-glycosylated flavonols, and proanthocyanins) and lignins, but had no significant effect on anthocyanin accumulation in tea leaves (cv. SCZ). In particular, the concentration of proanthocyanins and *O*-glycosylated flavonols in shaded leaves changed more than that of other flavonoid compounds, as compared to unshaded leaves. The authors speculated, therefore, that proanthocyanins and *O*-glycosylated flavonols, in addition to catechins, might be the key metabolites responsible for the improvement of tea quality promoted by the shade treatment, leading to a reduced astringency of the leaf powder. Unexpectedly, the expression levels of all the detected genes involved in the biosynthesis of different flavonoids were not correlated with the concentration of catechins in the leaf. However, expression of *phenylalanine ammonia lyase* (*PAL*), *flavanone 3-hydroxylase* (*F3H*), *flavonoid 3'-hydroxylase* (*F3'H*), *dihydroflavonol 4 reductase* (*DFR*), and *anthocyanidin reductase (ANR) -1* (*ANR1*) was notably correlated with the concentration of proanthocyanins in leaves; and expression of *chalcone synthase* (*CHS*) and *flavonoid 3',5'-hydroxylase* (*F3'5'H*) was remarkably correlated with the concentration of *O*-glycosylated flavonols. This confirms that proanthocyanins and *O*-glycosylated flavonols might be key metabolites influenced by shade treatment in tea leaves.

Conversely to the other phenylpropanoids, phenolic acids, especially galloyl and galloylquinic acids, were significantly increased due to shade, and the concentration of phenolic acids was negatively correlated with lignin content. This was probably due to a competition for the same substrate as these compounds are synthesized in the same branch of the upstream pathway. Similarly, always on the cv. SCZ, Lu et al. (2014) reported that the accumulation of phenolic acids, flavanol-3-ols, and flavonols in *C. sinensis* seedlings did not follow a consistent pattern under low light (45–55 μmol m^{-2} s^{-1} of white light for 10 days) and dark conditions or during de etiolation (10-day-old seedling exposed to the dark was illuminated for 7 days under low-intensity white light). The hydroxycinnamic acid derivatives and nongalloylated catechin (EGC) were promoted by low-intensity white light. However, the hydroxybenzoic acid derivatives (*β*-glucogallin and galloquinic acid) and galloylated catechins (EGCG and ECG) were restrained by low-intensity white light. Interestingly, the qRT-PCR analysis showed that the expression levels of *3 dehydroquinate dehydratase/shikimate 5-dehydrogenase* (*DHD/SDH*) and *uridine diphosphate glycosyltransferase* (*UGT*) genes were consistent with the patterns of accumulation of *β*-glucogallin, EGCG, and ECG in tea seedlings during de-etiolation.

The study by Ku et al. (2010) aimed to understand the effect of shading treatment on chemical composition changes in green tea (cv. Yabulita, 20-year-old) using a metabolomic approach. The total phenol and flavonoid contents (TPC and TFC, respectively) of green tea were significantly higher than those of tencha grown under 80% shade by directly covering for 10 days with black polyethylene. Ku et al. (2010) supposed that these decreases due to shading cultivation were probably due to lower biosynthesis of catechins through decreased expression of the genes encoding for CHS and PAL in the tea leaves. However, a different molecular regulation of catechin-related genes under low light was reported by Hong et al. (2014) by studying tea leaves (cv. Dangui) kept under dark conditions for 72 h. Although catechin accumulation in dark-treated plants decreased, the early genes of the flavonoid biosynthesis pathway, *CHS*, *F3H*, and *DFR*, displayed steady expression patterns in response to dark treatment. This indicated that the substrates and precursors of catechin biosynthesis were still sufficient. More importantly, the expression of *leucoanthocyanidin 4-reductase* (*LAR*) increased and that of *anthocyanidin synthase* (*ANS*) decreased under dark conditions.

LAR and ANS are the key enzymes at branch points of catechin biosynthesis; LAR catalyzes the reaction from leucoanthocyanidin to catechins (C, GC, CG, GCG), while ANS results involved epicatechin biosynthesis (EC, EGC, ECG, EGCG). Darkness promotes the downregulation of the expression level of ANS and reduced ECs buildup. Conversely, LAR transcripts increase as long as the concentration of Cs also in response to darkness. Ku et al. (2010) reported that in shade leaves, the concentrations of both TPC and TFC were well correlated with their antioxidant capacity; this may suggest a pivotal role of these compounds in the overall antioxidant activity of green tea, which was observed to be higher under sun than in tea under shade conditions. This is partially in agreement with Wu et al. (2016), who observed that superoxide dismutase and catalase activities increased significantly after a single day of shading, but they decreased after 2 and 3 days; on the other hand, peroxidase activity increased after 6 days of shading. Wu et al. (2016) concluded, therefore, that leaves of the yellow-leafed cultivar Bai Jiguan might be prone to a higher oxidative stress under high light. Differently, Topuz et al. (2014) did not find an increase in both TPC and the antioxidant activity of green tea powder as a consequence of a shade treatment. Ku et al. (2010) found that the level of galloylquinic acid, EGC, and EC decreased when green tea was grown under shade. Thus, the authors speculated that the lower antioxidant activity of tencha was due to the lower concentration of these compounds. Conversely, GC, strictinin, apigenin glucosyl arabinoside, quercetin *p*-coumaroylglucosylrhamnosylgalactoside, and kaempferol *p*-coumaroylglucosylrhamnosylgalactoside were found to decrease in the presence of shade. Furthermore, green tea showed higher levels of succinic acid and fructose together with lower levels of malic acid and pyroglutamic acid than shaded tencha.

Lee et al. (2013) carried out a metabolomic analysis of green teas (cv. Yabukita) grown with different shade periods (0, 15, 18, and 20 days; 5% light transmission) to monitor the nutritional status and the quality of tea liquors according to the shade duration. Among the phenolic compounds, the levels of quercetin galactosylrutinoside, kaempferol-glucosylrutinoside, ECG, and EGCG were elevated by the shade treatment, while those of kaempferol-glucoside, EGC, GC, myricetin-glucoside, and quercetin-glucosylrutinoside were conversely decreased. Overall, levels of TPC increased in the shaded plants; a phenomenon was in partial disagreement with the previous reports of Ku et al. (2010) and Wang et al. (2012). This discrepancy could be at least partly attributed to a difference in the experimental conditions, as Song et al. (2012) reported that the shade levels applied had an influence on the biosynthesis of catechins.

A metabolomic analysis was also performed on the albino tea cultivar YJX by Liu et al. (2017). Young albino or yellow leaves usually contain lower levels of catechins. Effectively, Liu et al. (2017) observed that the levels of total catechins, as well as of the four dominant catechins EGCG, EGC, ECG, and EC in both shaded and unshaded leaves of YJX, were all significantly lower than those of SCZ, a green cultivar, used as control. A 4-day shading treatment did not significantly affect the abundance of the tested catechins in YJX, except for ECG, which was reduced. However, in SCZ, shading resulted in a significant increase in galloylated catechins (EGCG and ECG) and a significant decrease in galloylated catechins (EC and EGC). Interestingly, high level of quercetin, a product branched off from the metabolic flux toward catechin biosynthesis, was found in unshaded leaves of YJX, followed by a significantly lower level in shaded ones, and both were dramatically higher in the two types of YJX leaves than in those of SCZ. The abundance of kaempferol in unshaded leaves was also higher than that in shaded ones. Similarly, shading led to a significant reduction in kaempferol in SCZ leaves. Since quercetin was enhanced and catechins decreased in YJX, flavonoid pathway gene expression was also examined with a focus on metabolic flux redirection in YJX. Results from qPCR showed that the expression of many upstream flavonoid genes, such as *CHS*, *CHI*, *F3H*, and *F3′ 5′ H*, was significantly lower in shaded leaves. Similarly, Wu et al. (2016) found that the expression of genes involved in flavonoid biosynthesis was also repressed in the yellow-leaf phenotype cultivar Bai Jiguan following shade treatments. However, other secondary metabolic processes, such as phenylpropanoid biosynthesis and terpenoid backbone biosynthesis, were also significantly enriched. Liu et al. (2017) found higher expression levels of *F3H* and *F3′5′H* in unshaded leaves; genes which are responsible for dihydroflavonol and flavonol quercetin biosynthesis were also consistent with an enhanced level of quercetin. A lower expression of *ANR1* and *ANR2*, *ANS*, and *LAR1* and *− 2* in YJX than in SCZ also correlated with decreased levels of different catechins as found in YJX. Furthermore, high expression of *flavonol synthase* (*FLS*) and low expression of *DFR* in YJX compared to SCZ was consistent with the difference in flavonol and flavan-3-ols between the two cultivars. These data suggested that these genes could be "turning points" for the metabolic flux redirection in YJX.

Together with phenylpropanoids, astringency of tea is strictly related also to the concentration of caffeine, which is probably best known for its ability to stimulate the central nervous system (Song et al. 2012). For this compound, there is general agreement that its presence decreases due to shade (Song et al. 2012; Lee et al.

Table 5.1 Characteristic flavor compounds in tea infusions and their odor descriptions and thresholds

Volatile compound	Odor description	Odor threshold
Carotenoid-derivative volatiles		
β-Ionone	Woody, violet-like	0.2 μg L^{-1}
Damascenone	Honey-like, fruity	0.004 μg L^{-1}
Fatty acid-derived volatiles		
Cis-3-Hexenol	Fresh, fruity green	13 μg L^{-1}
Hexanal	Green, grassy, metallic	10 μg L^{-1}
Hexanol	Green, grassy	92–97 μg L^{-1}
Methyl jasmonate	Sweet, floral	n.d.
Trans-2-hexanal	Green, fruity	190 μg L^{-1}
Volatile phenylpropanoids/benzenoids		
2-Phenylethanol	Flowery, rose-like, honey-like	1000 μg L^{-1}
Benzyl alcohol	Sweety, fruity	n.d.
Coumarin	Sweet, camphoraceous	0.02 μg L^{-1}
Phenylacetaldehyde	Honey-like	n.d.
Volatile terpenes		
Geraniol	Floral, rose-like	3.2 μg L^{-1}
Linalool	Floral, citrus-like	0.6 μg L^{-1}
Linalooloxides	Sweet floral, citrus, fruity	n.d.

Modified from Yang et al. (2013)
n.d. not defined

2013; Yang et al. 2014). However, Liu et al. (2017) did not found differences in caffeine content between shaded and unshaded leaves in both SCZ (green) and YJX (yellow) tea cultivars.

While phenylpropanoids are responsible for the color and taste, volatile compounds are essential for tea odor and aroma (Yang et al. 2013). Aroma is one of the main sensory properties which are crucial in tea quality. Table 5.1 summarizes some major tea volatiles, with their odor description and threshold. These compounds are derived from either the shikimate and terpenoid pathways or by oxidation of fatty acids and carotenoids. Simplified proposed biosynthetic pathways of volatile compounds are reported in Fig. 5.2.

Many experiments have been done to modify and improve volatile compounds of tea either by preharvest or by postharvest treatments, and it has been reported that tea volatile content can be improved by shading (Yang et al. 2013; Fu et al. 2015). For example, Yang et al. (2012) indicated that field-grown plants (cv. Yabukita) kept under shading conditions (50%, 98%, and 100% shade) for 3 weeks developed etiolated leaves with significantly increased levels of volatile fatty acid derivatives (VFADs) and even more of phenylpropanoids/benzenoids (VPBs), while volatile terpenes (VTs) were not significantly affected. Although many aroma compounds occur as their glycosidic precursors in plant cells, the shading treatments did not significantly affect the amounts of glycosides volatiles. This leaded the authors to

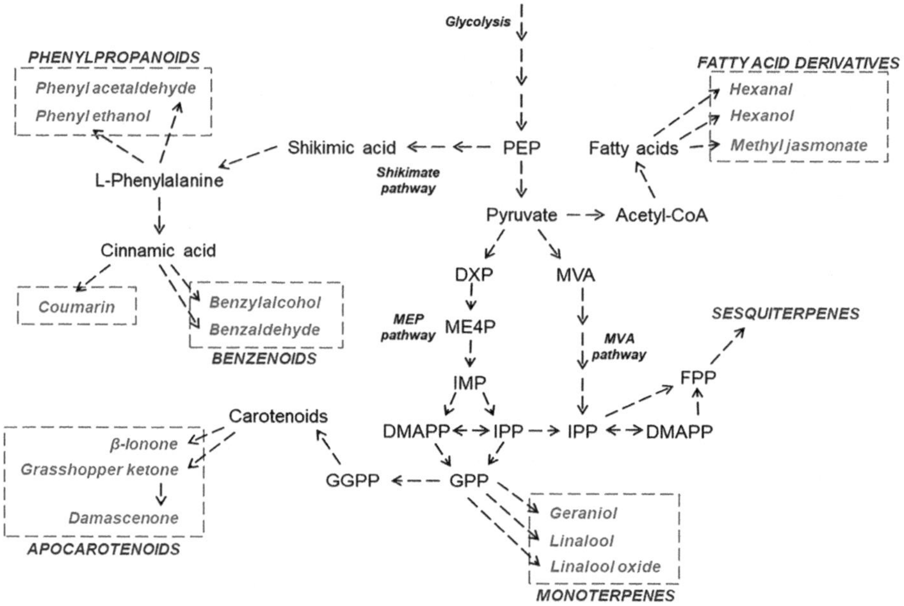

Fig. 5.2 Simplified biosynthetic pathways of tea volatiles, modified from Yang et al. (2013). Abbreviations: DMAPP, dimethylallyl pyrophosphate; DXP, deoxyxylulose phosphate; FPP, farnesyl pyrophosphate; GGPP, geranylgeranyl pyrophosphate; GPP, geranyl pyrophosphate; IMP, inosine monophosphate; IPP, isopentenyl pyrophosphate; ME4P, methylerythritol 4-phosphate; MEP, methylerythritol phosphate; MVA, mevalonic acid; PEP, phosphoenolpyruvate

suppose that the raise in free volatiles under dark treatments may relate to their upstream metabolism. Tontul et al. (2013) also observed increased amounts of volatile compounds in green tea powders from mature tea plants of two Turkish clones (Derepazari 7 and Fener) grown under different shade levels [control, light (50%) and dark (90%)]. Interestingly, Tontul et al. (2013) proposed phytol and tridecane as indicators of shading treatment, since these volatiles were present only in leaves grown under shade. Furthermore, to avoid the potential reduction of biomass of tea plants due to prolonged shading (Mohotti 2004), Fu et al. (2015) demonstrated that specific wavelengths could be usefully considered for the regulation of volatiles. In contrast to the natural light and dark treatment, blue light (470 nm) and red light (660 nm) applied by short-term (3 days) and long-term (14 days) LED treatments significantly increased most endogenous volatiles including VFADs, VPBs, and VTs in the pre-harvest leaves (cv. Jinxuan). VTs level was upregulated only under the short-term LED treatment, in agreement with Yang et al. (2012). Blue and red light significantly upregulated the expression levels of 9/13-lipoxygenases involved in VFADs formation, phenylalanine ammonia lyase involved in VPBs formation, and terpene synthases involved in VTs formation. Of note, biomass of tea plants were not affected by blue and red lights. These results suggest that the regulation of light intensity can be a promising approach for remodeling the aroma of tea leaves.

Many studies have shown that free amino acids also contribute to the quality of tea leaves and related liquors. Often, the amount of free amino acids is used as an index of tea quality, especially for green teas (Deng et al. 2013). Aromatic amino acids significantly affect aroma quality of leaves and are key precursors of tea aroma compounds (Chen et al. 2017). The main and characteristic amino acid in tea leaves is theanine, which not only influences tea quality and taste with its unique taste characteristic known as "umami" but also has several beneficial functions such as effects of antitumor agent and relaxation effect in humans (Deng et al. 2013).

Due to the relative low amino acid content in teas, many studies have been focused on improving or modifying the amino acids of tea leaves or beverages. It is generally accepted that dark treatments lead to accumulation of amino acids in leaves, regardless of cultivars and locations of tea plants. For example, Lee et al. (2013) observed that the levels of amino acids associated with the taste quality of green tea (tryptophan, phenylalanine, theanine, glutamine, and glutamate) were elevated by the shade treatment. Although tryptophan and phenylalanine are known to increase the intensity of bitter and the astringent taste of tea plants, a sensory evaluation dataset showed that the intensity of these two tea sensory features was not increased but rather slightly decreased, which may have been caused by the buildup of umami and sweet taste compounds, such as theanine, glutamine, and glutamate, because of the shade treatment (Lee et al. 2013). Thus, shading is nowadays extensively applied to tea cultivation for producing raw materials of teas with high levels of amino acids (Ku et al. 2010; Lee et al. 2013; Yang et al. 2012; Zhang et al. 2004; Zhang et al. 2014; Chen et al. 2017; Liu et al. 2017).

The intimate understanding for accumulation of free amino acids in dark-shade treated tea leaves is still unexplained. Two possible mechanisms may be related to this phenomenon: the first one is that the biosynthetic pathways of free amino acids may be activated and upregulated in dark-shade-treated plants; the alternative one is that proteins may be hydrolyzed to produce free amino acids when leaves are dark treated (Chen et al. 2017). These last authors also observed that the total content of amino acids significantly increased in the leaves of tea plants (cv. Jinxuan) dark treated for 2 weeks, whereas the content of soluble proteins was reduced after a dark treatment. Quantitative proteomics analysis showed that phospho-2-dehydro-3-deoxyheptonatealdolase 1 (DHAPS-1), glyceraldehyde-3-phosphatedehydrogenase 1 (GAPCP1), and D-3-phosphoglycerate dehydrogenase 2 (PGDH2), three key enzymes in the upstream biosynthetic pathways of amino acids (Fig. 5.3), were down-accumulated, suggesting that an accumulation of amino acids in dark-treated leaves may not be related to a direct biosynthesis of amino acids. Moreover, it was observed that (i) the protein content was decreased by dark treatment in chloroplasts, but not in other organelles; (ii) the amount of Rubisco decreased even in chloroplasts of dark-treated tea leaves; (iii) the dark treatment significantly decreased the presence of chloroplasts, although their structure was not modified; and (iv) most proteins involved in chloroplast formation and photosynthesis were down accumulated in dark-treated leaves. These observations suggested that the accumulation of amino acids was related to proteolysis in dark-treated leaves, which prevalently takes place in chloroplasts. Finally, most proteases showing differences

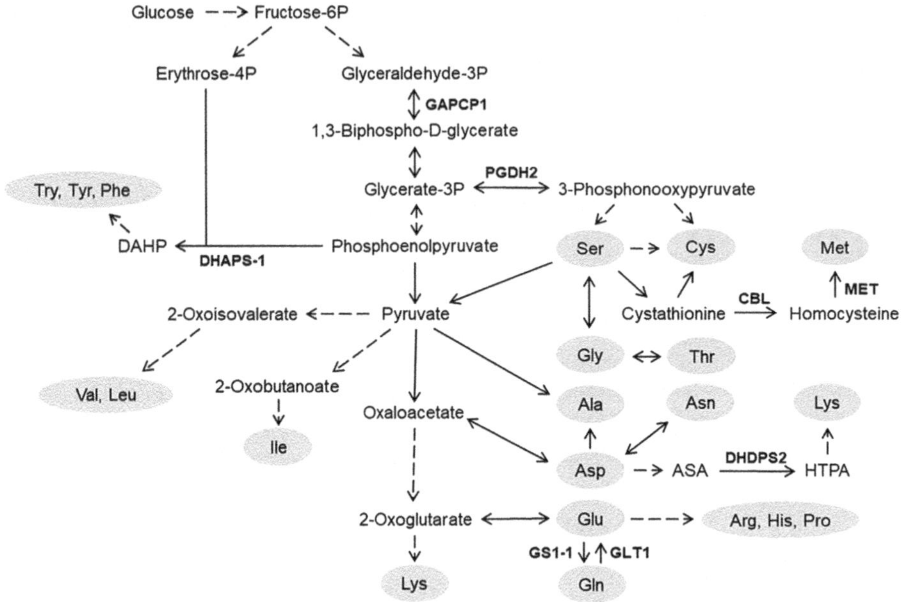

Fig. 5.3 Amino acid biosynthesis pathway in dark-treated tea leaves modified from Chen et al. (2017). Proteins up- (CBL, DHAPS-1, DHDPS2, GAPCP1, GLT1, PGDH2) and down accumulated (GS1–1, MET) in dark-treated leaves compared to control are shown. Abbreviations: Ala, alanine; Arg, arginine; ASA, L-aspartate 4-semialdehyde; Asn, asparagine; Asp, aspartic acid; CBL, cystathionine beta-lyase; Cys, cysteine; DAHP, 3-deoxy-D-arabinoheptulosonate 7 phosphate; DHAPS-1, phospho-2-dehydro-3-deoxyheptonate aldolase 1; DHDPS2, 4-hydroxy tetrahydrodipicolinate synthase 2; GAPCP1, glyceraldehyde-3-phosphate dehydrogenase; Gln, glutamine; GLT1, glutamate synthase [NADH]; GS1-1, glutamine synthetase; Glu, glutamate; Gly, glycine; His, histidine; HTPA, 4-hydroxy-2,3,4,5-tetrahydro-L,L-dipicolinic acid; Ile, isoleucine; Leu, leucine; Lys, lysine; Met, methionine; MET, 5-methyltetrahydropteroyltriglutamate homocysteine methyltransferase; PGDH2, D-3-phosphoglycerate dehydrogenase 2; Phe, phenylalanine; Pro, proline; Ser, serine; Thr, threonine; Try, tryptophan; Tyr, tyrosine; Val, valine

between control and dark-treated tea leaves were down-accumulated, whereas two proteases located in chloroplasts, ATP-dependent Clp protease proteolytic subunit 3 (ClpP3) and protease Do-like 2 (DegP2), were up-accumulated in dark-treated tea leaves. Thus, ClpP3 and DegP2 were involved in proteolysis of chloroplast proteins and synthesis of free amino acids. Similarly, Yang et al. (2012) hypothesized that an increase of amino acids in dark-treated leaves may be due to the degradation of proteins under dark-induced carbohydrate starvation. A different explanation was given for 6-month-old seedlings of tea (cv. Longjing 43) sheltered for 22 days by black shading nets (~20% light transmitted), where the levels of theanine (and his direct substrate, glutamic acid) and total free amino acids gradually increased in both shoots and roots, reaching their maximum at the end of the treatment (Deng et al. 2013). Since theanine is not incorporated into the proteins and an increase of the theanine synthetates was observed, the authors hypothesized that the increased content of theanine under shading could be related to increased N assimilation and reduced theanine.

Recently, an increment of theanine was also observed under shade in the albino cv. YJX (Liu et al. 2017). Transcriptomic analysis revealed the involvement of differentially expressed genes in many biosynthetic pathways, among which amino acid biosynthesis, as well as theanine biosynthesis. Theanine is synthesized in tea plants by theanine synthetase from glutamic acid and ethylamine, which is derived from L-alanine by the action of alanine decarboxylase, arginine decarboxylase, or S-adenosylmenthionine decarboxylase. However, out of these three decarboxylases, only the latter was mapped, and its higher expression was coincident with the variation of theanine abundance. Furthermore, all the genes involved in theanine metabolism were overexpressed in shaded leaves of the green cultivar SCZ (used as control in the comparison with the albino cv. YJX) than in non-shaded ones. This feature was not confirmed in YJX. Thus, the underlying biological mechanisms underlying theanine biosynthesis in YJX remained unclear.

Finally, shade-acclimated leaves are known to contain more Chl per unit mass than sunny ones. This is due to increased allocation of resources to light harvesting functions as compared to those involved in electron transport and CO_2 fixation capacity (Wijeratne et al. 2008). The increased amount of Chl due to shade treatment plays a key role in the chloroplast to maximize light absorption and causes an increase in the green color value of the tea (Lee et al. 2013). This response to shade has been largely reported in both young and mature plants (Ku et al. 2010; Wang et al. 2012; Lee et al. 2013; Liu et al. 2017). Interestingly, this shade adaptation has been also observed in albino cultivars of tea (Wu et al. 2016, Liu et al. 2017), which have commonly low Chl content (as compared with the green ones). Consistent with the alterations in leaf color and chloroplast ultrastructure in the albino YJX clone due to shade, levels of Chl *a* and *b*, their sum (Chl *a* + *b*), and the ratio of Chl *a*/*b* were higher in shaded leaves than pale leaves of YJX grown under full sunlight but still lower than in leaves of the green-leafed tea clone SCZ (whether grown under shade or full sunlight, Liu et al. 2017). Liu et al. (2017) found similar responses in terms of total carotenoids, β-carotene (according to Zhang et al. (2014) in green tea), cryptoxanthin, and lutein (according to Lee et al. (2013), whereas Zhang et al. (2014) did not find differences in green tea). However, a remarkable decrease in zeaxanthin was found in shaded leaves compare to pale ones, while violaxanthin was at the same level in both treatments. Furthermore, zeaxanthin, a crucial component of the violaxanthin cycle in plant photoprotection, was higher in YJX than in SCZ, conversely to all the other pigments. Zhang et al. (2014) also found an increase of neoxanthin due to the shade treatment, even though they did not observe an increase in Chl *a*. Under shade conditions, the leaves of the yellow-leaf phenotype Bai Jiguan rapidly turned green, coinciding with the biosynthesis of Chl *a* and Chl *b*. Transcriptome analysis suggested that the increment of Chl content under shade might be due to the upregulation of *protochlorphyllide oxidoreductase*. In contrast, carotenoid and lutein content decreased under shade. Indeed, all identified genes involved in carotenoid biosynthesis were downregulated (Wu et al. 2016). Different to all these experiments, Wijeratne et al. (2008) reported that none of the photosynthetic pigments analyzed showed significant differences among the shade treatments imposed therein. This confirms that more research is needed to understand how leaf pigments pattern could change in tea plants exposed to different irradiance conditions and how these changes can influence the status of the tea photosynthetic apparatus and, in turn, tea productivity and quality (De Costa et al. 2007).

3 Ultraviolet-B Radiation in *Camellia sinensis*: Stressor or Enhancer?

The UV region of sunlight is the portion of the electromagnetic spectrum that is conventionally divided into UV-A (315–400 nm), UV-B (280–315 nm), and UV-C (100–280 nm) radiation. UV-C radiation is completely absorbed by the stratospheric ozone (O_3) layer, whereas about 70% of UV-A and a smaller proportion of UV-B radiation (~20%) reach the Earth's surface (Sharma et al. 2017). Levels of UV-B radiation reaching the Earth's surface are dependent mainly on latitude and averages 0 to 12 kJ m^{-2} day^{-1} close to the equator, whereas mid-latitude areas and much more Arctic and Antarctic regions are exposed to higher UV-B irradiances (McKenzie et al. 2011; Zlatev et al. 2012).

Changes in atmosphere chemical composition, and mainly a substantial reduction of the protective stratospheric O_3 layer due to the abuse of chlorofluorocarbons (CFCs) protracted until to the 1980s, have led to serious increases in the solar UV-B reaching the Earth's surface (Bais et al. 2015). Since 1980, increases of UV-B ranging from 6% to 14% have indeed been found worldwide (Forster et al. 2011). The Montreal protocol (1989) has been successful in phasing out ozone-depleting CFCs, and, as a consequence, stratospheric O_3 concentrations are (slowly) recovering at low and mid-latitudes (Wu et al. 2013). However, whereas mid-latitude O_3 is expected to return close to the pre-1980 levels at mid-century, O_3 depletion is expected to continue at polar latitudes for several decades largely due to climate change events (Manney et al. 2011).

Though UV-B is a relatively minor component of solar spectrum, which accounts for less than 1% of total light energy reaching the Earth's surface, it is the most energetic waveband and thus can exert a pleiotropic effect on living organisms because of many organic compounds and cellular components (e.g., proteins, lipids, nucleic acids) absorb directly UV-B radiation (Wargent and Jordan 2013; Ulm and Jenkins 2015). Increased levels of UV-B radiation can also seriously alter some processes related to biogeochemical cycles, thus further impacting the biosphere (Erickson et al. 2015).

A huge body of experimental evidences has shown that enhanced UV-B radiation affects significantly morphological, physiological, and biochemical processes of many plant species (reviewed by Wargent and Jordan 2013; Kataria et al. 2014; Sharma et al. 2017). Plants are equipped to sense UV wavebands, such as UV-B light which is perceived by a specific UV-B photoreceptor, the so-called UVR8 (Rizzini et al. 2011). Variability of plant responses to high UV-B is principally dependent on the different sensitivity between Al-Oudat et al. 1998 and within plant species (Fedina et al. 2010; Kataria and Guruprasad 2012), as also observed for the tea cultivar Yulan (more UV-B tolerant) and Fuding Dabai (more UV-B sensitive) (Zheng et al. 2008). Other experimental factors, including UV-B doses (Robson et al. 2015), climatic conditions and geographical location of cultivation site (Teramura et al. 1991), and plant developmental stage (Al-Oudat et al. 1998), can be responsible for differential responses of plant species/genotypes against high UV irradiances. In tea plant, for example, the concentration of total foliar catechins and polyphenols declined with the leaf age, but also changes of individual catechins (decrease in catechin/epigallocatechin gallate and increase in epigallocatechin and epicatechin gallate) were found with the progression of leaf maturity (Mamati et al. 2006). Curiously, in some cases the plant's sex can also influence the sensitivity to UV-B as demonstrated for *Salix myrsinifolia* whose female individuals resulted more tolerant than males (Randriamanana et al. 2015).

Generally, the main changes induced by enhanced UV-B on aboveground plant morphology encompass the reduction of plant height (Reddy et al. 2013) with a parallel increase in plant compactness (Kakani et al. 2003; Reddy et al. 2013), the reduction of leaf surface area (Searles et al. 2001), as well as the increase of leaf thickness (Bornman and Vogelmann 1991) and leaf mass per area (Laposi et al. 2009). Little information is available on the effect of high UV-B levels on the allometry and plant morphology of tea, and, to the best of our knowledge, Mamati et al. (2004) only report data related to phenotypical changes of tea plant under high UV-B. The authors found that leaves of *C. sinensis* cv. Anhui 1 were scorched earlier compared to that of cv. Fuding Dabai Cha over an extended period of UV-B treatment (1.2 kJ m^{-2} h^{-1} UV-B). Anhui 1 started to shade off older leaves after 48 h of treatment. Fuding Dabai Cha on the other hand started showing signs of scorching stress on the leaves after 72 h of treatment. Dry matter content initially decreased in both the clonal tea cultivars but later increased with increasing time of exposure to UV-B. Similarly, reduction of dry matter was found in tea plant callus grown under elevated UV-B irradiances (Zagoskina et al. 2003). Conversely, more information exists on the biochemical and molecular effects induced by high UV-B, in particular on the stimulation of flavonoid metabolism.

Different classes of flavonoids possess protective roles in plant tissues (Grotewold 2006; Landi et al. 2015), and the stimulation of flavonoid biosynthesis is a typical response of plants to several environmental stressors (Winkel-Shirley 2002), such as excessive UV irradiances (Agati et al. 2009; Agati and Tattini 2010). Plants have indeed proficiently adapted to synthetize sunscreening molecules (e.g., flavonoids) when subjected to enhanced UV irradiances (Agati and Tattini 2010). Both induced and constitutively produced accumulation of such compounds has been shown to reduce the penetration of UV wavelengths deeper into leaves, protecting principally the photosynthetic apparatus from UV-triggered damages (Bornman and Vogelmann 1991; Tattini et al. 2014). In addition, UV-B-responsive flavonoids have the potential to reduce the oxidative damage by directly scavenging UV-B-triggered reactive oxygen species (Owens et al. 2008; Agati et al. 2009).

Catechins are principal protective sunscreen compounds of tea plant, and their UV-B-protecting ability was verified experimentally in leaves of this species (Mamati et al. 2004; Mamati 2005). In tea plants exposed to increased concentration of UV-B (0.2, 0.4, 1.0, and 1.7 μmol m^{-2} s^{-1} for 6 h), it was found that catechins accumulated only at lower doses of UV-B (0.2 and 0.4 μmol m^{-2} s^{-1}), whereas the level of catechins dropped down at higher UV-B irradiances (Zheng et al. 2008). This suggests that low UV-B doses stimulate the accumulation of major tea catechins, whereas excessive UV-B irradiation suppresses the accumulation of these compounds. Accordingly, when 1.7 μmol of UV-B m^{-2} s^{-1} (the highest UV-B rate) were applied, the levels of catechins increased 30 min after the beginning of the treatment, while their content declined following the duration of UV-B exposure (Zheng et al. 2008). Consistently, CHS, the first enzyme of the phenylpropanoid pathway's branch leading to the biosynthesis of catechins and other flavonoids (Fig. 5.1), was found to be overexpressed under elevated levels of UV-B in an *Arabidopsis* mutant tolerant to lethal UV-B (Bieza and Lois 2001). Similarly, Mamati (2005) observed a general increase in the accumulation of CHS mRNA in *Arabidopsis* leaves over the time of exposure to UV-B.

The UV-B-protecting ability of catechins was also observed in leaves of *Arabidopsis* sprayed with 250, 500, or 1000 mg L^{-1} tea catechins and irradiated with 1.2 kJ m^{-2} h^{-1} UV-B for 24 h (Lee et al. 2008). In that experiment, catechin sprayed leaves had a faster recovery by UV-B-induced wilting and curling of the leaves, and the overexpression of CHS was suppressed in comparison to unsprayed UV-B treated plants (Lee et al. 2008). Curiously, tea catechins have also demonstrated the ability to protect animal skins against UV-B, and it was found that treatment of human skin with green tea polyphenols reduced penetration of UV-B into deeper skin layers and prevents from skin cancer (Katiyar et al. 2001). At the same time, catechins are also major components conferring quality attributes and health benefits to processed tea (Mamati et al. 2006; Zheng et al. 2008), especially epigallocatechin gallate (Friedman et al. 2007), due to their strong anticarcinogenic, antioxidative, and radical scavenging capacity (Cabrera et al. 2006). Catechins and their enzymatic derivatives of tea leaves are also able to form complexes with other molecules, thereby acting as detoxifying compounds (reviewed by Crespy and Williamson 2004).

Beside the effect of UV-B on tea plant, changes in phenylpropanoid metabolism due to UV-B treatment were observed also in *C. sinensis* callus culture grown for 45 days with supplemented UV-B radiation (0.74 W m^{-2} h^{-1} for 2 h daily). Under control conditions, the total phenols and flavans peaked on the 25th day of culturing and then decreased, whereas UV-treated callus reached the maximum phenolic content only on the 45th day (Zagoskina et al. 2005). Supplemental UV radiation influenced also the fraction of the major phenolic moiety in tea leaves, namely, flavan, which represented 50–60% of the total phenolic fraction under control conditions and 80% in calluses exposed to UV rays (Zagoskina et al. 2005). It was also demonstrated that the stimulation of phenolic compounds was essential for photoprotecting tea plant callus against high UV-B (Zagoskina et al. 2003). The chitin synthase (ChS)-2 strain of tea plant callus, which accumulated higher level of phenolic compounds than ChS-1, was more resistant to UV-B radiation and exhibited a lower decrease in the mitotic activity of cells and, consequently, the cessation of culture growth.

4 Conclusions

The productivity of tea plants and health functions, economic value, and sensory quality of tea are related to physiological behavior and to the regulation of several compounds such as polyphenols, caffeine, amino acids, and aroma compounds. All these metabolites, as well as the physiology, are significantly altered by environmental light-related factors such as shade and UV stress.

The whole photosynthetic apparatus of tea is likely adapted to operate with maximum capacity under shade, as shown by the reactions of the different components of the photosynthetic apparatus (e.g., light capture, electron transport, photochemical- and non-photochemical quenching of energy, and carboxylation). Indeed, photoinhibition may be decreased by shading. Tea leaves grown under low light intensities also show significant anatomical ultrastructural changes. Usually, tea shade leaves are thinner than non-shaded ones and lack a well-organized palisade layer, and the mesophyll cells are primarily spongy; chloroplasts are reduced in number although their structure is not affected. Furthermore, shading influences leaf color of tea providing the characteristic vivid greenness that is the mark of high quality green tea powders. Shade can also effectively improve the quality of tea beverages by causing reduction of the concentration of phenylpropanoids (especially catechins) and caffeine, the main compounds that contribute to astringency, while increasing the presence of amino acids, the compounds that contribute to sweetness, and Chl contents in the leaves. However, due to the complexity of the pathways and the regulation mechanisms, the molecular and biochemical regulation of low light on the biosynthesis of these tea-quality metabolites remains unclear and sometimes controversial. A better understanding of the diversion of these metabolic fluxes regulations under shade would provide insights into the development of innovative strategies to enhance tea quality and taste. In conclusion, shade can be considered as beneficial for tea, although managing the shade correctly is essential, considering the prevailing environmental conditions in a given site.

Furthermore, experimental evidences reveal that exposition of tea plants to UV-B can alter the plant morphology and affect several biochemical and physiological processes, even though plant responses appear UV dose-dependent. Protracted exposition to low UV-B doses (similar to UV peaks actually perceive by plant in the ambient) leads to an enhancement of phenylpropanoid metabolism and in particular stimulate the biosynthesis of UV-protecting flavonoids (mainly catechins). Conversely, higher UV doses (fairly higher than those actually experienced by plants in natural environments) can seriously threaten the plant by hampering key metabolic processes (i.e., photosynthesis) and/or inducing irreversible (at some extent) damages to DNA. Likely as a consequence, under high UV the stimulation of phenylpropanoid metabolism is suppressed. Both these conditions, low and high UV radiation, worth to be investigated much more in detail in tea plant for a dual purpose. Indeed, changes in leaf phenylpropanoid metabolism take place principally for protecting the leaf from high UV-B irradiances (or at least as a first attempt to make the leaf equipped to deal with high UV), but, at the same time, those changes can sensibly alter the profile of "human-desired" flavonoids of tea leaves.

Acknowledgments We are thankful to Prof. Lucia Guidi, Prof. Cristina Nali, and Dr. Elisa Pellegrini for their critical review of the present manuscript and useful suggestions.

References

Agati G, Tattini M (2010) Multiple functional roles of flavonoids in photoprotection. New Phytol 186:786–793. https://doi.org/10.1111/j.1469-8137.2010.03269.x

Agati G, Biricolti S, Tattini M (2009) Mesophyll distribution of 'antioxidant' flavonoid glycosides in *Ligustrum vulgare* leaves under contrasting sunlight irradiance. Ann Bot 104:853–861. https://doi.org/10.1093/aob/mcp177

Al-Oudat M, Baydoun SA, Mohammad A (1998) Effects of enhanced UV-B on growth and yield of two Syrian crops wheat (*Triticum durum* var. Horani) and broad beans (*Vicia faba*) under field conditions. Environ Exp Bot 40:11–16. https://doi.org/10.1016/S0098-8472(98)00014-8

Anandacoomaraswamy A, De Costa WAJM, Shyamile HW et al (2000) Factors controlling transpiration of mature field-grown tea and its relationship with yield. Agric For Meteorol 103:375–386. https://doi.org/10.1016/S0168-1923(00)00134-9

Arts ICW, van de Putte B, Hollman PCH (2000) Catechin contents of foods commonly consumed in The Netherlands. 1. Fruits, vegetables, staple foods, and processed foods. J Agric Food Chem 48:1746–1751. https://doi.org/10.1021/jf000025h

Bais AF, McKenzie RL, Aucamp PJ et al (2015) Ozone depletion and climate change: impacts on UV radiation. Photochem Photobiol Sci 14:19–52. https://doi.org/10.1039/C4PP90032D

Ballaré CL, Pierik R (2017) The shade-avoidance syndrome: multiple signals and ecological consequences. Plant Cell Environ. https://doi.org/10.1111/pce.12914

Ballaré CL, Caldwell MM, Flint SD et al (2011) Effects of solar ultraviolet radiation on terrestrial ecosystems. Patterns, mechanisms and interactions with climate change. Photochem Photobiol Sci 10:226–241. https://doi.org/10.1039/C0PP90035D

Bieza K, Lois R (2001) An Arabidopsis mutant tolerant to lethal ultraviolet-B levels shows constitutively elevated accumulation of flavonoids and other phenolics. Plant Physiol 126:1105–1115. https://doi.org/10.1104/pp.126.3.1105

Bornman JF, Vogelmann TC (1991) Effect of UV-B radiation on leaf optical properties measured with fiber optics. J Exp Bot 42:547–554. https://doi.org/10.1093/jxb/42.4.547

Cabrera C, Artacho R, Gimenez R (2006) Beneficial effects of green tea – a review. J Am Coll Nutr 25:79–99. https://doi.org/10.1080/07315724.2006.10719518

Caldwell MM, Robberecht R, Flint ST (1983) Internal filters: prospects for up acclimation in higher plants. Physiol Plant 58:445–450. https://doi.org/10.1111/j.1399-3054.1983.tb04206.x

Carr MKV, Stephens W (1992) Climate, weather and the yield of tea. In: Willson KC, Clifford MN (eds) Tea: cultivation to consumption. Chapman & Hall, London, pp 87–135. https://doi.org/10.1007/978-94-011-2326-6_4

Casal JJ (2000) Phytochromes, cryptochromes, phototropin: photoreceptor interactions in plants. Photochem Photobiol 71:1–11. https://doi.org/10.1562/0031-8655(2000)071<0001:PCPPII>2.0.CO;2

Casal JJ (2013) Photoreceptor signaling netweorks in plant responses to shade. Annu Rev Plant Biol 64:403–427. https://doi.org/10.1146/annurev-arplant-050312-120221

Chen Y, Fu X, Mei X et al (2017) Proteolysis of chloroplast protein is responsible for accumulation of free amino acids in dark-treated tea (*Camellia sinensis*) leaves. J Proteomics 157:10–17. https://doi.org/10.1016/j.jprot.2017.01.017

Costa Galvão V, Fankhause C (2015) Sensing the light environment in plants: photoreceptors and early signaling steps. Curr Opin Neurobiol 34:46–53. https://doi.org/10.1016/j.conb.2015.01.013

Crespy V, Williamson G (2004) A review of the health effects of green tea catechins in in vivo animal models. J Nutr 134:3431–3440 0022-3166/0

De Costa WALM, Mohotti AJ, Wijeratne MA (2007) Ecophysiology of tea. Braz J Plant Physiol 19:299–332. https://doi.org/10.1590/S1677-04202007000400000

Deng W-W, Fei Y, Wang S et al (2013) Effect of shade treatment on theanine biosynthesis in *Camellia sinensis* seedlings. Plant Growth Regul 71:295–299. https://doi.org/10.1007/s10725-013-9828-1

Erickson DJ, Sulzberger B, Zepp RG et al (2015) Effects of stratospheric ozone depletion, solar UV radiation, and climate change on biogeochemical cycling: interactions and feedbacks. Photochem Photobiol Sci 14:127–148. https://doi.org/10.1039/C4PP90036G

Fedina I, Hidema J, Velitchkova M et al (2010) UV-B induced stress responses in three rice cultivars. Biol Plant 54:571. https://doi.org/10.1007/s10535-010-0102-3

Forster P, Thompson D, Baldwin MP (2011) Stratospheric changes and climate. In: Scientific assessment of ozone depletion: 2010, Global Ozone Research and Monitoring Project–Report No. 52. World Meteorological Organization, Geneva, Switzerland

Friedman M, Mackey BE, Kim HJ et al (2007) Structure–activity relationships of tea compounds against human cancer cells. J Agric Food Chem 55:243–253. https://doi.org/10.1021/jf062276h

Fu X, Chen Y, Mei X et al (2015) Regulation of formation of volatile compounds of tea (*Camellia sinensis*) leaves by single light wavelength. Sci Rep 5:16858. https://doi.org/10.1038/srep16858

Gamage AJ, Wijeratne MA, De Costa WAKM (2007) Role of shade trees in tea: a low-country perspective. In: Proceedings of the 215th meeting of the experiments and extension forum. Tea Research Institute of Sri Lanka, Talawakelle, Sri Lanka

Gee GW, Sandanam S, Kulasegaram S et al (1982) Effect of shade on leaf-water diffusion resistance in clonal tea (*Camellia sinensis* L.). Tea Quart 51:12–20

Gommers CM, Visser EJ, Onge KR et al (2013) Shade tolerance: when growing tall is not an option. Trends Plant Sci 18:65–71. https://doi.org/10.1016/j.tplants.2012.09.008

Gong WZ, Jiang CD, Wu YS et al (2015) Tolerance vs. avoidance: two strategies of soybean (*Glycine max*) seedlings in response to shade in intercopping. Photosynthetica 53:259–268. https://doi.org/10.1007/s11099-015-0103-8

Grotewold E (2006) The science of flavonoids. Springer, New York

Guidi L, Penella C, Landi M (2015) Anthocyanins in Mediterranean diet: common and innovative sources. In: Warner LM (ed) Handbook of anthocyanins: food sources, chemical application and health benefit. Nova Publisher, New York, pp 1–50

Haslam E (2003) Thoughts on thearubigins. Phytochem 64:61–73. https://doi.org/10.1016/S0031-9422(03)00355-8

Hong G, Wang J, Zhang Y et al (2014) Biosynthesis of catechin components is differentially regulated in dark-treated tea (*Camellia sinensis* L.). Plant Physiol Biochem 78:49–52. https://doi.org/10.1016/j.plaphy.2014.02.017

Jansen MAK, Hectors K, O'Brien NM et al (2008) Plant stress and human health: do human consumers benefit from UV-B acclimated crops? Plant Sci 175:449–458. https://doi.org/10.1016/j.plantsci.2008.04.010

Kakani VG, Reddy KR, Zhao D et al (2003) Field crop responses to ultraviolet-B radiation: a review. Agric Forest Meteorol 120:191–218. https://doi.org/10.1016/j.agrformet.2003.08.015

Karunaratne PMAS, Mohotti AJ, Nissanka SP et al (2003) Effect of shade in minimizing photoinhibition of photosynthesis of high grown tea (*Camellia sinensis* L.O. Kuntze) in Sri Lanka. Trop Agric Res 15:133–143

Kataria S, Guruprasad KN (2012) Solar UV-B and UV-A/B exclusion effects on intraspecific variations in crop growth and yield of wheat varieties. Field Crops Res 125:8–13. https://doi.org/10.1016/j.fcr.2011.08.011

Kataria S, Jajoo A, Guruprasad KN (2014) Impact of increasing ultraviolet-B (UV-B) radiation on photosynthetic processes. J Photochem Photobiol B Biol 137:55–66. https://doi.org/10.1016/j.jphotobiol.2014.02.004

Katiyar SK, Bergamo BM, Vyalil PK et al (2001) Green tea polyphenols: DNA photodamage and photo immunology. J Photochem Photobiol B 65:109–114. https://doi.org/10.1016/S1011-1344(01)00248-2

Kaur L, Jayasekera S, Moughan PJ (2014) Antioxidant quality of tea (*Camellia sinensis*) as affected by environmental factors. In: Preedy V (ed) Processing and impact on antioxidants in beverages, 1st edn. Academic, Cambridge, MA, pp 121–129. https://doi.org/10.1016/B978-0-12-404738-9.00013-1

Ku KM, Choi JN, Kim J et al (2010) Metabolomics analysis reveals the compositional differences of shade grown tea (*Camellia sinensis* L.). J Agric Food Chem 58:418–426. https://doi.org/10.1021/jf902929h

Landi M, Tattini M, Gould KS (2015) Multiple functional roles of anthocyanins in plant environment interactions. Environ Exp Bot 119:4–17. https://doi.org/10.1016/j.envexpbot.2015.05.012

Laposi R, Veres S, Lakatos G et al (2009) Responses of leaf traits of European beech (*Fagus sylvatica* L.) saplings to supplemental UV-B radiation and UV-B exclusion. Agric For Meteorol 149:745–755. https://doi.org/10.1016/j.agrformet.2008.10.023

Lawlor DW (2001) Photosynthesis: molecular, physiological and environment processes, 3rd edn. Bios Scientific Publisher, Oxford

Lee XZ, Liang YR, Lu JL et al (2008) Alleviation of UV-B stress in Arabidopsis using tea catechins. Afr J Biotechnol 7:4111–4115. https://doi.org/10.5897/AJB08.784

Lee L-S, Choi JH, Son N et al (2013) Metabolomic analysis of the effect of shade treatment on the nutritional and sensory qualities of green tea. J Agric Food Chem 61:332–338. https://doi.org/10.1021/jf304161y

Liang Y, Ma W, Lu J et al (2001) Comparison of chemical composition of *Ilex latifolia* Thumb. and *Camellia sinensis* L. Food Chem 75:339–343. https://doi.org/10.1016/S0308-8146(01)00209-6

Liu G-F, Han Z-X, Feng L et al (2017) Metabolic flux redirection and transcriptomic reprogramming in the albino tea cultivar 'Yu-Jin-Xiang' with and emphasis on catechin production. Sci Rep 7:45062. https://doi.org/10.1038/srep45062

Lu Z, Liu Y, Zhao L et al (2014) Effect of low-intensity white light mediated de-etiolation on the biosynthesis of polyphenols in tea seedlings. Plant Physiol Biochem 80:328–336. https://doi.org/10.1016/j.plaphy.2014.04.016

Mamati EG (2005). Tea polyphenols biosynthesis and genes expression control in tea (*Camellia sinensis*). PhD dissertation, Zhejiang University, Hangzhou, China

Mamati EG, Liang YR, Lu JL et al (2004) The role of tea polyphenols in protection of plants against UV-B. In: Proceedings of international conference on O-Cha (tea) Culture and Science (ICOS), Shizuoka, Japan, 4–6 November 2004

Mamati GE, Liang YR, Lu JL (2006) Expression of basic genes involved in tea polyphenol synthesis in relation to accumulation of catechins and total tea polyphenols. J Sci Food Agric 86:459–464. https://doi.org/10.1002/jsfa.2368

Manney GL, Santee ML, Rex M et al (2011) Unprecedented Arctic ozone loss in 2011. Nature 478:469–475. https://doi.org/10.1038/nature10556

McKenzie RL, Aucamp PJ, Bais AF et al (2011) Ozone depletion and climate change: impacts on UV radiation. Photochem Photobiol Sci 10:182–198. https://doi.org/10.1039/C0PP90034F

Mohotti AJ (2004) Shade in tea: is it beneficial? SLJ Tea Sci 69:27–39

Mohotti AJ, Lawlor DW (2002) Diurnal variation of photosynthesis and photoinhibition in tea: effects of irradiance and nitrogen supply during growth in the field. J Exp Bot 53:313–322. https://doi.org/10.1093/jexbot/53.367.313

Mohotti AJ, Dennett MD, Lawlor DW (2000) Electron transport as a limitation of photosynthesis of tea (*Camellia sinensis* (L.) O. Kuntz): a comparison with sunflower (*Heliantus annuus* L.) with special reference to irradiance. Trop Agric Res 12:1–10

Murthy SBK (1996) Shade in tea plantations. The Plant Chron:591–595

Nithia SMJ, Shanthi N, Kulandaivelu G (2005) Different responses to UV-B enhanced solar radiation in radish and carrot. Photosynthetica 43:307–311. https://doi.org/10.1007/s11099-005-0051-9

Okano K, Komaki S, Matsuo K et al (1995) Analysis of canopy photosynthesis in mature tea (*Camellia sinensis* L.) bush at late autumn. Jpn J Crop Sci 64:310–316. https://doi.org/10.1626/jcs.64.310

Owens DK, Alerding AB, Crosby KC et al (2008) Functional analysis of a predicted flavonol synthase gene family in Arabidopsis. Plant Physiol 147:1046–1061. https://doi.org/10.1104/pp.108.117457

Randriamanana TR, Nissinen K, Moilanen J et al (2015) Long-term UV-B and temperature enhancements suggest that females of *Salix myrsinifolia* plants are more tolerant to UV-B than males. Environ Exp Bot 109:296–305. https://doi.org/10.1016/j.envexpbot.2014.06.007

Reddy KR, Singh SK, Koti S (2013) Quantifying the effects of corn growth and physiological responses to ultraviolet-B radiation for modeling. Agro J 105:1367–1377. https://doi.org/10.2134/agronj2013.0113

Rizzini L, Favory J, Faggionato D et al (2011) Perception of UV-B by the Arabidopsis UVR8 protein. Science 332:103–106. https://doi.org/10.1126/science.1200660

Roberts GR, Keys AJ (1978) The mechanism of photosynthesis in the tea plant (*Camellia sinensis* L.). J Exp Bot 29:1403–1407. https://doi.org/10.1093/jxb/29.6.1403

Robson TM, Klem K, Urban O et al (2015) Re-interpreting plant morphological responses to UV-B radiation. Plant Cell Environ 38:856–866. https://doi.org/10.1111/pce.12374

Ruberti I, Sessa G, Ciolfi A et al (2012) Plant adaptatation to dynamically changing environment: the shade avoidance response. Biotechnol Adv 30:1047–1058. https://doi.org/10.1016/j.biotechadv.2011.08.014

Searles PS, Flint SD, Caldwell MM (2001) A meta analysis of plant field studies simulating stratospheric ozone depletion. Oecologia 127:1–10. https://doi.org/10.1007/s004420000592

Seemann JR (1989) Light adaptation/acclimation of photosynthesis and the regulation of Ribulose 1,5-Bisphosphate carboxylase activity in sun and shade plants. Plant Physiol 91:379–386. https://doi.org/10.1104/pp.91.1.379

Sharma S, Chatterjee S, Kataria S et al (2017) A review of stress and responses of plants to UV-B radiation. In: Singh VP, Singh S, Prasad MS, Parihar P (eds) UV-B radiation: from environmental stressor to regulator of plant growth. Wiley-Blackwell, New York, pp 75–99

Smith BG, Stephens W, Burgess PJ et al (1993) Effect of light, temperature, irrigation and fertilizer on photosynthetic rate in tea (*Camellia sinensis*). Expl Agric 29:291–306. https://doi.org/10.1017/s001447970002056x

Smith BG, Burgess PJ, Carr MKV (1994) Effects of clone and irrigation on the stomatal conductance and photosynthetic rate of tea (*Camellia sinensis*). Expl Agric 30:1–16. https://doi.org/10.1017/s0014479700023802

Song R, Kelman D, Johns KL et al (2012) Correlation between leaf age, shade levels, and characteristic beneficial natural constituents of tea (*Camellia sinensis*) grown in Hawaii. Food Chem 133:707–714. https://doi.org/10.1016/j.foodchem.2012.01.078

Squire GR (1977) Seasonal changes in photosynthesis of tea (*Camellia sinensis* L.). J Appl Ecol 14:303–316. https://doi.org/10.2307/2401847

Tattini M, Landi M, Brunetti C et al (2014) Epidermal coumaroyl anthocyanins protect sweet basil against excess light stress: multiple consequences of light attenuation. Physiol Plant 152:585–598. https://doi.org/10.1111/ppl.12201

Teramura AH, Ziska LH, Sztein AE (1991) Changes in growth and photosynthetic capacity of rice with increased UV-B radiation. Physiol Plant 83:373–380. https://doi.org/10.1111/j.1399-3054.1991.tb00108.x

Tontul I, Torun M, Dincer C et al (2013) Comparative study on volatile compounds in Turkish green tea powder: impact of clone, shading level and shooting period. Food Res Int 53:744–750. https://doi.org/10.1016/j.foodres.2012.12.026

Topuz et al (2014) Physicochemical properties of Turkish green tea powder: effects of shooting period, shading, and clone. Turk J Agric For 38:233–241. https://doi.org/10.3906/tar-1307-17

Ulm R, Jenkins GI (2015) Q&A: how do plants sense and respond to UV-B radiation? BMC Biol 13:45. https://doi.org/10.1186/s12915-015-0156-y

Valladares F, Niinemets Ü (2008) Shade tolerance, a key plant feature of complex nature and consequences. Annu Rev Ecol Evol Syst 39:237–257. https://doi.org/10.1146/annurev.eclsys.39.110707.173506

Wang H, Provan G, Helliwell K (2000) Tea flavonoids: their function, utilization and analysis. Trends Food Sci Technol 11:152–160. https://doi.org/10.1016/S0924-2244(00)00061-3

Wang Y, Gao L, Shan Y et al (2012) Influence of shade flavonoid biosynthesis in tea (*Camellia sinensis* (L.) O. Kuntze. Sci Horticult 141:7–16. https://doi.org/10.1016/j.scienta.2012.04.013

Wargent JJ, Jordan BR (2013) From ozone depletion to agriculture: understanding the role of UV radiation in sustainable crop production. New Phytol 197:1058–1076. https://doi.org/10.1111/nph.12132

Wei CL, Jiang CJ, Tao HZ et al (2004) Analysis of the xanthophyll cycle component in fresh leaves of tea plant by high performance liquid chromatography ad its photo-protection. J Tea Sci 24:60–64

Wijeratne TL, Mohotti AJ, Nissanka SP (2008) Impact of long term shade on the physiological, anatomical and biochemical changes in tea (*Camellia sinensis* (L.) O. Kuntz). Trop Agric Res 20:376–387

Winkel-Shirley B (2002) Biosynthesis of flavonoids and effect of stress. Curr Opin Plant Biol 5:218–223. https://doi.org/10.1016/S1369-5266(02)00256-X

Wu Y, Polvani LM, Seager R (2013) The importance of the Montreal Protocol in protecting Earth's hydroclimate. J Climate 26:4049–4068. https://doi.org/10.1175/JCLI-D-12-00675.1

Wu et al (2016) De novo sequencing of the leaf transcriptome reveals complex light-responsive regulatory networks in camellia sinensis cv. Baijiguan. Front Plant Sci 7. https://doi.org/10.3389/fpls.2016.00332

Xue J, Tang R (1998) Practices and investigations of tree-tea intercropping systems in China. Int Tree Crops J 9:179–185. https://doi.org/10.1080/01435698.1998.9752973

Yang Z, Kobayashi E, Katsuno T et al (2012) Characterisation of volatile and non-volatile metabolites in etiolated leaves of tea (*Camellia sinensis*) plants in the dark. Food Chem 135:2268–2276. https://doi.org/10.1016/j.foodchem.2012.07.066

Yang Z, Baldermann S, Watanabe N (2013) Recent studies of the volatile compounds in tea. Food Res Int 53:585–599. https://doi.org/10.1016/j.foodres.2013.02.011

Yang et al (2014) Comparison of in vitro antiviral activity of tea polyphenols against influenza A and B viruses and structure-activity relationship analysis. Fitoterapia 93:47–53. https://doi.org/10.1016/j.fitote.2013.12.011

Zagoskina NV, Dubravina GA, Alyavina AK (2003) Effect of ultraviolet (UV-B) radiation on the formation and localization of phenolic compounds in tea plant callus cultures. Russ J Plant Physiol 50:270–275. https://doi.org/10.1023/A:1022945819389

Zagoskina NV, Alyavina AK, Gladyshko OT (2005) Ultraviolet rays promote development of photosystem II photochemical activity and accumulation of phenolic compounds in the tea callus culture (*Camellia sinensis*). Russ J Plant Physiol 52:830–838. https://doi.org/10.1007/s11183-005-0109-3

Zhang WJ, Liang YR, Zhang FZ et al (2004) Effects of yield and quality of Oolong tea by covering with shading net. J Tea Sci 24:2786–2282

Zhang Q, Shi Y, Ma L et al (2014) Metabolomic analysis using Ultra-Performance Liquid Chromatography-Quadrupole-Time of Flight Mass Spectrometry (UPLC-Q-TOF MS) uncovers the effects of light intensity and temperature under shading treatments in the metabolites in tea. PLoS ONE 9:e112572. https://doi.org/10.1371/journal.pone.0112572

Zheng XQ, Jin J, Chen H et al (2008) Effect of ultraviolet B irradiation on accumulation of catechins in tea (*Camellia sinensis* (L) O. Kuntze). Afr J Biotechnol 7:3283–3287. https://doi.org/10.5897/AJB08.509

Zlatev ZS, Lidon FJC, Kaimakanova M (2012) Plant physiological responses to UV-B radiation. Emir J Food Agric 24:481–450. https://doi.org/10.9755/ejfa.v24i6.14669

Chapter 6
UV-B Radiation-Induced Changes in Tea Metabolites and Related Gene Expression

Fang-yuan Fan, Chun-lin Li, Zhou Luo, and Gui-zhen Tang

Abstract UV-B radiation is an inevitable abiotic stress, which could induce a series of changes in metabolites and related metabolisms in plants. UV-B-induced metabolic changes in leaves of *Camellia sinensis* affect the tea quality. This review summarizes the recent investigations into UV-B radiation-induced changes in tea metabolites and their related gene expression, involving in flavonoids, amino acids, and volatile compounds. UV-B radiation induces flavonoid accumulation by increasing expression of key genes in general phenylpropanoid pathway and flavonoid pathway. The UV-B radiation-induced gene expressions in flavonoid biosynthesis pathway also are affected by transcription factors and endogenous phytohormones signaling pathway. Changes of individual amino acids under UV-B radiation exhibit significant variation among different plants, and their responses to UV-B radiation dose are different. These regulations involve in modulation of gene expressions related to GABA shunt and tricarboxylic acid cycle (TCA). Volatile compounds in *Camellia sinensis* under UV-B radiation are regulated by both metabolites biosynthesis and volatile glycosidic-precursors hydrolysis. In a word, UV-B radiation influences metabolisms in tea in a rather complex way. More researches on UV-B-induced transcriptional regulation, endogenous-phytohormone signal regulation, metabolisms diversions regulation, etc. are needed in the future.

Keywords UV-B radiation · Tea polyphenols · Amino acids · Volatile compounds · Signal regulation · Gene expression

F.-y. Fan (✉) · C.-l. Li · Z. Luo · G.-z. Tang
Zhejiang University Tea Research Institute, Hangzhou, China
e-mail: FangyFan@zju.edu.cn

W.-Y. Han et al. (eds.), *Stress Physiology of Tea in the Face of Climate Change*,
https://doi.org/10.1007/978-981-13-2140-5_6

1 Introduction

Sunlight is necessary for photosynthetic plant life on Earth. Ultraviolet B radiation (UV-B) has short wavelengths ranging from 280 nm to 315 nm, and most of the UV-B in solar radiation is absorbed by ozone (O_3) in the Earth's atmosphere before reaching Earth's surface. In recent years, increasing UV-B radiation reaches the Earth's surface due to stratospheric ozone depletion, which largely were attributed to climatic variation, land-use patterns alteration, and new O_3-depleting substances (Anderson et al. 2012; Laube et al. 2014). The effects of UV-B on plant growth and metabolism have been widely reported for crop plants (Kakani et al. 2003), vegetables (Giuntini et al. 2005), and fruits (Martínez-Lüscher et al. 2014). High levels of UV-B radiation will lead to serious distress like DNA damage and decrease of the photosynthetic performance (Inostroza-Blancheteau et al. 2016; Rastogi et al. 2010), while moderate ambient UV-B radiation will promote secondary metabolism to elevate UV-B tolerance without substantial damage (Hectors et al. 2007; Ballaré et al. 2011). ROS metabolism and oxidative defense occur under both mild and severe stress induced by UV-B radiation, involving complex signal transduction pathways and different signaling molecules (Mackerness 2000; Eva et al. 2013). Metabolic regulation is an important strategy for resistance to UV-B radiation. In plants, UV-B stress promotes biosynthesis of enzymatic antioxidants like superoxide dismutase (SOD), ascorbate peroxidase (APX), glutathione reductase (GR) (Mittler et al. 2004), as well as the production of secondary metabolites which function as antioxidants like flavanols and flavonoids (Bieza and Lois 2001; Wen et al. 2015).

Tea (*Camellia sinensis* (L.) O. Kuntze) is an economically important evergreen plant whose photosynthesizing leaves are harvested for tea beverages, one of the most popular drinks in the world. The metabolites in tea plants, especially secondary metabolites disturbed by environmental changes, are not only known to be beneficial to human health and rich flavors of brewed tea but also play important roles in defense and acclimation. Leaves and buds used as raw materials for tea production are rich in flavonoids, caffeine, theanine, and volatile aromatic compounds. Some studies have shown that tea plants respond to UV-B radiation by activating or inhibiting genes responsible for metabolism and accumulation of secondary metabolites. The objective of this review is to investigate the effect of UV-B radiation on the tea metabolites and related gene expression.

2 Phenolic Compound Metabolism and Related Gene Expression

2.1 UV-B-Induced General Phenylpropanoid Pathway in Polyphenol Metabolism

Tea polyphenols are important secondary metabolites, not only in brewed tea as contributors to human health and sensory quality but also as protectors against abiotic and biotic stress in *Camellia sinensis*. Phenolic compounds enhance plant resistance to UV-B due to their capacity to absorb short wavelength solar radiation (Kusano et al. 2011). Absorbance of UV-B by phenolic compounds protects leaf tissue from UV-B radiation before it reaches to UV-sensitive targets such as chloroplasts and other organelles (Zagoskina et al. 2005). UV-B radiation induces the accumulation of phenolic compounds in tea, including soluble and polymeric forms like phenylpropanoids, flavonoids, and lignin (Zagoskina et al. 2005).

Accumulation of flavonoids induced by UV-B treatment has been found to play an important role in resistance to UV-B radiation in many plant species and various plant tissues such as the skin of apple fruits and leaves of seedlings, in which flavonoid concentrations are enhanced by UV-B exposure (Solovchenko and Schmitzeiberger 2003). Li et al. (1993) investigated UV-B protective mechanisms in the model plant *Arabidopsis* using mutants with defects in the synthesis of UV-absorptive products, which are derived from phenylalanine. Flavonoids are a group of plant polyphenol compounds including flavones, flavonols, isoflavones, flavanones, flavanols, and anthocyanidins, which are synthesized in the general phenylpropanoid pathway (Fig. 6.1), a highly conserved pathway which has been well characterized (Rani et al. 2012; Petrussa et al. 2013). Phenylalanine ammonia lyase (PAL) catalyzes the first step in the phenylpropanoid biosynthetic pathway, which is regulated by UV-B radiation via a putative UV-B photoreceptor (Takeda et al. 1993; Takeda et al. 1994). Action spectrum analysis for promoter activity of the *PAL* gene in carrot showed that the activation of the *PAL* promoter is categorizable as a UV-B light response (Takeda et al. 1997). The study also showed that nitric oxide (NO)-independent pathways in UV-B signaling networks mediated *PAL* activation and flavonoids synthesis (Hao et al. 2009). Chalcone synthase (CHS) is a key enzyme for the first committed step in the flavonoid biosynthetic pathway (Fig. 6.1). Constitutively elevated *CHS* gene expression and consequent flavonoid accumulation in *Arabidopsis* mutant *uvt1* confer a remarkable tolerance to UV-B radiation conditions that would kill wild-type plants (Bieza and Lois 2001). The *CHS* expres-

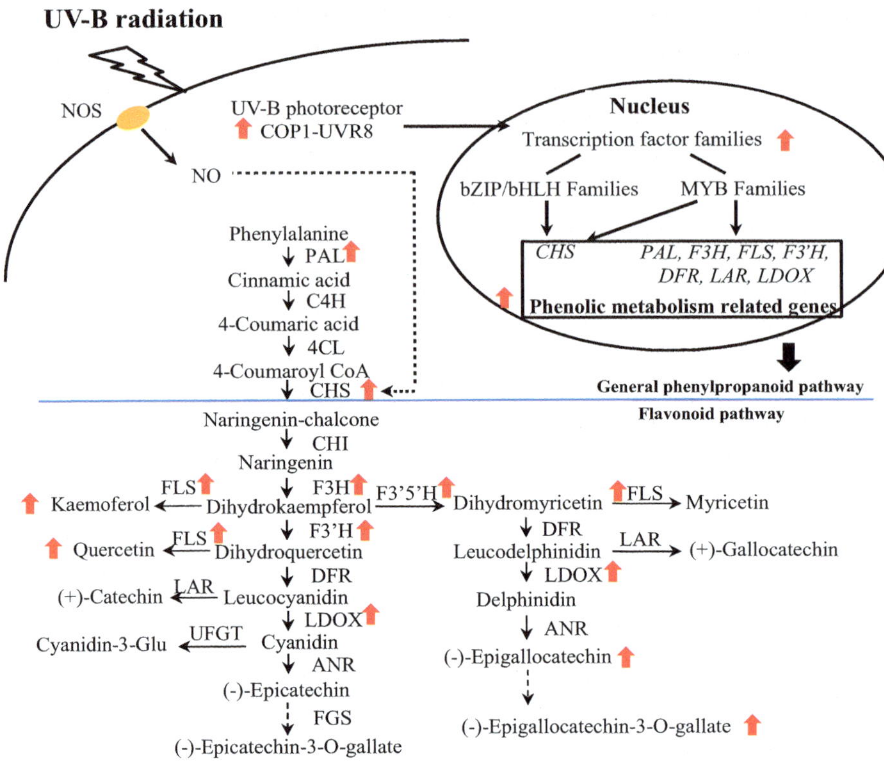

Fig. 6.1 Potential effect of UV-B radiation on phenolic metabolism
General phenylpropanoid pathway combined with flavonoid pathway produce the major phenolic compounds in *Camellia sinensis*. *COP1-UVR8* constitutively photomorphogenic 1-UV response locus 8, *PAL* phenylalanine ammonialyase, *C4H* cinnamate 4-hydroxylase, *4CL* 4-coumaroyl-CoA ligase, *CHS* chalcone synthase, *CHI* chalcone isomerase, *FLS* flavonol synthase, *F3H* flavanone 3′-hydroxylase, *F3′H* flavonoid 3′ -hydroxylase, *F3′5′H* flavonoid 3′, 5′-hydroxylase, *DFR* dihydroflavonol 4-reductase, *LDOX* leucoanthocyanidin oxidase, *LAR* leucoanthocyanidin reductase, *ANR* anthocyanidin reductase, *FGS* flavan-3-ol gallate synthase

sion induced by UV-B radiation has been reported to involve in reversible protein phosphorylation, regulated by a phytochrome signal transduction pathway (Christie and Jenkins 1996). For instance, phytochrome B (phyB) acts as a negative regulator for the *CHS* gene expression of *Arabidopsis* leaf tissue that is induced by UV-B radiation (Wade et al. 2001). Like *PAL* activation, NO has been also considered as an important signaling compound synthesized by increased nitric oxide synthase (NOS) activity to induce the upregulation of *CHS* expression under UV-B radiation (Fig. 6.1) (Mackerness et al. 2001). In *Camellia sinensis*, three full-length cDNAs (*CHS1*, *CHS2*, and *CHS3*) encoding chalcone synthase (CHS; EC 2.3.1.74) have been isolated, with 93–96% of amino acid residues identity to one another. It was reported that the light induced an increase in transcript levels of *CHS* in young tea leaves, possibly indirectly by regulating sugar (glucose, sucrose, fructose, maltose, etc.) levels (Takeuchi et al. 1994).

2.2 UV-B-Induced Changes of Flavonols/Flavonol Glycosides and Related Gene Expression

Flavonols (including flavonol glycosides) are one of the major flavonoid subclasses found in *Camellia sinensis* and offer silky, mouth-drying, mouth-coating, and bitter-amplifying sensations to tea infusions at very low threshold concentrations (Scharbert et al. 2004; Scharbert and Hofmann 2005). Flavonols and flavonol glycosides are biosynthesized by general phenylpropanoid pathway and flavonoid pathway (Fig. 6.1) and accumulated in the form of O-glycosides via glycosylation. The influence of UV-B radiation on metabolism of flavonols and their glycosides depends on UV-B radiation dose and flavonol structure and involves key genes in flavonoids pathway including flavanone 3-hydroxylase (*F3H*), flavonoid 3′-hydroxylase (*F3′H*), flavonoid 3′,5′-hydroxylase (*F3′5′H*), and flavonol synthase (*FLS*). It has been reported that moderate UV-B radiation (0.22–0.88 kJ m^{-2} d^{-1}) decreased quercetin and kaempferol derivatives concentration, and the sugar moieties at 7-O-position of kaempferol glycosides influenced the decline induced by UV-B radiation (Neugart et al. 2012). Under controlled environment rooms with supplemental UV-B radiation of 13.3 kJ m^{-2} d^{-1} for 12 weeks, quercetin glycoside in white clover (*Trifolium repens* L.) specifically accumulated on average by 200% (Hofmann et al. 2000). In silver birch (*Betula pendula*) seedlings, quercetin glycosides (i.e., quercetin-3-galactoside and quercetin-3-rhamnoside) and kaempferol glycosides (i.e., kaempferol-3-rhamnoside) significantly increased by UV-B radiation (7.3–8.5 kJ m^{-2} d^{-1} for 10 days), while the myricetin derivatives were unaffected (Tegelberg et al. 2004). The similar results also showed in grape berries that UV-B radiation (5.98 and 9.66 kJ m^{-2} d^{-1}) could increase the proportion of monosubstituted flavonols (i.e., kaempferol-3-O-galactoside and kaempferol-3-O-glucoside) and disubstituted flavonols (i.e., isorhamnetin-3-O-glucoside and quercetin derivatives), while reducing that of trisubstituted flavonols (i.e., myricetin derivatives). This behavior under UV-B radiation is performed by increasing FLS activity to regulate biosynthetic flux (Martínez-Lüscher et al. 2014). Biosynthetic conversion of monosubstituted flavonols (kaempferol derivatives) to their ortho-disubstituted flavonols (quercetin derivatives) could be considered as an important plant protection mechanism against UV-B stress. This conversion is catalyzed by flavonoid 3′-hydroxylase (*F3′H*). Ryan et al. (2002) indicated that UV-B radiation upregulated the expression of *F3′H* gene and other relevant genes in the phenylpropanoid pathway, leading to an increase in total flavonols and a higher ratio of quercetin to kaempferol, while *F3′H*-deficient mutants in Petunia showed low levels of quercetin and kaempferol as predominant flavonoid. In addition, *F3H* gene, which was cloned from soybean (*GmF3H*), could be induced by UV-B radiation and shown maximal expression by 2 h post treatment. The flavonoids content had no change immediately after UV-B radiation and increased after treatment with UV-B radiation for 23 h (Kim et al. 2008). In *Camellia sinensis*, flavonols usually are accumulated as O-glycosylated forms by glycosylation. Ohgami et al. (2014) have isolated a candidate UGT gene from *Camellia sinensis* var. sinensis cv Yabukita, designated as UGT73A17, whose preferential expression is in mature leaves. The recombinant UGT73A17 protein in vitro showed high activity toward quercetin and UDP-glucose. Although research on the metabolic changes of flavonols/flavonol glycosides in response to UV-B in *Camellia sinensis* is lacking, the metabolic trends of flavonols/flavonol glycosides in other plant species and plant tissues are speculated to apply to *Camellia sinensis*.

2.3 UV-B-Induced Flavanols Metabolism and Related Gene Expression

Flavanols are another flavonoid subclass which mainly refers to catechins in *Camellia sinensis*, consisting of epigallocatechin gallate (EGCg), gallocatechin gallate (GCg), epigallocatechin (EGC), gallocatechin (GC), epicatechin gallate (ECg), catechin gallate (Cg), epicatechin (EC), and catechin (C), which are key substances that influence tea infusion quality. UV-B radiation could affect tea sensory quality by modifying catechins content and catechins metabolism. Zheng et al. (2008) investigated the accumulation of catechins in tea leaves affected by UV-B radiation time and irradiation dose-rate, which showed that short period (within 30 min) and low irradiation dose-rate (below 1.0 μmol m^{-2} s^{-1}) of UV-B radiation stimulated accumulation of major catechins such as EGCg and EGC, whereas the accumulation of tea catechins was suppressed by excessive UV-B radiation including high irradiation dose-rate and long period radiation. Among eight catechins, EGCg increased more quickly than others under moderate UV-B radiation. In addition, tea cultivars differed in their sensitivity to UV-B damage because of differences in their leaf structure such as thickness of the epidermal wax layer.

The metabolism of flavanols in *Camellia sinensis* also involves in general phenylpropanoid pathway and flavonoid pathway (Fig. 6.1). As shown in Fig. 6.1, *F3H*, *F3′H*, and *F3′5′H* are early key genes in the flavonoid pathway and are also key controllers of catechin synthesis in tea plants. F3′H and F3′5′H are key cytochrome P450-dependent monooxygenases (P450s) involved in the formation of dihydroxylated catechins (C, EC, and ECg) and trihydroxylated catechins (GC, EGC, and EGCg), the ratio between which is an important indicator of tea quality and a biochemical marker for genetic diversity in *Camellia sinensis* (Wei et al. 2015). The combined actions of both F3′H and F3′5′H enzymes control the hydroxylation position of dihydrokaempferol, and dihydroxylated and trihydroxylated flavonoids are formed by downstream enzymes (Fig. 6.1). *F3′5′H* gene in *Camellia sinensis* (*CsF3′5′H*) has been isolated, and its expression could be enhanced by light and sucrose (Wang et al. 2014). In Antarctic moss, UV-B radiation accumulated F3′H and F3′5′H transcripts for acclimation to polar environment (Liu et al. 2014). In addition, flavanone 3-hydroxylase (F3H) also is a key enzyme during flavonoid metabolism, which catalyzes naringenin to dihydroflavonols (Fig. 6.1). In *Reaumuria soongorica*, UV-B radiation induced an increase in *F3H* gene expression and *F3H* enzyme activity (Liu et al. 2013). The *Camellia sinensis F3H* (*CsF3H*) has been isolated, which comprised a full-length cDNA sequence of 1365 bp with an open reading frame of 1107 nt encoding a polypeptide of 368 amino acids (Singh et al. 2008). Downregulation of *CsF3H* expression caused by exposure of tea shoots to 50–100 mM catechins suggested the substrate-mediated feedback regulation of *F3H* and its critical role in catechin biosynthesis. The study also showed that *CsF3H* expression was downregulated in response to drought, abscisic acid, and gibberellic acid treatment but upregulated in response to wounding.

Anthocyanidin reductase (*ANR*), leucoanthocyanidin reductase (*LAR*), leucoanthocyanidin oxidase (*LDOX*), and flavan-3-ol gallate synthase (*FGS*) are downstream genes involved in the production of individual catechins in the flavonoid biosynthesis pathway of *Camellia sinensis*. Leucoanthocyanidins are the direct precursors of catechins (Fig. 6.1). Epicatechins are produced through a two-step reaction from leucoanthocyanidin by *LDOX* and *ANR*, and the conversion from epi(gallo)-catechin to epi(gallo)-catechin-3-O-gallate is performed by FGS (Li et al. 2015). *LDOX* gene expression could be induced by light and sucrose, with involvement of a UV receptor signal transduction pathway and a sugar-mediated signal transduction pathway (Gollop et al. 2001). ANR catalyzes anthocyanidin to epicatechin and LAR catalyzes the conversion of leucocyanidin to catechin. An ANR gene encoding anthocyanidin reductase has been cloned from *Camellia sinensis* (*CsANR*, Accession No. AY641729), which was downregulated by drought, ABA and GA_3 treatments, and was upregulated by wounding (Singh et al. 2009). A functional *CsANR* protein yielded by expression in an *Escherichia coli* expression vector catalyzed the conversion of cyanidin to epicatechin in the presence of NADPH (Singh et al. 2009). Over all, almost all of flavonoid-pathway gene expression could be induced in response to UV-B radiation, leading to accumulation of catechins in *Camellia sinensis*.

2.4 UV-B-Induced Transcription Factors Involving in Flavonoid Biosynthesis

It was found that the UV-B-induced gene expression involved in the phenylpropanoid biosynthesis pathway is mediated by light-responsive members of the bZIP/bHLH and MYB transcription factor families (Fig. 6.1). ELONGATED HYPOCOTYL5 (HY5) is a bZIP transcriptional regulator, which has been reported to link to *CHS* gene activation and flavonoid accumulation under UV-B radiation (Oravecz et al. 2006). The activation of *HY5* gene expression and HY5 protein accumulation in *Arabidopsis* under UV-B radiation require functional UV RESISTANCE LOCUS8 (UVR8) and CONSTITUTIVE PHYOTOMORPHOGENIC1 (COP1), which increase flavonoid concentrations in order to protect against UV-B radiation (Oravecz et al. 2006; Favory et al. 2009). MYB transcription factors, a large family of proteins, contain a conserved MYB DNA-binding domain. It was found that

MYB transcription factors involved in flavonoid biosynthesis could be upregulated by sun exposure, which increases flavonoid content in berries after grape veraison (Matus et al. 2009). One subgroup in R2R3-type MYB proteins is involved in regulation about flavonoid biosynthesis during plant development, namely, the *SG7/ PRODUCTION OF FLAVONOL GLYCOSIDES* (PFG) family including PFG1/ MYB12, PFG2/MYB11, and PFG3/MYB111 (Stracke et al. 2010). Among these, PFG1/MYB12 has been found to be a transcriptional regulator of *CHS* and *FLS* gene in planta (Mehrtens et al. 2005). The *VvMYBF1* transcription factor with characteristic of light inducible, also an R2R3-type MYB member, is reported to be a specific activator of *VvFLS1* involving flavonol synthesis (Czemmel et al. 2009). In tea, 26 transcription factors belonging to the MYB4 family have been identified. These transcription factors have been predicted to regulate the expression of PAL, F3H, F3′H, and FLS (Li et al. 2015). The expression of *AtMYB12* in *Arabidopsis thaliana* (Mehrtens et al. 2005) is found to be regulated by UV-B radiation, through UV RESISTANCE LOCUS 8 protein (UVR8) (Cloix and Jenkins 2008; Rizzini et al. 2011). The photoreceptor *VvUVR8* in grape skin could mediate UV-B-induced flavonol accumulation by upregulating secondary metabolism-related transcripts (Carbonell-bejerano et al. 2014). Thus, both *HY5* and *MYB12* bind to promoters of common target genes like *CHS*, which indicates that HY5-MYB12 cooperativity might act at the promoter site. Stracke et al. (2010) indicate that HY5 is indeed required for UV-B-induced regulation of related genes, and HY5-MYB12 cooperation is required for optimal regulation of flavonoid biosynthesis. The precise regulatory mechanisms of different transcriptional regulators at common or different promoters of genes involved in flavonoid biosynthesis pathways need to be further studied.

2.5 Phytohormone and Flavonoid Biosynthesis

ABA and ethylene are abiotic stress-related phytohormones, which are also associated with plant tolerance to UV-B radiation through the regulation of signaling pathways (Cassia 2009; Berli et al. 2010; Wang et al. 2006). The effect of solar UV-B and exogenous ABA on grape berry has been explored and indicates exogenous ABA could strongly increase flavonol accumulation under UV-B radiation through activation of phenylpropanoid and flavonoid biosynthetic pathway enzymes like FLS (Berli et al. 2011). Sun et al. (2011) have explored the effect of *ethylene-insensitive 2* (*EIN2*), a central gene component of the ethylene signaling pathway, on mediating the UV-B response in *Arabidopsis*, by *EIN2* mutant. It has been found that *EIN2* gene with repression in response to UV-B radiation mediates flavonoids biosynthesis through modulation of *CHS* and *C4H* gene expression under UV-B exposure. Above studies indicate that the effect of endogenous phytohormones on flavonoid biosynthesis regulation and their mechanisms under UV-B radiation should be explored in *Camellia sinensis* in future research.

3 UV-B-Induced Amino Acids Metabolism and Related Gene Expression

Free amino acids in tea contribute to tea taste sensory, especially glutamic acid and theanine to provide basis for umami taste (Alasalvar et al. 2012; Hsieh and Chen 2007). Changes in free amino acid concentrations in response to UV-B radiation have been discussed in the model plant *Arabidopsis*. Levels of individual free amino acids varied significantly following 24 h and 96 h UV-B treatment, among which glutamine, histidine, proline, lysine, asparagine, aspartate, threonine, methionine, and glycine were dramatically increased, whereas glutamate and γ-amino butyrate acid (GABA) were moderately elevated following UV-B treatment. In addition, the levels of tyrosine and phenylalanine were obviously increased by 24 h of UV-B treatment, whereas the changes in levels were considerably reduced in the extended UV-B treatment (Kusano et al. 2011). Diverse responses of individual amino acid concentrations in grape berries under UV-B radiation resulted in total free amino acid content remaining unaffected, despite increased concentration of GABA and decreased concentrations of threonine, isoleucine, methionine, serine, and glycine (Martínez-Lüscher et al. 2014). It is speculated that the UV-B-induced diverse changes in amino acids might be caused by UV-B radiation dose. Above studies indicate that biphasic metabolism responses exist in amino acid metabolism induced by UV-B radiation, and amino acid-mediated signaling pathways are triggered in response to UV-B stress, which cause amino acid level variation according to UV-B radiation dose. It has been reported that UV-B radiation decreased stomatal conductance and sub-stomatal CO_2 concentration to reduce net carbon fixation (Martínez-Lüscher et al. 2013) and slow nitrogen metabolism by nitrogen-assimilating enzymes (Singh et al. 2012; Bashri et al. 2018). The glutamine synthetase/glutamate synthase (GS/GOGAT) cycle and glutamate dehydrogenase (GDH) alternative pathway play important roles in inorganic nitrogen assimilation and reassimilation into amino acids (Fig. 6.2) (Valadier et al. 2010). UV-B radiation could inhibit GS/GOGAT activities evidenced by inhibiting activities of nitrate reductase, nitrite reductase, glutamine synthetase, and glutamate synthase (Bashri et al. 2018), which might have been attributed to UV-B-induced reactive oxygen species (ROS) accumulation (Mishra et al. 2011) and subsequently led to ammonium accumulation (Bashri et al. 2018). The activity of GDH has been shown to be enhanced by UV-B stress (Bashri et al. 2018), which may relieve the pressure of accumulating toxic ammonium and provide the function of glutamate oxidation (Robinson et al. 1991). Increases in GABA concentrations induced by UV-B radiation has been researched in various plants. Studies show that GABA in plants functions not only as a metabolite in response to stress but also as a signal molecule playing an important role in the GABA metabolic pathway and carbon-nitrogen metabolism (Bouché and Fromm 2004). GABA is mainly metabolized via the GABA shunt pathway involving the cytosolic enzyme glutamate decarboxylase (GAD) and the mitochondrial enzymes GABA transaminase (GABA-TA) and succinate semialdehyde dehydrogenase (SSADH). GABA, alanine, and glutamate are GABA shunt-related

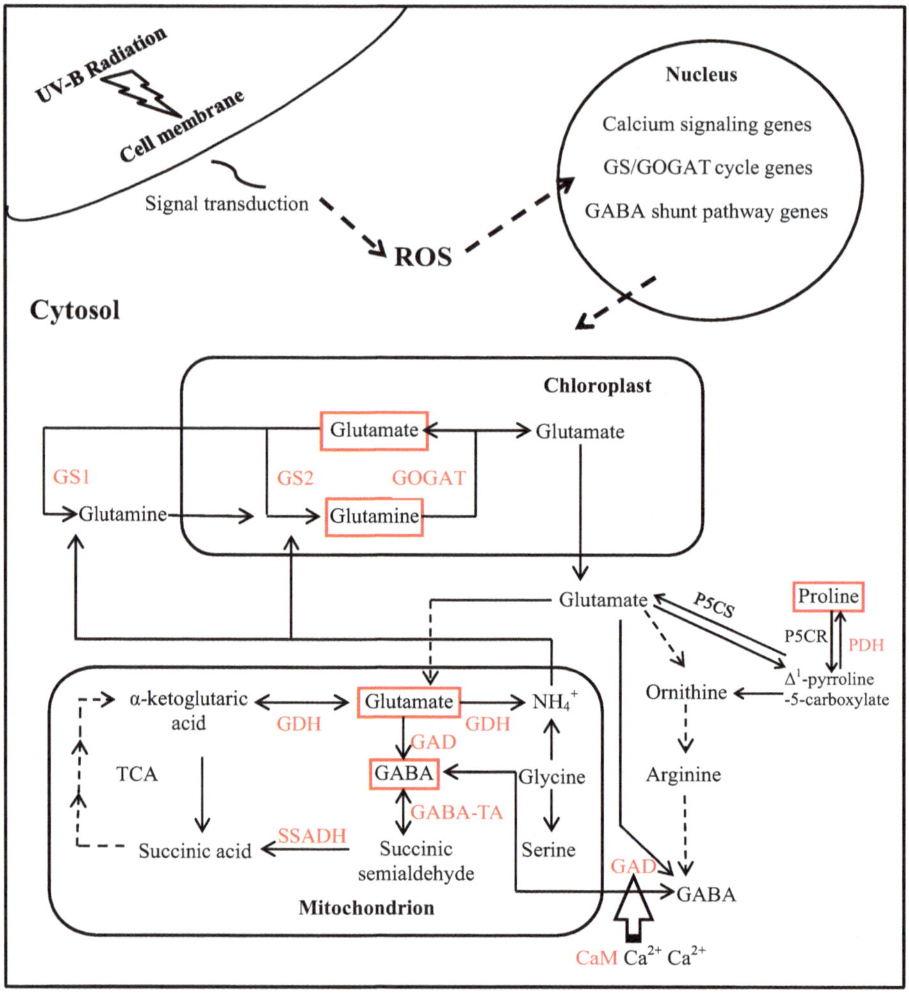

Fig. 6.2 Potential amino acids metabolism under UV-B radiation
UV-B radiation leading to ROS generation, regulates the genes expression involving in GS/GOGAT cycle, GABA shunt pathway and Calcium signaling pathway, which participates in amino acids metabolism. *TCA* tricarboxylic acid cycle, *GS1* cytosolic glutamine synthetase, *GS2* chloroplastic glutamine synthetase, *GOGAT* glutamate synthase, *GDH* glutamate dehydrogenase, *GAD* glutamate decarboxylase, *GABA-TA* GABA transaminase, *SSADH* succinate semialdehyde dehydrogenase, *PDH* proline dehydrogenase, *P5CS* Δ^1-pyrroline-5-carboxylate synthetase, *P5CR* Δ^1-pyrroline-5-carboxylate reductase

metabolites (Fig. 6.2) that are accumulated in response to oxidative stress induced by UV-B radiation (Al-Quraan 2015). Al-Quraan (Al-Quraan 2015) investigated the effect of GABA shunt on ROS accumulation under UV-B stress via *Arabidopsis thaliana* calmodulin mutants, which indicated that GABA shunt pathway, GABA metabolites accumulation, and Ca^{2+}/CaM-mediating signaling mechanisms, acting as major components of antioxidant machinery, were involved in ROS scavenging, H_2O_2 equilibrium, maintaining balance of cellular redox state, and enhancing tolerance in cellular signaling in response to UV stress. Proline, also an antioxidant, has been reported to be induced in different species under UV-B exposure (Saradhi 1995). Proline is synthesized from glutamate by a series of reduction reactions (Fig. 6.2). During this synthesis, redox and hydride ion-mediated stimulation of the pentose phosphate pathway may be regulated by proline (Hagedorn and Phang 1983; Phang 1985). Increase in total proline content in turfgrass under UV-B radiation also indicates that carbon flux is diverted from the TCA cycle to proline synthesis from glutamate through α-aketoglutarate. This finding is also supported by increased activity of glucose-6-phosphate dehydrogenase (G6PDH) and decrease in that of succinate dehydrogenase (SDH) (Sarkar and Bhowmik 2011). In addition, UV-B radiation could also induce the activity of proline dehydrogenase (PDH) activity, an enzyme involved in proline oxidation (Sarkar and Bhowmik 2011).

4 UV-B Induced Changes of Volatile Compounds and Related Gene Expression

Volatile organic compounds are considered not only to be fundamental substances for tea aroma influencing tea quality and tea characteristic but also to be chemical defense compounds increasing tolerance to biotic and abiotic stress (Dong et al. 2016). So far, more than 200 and 400 volatile compounds have been identified in green and black tea, respectively. Volatile profiles of different kinds of tea vary with tea cultivar and tea processing. The volatile compounds produced in live tea plants serve as precursors for aroma development during tea processing. UV-B radiation influences biosynthesis of volatiles by impacting carotenoid pathway, terpenoid pathway, and hydrolysis of glycosidic-bound volatiles.

4.1 UV-B-Induced Carotenoid-Derived Volatiles and Related Gene Expression

Carotenoids, a group of pigments and antioxidants in plants, play an important role in response to various environmental stresses (Strzałka et al. 2003) and are precursors to many volatile compounds produced by plants. In field-grown grapevine berries, Joubert et al. (2016) reports that UV-B radiation could induce accumulation of carotenoids and xanthophylls, which provide substrates for formation of volatile aroma compounds like norisoprenoids. Giuntini et al. (2005) showed the effect of UV-B radiation on carotenoid levels differed among three tomato genotypes. The HP1 genotype showed negligible induction of carotenoids and was therefore susceptible to the detrimental effects of UV-B, while the DRW genotype showed carotenoid accumulation under sunlight conditions and a further promotion by UV-B treatment. Overexpression of bacterial β-carotene hydroxylase gene *CrtZ*, an enzyme responsible for the conversion of β-carotene into zeaxanthin, led to carotenoid accumulation and higher UV-B tolerance, which demonstrates the relationship between the carotenoid pathway and UV-B stress protection (Götz et al. 2002). Ethylene has been found to be regulatory factors of carotenoid metabolism. Becatti et al. (2009) have shown that UV-B radiation regulates carotenoid metabolism in a rather complex way, involving ethylene-dependent and ethylene-independent mechanisms. Carotenoids serve as precursors to terpenoid volatiles which contribute to the flavor and aroma of tea (Fig. 6.3) (Simkin et al. 2004). Carotenoid-derived volatile compounds with low human odor perception thresholds, such as β-ionone and damascenone, contribute to black tea flavor quality (Schuh and Schieberle 2006). Carotenoid cleavage enzymes (CCDs) and CCDx-like enzymes are important specific carotenoid oxidation enzymes involved in apocarotenoid formation (Fig. 6.3) that have been identified and partially characterized in *Camellia sinensis* (Baldermann et al. 2009). In addition, changes in the phosphorylation status of CCDs from *Camellia sinensis* with seasonal variation are reported to influence enzyme activities (Felfe et al. 2011). A carotenoid cleavage enzyme identified from petunia flowers (*PhCCD1*) catalyzes β-carotene to generate β-ionone (Schwartz et al. 2001) with property of light-induced expression variation, which lead to β-ionone emission occurred principally during daylight paralleling *PhCCD1* gene expression (Simkin et al. 2004).

4.2 UV-B-Induced Terpene Volatiles and Related Gene Expression

Terpenoids are the dominant classes in volatile organic compounds in tea, which contribute to tea flavor and fragrance. Linalool and linalool oxides, important terpenoids, are estimated to be about 50% of volatile concentrations in oxidized and semi-oxidized teas (Wang et al. 1994). The effects of UV-B radiation on terpenoids concentration have been studied in many plants. High UV-B exposure increased the content of α-pinene, 3-carene, terpinolene, and nerolidol in grapevine leaves, which were not detected in control plants suggesting that they were synthesized de novo as a result of UV-B elicitation (Gil et al. 2012). Monoterpenes, aldehydes, alcohols, and ketones were markedly increased by UV-B treatment on grape berries (Gil et al. 2013). In grapevine flowers, UV-B radiation did not change total amount of volatiles but increased the proportion of valencene, β-farnesene and α-panasinsene (Gil et al. 2014). α-pinene, β-pinene, myrcene, ocimene, and limonene were increased in peach fruit during UV-B radiation exposure (Liu et al. 2017), which have been reported to contribute to thermotolerance in leaves of some non-isoprenoid emitting oak species (Delfine et al. 2010).

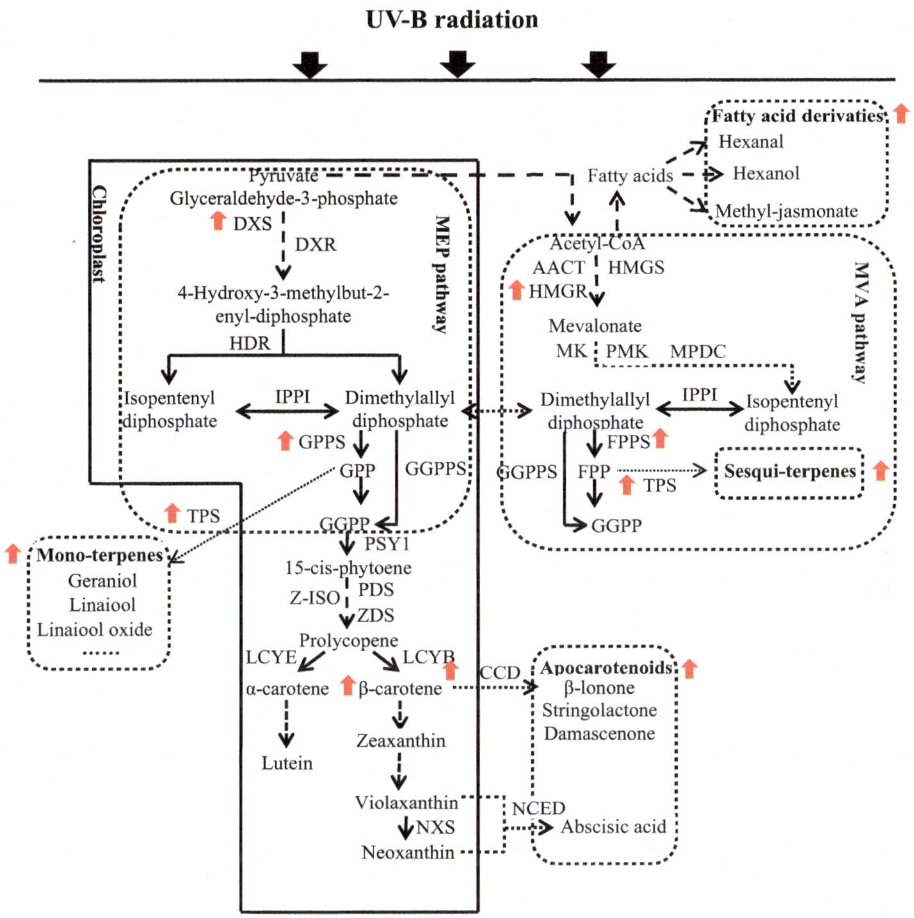

Fig. 6.3 Potential effect of UV-B radiation on major volatiles metabolism
Carotenoid-derived volatiles and terpene volatiles arise from 5-carbon precursor isopentenyl diphosphate and dimethylallyl diphosphate, which are products of the 2-C-methyl-D-erythritol-4-phosphate (MEP) pathway and mevalonate (MVA) pathway. *DXS* 1-deoxy-D-xylulose-5-phosphate synthase, *DXR* 1-deoxy-D-xylulose 5-phosphate reductosiomerase, *HDR* 4-hydroxy-3-methylbut-2-enyldiphosphate reductase, *IPPI* isopentenyl diphosphate isomerase, *GPPS* geranyl diphosphate synthase, *GPP* GGPPS, geranylgeranyl pyrophosphate synthase, *GGPP* geranylgeranyl diphosphate, *PSY1* phytoene synthase, *PDS* Z-ISO, ζ-carotene isomerase, *ZDS* ζ-carotene desaturase, *LCYE* lycopene ε-cyclase, *LCYB* lycopene β-cyclase, *CCD* carotenoid cleavage dioxygenases, *NXS* neoxanthin synthase, *NCED* 9-cis-expoxycarotenoid dioxygenases, *AACT* acetoacetyl-CoA thiolase, *HMGS* 3-methylglutaryl-CoA synthase, *HMGR* 3-hydroxy-3-methylglutaryl synthase, *MK* MVA kinase, *PMK* phosphor-MVA kinase, *MPDC* MVA diphosphate decarboxylase, *FPPS* farnesyl diphosphate synthase, *FPP* farnesyl diphosphate, *TPS* terpenoid synthase

Terpenoids are biosynthesized via two pathways including the cytosolic mevalonic acid (MVA) pathway and the plastid methylerythritol phosphate (MEP) pathway (Fig. 6.3), which both lead to the formation of the C5 compounds isopentenyl pyrophosphate (IPP) and dimethylallyl pyrophosphate (DMAD) (Yang et al. 2013). *DXS* and *GPPS* in the MEP pathway and *HMGR* and *FPPS* in the MAV pathway have been reported to be upregulated under UV-B radiation in peppermint, which contribute to essential terpenoid accumulation (Dolzhenko et al. 2010; Wang et al. 2016). The C5 compounds are condensed by prenyltransferases to synthesize prenyl diphosphates such as geranyl pyrophosphate (GPP), farnesyl pyrophosphate, and geranylgeranyl pyrophosphate, which are precursors to various mono-, sesqui-, and diterpenes formed by cyclizations and other transformations catalyzed by terpene synthases (TPS) (Fig. 6.3) (Hampel and Mosandl 2005; Tholl 2006; Martin et al. 2010). It has been reported that there were more than 100 known *TPS* genes with seven subfamilies: Tps-a to Tps-g (Dudareva et al. 2006). Monoterpenoid biosynthesis in tea is predicted based on the knowledge of the pathways elucidated in other plants using sequence comparisons (Yang et al. 2013). TPS activity in vegetative grapevine tissues was induced by low and high UV-B radiation, resulting in increasing membrane content of sitosterol, stigmasterol, and lupeol to enhance the membrane stability (Gil et al. 2012). Liu et al. (2017) investigated regulation of terpene synthases and terpene content in peach fruit and leaves by UV-B radiation, which showed that UV-B radiation reduced monoterpene linalool and increased (*E, E*)-, farnesene involving in *TPS* gene-regulation mechanism. *PpTPS1*, as a member of Tps-g subfamily involving in formation of linalool, geraniol, ocimene, and β-myrcene, was inhibited under UV-B radiation, while the transcript levels of *PpTPS2*, a Tps-b subfamily member responsible for (*E, E*)-α-farnesene, (*Z, E*)-α-farnesene, and nerolidol, increased by UV-B treatment.

The biosynthesis of terpenoids and expression of genes related to metabolism are also regulated by transcription factors (Li et al. 2015; Schluttenhofer and Yuan 2015). JAZs are jasmonate ZIM-domain proteins involving the jasmonic acid signaling pathway at the transcriptional level. It has been reported that *JAZ* interacts with AP2/ERF, bHLH, and WRKY to regulate plant secondary metabolism (Moerkercke et al. 2015; Shen et al. 2016; Spyropoulou et al. 2014); among these *JAZs* are negative regulators involved in linalool biosynthesis (Taniguchi et al. 2014). During UV-B radiation, a significant increase in *JAZs* transcripts and reduction in transcript levels of AP2/ERF, bHLH, and WRKY were observed in peach fruit and leaves (Liu et al. 2017).

4.3 UV-B-Induced Hydrolysis of Glycosidic-Bound Volatiles and Related Gene Expression

Many volatile compounds including ABA, (Z)-3-hexenol, (S)-linalool, and so on are stored as glycosidic precursors in tea plant cells (Yang et al. 2013) such as β-primeverosides (Guo et al. 1993; Guo et al. 1994; Moon et al. 1996; Nishikitani et al. 1999), β-D-glucopyranside (Ma et al. 2001), vicianoside (Nishikitani et al. 1996), and acuminoside (Mizutani and Sakata 2002). Both enzymatic and nonenzymatic hydrolysis of glycosidic-bound volatiles are major metabolic pathways of volatile formation, following release of aromatic volatile compounds further to enhance tea flavor quality or to resist environmental stresses (Yang et al. 2013). Increase in concentration of the sesquiterpene hormone, abscisic acid (ABA), has been found under UV-B radiation (Cassia 2009; Gil et al. 2012) following a dose-response behavior according to UV-B irradiation dose-rate (Gil et al. 2012). However, previous studies showed there was no increase in the expression of genes related to ABA biosynthesis by UV-B treatment (Martinez-Zapater et al. 2010). Thus, Gil et al. explain increase in ABA levels under UV-B treatment by UV-B-induced hydrolysis of glycosylated ABA from vacuoles (Gil et al. 2012).

In tea, Jang et al. (2010) found that UV-B radiation could induce new volatiles to increase the number of volatiles in freshly picked tea shoots. Total volatile concentration increased under UV-B radiation (3.0 μmol m^{-2} s^{-1}) for 2 h and then decreased when UV-B radiation was applied for 6 h. This means that the degradation of volatiles and release of new volatiles takes place simultaneously under UV-B treatment, and regulation of volatile concentration depends on hydrolysis of precursor substance and volatiles degradation induced by UV-B radiation dose. β-primeverosidase and β-glucosidase, two important glycoside hydrolases (Fig. 6.4), were confirmed to be important enzymes hydrolyzing glycosidic precursors to release volatiles (Wang et al. 2001; Seungjin et al. 2001). A β-primeverosidase has been purified from fresh leaves of *Camellia sinensis* var. *sinensis* cv. Yabukita, and its partial protein sequence was classified into a family of glycosyl hydrolases. Previous studies have indicated most plant glycosyl hydrolases in this family were involved in

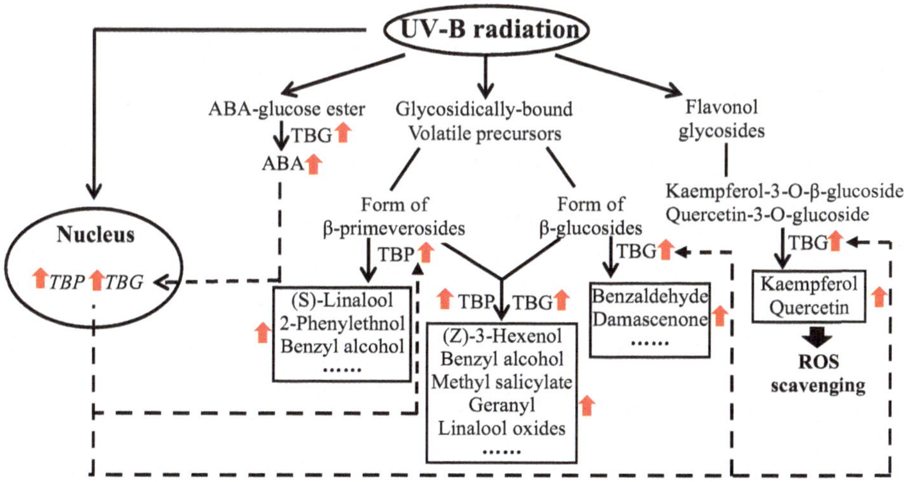

Fig. 6.4 Potential effect of UV-B radiation on glycosidically-bound compounds
ABA abscisic acid, *TBP* β-primeverosidase, *TBG* β-glucosidase, *ROS* reactive oxygen species

defense mechanisms (Mizutani and Sakata 2002). Li et al. (2005) purified a β-glucosidase from *Camellia sinensis* cv. *Fuding da bai cha*, which showed optimum activity at 50 °C with good stability below 40 °C and was active between pH 4.0 and pH 7.0. Jang et al. (2010) indicated UV-B radiation induced increase in volatile types by action of β-primeverosidase and β-glucosidase on glycosidically bound volatiles, which was confirmed by elevated expression of the β-primeverosidase gene (*TBP*) and the β-glucosidase gene (*TBG*) stimulated by UV-B radiation. These two genes' expression profiles also follow a dose-response behavior according to UV-B irradiation dose-rate. A previous study showed that the expressions of *TBP* and *TBG* in tea shoots were enhanced by UV-B treatment for 2 h, while decreased as UV-B treatment was extended to 6 h, which indicated that appropriate UV-B dose increased expression of *TBP* and *TBG* genes promote biosynthesis of β-primeverosidase and β-glucosidase, but excessive UV-B radiation suppressed gene expression and related biosynthesis (Jang et al. 2010). β-glucosidases in many plants have been reported to exhibit high affinity for abiotic stress tolerance. The expression of the β-glucosidase gene (*CsBGlu*) in *Crocus sativus* could be significantly induced in response to dehydration, NaCl, UV-B treatments, and other phytohormone treatments like methyl jasmonate and ABA treatments. The fact that increase in *CsBGlu* expression accompanying flavonols (such as kaempferol and quercetin) accumulation suggests glucosidase catalyzing deglycosylation of flavonol β-glucosides (Fig. 6.4) confer tolerance to various abiotic stresses such as UV-B radiation (Baba et al. 2017).

5 Conclusion and Perspectives

This chapter provides a comprehensive review of tea metabolites and related gene expression induced by UV-B radiation. UV-B radiation-induced metabolic changes mainly involve in flavonoids, amino acids, and volatile compounds. Flavonoid compounds have been reported to protect against UV-B radiation in several plant species. In *Camellia sinensis*, total polyphenol content, including soluble and polymeric forms, obviously accumulate under UV-B exposure. Catechins, important metabolites for tea quality and human health, could be increased by exposure to UV-B radiation with differing responses due to UV-B radiation dose and cultivar variation. UV-B radiation could influence gene expression of flavonoid biosynthesis including the general phenylpropanoid pathway and the flavonoid pathway. Most of the flavonoid biosynthesis-associated gene expressions are enhanced by moderate UV-B exposure, leading to flavonoid accumulation. In addition, UV-B-induced expression of genes involved in the flavonoid biosynthetic pathway is mediated by light-responsive members of the bZIP/bHLH and MYB transcription factor families. Twenty-six transcription factors in MYB family have been identified in *Camellia sinensis*, involving gene expression regulation of *PAL*, *F3H*, *F3′H*, and *FLS* in flavonoid biosynthesis. There is considerably less systematic research on UV-B-induced changes in amino acid concentrations. UV-B radiation regulates inorganic nitrogen assimilation into amino acids by regulation of the glutamine synthetase/glutamate synthase cycle (GC/GOGAT). GABA is a metabolite produced via the GABA shunt pathway in response to stress, and it also is a signal molecule affecting carbon-nitrogen metabolism. UV-B exposure induces changes of volatile aroma compounds including alcohols, ketones, aldehydes, terpenoids, and so on. Changes in gene expression related to UV-B radiation occur mainly in terpenoid biosynthesis and volatile glycosidic-precursors hydrolysis. In addition, there are several aspects involving UV-B-induced metabolism in *Camellia sinensis* that should be addressed in future studies:

(a) The precise regulatory mechanism of different transcriptional regulators at common/different promoters of genes involving flavonoid biosynthetic pathway needs to be further studied.
(b) The knowledge of changes of UV-B-induced amino acids and related gene expression is most based on other plant species rather than *Camellia sinensis*. Direct investigation into amino acid metabolism under UV-B radiation remains to be done.
(c) More research into the mechanism of endogenous phytohormone signal-induced regulation of metabolites and related metabolic pathways under UV-B radiation is needed.
(d) UV-B-induced metabolic diversions and their regulatory mechanisms among metabolite species need to be discussed in future.

Acknowledgments We thank Eric Scott (Tufts University, USA) for improvements and corrections on the manuscript.

References

Alasalvar C, Topal B, Serpen A et al (2012) Flavor characteristics of seven grades of black tea produced in Turkey. J Agric Food Chem 60(25):6323

Al-Quraan NA (2015) GABA shunt deficiencies and accumulation of reactive oxygen species under UV treatments: insight from Arabidopsis thaliana calmodulin mutants. Acta Physiol Plant 37(4):86

Anderson JG, Wilmouth DM, Smith JB et al (2012) UV dosage levels in summer: increased risk of ozone loss from convectively injected water vapor. Science 337(6096):835

Baba SA, Vishwakarma RA, Ashraf N (2017) Functional characterization of CsBGlu12, a β-glucosidase from Crocus sativus provides insights into its role in abiotic stress through accumulation of antioxidant flavonols. J Biol Chem 292(11):jbc.M116.762161

Baldermann S, Fleischmann P, Bolten M et al (2009) Centrifugal precipitation chromatography – a powerful technique for the isolation of active enzymes from tea leaves (Camellia sinensis). J Chromatogr A 1216(19):4263–4267

Ballaré CL, Caldwell MM, Flint SD et al (2011) Effects of solar ultraviolet radiation on terrestrial ecosystems. Patterns, mechanisms, and interactions with climate change. Photochem Photobiol Sci 10(2):226–241

Bashri G, Singh M, Mishra RK et al (2018) Kinetin regulates UV-B-induced damage to growth, photosystem II photochemistry, and nitrogen metabolism in tomato seedlings. J Plant Growth Regul 37(1):233–245

Becatti E, Petroni K, Giuntini D et al (2009) Solar UV-B radiation influences carotenoid accumulation of tomato fruit through both ethylene-dependent and -independent mechanisms. J Agric Food Chem 57(22):10979

Berli FJ, Moreno D, Piccoli P et al (2010) Abscisic acid is involved in the response of grape (Vitis vinifera L.) cv. Malbec leaf tissues to ultraviolet-B radiation by enhancing ultraviolet-absorbing compounds, antioxidant enzymes and membrane sterols. Plant Cell Environ 33(1):1

Berli FJ, Fanzone M, Piccoli P (2011) Solar UV-B and ABA Are Involved in Phenol Metabolism of Vitis vinifera L Increasing Biosynthesis of Berry Skin Polyphenols. J Agric Food Chem 59(9):4874–4884

Bieza K, Lois R (2001) An Arabidopsis mutant tolerant to lethal ultraviolet-B levels shows constitutively elevated accumulation of flavonoids and other phenolics. Plant Physiol 126(3):1105–1115

Bouché N, Fromm H (2004) GABA in plants: just a metabolite? Trends Plant Sci 9(3):110–115

Carbonell-bejerano P, Diago MP, Martínezabaigar J et al (2014) Solar ultraviolet radiation is necessary to enhance grapevine fruit ripening transcriptional and phenolic responses. BMC Plant Biol 14(1):183

Cassia R (2009) An increase in the concentration of abscisic acid is critical for nitric oxide-mediated plant adaptive responses to UV-B irradiation. New Phytol 181(4):871–879

Christie JM, Jenkins GI (1996) Distinct UV-B and UV-A/blue light signal transduction pathways induce chalcone synthase gene expression in Arabidopsis cells. Plant Cell 8(9):1555–1567

Cloix C, Jenkins GI (2008) Interaction of the Arabidopsis UV-B-specific signaling component UVR8 with chromatin. Mol Plant 1:118–128

Czemmel S, Stracke R, Weisshaar B et al (2009) The grapevine R2R3-MYB transcription factor VvMYBF1 regulates flavonol synthesis in developing grape berries. Plant Physiol 151(3):1513–1530

Delfine S, Csiky O, Seufert G et al (2010) Fumigation with exogenous monoterpenes of a non-isoprenoid-emitting oak (Quercus suber): monoterpene acquisition, translocation, and effect on the photosynthetic properties at high temperatures. New Phytol 146(1):27–36

Dolzhenko Y, Bertea CM, Occhipinti A et al (2010) UV-B modulates the interplay between terpenoids and flavonoids in peppermint (piperita L.). J Photochem Photobiol B Biol 100(2):67

Dong F, Fu X, Watanabe N et al (2016) Recent advances in the emission and functions of plant vegetative volatiles. Molecules 21(2):124

Dudareva N, Negre F, Nagegowda DA et al (2006) Plant volatiles: recent advances and future perspectives. Crit Rev Plant Sci 25(5):417–440

Eva H, Jansen MAK, Ake S (2013) UV-B exposure, ROS, and stress: inseparable companions or loosely linked associates? Trends Plant Sci 18(2):107–115

Favory J, Stec A, Gruber H et al (2009) Interaction of COP1 and UVR8 regulates UV-B-induced photomorphogenesis and stress acclimation in Arabidopsis. EMBO J 28(5):591–601

Felfe C, Schemainda M, Baldermann S et al (2011) Metabolism of carotenoid degradation in leaves of Camellia sinensis—Functional and biochemical modifications. J Food Compos Anal 24(6):821–825

Gil M, Pontin M, Berli F et al (2012) Metabolism of terpenes in the response of grape (Vitis vinifera, L.) leaf tissues to UV-B radiation. Phytochemistry 77(15–16):89–98

Gil M, Bottini R, Berli F et al (2013) Volatile organic compounds characterized from grapevine (Vitis vinifera, L. cv. Malbec) berries increase at pre-harvest and in response to UV-B radiation. Phytochemistry 96(12):148

Gil M, Bottini R, Pontin M et al (2014) Solar UV-B radiation modifies the proportion of volatile organic compounds in flowers of field-grown grapevine (Vitis vinifera, L.) cv. Malbec. Plant Growth Regul 74(2):193–197

Giuntini D, Graziani G, Lercari B et al (2005) Changes in carotenoid and ascorbic acid contents in fruits of different tomato genotypes related to the depletion of UV-B radiation. J Agric Food Chem 53(8):3174–3181

Gollop R, Farhi S, Perl A (2001) Regulation of the leucoanthocyanidin dioxygenase gene expression in Vitis vinifera. Plant Sci 161(3):579–588

Götz T, Sandmann G, Römer S (2002) Expression of a bacterial carotene hydroxylase gene (crtZ) enhances UV tolerance in tobacco. Plant Mol Biol 50(1):127–140

Guo W, Sakata K, Watanabe N et al (1993) Geranyl 6-O-β-d-xylopyranosyl-β-d-glucopyranoside isolated as an aroma precursor from tea leaves for oolong tea. Phytochemistry 33(6):1373–1375

Guo W, Hosoi R, Sakata K et al (1994) (S)-linalyl, 2-phenylethyl, and benzyl disaccharide glycosides isolated as aroma precursors from oolong tea leaves. Biosci Biotechnol Biochem 58(8):1532–1534

Hagedorn CH, Phang JM (1983) Transfer of reducing equivalents into mitochondria by the interconversions of proline and delta 1-pyrroline-5-carboxylate. Arch Biochem Biophys 225(1):95–101

Hampel D, Mosandl A (2005) Wüst, A, M. Induction of de Novo Volatile Terpene Biosynthesis via Cytosolic and Plastidial Pathways by Methyl Jasmonate in Foliage of Vitis vinifera L. J Agric Food Chem 53(7):2652–2657

Hao G, Du X, Zhao F et al (2009) Role of nitric oxide in UV-B-induced activation of PAL and stimulation of flavonoid biosynthesis in Ginkgo biloba callus. Plant Cell Tissue Org Cult 97(2):175–185

Hectors K, Prinsen E, Coen WD et al (2007) Arabidopsis thaliana plants acclimated to low dose rates of ultraviolet B radiation show specific changes in morphology and gene expression in the absence of stress symptoms. New Phytol 175(2):255–270

Hofmann RW, Swinny EE, Bloor SJ et al (2000) Responses of nine Trifolium repens L. populations to ultraviolet-B radiation: differential flavonol glycoside accumulation and biomass production. Ann Bot 86(3):527–537

Hsieh MM, Chen SM (2007) Determination of amino acids in tea leaves and beverages using capillary electrophoresis with light-emitting diode-induced fluorescence detection. Talanta 73(2):326–331

Inostroza-Blancheteau C, Acevedo P, Loyola R et al (2016) Short-term UV-B radiation affects photosynthetic performance and antioxidant gene expression in highbush blueberry leaves. Plant Physiol Biochem 107:301–309

Jang J, Yang YC, Zhang GH et al (2010) Effect of ultra-violet B on release of volatiles in tea leaf. Int J Food Prop 13(3):608–617

Joubert C, Young PR, Eyéghé-Bickong HA et al (2016) Field-grown grapevine berries use carotenoids and the associated xanthophyll cycles to acclimate to UV exposure differentially in high and low light (shade) conditions. Front Plant Sci 7(286):786

Kakani VG, Reddy KR, Zhao D et al (2003) Field crop responses to ultraviolet-B radiation: a review. Agric Forest Meteorol 120(1–4):191–218

Kim BG, Kim JH, Kim J et al (2008) Accumulation of flavonols in response to ultraviolet-B irradiation in soybean is related to induction of flavanone 3-beta-hydroxylase and flavonol synthase. Mol Cells 25(2):247–252

Kusano M, Tohge T, Fukushima A et al (2011) Metabolomics reveals comprehensive reprogramming involving two independent metabolic responses of Arabidopsis to UV-B light. Plant J Cell Mol Biol 67(2):354–369

Laube JC, Newland MJ, Hogan C et al (2014) Newly detected ozone-depleting substances in the atmosphere. Nat Geosci 7(4):266–269

Li J, Ou-Lee TM, Raba R et al (1993) Arabidopsis flavonoid mutants are hypersensitive to UV-B irradiation. Plant Cell 5(2):171–179

Li YY, Jiang CJ, Wan XC et al (2005) Purification and partial characterization of β-glucosidase from fresh leaves of tea plants (Camellia sinensis (L.) O. Kuntze). Acta Biochim 37(6):363–370

Li CF, Zhu Y, Yu Y et al (2015) Global transcriptome and gene regulation network for secondary metabolite biosynthesis of tea plant (Camellia sinensis). BMC Genomics 16(1):560

Liu M, Li X, Liu Y et al (2013) Regulation of flavanone 3-hydroxylase gene involved in the flavonoid biosynthesis pathway in response to UV-B radiation and drought stress in the desert plant, Reaumuria soongorica. Plant Physiol Biochem 73(6):161–167

Liu S, Ju J, Xia G (2014) Identification of the flavonoid 3′-hydroxylase and flavonoid 3′,5′-hydroxylase genes from Antarctic moss and their regulation during abiotic stress. Gene 543(1):145–152

Liu H, Cao X, Liu X et al (2017) UV-B irradiation differentially regulates terpene synthases and terpene content of peach. Plant Cell Environ 40(10):2261–2275

Ma SJ, Watanabe N, Yagi A et al (2001) The (3R,9R)-3-hydroxy-7,8-dihydro-beta-ionol disaccharide glycoside is an aroma precursor in tea leaves. Phytochemistry 56(8):819–825

Mackerness SAH (2000) Plant responses to ultraviolet-B (UV-B: 280–320 nm) stress: what are the key regulators? Plant Growth Regul 32(1):27–39

Mackerness SAH, John CF, Jordan B et al (2001) Early signaling components in ultraviolet-B responses: distinct roles for different reactive oxygen species and nitric oxide. FEBS Lett 489(2–3):237

Martin DM, Aubourg S, Schouwey MB et al (2010) Functional annotation, genome organization and phylogeny of the grapevine (Vitis vinifera) terpene synthase gene family based on genome assembly, FLcDNA cloning, and enzyme assays. BMC Plant Biol 10(1):226

Martínez-Lüscher J, Morales F, Delrot S et al (2013) Short- and long-term physiological responses of grapevine leaves to UV-B radiation. Plant Sci 213(4):114

Martínez-Lüscher J, Torres N, Hilbert G et al (2014) Ultraviolet-B radiation modifies the quantitative and qualitative profile of flavonoids and amino acids in grape berries. Phytochemistry 102(6):106–114

Martinez-Zapater JM, Ruben B, Rita F et al (2010) Transcriptome changes in grapevine (Vitis vinifera L.) cv. Malbec leaves induced by ultraviolet-B radiation. BMC Plant Biol 10(1):224

Matus JT, Loyola R, Vega A et al (2009) Post-veraison sunlight exposure induces MYB-mediated transcriptional regulation of anthocyanin and flavonol synthesis in berry skins of Vitis vinifera. J Exp Bot 60(3):853

Mehrtens F, Kranz H, Bednarek P et al (2005) The Arabidopsis transcription factor MYB12 is a flavonol-specific regulator of phenylpropanoid biosynthesis. Plant Physiol 138(2):1083–1096

Mishra V, Mishra P, Srivastava G et al (2011) Effect of dimethoate and UV-B irradiation on the response of antioxidant defense systems in cowpea (Vigna unguiculata L.) seedlings. Pestic Biochem Physiol 99(2):118–123

Mittler R, Vanderauwera S, Gollery M et al (2004) Reactive oxygen gene network of plants. Trends Plant Sci 9(10):490

Mizutani M, Sakata K (2002) Cloning of β-primeverosidase from tea leaves, a key enzyme in tea aroma formation. Plant Physiol 130(4):2164–2176

Moerkercke AV, Steensma P, Schweizer F et al (2015) The bHLH transcription factor BIS1 controls the iridoid branch of the monoterpenoid indole alkaloid pathway in Catharanthus roseus. Proc Natl Acad Sci U S A 112(26):8130–8135

Moon JH, Watanabe NY, Yagi A, Sakata K (1996) Studies on aroma formation mechanism in oolong tea. 6. cis- and trans-linalool 3,7-oxides and methyl salicylate glycosides and (z)-3-hexenyl beta-d-glucopyranoside as aroma precursors from tea leaves for oolong tea. Biosci Biotechnol Biochem 60(11):1815–1819

Neugart S, Zietz M, Schreiner M et al (2012) Structurally different flavonol glycosides and hydroxycinnamic acid derivatives respond differently to moderate UV-B radiation exposure. Physiol Plant 145(4):582–593

Nishikitani M, Kubota K, Kobayashi A et al (1996) Geranyl 6-O-alpha-L-Arabinopyranosyl-beta-D -glucopyranoside isolated as an aroma precursor from leaves of a green tea cultivar. Biosci Biotechnol Biochem 60(5):929–931

Nishikitani M, Wang D, Kubota K et al (1999) (Z)-3-hexenyl and trans-linalool 3,7-oxide beta-primeverosides isolated as aroma precursors from leaves of a green tea cultivar. Biosci Biotechnol Biochem 63(9):1631–1633

Ohgami S, Ono E, Toyonaga H et al (2014) Identification and characterization of Camellia sinensis glucosyltransferase, UGT73A17: a possible role in flavonol glucosylation. Plant Biotechnol 31(5):573–578

Oravecz A, Baumann A, Máté Z et al (2006) CONSTITUTIVELY PHOTOMORPHOGENIC1 is required for the UV-B response in Arabidopsis. Plant Cell 18(8):1975–1990

Petrussa E, Braidot E, Zancani M et al (2013) Plant flavonoids—Biosynthesis, transport and involvement in stress responses. Int J Mol Sci 14(7):14950

Phang JM (1985) The regulatory functions of proline and pyrroline-5-carboxylic acid. Curr Top Cell Regul 25(4):91–132

Rani A, Singh K, Ahuja PS et al (2012) Molecular regulation of catechins biosynthesis in tea [*Camellia sinensis* (L.) O. Kuntze]. Gene 495(2):205–210

Rastogi RP, Richa KA et al (2010) Molecular mechanisms of ultraviolet radiation-induced DNA damage and repair. J Nucleic Acids 2010(6551):592980

Rizzini L, Favory JJ, Cloix C et al (2011) Perception of UV-B by the Arabidopsis UVR8 protein. Science 332:103–106

Robinson SA, Slade AP, Fox GG et al (1991) The role of glutamate dehydrogenase in plant nitrogen metabolism. Plant Physiol 95(2):509–516

Ryan KG, Swinny EE, Markham KR et al (2002) Flavonoid gene expression and UV photoprotection in transgenic and mutant Petunia leaves. Phytochemistry 59(1):23–32

Saradhi PP (1995) Alia, Arora S, et al. Proline accumulates in plants exposed to UV radiation and protects them against UV induced peroxidation. Biochem Biophys Res Commun 209(1):1–5

Sarkar D, Bhowmik PC (2011) Young-In-Kwon, et al. The role of proline-associated pentose phosphate pathway in cool-season turfgrasses after UV-B exposure. Environ Exp Bot 70(2–3):251–258

Scharbert S, Hofmann T (2005) Molecular definition of black tea taste by means of quantitative studies, taste reconstitution, and omission experiments. J Agric Food Chem 53(13):5377–5384

Scharbert S, Holzmann N, Hofmann T (2004) Identification of the astringent taste compounds in black tea infusions by combining instrumental analysis and human bioresponse. J Agric Food Chem 52(11):3498

Schluttenhofer C, Yuan L (2015) Regulation of specialized metabolism by WRKY transcription factors. Plant Physiol 167(2):295–306

Schuh C, Schieberle P (2006) Characterization of the key aroma compounds in the beverage prepared from Darjeeling black tea: quantitative differences between tea leaves and infusion. J Agric Food Chem 54(3):916–924

Schwartz SH, Qin X, Zeevaart JA (2001) Characterization of a novel carotenoid cleavage dioxygenase from plants. J Biol Chem 276(27):25208

Seungjin MA, Mizutani M, Hiratake J et al (2001) Substrate specificity of β-primeverosidase, a key enzyme in aroma formation during oolong tea and black tea manufacturing. Biosci Biotechnol Biochem 65(12):2719–2729

Shen SL, Yin XR, Zhang B et al (2016) CitAP2.10 activation of the terpene synthase CsTPS1 is associated with the synthesis of (+)-valencene in 'Newhall' orange. J Exp Bot 67(14):4105–4115

Simkin AJ, Underwood BA, Auldridge M et al (2004) Circadian regulation of the PhCCD1 carotenoid cleavage dioxygenase controls emission of beta-ionone, a fragrance volatile of petunia flowers. Plant Physiol 136(3):3504

Singh K, Rani A, Kumar S et al (2008) An early gene of flavonoid pathway, flavanone 3-hydroxylase, exhibits a positive relationship with catechins content in tea (Camellia sinensis (L.) O. Kuntze). Tree Physiol. Tree Physiol 28(9):1349–1356

Singh K, Rani A, Paul A et al (2009) Differential display mediated cloning of anthocyanidin reductase gene from tea (Camellia sinensis) and its relationship with the concentration of epicatechins. Tree Physiol 29(6):837

Singh VP, Srivastava PK, Prasad SM (2012) UV-B induced differential effect on growth and nitrogen metabolism in two cyanobacteria under copper toxicity. Cell Mol Biol (Noisy-le-Grand) 58(1):85–95

Mackerness SAH (2000) Plant responses to ultraviolet-B (UV-B: 280–320 nm) stress: what are the key regulators? Plant Growth Regul 32(1):27–39

Solovchenko A, Schmitzeiberger M (2003) Significance of skin flavonoids for UV-B-protection in apple fruits. J Exp Bot 54(389):1977–1984

Spyropoulou EA, Haring MA, Schuurink RC (2014) RNA sequencing on Solanum lycopersicum trichomes identifies transcription factors that activate terpene synthase promoters. BMC Genomics 15(1):402

Stracke R, Favory JJ, Gruber H et al (2010) The Arabidopsis bZIP transcription factor HY5 regulates expression of the PFG1/MYB12 gene in response to light and ultraviolet-B radiation. Plant Cell Environ 33(1):88–103

Strzałka K, Kostecka-Gugała A, Latowski D (2003) Carotenoids and environmental stress in plants: significance of carotenoid-mediated modulation of membrane physical properties. Russ J Plant Physiol 50(2):168–173

Sun Z, Chen W, Chen Z et al (2011) A role for Ethylene – Insensitive 2, gene in the regulation of the ultraviolet-B response in Arabidopsis. Acta Physiol Plant 33(3):1025–1030

Takeda J, Abe S, Hirose Y et al (1993) Effect of light and 2,4-dichlorophenoxyacetic acid on the level of mRNAs for phenylalanine ammonia-lyase and chalcone synthase in carrot cells cultured in suspension. Physiol Plant 89(1):4–10

Takeda J, Obi I, Yoshida K (1994) Action spectra of phenylalanine ammonia-lyase and chalcone synthase expression in carrot cells in suspension. Physiol Plant 91(3):517–521

Takeda J, Ozeki Y, Yoshida K (1997) Action spectrum for induction of promoter activity of phenylalanine ammonia-lyase gene by UV in carrot suspension cells. Photochem Photobiol 66(4):464–470

Takeuchi A, Matsumoto S, Hayatsu M (1994) Chalcone synthase from Camellia sinensis: isolation of the cDNAs and the organ-specific and sugar-responsive expression of the genes. Plant Cell Physiol 35(7):1011–1018

Taniguchi S, Hosokawa-Shinonaga Y, Tamaoki D et al (2014) Jasmonate induction of the monoterpene linalool confers resistance to rice bacterial blight and its biosynthesis is regulated by JAZ protein in rice. Plant Cell Environ 37(2):451

Tegelberg R, Julkunen-Tiitto R, Aphalo PJ (2004) Red : far-red light ratio and UV-B radiation: their effects on leaf phenolics and growth of silver birch seedlings. Plant Cell Environ 27(8):1005–1013

Tholl D (2006) Terpene synthases and the regulation, diversity and biological roles of terpene metabolism. Curr Opin Plant Biol 9(3):297–304

Valadier MH, Yoshida A, Grandjean O et al (2010) Implication of the glutamine synthetase/glutamate synthase pathway in conditioning the amino acid metabolism in bundle sheath and mesophyll cells of maize leaves. FEBS J 275(12):3193–3206

Wade HK, Bibikova TN, Valentine WJ et al (2001) Interactions within a network of phytochrome, cryptochrome and UV-B phototransduction pathways regulate chalcone synthase gene expression in Arabidopsis leaf tissue. Plant J Cell Mol Biol 25(6):675–685

Wang D, Ando K, Morita K, et al (1994) Optical isomers of linalool and linalool oxides in tea aroma. Validation of Shipmo7 & Precal with the Cpf Hydroelastic Model 58(11):2050–2053

Wang D, Kurasawa E, Yamaguchi Y et al (2001) Analysis of glycosidically bound aroma precursors in tea leaves. 2. Changes in glycoside contents and glycosidase activities in tea leaves during the black tea manufacturing process. J Agric Food Chem 49(4):1900

Wang Y, Feng H, Qu Y et al (2006) The relationship between reactive oxygen species and nitric oxide in ultraviolet-B-induced ethylene production in leaves of maize seedlings. Environ Exp Bot 57(1–2):51–61

Wang YS, Xu YJ, Gao LP et al (2014) Functional analysis of Flavonoid 3′,5′-hydroxylase from Tea plant (Camellia sinensis): critical role in the accumulation of catechins. BMC Plant Biol 14(1):347

Wang CH, Zheng LP, Tian H et al (2016) Synergistic effects of ultraviolet-B and methyl jasmonate on tanshinone biosynthesis in Salvia miltiorrhiza, hairy roots. J Photochem Photobiol B Biol 159:93–100

Wei K, Wang L, Zhang C et al (2015) Transcriptome analysis reveals key flavonoid 3′-hydroxylase and flavonoid 3′,5′-hydroxylase genes in affecting the ratio of dihydroxylated to trihydroxylated catechins in Camellia sinensis. PLoS One 10(9):e0137925

Wen PF, Ji W, Gao MY et al (2015) Accumulation of flavanols and expression of leucoanthocyanidin reductase induced by postharvest UV-C irradiation in grape berry. Genet Mol Res Gmr 14(3):7687

Yang Z, Baldermann S, Watanabe N (2013) Recent studies of the volatile compounds in tea. Food Res Int 53(2):585–599

Zagoskina NV, Alyavina AK, Gladyshko TO et al (2005) Ultraviolet rays promote development of photosystem II photochemical activity and accumulation of phenolic compounds in the tea callus culture (Camellia sinensis). Russ J Plant Physiol 52(6):731–739

Zheng XQ, Jin J, Chen H, et al (2008) Effect of ultraviolet B irradiation on accumulation of catechins in tea (Camellia sinensis (L) O. Kuntze. Afr J Biotechnol 7(18):3283–3287

Chapter 7
Elevated Carbon Dioxide-Induced Perturbations in Metabolism of Tea Plants

Xin Li, Golam Jalal Ahammed, Lan Zhang, Peng Yan, Liping Zhang, and Wen-Yan Han

Abstract Rising CO_2 on the earth's surface is the single most influential driving force for climate change. A CO_2-enriched environment impacts leaf chemistry and metabolism, leading to remarkable changes in plant productivity. While effect of elevated CO_2 on the yield of economic crops has been well addressed, its impact on important beverage crops such as tea (*Camellia sinensis* L.) has received less attention. In tea plants, elevated CO_2 increases photosynthesis and the production of photoassimilates, whereas increased photosynthesis necessitates increased respiration to provide more energy to drive photosynthesis. Furthermore, elevated CO_2 increases total carbon concentration and decreases total nitrogen concentration, resulting in an increased ratio of carbon to nitrogen in tea leaves. Despite the discrepancies in research reports, elevated CO_2 improves tea quality by causing enhancements in the concentrations of tea polyphenol, free amino acid, including theanine and a reduction in that of caffeine. A proper balance between photosynthesis and respiration under elevated CO_2 increases net productivity in tea; however, tea plant defense response to necrotrophic pathogens is attenuated in a CO_2-enriched environment. In this chapter, we aim to review and summarize available literature on the changes in plant physiological processes, primary and secondary metabolite synthesis, nutraceutical properties, overall tea quality, tea yield, and plant defense in response to elevated CO_2.

Keywords Elevated CO_2 · Tea quality · Photosynthesis · Respiration · Tea yield · Defense response

X. Li · W.-Y. Han (✉)
Chinese Academy of Agricultural Sciences, Tea Research Institute, Hangzhou, PR China
e-mail: lixin@tricaas.com; hanwy@tricaas.com

G. J. Ahammed
Department of Horticulture, College of Forestry, Henan University of Science and Technology, Luoyang, PR China
e-mail: ahammed@haust.edu.cn

L. Zhang · P. Yan · L. Zhang
Key Laboratory of Tea Quality and Safety Control, Ministry of Agriculture, Tea Research Institute, Chinese Academy of Agricultural Sciences, Hangzhou, People's Republic of China

W.-Y. Han et al. (eds.), *Stress Physiology of Tea in the Face of Climate Change*,
https://doi.org/10.1007/978-981-13-2140-5_7

1 Introduction

Rising CO_2 on the earth's surface is the single most influential driving force for global warming as well as climate change. Recent global atmospheric CO_2 concentration is 407.54 parts per million (ppm) by volume (recorded in January 2018 http://www.esrl.noaa.gov/gmd/ccgg/trends/) with an annual mean growth rate of 2.34 ppm. However, the concentration of atmospheric CO_2 was only 270 ppm before the industrial revolution, around 200 years ago. This huge upsurge in atmospheric CO_2 is mainly due to human activities including large-scale deforestation, fossil fuel combustion, urbanization, and industrialization (Ahuja et al. 2010). Despite the implementation of numerous strategies to reduce global carbon emission, the concentration of atmospheric CO_2 is increasing, and it will further increase in the coming decades and possibly will reach 550 ppm by 2050 and 800 ppm by 2100, respectively. Since natural calamities are linked to climate change, the severity and frequency of heat waves, cold, drought, and flooding will increase with rising atmospheric CO_2 (Field et al. 2014).

Tea is the most widely consumed nonalcoholic beverage in the world after water (Macfarlane and Macfarlane 2004). In principle, a bud and the adjacent two young leaves of *Camellia sinensis* (L.) Kuntze are used to manufacture green tea through rapid roasting of fresh harvest to avoid oxidation (Han et al. 2016). In recent years, the popularity of green tea is rising across the world due to its pleasant flavor and numerous health benefits such as anticancer, anti-inflammatory, anti-allergic, and anti-obesity effects on human (Kim et al. 2009; Siamwala et al. 2013; Mancini et al. 2017). To meet the increasing demand, tea cultivation has been expanding in many countries of tropical and subtropical regions including China (Han et al. 2016; Han et al. 2018). Health benefits of tea are mainly attributed to some bioactive compounds, and a few of them exclusively occur in tea (Kim et al. 2009). While effect of climate change on the yield of economic crops has been well addressed, its impact on important beverage crops such as tea has received less attention (Ahmed et al. 2014). Unlike annual crops, tea plants remain in active production for a long period of time, even for a hundred years, which may allow them to experience climate change over the century (Larson 2015; Li et al. 2017). It is believed that long life span of tea plants may lead them to operate massive physiological adaptation instead of genetic modification (Li et al. 2017). Studies show that climate change factors, especially increasing temperatures, differentially affect tea yield and quality at various altitudes (Wijeratne et al. 2007; Han et al. 2016). However, compared to other climatic parameters, information relating to the effects of rising atmospheric CO_2 on tea yield and quality is scanty. In fact, this area of tea science is still in its infancy. However, recently the issue has received new momentum, and intergovernmental initiatives have been undertaken to explore climate change effect on tea system (Han et al. 2018). Several papers on the effects of elevated CO_2 on tea yield, tea quality, and defense response were published in the last 2 years (Hui et al. 2016; Li et al. 2016b; Li et al. 2017). In this chapter, we intend to review and summarize available literature on the changes in plant physiological processes, primary and secondary metabolite synthesis, nutraceutical properties, overall tea quality, tea yield, and plant defense in response to elevated CO_2.

2 Effects of Elevated CO_2 on Photosynthesis and Photosystems

2.1 *Gas Exchange and CO_2 Assimilation*

Photosynthesis is tied to atmospheric CO_2 and plays a key role in driving global carbon cycle (Amthor 1995). Moreover, plant productivity is primarily dependent on photosynthesis. In brief, C is fixed from atmospheric CO_2 through the process of photosynthesis, and the photoassimilates provide C-skeleton for multiple metabolic processes including synthesis of numerous structural and nonstructural compounds required for plant growth, development, and stress response. Photosynthetic CO_2 assimilation largely depends on a key enzyme called ribulose-1,5-bis-phosphate (RuBP) carboxylase/oxygenase (RuBisCO) which can catalyze both carboxylation and oxygenation of RuBP utilizing CO_2 and O_2 as substrate, respectively (Li et al. 2017). Since CO_2 and O_2 compete for the active site of the RuBisCO, elevated CO_2 promotes internal CO_2 concentration and the ratio of CO_2:O_2 at the RuBisCO site, favoring the carboxylation over the oxygenation of RuBP. In fact, oxygenation is a respiratory (photorespiration) loss of CO_2 which has detrimental effect on yield (Amthor 1997; Li et al. 2013).

In response to elevated CO_2, plants increase photosynthesis, and such photosynthetic response is common in both C_3 and C_4 plants (Ainsworth and Rogers 2007); however, the mechanisms may differ to a great extent. In C_3 plants, elevated CO_2 increases activity of RuBisCO to increase the photosynthesis, while in C_4 plants, elevated CO_2 decreases stomatal conductance to minimize associated stress (such as drought) effects on photosynthetic process (Ainsworth and Long 2005; Ahuja et al. 2010). Similar to other C_3 plants, tea plants elevate the net photosynthetic rate in response to high CO_2 concentration (Li et al. 2017). However, maximum light-saturated photosynthetic rates in tea plants are lower than the average for C_3 plants (De Costa et al. 2007). Furthermore, elevated CO_2 also influences plant respiration that utilizes photoassimilates to generate usable energy such as ATP (Fig. 7.1). Recent studies revealed that increased carbohydrate availability and energy demand under elevated CO_2 also enhance respiration rate in tea plants, which helps plant to optimize the allocation of carbon and nutrient for maximizing photosynthesis and plant growth (Li et al. 2013; Li et al. 2017). This implies that CO_2-enriched environments greatly impact plant physiology, biochemistry, and metabolism, leading to remarkable changes to plant productivity (Dietterich et al. 2015).

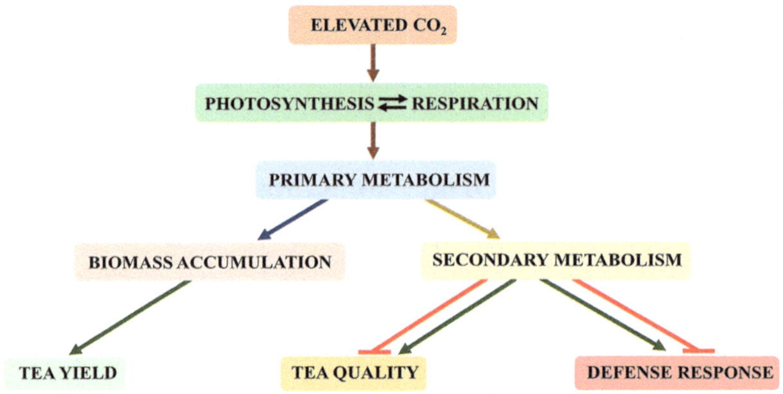

Fig. 7.1 Schematic representation of the effect of elevated CO_2 on physiological processes, primary and secondary metabolite synthesis, tea yield, tea quality, and defense response in tea plants → indicates "enhancement", while —⊣ represents "inhibition." For description, readers are referred to the main text

Like food crops, tea plants, which are considered as the most popular beverage crop in the world, increase instantaneous photosynthesis in response to CO_2 enrichment. In mature field-grown tea, photosynthetic rate can reach maximum 30 μmol $m^{-2}s^{-1}$ at 1000–1200 μmol mol^{-1} sub-stomatal CO_2 concentration, while that in greenhouse-grown young tea plants is only 7–10 μmol m^{-2} s^{-1} at 500–600 μmol mol^{-1} sub-stomatal CO_2 concentration (Mohotti 1998; De Costa et al. 2007). However, intercellular CO_2 is largely dependent on stomatal conductance, a gas exchange parameter, which is highly responsive to water availability. Thus, elevated CO_2-induced photosynthetic enhancement also depends on the availability of water (Hajiboland 2017).

In general, elevated CO_2 causes partial stomatal closure and reduces stomatal conductance, leading to a lower transpiration rate (Kimball 2016). However, the effects of elevated CO_2 on stomatal conductance become more pronounced when nitrogen (N) supply is limited (Ainsworth and Long 2005; Ainsworth and Rogers 2007) due to the fact that N limitation reduces availability of RuBisCO concentration as N is a constituent of RuBisCO. Thus plants decrease stomatal conductance to balance the CO_2 ratio between internal leaf CO_2 and outside air under elevated CO_2 environment (Kimball 2016).

Since plants reduce stomatal conductance and transpiration rate in response to elevated CO_2, the natural leaf cooling mechanism is hampered, which results in approximately 0.4–1.7 °C temperature rise at the canopy level (Asseng et al. 2014). Elevated CO_2-induced rise in canopy temperature varies largely depending on the plant species; however, it is on an average 0.6 °C in C_3 plants. A study on tea plants showed that 550 and 700 μmol mol^{-1} CO_2 can increase canopy temperature by 1 ~ 2 °C and 3 ~ 4 °C, respectively, at 800 μmol m^{-2} s^{-1} photosynthetic active radiation (Yang et al. 2005). It is suggested that a 0.6 °C increase in temperature would cause 3.6% yield reduction in wheat (Asseng et al. 2014); however, relevant data on tea is currently unavailable.

It is well conceived that exposure of plants to elevated CO_2 alters leaf gas exchange as reflected by increased photosynthesis and decreased stomatal conductance and transpiration rate in a range of plant species (Ainsworth and Rogers 2007; Zavala et al. 2013; Sun et al. 2015). Similarly, a 45-day CO_2 enrichment study conducted in open-top chambers showed that elevated CO_2 (648 ~ 658 μmol mol^{-1}) increases net photosynthetic rate and water use efficiency but decreases stomatal conductance and transpiration rate in leaves of Longjingchangye tea cultivar. In addition, a combined treatment of elevated CO_2 and +0.57 °C temperature has additive effect on photosynthesis compared with that of their only treatments (Hui et al. 2016). Another study on tea plants showed that elevated CO_2 (550/750 μmol mol^{-1}) increases photosynthetic rate and water use efficiency by 17.9%/25.8% and 21.6%/35.8% but decreases stomatal conductance by 7.6%/13.0%, respectively (Jiang et al. 2005). A study conducted on tea plants in open-top chambers showed that warm temperature (above 34 °C)-induced reduction in photosynthetic rate is greatly attenuated by elevated CO_2 at 800 μmol m^{-2} s^{-1} PAR. While the photosynthetic rate drastically declines at low PAR (below 400 μmol m^{-2} s^{-1}) and a temperature above 30 °C, elevated CO_2 attenuates the reduction in the photosynthetic rate (Yang et al. 2005).

Photosynthetic acclimation is an important aspect of the effect of elevated CO_2 on plant photosynthesis. Over the time, the expression levels of RuBisCO and other photosynthesis-related genes decline under elevated CO_2 which potentially cause photosynthesis acclimation. The fact that elevated CO_2 increases accumulation of soluble sugars, the expression of photosynthesis-related genes declines, leading to a reduction in photosynthesis (Feng et al. 2014). In addition, accelerated leaf senescence and limited growth rate may also cause downregulation of photosynthesis, which is highly tissue and species/ecotype specific. In a 24-day-long CO_2 enrichment (800 μmol mol^{-1}) study in tea plants, we recorded net photosynthetic rate at five time points using an open-flow infrared gas analyzer adapted with light and temperature control systems (Li et al. 2017). We found that elevated CO_2 rapidly increases the net photosynthetic rate, which gradually reaches the peak at 12 days. Afterward no further increase in the photosynthetic rate occurs up to 24 days. This implies that a photosynthetic acclimation may occur after 12 days in tea plants under elevated CO_2 conditions. The trend of changes in maximum carboxylation rate of RuBisCO (Vcmax) and maximum rates of RuBP regeneration (Jmax) follows that of CO_2 assimilation rate, implying that effect of elevated CO_2 on RuBisCO potentially causes photosynthetic acclimation. Indeed, many C_3 plant species respond to elevated atmospheric CO_2 concentration with an initial stimulation of photosynthesis, which subsequently declines to levels similar to that of plants grown under ambient conditions due to the photosynthetic acclimation (Sage 1994).

2.2 *Chlorophyll Content and Chlorophyll Fluorescence*

Chlorophylls are important light-harvesting molecules in photosystem. A decrease in chlorophyll content, which can be recognized by leaf yellowing due to either the chlorophyll degradation or reduced synthesis, is often regarded as an important sign of leaf senescence. When tea plants are exposed to elevated CO_2 concentration, chlorophyll contents may either increase or remain stable. Jing et al (2005) found that CO_2 enrichments (550/750 μmol mol^{-1}) for 3 months could increase chlorophyll a, chlorophyll b, and carotenoid contents by 12.8%/18.4%, 14.0%/22.0%, and 17.2%/20.1% in shoots of tea plants (Jiang et al. 2005), whereas the ratio of chlorophyll a to chlorophyll b is decreased by elevated CO_2. However, Hui et al. (2016) found only slight, but not significant, increase in chlorophyll a and b contents in tea leaves following a 45-day-long exposure of tea plants to 648–658 μmol mol^{-1} CO_2 in open-top chambers. This implies that elevated CO_2 favors increased photosynthetic pigment concentration and potentially delay leaf senescence in tea plants.

Chlorophyll a fluorescence measurements are rapid, noninvasive, and easy to use methods for assessing the status of photosynthetic apparatus under environmental perturbation (Ahammed et al. 2015). The maximal photochemical efficiency of PSII (Fv/Fm) is the most commonly used chlorophyll fluorescence parameter that represents amount of energy trapped in PSII reaction centers with regard to energy absorbed. The Fv/Fm value also represents the occurrence of photoinhibition. Compared to other C_3 plant species, tea plants are sensitive to photoinhibition in response to high light intensities (De Costa et al. 2007). Analysis of chlorophyll fluorescence parameters showed that elevated CO_2 causes significant increases in Fv/Fm and photosynthetic light absorption (Hui et al. 2016). In one of our previous studies, we also analyzed the quantum efficiency of PSII photochemistry (Φ_{PSII}) that represents photosynthetic efficiency of tea leaves (Li et al. 2017). We noticed a gradual increase in Φ_{PSII} up to 18 days following exposure of tea plants to 800 μmol mol^{-1} CO_2. After that no noticeable change in Φ_{PSII} was found, implying that response of PSII to elevated CO_2 is saturated possibly due to photosynthetic acclimation.

3 Effects of Elevated CO_2 on Respiration

In plant cells, respiration oxidizes the carbohydrate to CO_2 and O_2, while leaves, buds, shoots, and roots consume part of the carbon and the rest is released to the atmosphere as CO_2 (Huang et al. 2007). Eventually, respiration provides energy in a form of ATP to cells. An optimum balance between photosynthesis and respiration is required for proper biomass accumulation (Amthor 1995; Amthor 1997; Li et al. 2017). There is huge discrepancy among study reports regarding the effect of elevated CO_2 on respiration. Some studies show that respiration is triggered under elevated CO_2, while others show inhibition of respiration at CO_2-enriched environments (Li et al. 2013). The fact that elevated CO_2 increases production of photoassimilates, the increased photosynthesis necessitates increased respiration to provide more energy to drive photosynthesis (Fig. 7.1). However, a proper balance between photosynthesis and respiration can only increase net productivity by providing more energy to export the photoassimilates from source leaves to the sink tissues. There is scanty of information regarding the effect of elevated CO_2 on respiration in tea plants. Li et al. (2017) showed that elevated CO_2 (800 μmol mol^{-1}) increases total respiration rate over a 24-day-long study period. This total respiration is attributed to simultaneous increases in SHAM-resistant and CN-resistant respiration in tea leaves. However, a respiratory acclimation response to elevated CO_2 can be observed in tea plants after 18-day exposure to 800 μmol mol^{-1} CO_2.

4 Effects of Elevated CO_2 on Primary and Secondary Metabolites in Tea Plants

Plants produce quite thousands numbers of metabolites that are diverse both structurally and functionally. Among them, only a small portion of metabolites (primary metabolites) are essential for plant growth and development, while the rest (secondary metabolites) are used for multiple other functions such as plant defense against biotic and abiotic stressors (Zhao et al. 2013). Recently, human awareness of the benefits from phytochemicals has increased tremendously based on epidemiological studies (Pérez-López et al. 2018). In many cases, mild stresses that perturb plant metabolism eventually improve the nutraceutical quality of plant-derived food (Han et al. 2016; Li et al. 2016a). Moreover, enhanced secondary metabolism correlates well to the improvement of the nutraceutical quality (Pérez-López et al. 2018). However, such increase often associates a certain level of yield penalty (Han et al. 2016). Elevated CO_2 increases the photosynthetic rate and also the demand of ATP and NADPH. Higher photosynthesis results in increased availability of carbohydrates, which promotes the synthesis of secondary metabolites (Fig. 7.1). With the advancement in the field of metabolomics, novel and complex information is continuingly accumulating in the chemistry of plant secondary metabolism (Zhao et al. 2013). High-throughput Illumina RNA-seq of *Camellia sinensis* discovered novel candidate genes that are related to plant secondary metabolism and specific to tea plants (Shi et al. 2011). In addition, majority of other genes that are associated with natural product pathways, such as biosynthesis of flavonoids, theanine, and caffeine, are important to tea quality. In the following subheadings, we discuss the effect of elevated CO_2 on some key metabolites in tea plants.

4.1 Sugar and Saccharides

Carbohydrates are the primary product of photosynthesis. Since elevated CO_2 dramatically increases CO_2 assimilation rate in tea plants, significant increases in sugar and starch accumulation occur in tea leaves in response to elevated CO_2 condition.

CO_2 enrichment treatments for 6 months with 500 and 750 μmol mol^{-1} CO_2 increased the soluble saccharide contents by 8.4 ~ 14.4% and 18.1 ~ 28.2%, respectively, in spring tea (Jiang et al. 2006). In our own study, relatively short duration (24-day) exposure of tea plants to 800 μmol mol^{-1} CO_2 resulted in a significant increase in soluble sugar, sucrose, and starch contents in Longjing tea leaves.

4.2 Amino Acids

Amino acids are one of the key determinants of tea quality. However, there are discrepancies in research reports regarding the effect of elevated CO_2 on amino acid content in tea leaves. Jiang et al. (2006) found that 500 and 750 μmol mol^{-1} CO_2 treatments for 6 months result in 1.7 ~ 4.5% and 6.7 ~ 12.2% decreased amino acid concentration in spring tea compared to the ambient air CO_2 (350 μmol mol^{-1}) concentration, respectively (Jiang et al. 2006). By contrast, Li et al. (2017) reported an increased amino acid content when tea plants are grown under 800 μmol mol^{-1} CO_2 for 24 days compared with that in plants grown under 400 μmol mol^{-1} CO_2. Moreover, they found relatively increased concentration of aspartic acid, theanine, proline, alanine, and phenylalanine, while a decreased concentration of threonine and serine following exposure of tea plants to elevated CO_2. Meanwhile, the concentrations of glutamic acid, glycine, valine, isoleucine, tyrosine, histidine, lysine, and arginine remained unaltered following CO_2 enrichment treatment. In an ongoing free-air CO_2 enrichment (FACE) study, we noticed that amino acids tended to decline following exposure of tea plants to elevated CO_2 for 3 months (unpublished). This implies that elevated CO_2-induced fluctuations in amino acid biosynthesis are responsive to the duration of CO_2 treatment, possibly attributed to the associated acclimation responses.

4.3 Tea Polyphenol and Flavonoids

In general, high concentrations of flavonoid components relate well with the health benefits of tea but can increase astringency and make the tea bitter (Tounekti et al. 2013). Therefore, an optimal balance between the tea polyphenol and the amino acids is vital for the quality of tea (Han et al. 2016; Li et al. 2016a). Hui et al. (2016) found that elevated CO_2 increased tea polyphenol content but decreased amino acid (including theanine) content, leading to an increased TP to AA ratio (TP/AA) in tea leaves (Hui et al. 2016). Likewise, Jiang et al. (2006) showed that 500 and 750 μmol mol^{-1} CO_2 treatments for 6 months increased tea polyphenol contents by 3.8 ~ 6.0% and 6.9 ~ 11.3% %, respectively, in spring tea (Jiang et al. 2006). We also found increased total polyphenol contents in tea leaves, when grown under elevated CO_2

(800 μmol mol^{-1}) for 24 days. In addition, total catechin content was high in elevated CO_2-grown plants, which was attributed to significant increases in (-)-epigallocatechin (EGC) and (-)-epigallocatechin-3-gallate (EGCG) concentrations. However, individual catechins such as (-)-gallocatechin (GC) and (-)-catechin (C) concentrations were not altered by elevated CO_2 (800 μmol mol^{-1}) treatment for 24 days. This implies that elevated CO_2 induces gallylation of epicatechins leading to an increased accumulation of EGCG and ECG under elevated CO_2 conditions in tea leaves. Given that epicatechins constitute about 90% of total catechins, an enhancement in epicatechin content ultimately increases total catechin content in tea leaves.

Catechin biosynthesis is mediated by several key enzymes involved in phenylpropanoid and flavonoid biosynthesis pathways. The first committed step in the biosynthesis of catechins is deamination of L-phenylalanine to trans-cinnamic acid, catalyzed by the enzyme phenylalanine ammonia-lyase (PAL). The PAL is encoded by *CsPAL* in tea (Rani et al. 2012), and elevated CO_2 treatment (800 μmol mol^{-1} for 24 d) upregulates the transcript level of *CsPAL* in tea leaves (Fig. 7.2a). Furthermore, elevated CO_2 upregulates the transcript levels of *4CL, CsCHS, CsCHI, CsF3H, CsDFR, CsANS, CsUFGT,* and *CsANR* involved in catechin biosynthetic pathway. The LAR functions in the final step of catechin biosynthesis, which catalyzes conversion of leucocyanidins into catechins (C, GC), whereas the ANR catalyzes conversion of anthocyanidins into epicatechins (EC, EGC). Thus elevated CO_2-induced suppression in *CsLAR* expression eventually causes slight reduction in GC concentration in tea leaves under elevated CO_2 conditions. On the other hand, elevated CO_2-inudced upregulation of *CsANR* ultimately increases EGC concentration (Li et al. 2017). Thus it is highly convincing that elevated CO_2 can directly or indirectly promote the transcription of the key genes involved in catechin biosynthetic pathway, leading to an increased accumulation of epicatechins and total catechins in tea leaves.

4.4 Caffeine

Caffeine is an important purine alkaloid (secondary metabolite) occurring in various plant species including *Camellia sinensis* (Filho and Mazzafera 2000; Ashihara et al. 2008). Previous studies have found an important role of caffeine in chemical defense and allelopathy (Ashihara et al. 2008). Our group revealed that decreased caffeine biosynthesis under elevated CO_2 condition increases plant susceptibility to *Colletotrichum gloeosporioides* (Hui et al. 2016; Li et al. 2016b; Li et al. 2017). Although a high caffeine content improves plant defense against pathogens, a high caffeine concentration in green tea is harmful to human health as it causes sleep deprivation, tachycardia, abortion, and miscarriages (Bermejo et al. 2015; Unno et al. 2017). In spring tea, 500 and 750 μmol mol^{-1} CO_2

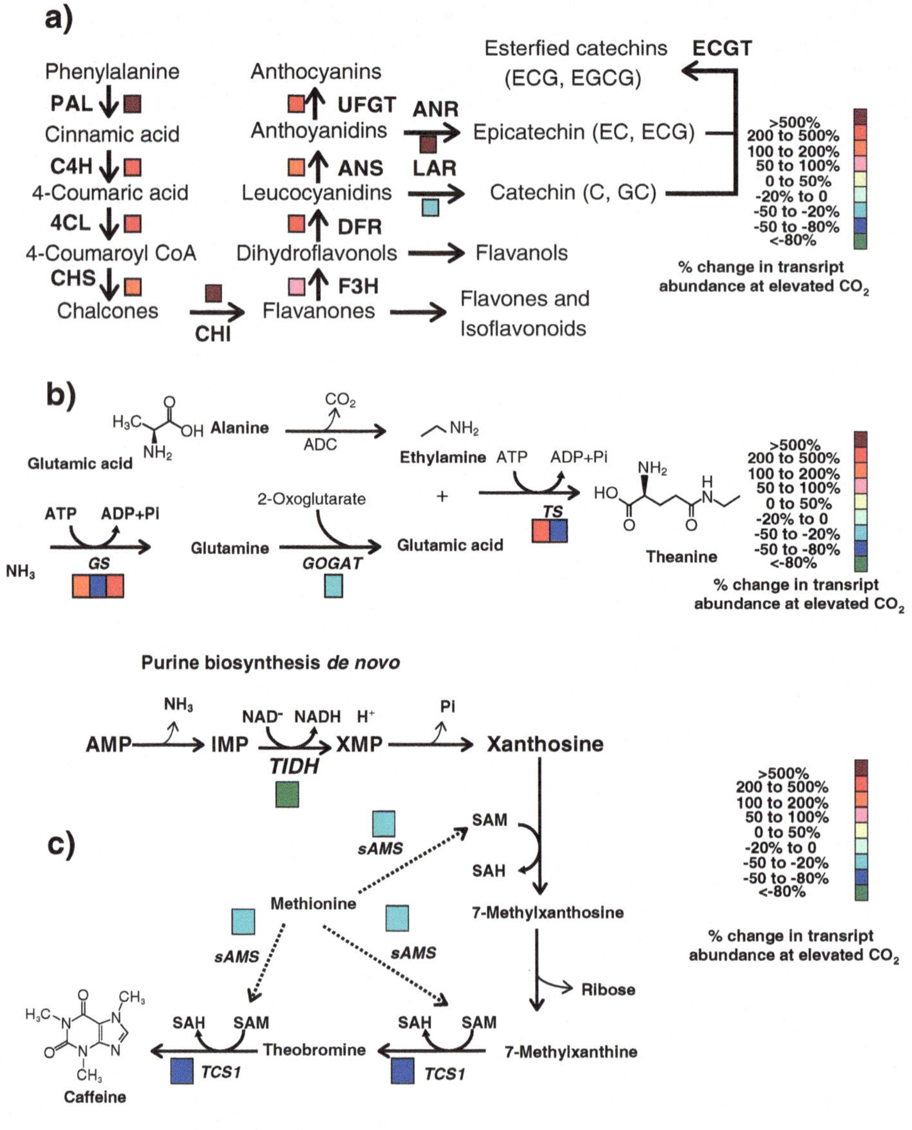

Fig. 7.2 Transcript levels of catechins, theanine, and caffeine synthetic pathway-related genes in tea leaves as influenced by ambient (380 μmol mol^{-1}) or elevated CO_2 concentration (800 μmol mol^{-1}). Changes in transcript levels of catechins (**a**), theanine (**b**), and caffeine (**c**) synthetic pathway-related genes in tea leaves. Leaf samples were harvested at 24 days following exposure of tea seedlings to different atmospheric CO_2 concentrations. Expression levels of genes were analyzed by qPCR using gene-specific primer pairs (Li et al. 2017). *PHENYLALANINE AMMONIA-LYASE (PAL), CINNAMATE 4-HYDROXYLASE (C4H), P-COUMARATE:COA LIGASE (4CL), CHALCONE SYNTHASE (CHS), CHALCONE ISOMERASE (CHI), FLAVANONE 3-HYDROXYLASE (F3H), DIHYDROFLAVONOL 4-REDUCTASE (DFR), ANTHOCYANIDIN SYNTHASE (ANS), UDP- GLUCOSE FLAVONOID 3-O-GLUCOSYL TRANSFERASE (UFGT), ANTHOCYANIDIN REDUCTASE (ANR), LEUACOANTHOCYANIDIN REDUCTASE (LAR), GLUTAMINE SYNTHETASE (GS), GLUTAMINE:2-OXOGLUTARATE AMINOTRANSFERASE (GOGAT), THEANINE SYNTHASE (TS), INOSINE 5'-MONOPHOSPHATE DEHYDROGENASE (TIDH), S-ADENOSYL-L-METHIONINE SYNTHASE (sAMS)*, and *TEA CAFFEINE SYNTHASE 1 (TCS1)*

treatments for 6 months reduced caffeine contents by 3.1 ~ 4.6% and 5.1 ~ 10.7%, respectively, compared to the ambient air CO_2 (350 μmol mol^{-1}) concentration (Jiang et al. 2006). Even a short period of CO_2 enrichment (800 μmol mol^{-1} for 24 days) can cause a significant reduction in caffeine content during summer (Li et al. 2017). *In planta* caffeine synthesis is dependent on several key genes, such as such as *INOSINE 5′-MONOPHOSPHATE DEHYDROGENASE (TIDH)*, *S-ADENOSYL-L-METHIONINE SYNTHASE (sAMS)*, and *TEA CAFFEINE SYNTHASE 1 (TCS1)*. Li et al. (2017) showed that elevated CO_2 treatment could substantially suppress the transcript levels of *TIDH*, *sAMS*, and *TCS1* in tea leaves (Fig. 7.2c), indicating that elevated CO_2-induced downregulations of caffeine biosynthetic genes eventually decrease caffeine content in tea leaves under elevated CO_2 condition.

4.5 Theanine

Theanine is the major tea amino acids accounting for about 50% of total free amino acid in tea leaves (Han et al. 2016). It is a unique nonprotein amino acids that imparts the so-called "umami" taste of tea (Vuong et al. 2011; Cheng et al. 2017). Studies have revealed that L-theanine of tea exhibits multiple positive effects on human health (Kim et al. 2009; Siamwala et al. 2013), such as improvement of memory and learning ability, reduction of blood pressure, mental stress and anxiety, and promotion of relaxation and concentration by inhibiting the negative effects of caffeine (Kim et al. 2009; Unno et al. 2017). To the best of our knowledge, only a few studies assessed the effect of elevated CO_2 on theanine concentration in tea leaves. Jiang et al. (2006) reported that exposure of tea plants to 500 and 750 μmol mol^{-1} CO_2 treatments for 6 months resulted in a decreased theanine concentration in spring tea compared to that in the ambient air CO_2 (350 μmol mol^{-1}) concentration (Jiang et al. 2006). However, Li et al. (2017) found that tea plants grown under 800 μmol mol^{-1} CO_2 for 24 days accumulated an increased theanine content compared with that in plants grown under 400 μmol mol^{-1} CO_2. The discrepancy can be explained with the difference in exposure duration as well as growing season since the later study was conducted in the summer with a short exposure period.

The concentration of theanine is largely dependent on the expression of its key biosynthetic genes. There are 17 genes that encode key enzymes involved in theanine metabolism pathway. However, *THEANINE SYNTHASE* (*CsTS*), *GLUTAMINE SYNTHETASE* (*CsGS*), and *GLUTAMINE: 2-OXOGLUTARATE AMINOTRANSFERASE* (*CsGOGAT*) have positive correlation with the theanine content in tea leaves (Liu et al. 2017). Real-time quantitative polymerase chain reaction (qPCR) data of theanine biosynthetic genes show that elevated CO_2 (800 μmol mol^{-1} for 24 days) upregulates transcript levels of *CsGS* and *CsTS* (Fig. 7.2b), indicating that an increased expression of theanine biosynthetic genes is well in accord with the content of theanine in tea leaves under elevated CO_2.

5 Effects of Elevated CO_2 on Nutrient and Element Contents

5.1 *Elements and Nutritional Values*

One important consequence of climate change includes its effect on nutritional values of plant-based food that directly affect human health (Myers et al. 2014). A large-scale meta-analysis of the nutritional contents showed that elevated CO_2 has profound effects on crop macro- and micronutrient concentrations of important food crops (Dietterich et al. 2015). Earlier studies showed that C_3 grains and legumes contain relatively lower concentrations of zinc and iron under elevated atmospheric CO_2 concentration (Myers et al. 2014). Moreover, C_3 crops (excluding legumes) also have decreased protein concentrations at elevated CO_2 compared with that at ambient CO_2 conditions. In rice, elevated CO_2 concentrations are expected to degrade the grain quality by decreasing grain protein content and increasing grain chalkiness (Zhang et al. 2015a). Due to the enhancement of plant growth under elevated CO_2, CO_2-enriched environments increase plant nutrient demands. While under limited iron supply, elevated CO_2 increases plant capacity to acquire iron from frugally soluble sources, which is mediated partially by nitric oxide-induced regulation of iron transporter genes in tomato (Jin et al. 2009). However, the nutrient acquisition mechanism in tea plants under elevated CO_2 conditions is largely unknown. There are only few studies of the effect of elevated CO_2 on element contents. These studies show that the N, K, Ca, P, and Na contents in tea declined to some extent following exposure of tea plants to 500/750 $\mu mol\ mol^{-1}$ CO_2 for 6 months, whereas the Zn, Mg, Fe, and Cu contents increased following CO_2 enrichment (Jiang et al. 2005; Jiang et al. 2006). Since tea is not consumed as a source of mineral and nutrients, changes in the element contents due to rise in atmospheric CO_2 concentration may not directly affect human health by causing malnutrition. However, elevated CO_2-induced decrease in N, P, and K is a crucial threat to tea plant health and overall tea cultivation. Therefore, the fertilizer management strategies are to be developed to sustain tea cultivation in the future climates.

5.2 *Carbon/Nitrogen Ratio (C:N)*

An optimal C:N ratio is critical for plant growth and productivity. The majority of plants that show a positive relationship between CO_2 enrichment and photosynthesis rate also exhibit an increased ratio of C:N in their leaves (Ainsworth and Rogers 2007; Terashima et al. 2014). According to the CO_2 fertilization hypothesis, elevated CO_2 exerts a positive effect on plant growth by increasing availability of carbon. The enhancement in primary productivity and water use efficiency in response to elevated CO_2 is commonly referred to the direct "fertilization effect"

(Huang et al. 2007). The C:N ratio is increased by elevated CO_2 as a result of increased accumulation of C assimilates and sugars due to enhanced photosynthesis under elevated CO_2. However, N supply plays a crucial role in determining the C:N under elevated CO_2. The reduced N availability under elevated CO_2 is likely to limit the stimulatory effects of high CO_2 on plant growth (Luo et al. 2004; Reich et al. 2006). In a deciduous forest dominated by *Liquidambar styraciflua*, elevated CO_2 increased C:N ratio up to 6 years before the appearance of N shortage in soils (Norby et al. 2010). However, in the next 6 years C:N drastically declined potentially due to limited N availability. Although data relating to such long-term study of the effect of elevated CO_2 on C:N ratio in tea plants are not available, similar effect of CO_2 enrichment on tea cannot be ignored after years of treatment.

The ratio of C:N is not only related to photosynthesis but to kinds and amounts of biomass products (Feng et al. 2014). Despite elevated CO_2-induced increased photosynthesis, specific metabolite synthesis may decrease. For instance, elevated CO_2 decrease N content in tea shoots which is attributed to decreased caffeine content in tea (Li et al. 2017). In fact, elevated CO_2 alters gene expression patterns associated with nitrogen metabolism which is more likely the patterns observed under N deficiency (Li et al. 2008), indicating that elevated CO_2 virtually creates a perturbed nitrogen-limited constraint that favors C metabolism but inhibits N metabolism in plants (Huang et al. 2007). In Table 7.1, changes in different physiological processes, primary/secondary metabolite contents, elements and defense compounds in response to elevated CO_2 levels are summarized.

6 Effect of Elevated CO_2 on Biomass Production and Yield

Elevated CO_2 importunately increases leaf area index, leaf number, and branches under optimal conditions in a range of plant species (Huang et al. 2007; Kimball 2016). When other variables are ample, elevated CO_2 causes about 17% biomass increase in C_3 cereal crops. However, under limited water supply, this increase is about 23% (Kimball 2016). Moreover, studies conducted in chambers and greenhouses have shown that plants grown under elevated CO_2 can accumulate higher biomass, leading to a 33% increased agricultural yield. However, studies using free-air CO_2 enrichment (FACE) showed 15% and 19% yield increment in wheat and rice, respectively, which are lower than that of chamber studies possibly due to the differences in solar and thermal radiation, wind flow, air temperature, and humidity among the systems (Kimball 2016). Elevated CO_2 causes about 19% and 16% grain yield increase in C_3 cereals and grain legumes, respectively, under ample N and water supply. Moreover, elevated CO_2 conditions increase the ratio of root mass to shoot mass under low nutrient supply (Huang et al. 2007). In deciduous sweet gum forest, the annual fine root production was more than double under 550 $\mu mol\ mol^{-1}$ CO_2 compared with that in ambient CO_2 in a closed-canopy FACE experiment (Norby et al. 2004; Huang et al. 2007). The yield of tea is closely associated with the shoot biomass production. In one of our previous studies, we found that elevated

Table 7.1 Effect of CO_2 enrichment on different physiological, ecophysiological, and biochemical parameters in tea leaves

Parameters	CO_2 concentrations ($\mu mol\ mol^{-1}$)	Duration	Methods	Effects	References
Net photosynthetic rate (Pn)	500/750	3 m	Open-top	Increased	Jiang et al. (2005)
	648–658	45 d	Open-top	Increased	Hui et al. (2016)
	800	6 d, 12 d, 18 d, 24 d	Closed	Increased	Li et al. (2017)
Maximum carboxylation rate of RuBisCO (Vc,max)	800	6 d, 12 d, 18 d, 24 d	Closed	Increased	Li et al. (2017)
Maximum rate of RuBP regeneration (Jmax)	800	6 d, 12 d, 18 d, 24 d	Closed	Increased	Li et al. (2017)
Stomatal conductance (Gs)	648–658	45 d	Open-top	Decreased	Hui et al. (2016)
	500/750	3 m		Decreased	Jiang et al. (2005)
Transpiration	648–658	45 d	Open-top	Decreased	Hui et al. (2016)
	500/750	3 m	Open-top	Decreased	Jiang et al. (2005)
Water use efficiency	648–658	45 d	Open-top	Increased	Hui et al. (2016)
	500/750	3 m	Open-top	Increased	Jiang et al. (2005)
Respiration	800	6 d, 12 d, 18 d, 24 d	Closed	Increased	Li et al. (2017)
SHAM-resistant respiration	800	6 d, 12 d, 18 d, 24 d	Closed	Increased	Li et al. (2017)
CN-resistant respiration	800	6 d, 12 d, 18 d, 24 d	Closed	Increased	Li et al. (2017)
Chlorophyll a	648–658	45 d	Open-top	No change	Hui et al. (2016)
	500/750	3 m	Open-top	Increased	Jiang et al. (2005)
Chlorophyll b	648–658	45 d	Open-top	No change	Hui et al. (2016)
	500/750	3 m	Open-top	Increased	Jiang et al. (2005)
Chl a: Chl b	500/750	3 m	Open-top	Decreased	Jiang et al. (2005)
Carotenoids	500/750	3 m	Open-top	Increased	Jiang et al. (2005)

(continued)

Table 7.1 (continued)

Parameters	CO_2 concentrations (μmol mol^{-1})	Duration	Methods	Effects	References
Maximum photochemical efficiency of PSII (Fv/Fm)	648–658	45 d	Open-top	Increased	Hui et al. (2016)
Quantum efficiency of PSII (ΦPSII)	800	6 d, 12 d, 18 d, 24 d	Closed	Increased	Li et al. (2017)
PI_{ABS}	648–658	45 d	Open-top	Increased	Hui et al. (2016)
Soluble sugar/saccharide	500/750	6 m	Open-top	Increased	Jiang et al. (2006)
	800	24 d	Closed	Increased	Li et al. (2017)
Sucrose	800	24 d	Closed	Increased	Li et al. (2017)
Starch	800	24 d	Closed	Increased	Li et al. (2017)
Key quality parameters					
Tea polyphenol	648–658	45 d	Open-top	Increased	Hui et al. (2016)
	500/750	6 m	Open-top	Increased	Jiang et al. (2006)
	800	24 d	Closed	Increased	Li et al. (2017)
Amino acids	500/750	6 m	Open-top	Decreased	Jiang et al. (2006)
	648–658	45 d	Open-top	Decreased	Hui et al. (2016)
	800	24 d	Closed	Decreased	Li et al. (2017)
TP/AA	648–658	45 d	Open-top	Increased	Hui et al. (2016)
Total catechins	800	24 d	Closed	Increased	Li et al. (2017)
(-)-epigallocatechin (EGC)	800	24 d	Closed	Increased	Li et al. (2017)
(-)-epigallocatechin-3-gallate (EGCG)	800	24 d	Closed	Increased	Li et al. (2017)
L-Theanine	800	24 d	Closed	Increased	Li et al. (2017)
Caffeine	500/750	6 m	Open-top	Decreased	Jiang et al. (2006)
	648–658	45 d	Open-top	Decreased	Hui et al. (2016)
	800	24 d	Closed	Decreased	Li et al. (2017)

(continued)

Table 7.1 (continued)

Parameters	CO_2 concentrations ($\mu mol\ mol^{-1}$)	Duration	Methods	Effects	References
Elements					
C:N ratio	800	24 d	Closed	Increased	Li et al. (2017)
Zn, Fe, Mg, Mn and Cu	500/750	6 m	Open-top	Increased	Jiang et al. (2005, 2006)
N, K, Ca, P and Na	500/750	6 m	Open-top	Decreased	Jiang et al. (2005, 2006)
Signaling molecule					
Jasmonic acid	800	24 d	Closed	Decreased	Li et al. (2016b)

CO_2 (800 μmol mol^{-1} for 24 days) could increase tea plant height, dry weight of shoot, and dry weight of root by 13.46%, 24.68, and 67.80%, respectively. Elevated CO_2-induced promotion in both shoot and root biomass accumulation ultimately improves the ratio of root to shoot by 27.66% compared with that in ambient CO_2. A study conducted in Sri Lanka to assess the impact of climate change on tea production showed that elevation of atmospheric CO_2 from 370 ppm to 600 ppm will likely increase the tea yield by approximately 33–37% depending on the altitudinal variation (Wijeratne et al. 2007). However, elevated CO_2-induced increases in tea yield and subsequent positive effects of CO_2 fertilization are abolished by climate warming-induced high temperatures at low elevations (Wijeratne et al. 2007).

7 Effect of Elevated CO_2 on Defense and Stress Tolerance in Tea Plants

The main drivers of climate change such as increasing temperature, declining precipitation, and rising atmospheric CO_2 level not only affect the yields and quality of tea directly but also indirectly by interaction with abiotic and biotic stressors such as insects and disease causing pathogens that eventually impact tea yield and quality (Ahmed et al. 2014; Terashima et al. 2014; AbdElgawad et al. 2016; Sun et al. 2016; Zavala et al. 2017). Previous studies showed that elevated CO_2 can ameliorate abiotic stresses in a range of plant species (Albert et al. 2011; Perez-Lopez et al. 2012; Shanmugam et al. 2013; Martins et al. 2014; Ahammed et al. 2015; Li et al. 2015). For instance, elevated CO_2 increases photosynthesis under heat stress, and the enhancement in photosynthesis under elevated CO_2 is attributed to an increase in the level of proteins involved in photosynthetic light reactions (chlorophyll a–b-binding protein), electron transport carrier molecules (ferredoxin), and carbon assimilation enzymes (aldolase), as well as higher carbon assimilation efficiency and carboxylation enzyme activities of the Benson-Calvin cycle (Li et al. 2015). In tea plants, open-top chamber study showed that elevated CO_2 can alleviate midday high-temperature-induced depression in photosynthesis (Jiang et al. 2005). However, elevated CO_2-induced increases in tea yield and subsequent positive effects of CO_2 fertilization are abolished by climate warming-induced high temperatures at low elevations (Wijeratne et al. 2007). Elevated CO_2 also improves leaf water potential under water stress, which is associated with an increase in chlorophyll and lipid content as well as a degradation of monogalactosyl diacylglycerol (Sgherri et al. 2000). Moreover, tomato plants grown under elevated CO_2 levels show increased salt tolerance, together with decreased leaf transpiration rates and lower sodium concentrations in the xylem sap, vascular tissues, and leaves. It is revealed that elevated CO_2 concentrations improve salt stress tolerance in an apoplastic H_2O_2-dependent manner, which suppresses transpiration to minimize the Na^+ delivery from the roots to the shoots in tomato (Yi et al. 2015). However, to date, in-depth physiological and molecular mechanisms of elevated CO_2-mediated response to abiotic stresses are largely unknown in tea plants and thus demand urgent elucidation to sustain tea production in the face of climate change.

While effects of elevated CO_2 on many crop diseases have extensively been investigated, research on the impact of elevated CO_2 on tea disease and pathogens still remains in its infancy. Despite the conflicting results regarding the effect of elevated CO_2 on plant-pathogen interactions, two key hormones, such as salicylic acid and jasmonic acid, have been implicated in the divergence of plant responses to pathogens under elevated CO_2 conditions (Vary et al. 2015). In general, the biosynthesis and signaling pathways of salicylic acid are stimulated, whereas that of jasmonic acid are inhibited by elevated CO_2 in plants. However, the responses greatly vary depending on the type, nature, and behavior of the pathogens (Noctor and Mhamdi 2017). Moreover, elevated CO_2-induced changes in photorespiration have been shown to affect plant defense in the future climates (Noctor and Mhamdi 2017; Ahammed et al. 2018). Broadly, severity and incidence of biotrophic pathogens such as tobacco mosaic virus and *Pseudomonas syringae* are suppressed in a CO_2-enriched environment which is associated with the increased biosynthesis and signaling of salicylic acid (Zhang et al. 2015b). However, virulence of necrotrophic pathogen *Botrytis cinerea* is increased by elevated CO_2, which is attributed to elevated CO_2-induced inhibition in jasmonic acid biosynthesis in tomato plants, implying that salicylic acid and jasmonic acid potentially function antagonistically on plant defense under elevated CO_2. In one of our previous studies, we found that elevated CO_2 enhanced susceptibility of tea plants to *Colletotrichum gloeosporioides* (Li et al. 2016b). The underlying mechanisms of elevated CO_2-induced suppression of tea defense involve decreased synthesis of caffeine and jasmonic acid via lipoxygenase (LOX) pathway.

8 Conclusions and Future Perspectives

Climate change-associated unusual weather events greatly impact global tea production (Ahmed et al. 2014). While the increases in temperature and CO_2 may favor tea production, the occurrence of frequent drought, heavy rainfall, frosts, pests, and disease abundance remains as a key challenge to sustain tea yield and quality. Under the natural conditions, future climate will affect fundamental plant physiological processes in a way which is difficult to assume from a single climate factor (e.g., elevated CO_2) study. In fact, environmental factors such as temperature, moisture availability, light intensity, light quality (ratio of red to far red light), nutrient availability, and elevated CO_2 potentially interact with each other, and their effects may vary depending on the plant species. Currently, little information is available on the effect of elevated CO_2, especially long duration free-air CO_2 enrichment (FACE), on tea plant physiology and the interaction with other abiotic and biotic stressors. Therefore, more research on the physiological and molecular adaptations in tea plants to elevated CO_2 is required to develop a climate resilient tea system.

Acknowledgments Research in the authors' laboratories is supported by the grants from Key Project of International Science & Technology Cooperation, National Key Research and Development Programme of China (2017YFE0107500), the Zhejiang Provincial Natural Science Foundation of China (Y19C160031), the Innovation Project of the Chinese Academy of Agricultural Sciences (CAAS-ASTIP-2015-TRICAAS), the National Natural Science Foundation of China (31600561), and Henan University of Science and Technology (HAUST) Research Start-up Fund for New Faculty (13480058).

References

AbdElgawad H, Zinta G, Beemster GT, Janssens IA, Asard H (2016) Future climate CO_2 levels mitigate stress impact on plants: increased defense or decreased challenge? Front Plant Sci 7:556

Ahammed GJ, Li X, Yu J, Shi K (2015) NPR1-dependent salicylic acid signaling is not involved in elevated CO_2-induced heat stress tolerance in Arabidopsis thaliana. Plant Signal Behav 10:e1011944

Ahammed GJ, Li X, Zhang G, Zhang H, Shi J, Pan C, Yu J, Shi K (2018) Tomato photorespiratory glycolate-oxidase-derived H2 O2 production contributes to basal defence against Pseudomonas syringae. Plant Cell Environ 41:1126–1138

Ahmed S, Stepp JR, Orians C, Griffin T, Matyas C, Robbat A, Cash S, Xue D, Long C, Unachukwu U, Buckley S, Small D, Kennelly E (2014) Effects of extreme climate events on tea (Camellia sinensis) functional quality validate indigenous farmer knowledge and sensory preferences in tropical China. PloS One 9:e109126

Ahuja I, de Vos RC, Bones AM, Hall RD (2010) Plant molecular stress responses face climate change. Trends Plant Sci 15:664–674

Ainsworth EA, Long SP (2005) What have we learned from 15 years of free-air CO_2 enrichment (FACE)? A meta-analytic review of the responses of photosynthesis, canopy properties and plant production to rising CO_2. New Phytol 165:351–371

Ainsworth EA, Rogers A (2007) The response of photosynthesis and stomatal conductance to rising [CO_2]: mechanisms and environmental interactions. Plant Cell Environ 30:258–270

Albert KR, Mikkelsen TN, Michelsen A, Ro-Poulsen H, van der Linden L (2011) Interactive effects of drought, elevated CO_2 and warming on photosynthetic capacity and photosystem performance in temperate heath plants. J Plant Physiol 168:1550–1561

Amthor JS (1995) Terrestrial higher-plant response to increasing atmospheric [CO_2] in relation to the global carbon cycle. Glob Chang Biol 1:243–274

Amthor JS (1997) Plant respiratory responses to elevated carbon dioxide partial pressure. In: Allen LH Jr, Kirkham MB, Olszyk DM, Whitman CE (eds) Advances in carbon dioxide effects research. American Society of Agronomy, Madison, pp 35–77

Ashihara H, Sano H, Crozier A (2008) Caffeine and related purine alkaloids: biosynthesis, catabolism, function and genetic engineering. Phytochemistry 69:841–856

Asseng S, Ewert F, Martre P, Rötter RP, Lobell DB, Cammarano D, Kimball BA, Ottman MJ, Wall GW, White JW, Reynolds MP, Alderman PD, Prasad PVV, Aggarwal PK, Anothai J, Basso B, Biernath C, Challinor AJ, De Sanctis G, Doltra J, Fereres E, Garcia-Vila M, Gayler S, Hoogenboom G, Hunt LA, Izaurralde RC, Jabloun M, Jones CD, Kersebaum KC, Koehler AK, Müller C, Naresh Kumar S, Nendel C, O'Leary G, Olesen JE, Palosuo T, Priesack E, Eyshi Rezaei E, Ruane AC, Semenov MA, Shcherbak I, Stöckle C, Stratonovitch P, Streck T, Supit I, Tao F, Thorburn PJ, Waha K, Wang E, Wallach D, Wolf J, Zhao Z, Zhu Y (2014) Rising temperatures reduce global wheat production. Nat Clim Chang 5:143

Bermejo DV, Mendiola JA, Ibáñez E, Reglero G, Fornari T (2015) Pressurized liquid extraction of caffeine and catechins from green tea leaves using ethyl lactate, water and ethyl lactate+water mixtures. Food Bioprod Process 96:106–112

Cheng S, Fu X, Wang X, Liao Y, Zeng L, Dong F, Yang Z (2017) Studies on the Biochemical Formation Pathway of the Amino Acid l-Theanine in Tea (Camellia sinensis) and Other Plants. J Agric Food Chem 65:7210–7216

De Costa WAJM, Mohotti AJ, Wijeratne MA (2007) Ecophysiology of tea. Braz J Plant Physiol 19:299–332

Dietterich LH, Zanobetti A, Kloog I, Huybers P, Leakey AD, Bloom AJ, Carlisle E, Fernando N, Fitzgerald G, Hasegawa T, Holbrook NM, Nelson RL, Norton R, Ottman MJ, Raboy V, Sakai H, Sartor KA, Schwartz J, Seneweera S, Usui Y, Yoshinaga S, Myers SS (2015) Impacts of elevated atmospheric CO(2) on nutrient content of important food crops. Scientific Data 2:150036

Feng G-Q, Li Y, Cheng Z-M (2014) Plant molecular and genomic responses to stresses in projected future CO_2 environment. Crit Rev Plant Sci 33:238–249

Field CB, Barros VR, Mach K, Mastrandrea M (2014) Climate change 2014: impacts, adaptation, and vulnerability. Cambridge University Press, New York

Filho O, Mazzafera P (2000) Caffeine does not protect coffee against the leaf miner Perileucoptera coffeella. J Chem Ecol 26:1447–1464

Hajiboland R (2017) Environmental and nutritional requirements for tea cultivation. Folia Hortic 29(2):199–220

Han W-Y, Huang J-G, Li X, Li Z-X, Ahammed GJ, Yan P, Stepp JR (2016) Altitudinal effects on the quality of green tea in east China: a climate change perspective. Eur Food Res Technol 243(2):1–8

Han W, Li X, Yan P, Zhang L, Jalal Ahammed G (2018) Tea cultivation under changing climatic conditions. Global Tea Science. pp 455–472

Huang J-G, Bergeron Y, Denneler B, Berninger F, Tardif J (2007) Response of forest trees to increased atmospheric CO_2. Crit Rev Plant Sci 26:265–283

Hui XU, Lei LI, Qinghui LI, Zhou L, Zhu X, Chen F, Wang Y, Fang W, University, N.A. (2016) Effects of elevated atmospheric CO_2 concentration and temperature on photosynthesis system and quality components in tea plant. J Nanjing Agric Univ 39:550–556

Jiang Y, Zhang Q, Zhang S (2006) Effects of atmospheric CO_2 concentration on tea quality. J Tea Sci 26:299–304

Jiang Y, Zhang S, Zhang Q (2005) Effects of elevated atmospheric CO_2 concentration on photophysiological characteristics of tea plant. J Tea Sci 25:43–48

Jin CW, Du ST, Chen WW, Li GX, Zhang YS, Zheng SJ (2009) Elevated carbon dioxide improves plant iron nutrition through enhancing the iron-deficiency-induced responses under iron-limited conditions in tomato. Plant Physiol 150:272–280

Kim TI, Lee YK, Park SG, Choi IS, Ban JO, Park HK, Nam SY, Yun YW, Han SB, Oh KW, Hong JT (2009) l-Theanine, an amino acid in green tea, attenuates beta-amyloid-induced cognitive dysfunction and neurotoxicity: reduction in oxidative damage and inactivation of ERK/p38 kinase and NF-kappaB pathways. Free Radic Biol Med 47:1601–1610

Kimball BA (2016) Crop responses to elevated CO_2 and interactions with H_2O, N, and temperature. Curr Opin Plant Biol 31:36–43

Larson C (2015) Reading the tea leaves for effects of climate change. Science 348:953–954

Li P, Ainsworth EA, Leakey AD, Ulanov A, Lozovaya V, Ort DR, Bohnert HJ (2008) Arabidopsis transcript and metabolite profiles: ecotype-specific responses to open-air elevated [CO_2]. Plant Cell Environ 31:1673–1687

Li X, Ahammed GJ, Li Z-X, Zhang L, Wei J-P, Shen C, Yan P, Zhang L-P, Han W-Y (2016a) Brassinosteroids improve quality of summer tea (Camellia sinensis L.) by balancing biosynthesis of polyphenols and amino acids. Front Plant Sci 7:1304

Li X, Ahammed GJ, Li Z, Tang M, Yan P, Han W (2016b) Decreased biosynthesis of jasmonic acid via lipoxygenase pathway compromised caffeine-induced resistance to Colletotrichum gloeosporioides under elevated CO_2 in tea seedlings. Phytopathology 106:1270–1277

Li X, Ahammed GJ, Zhang YQ, Zhang GQ, Sun ZH, Zhou J, Zhou YH, Xia XJ, Yu JQ, Shi K (2015) Carbon dioxide enrichment alleviates heat stress by improving cellular redox homeostasis through an ABA-independent process in tomato plants. Plant Biol 17:81–89

Li X, Zhang G, Sun B, Zhang S, Zhang Y, Liao Y, Zhou Y, Xia X, Shi K, Yu J (2013) Stimulated leaf dark respiration in tomato in an elevated carbon dioxide atmosphere. Sci Rep 3:3433

Li X, Zhang L, Ahammed GJ, Li Z-X, Wei J-P, Shen C, Yan P, Zhang L-P, Han W-Y (2017) Stimulation in primary and secondary metabolism by elevated carbon dioxide alters green tea quality in Camellia sinensis L. Sci Rep 7(1):7937

Liu ZW, Wu ZJ, Li H, Wang YX, Zhuang J (2017) L-Theanine content and related gene expression: novel insights into theanine biosynthesis and hydrolysis among different tea plant (Camellia sinensis L.) tissues and cultivars. Front Plant Sci 8:498

Luo Y, Su B, Currie WS, Dukes JS, Finzi A, Hartwig U, Hungate B, McMurtrie RE, Oren R, Parton WJ, Pataki DE, Shaw RM, Zak DR, Field CB (2004) Progressive nitrogen limitation of ecosystem responses to rising atmospheric carbon dioxide. Bioscience 54:731–739

Macfarlane A, Macfarlane I (2004) The empire of tea: the remarkable history of the plant that took over the world. Overlook Press, New York

Mancini E, Beglinger C, Drewe J, Zanchi D, Lang UE, Borgwardt S (2017) Green tea effects on cognition, mood and human brain function: a systematic review. Phytomedicine: Int J Phytother Phytopharmacol 34:26–37

Martins LD, Tomaz MA, Lidon FC, DaMatta FM, Ramalho JC (2014) Combined effects of elevated [CO_2] and high temperature on leaf mineral balance in Coffea spp. plants. Clim Chang 126:365–379

Mohotti AJ (1998) Effect of irradiance and N nutrition on photosynthesis of tea (*Camellia sinensis* (L.) O. Kuntze) in comparison with sunflower (*Helianthus annus* L.). University of Reading

Myers SS, Zanobetti A, Kloog I, Huybers P, Leakey ADB, Bloom AJ, Carlisle E, Dietterich LH, Fitzgerald G, Hasegawa T, Holbrook NM, Nelson RL, Ottman MJ, Raboy V, Sakai H, Sartor KA, Schwartz J, Seneweera S, Tausz M, Usui Y (2014) Increasing CO_2 threatens human nutrition. Nature 510:139–142

Noctor G, Mhamdi A (2017) Climate change, CO_2, and defense: the metabolic, redox, and signaling perspectives. Trends Plant Sci 22(10):857–870

Norby RJ, Ledford J, Reilly CD, Miller NE, O'Neill EG (2004) Fine-root production dominates response of a deciduous forest to atmospheric CO_2 enrichment. Proc Natl Acad Sci U S A 101:9689–9693

Norby RJ, Warren JM, Iversen CM, Medlyn BE, McMurtrie RE (2010) CO_2 enhancement of forest productivity constrained by limited nitrogen availability. Proc Natl Acad Sci 107:19368–19373

Perez-Lopez U, Robredo A, Lacuesta M, Mena-Petite A, Munoz-Rueda A (2012) Elevated CO_2 reduces stomatal and metabolic limitations on photosynthesis caused by salinity in Hordeum vulgare. Photosynth Res 111:269–283

Pérez-López U, Sgherri C, Miranda-Apodaca J, Micaelli F, Lacuesta M, Mena-Petite A, Quartacci MF, Muñoz-Rueda A (2018) Concentration of phenolic compounds is increased in lettuce grown under high light intensity and elevated CO_2. Plant Physiol Biochem 123:233–241

Rani A, Singh K, Ahuja PS, Kumar S (2012) Molecular regulation of catechins biosynthesis in tea [Camellia sinensis (L.) O. Kuntze]. Gene 495:205–210

Reich PB, Hobbie SE, Lee T, Ellsworth DS, West JB, Tilman D, Knops JMH, Naeem S, Trost J (2006) Nitrogen limitation constrains sustainability of ecosystem response to CO_2. Nature 440:922

Sage RF (1994) Acclimation of photosynthesis to increasing atmospheric CO_2: the gas exchange perspective. Photosynth Res 39:351–368
Sgherri CLM, Salvateci P, Menconi M, Raschi A, Navari-Izzo F (2000) Interaction between drought and elevated CO_2 in the response of alfalfa plants to oxidative stress. J Plant Physiol 156:360–366
Shanmugam S, Kjaer KH, Ottosen CO, Rosenqvist E, Kumari Sharma D, Wollenweber B (2013) The alleviating effect of elevated CO_2 on heat stress susceptibility of two wheat (Triticum aestivumL.) cultivars. J Agron Crop Sci 199:340–350
Shi C-Y, Yang H, Wei C-L, Yu O, Zhang Z-Z, Jiang C-J, Sun J, Li Y-Y, Chen Q, Xia T, Wan X-C (2011) Deep sequencing of the Camellia sinensis transcriptome revealed candidate genes for major metabolic pathways of tea-specific compounds. BMC Genomics 12:131
Siamwala JH, Dias PM, Majumder S, Joshi MK, Sinkar VP, Banerjee G, Chatterjee S (2013) L-theanine promotes nitric oxide production in endothelial cells through eNOS phosphorylation. J Nutr Biochem 24:595–605
Sun Y, Guo H, Ge F (2016) Plant-aphid interactions under elevated CO_2: some cues from aphid feeding behavior. Front Plant Sci 7:502
Sun Y, Guo H, Yuan L, Wei J, Zhang W, Ge F (2015) Plant stomatal closure improves aphid feeding under elevated CO_2. Glob Chang Biol 21(7):2739–2748
Terashima I, Yanagisawa S, Sakakibara H (2014) Plant responses to CO_2: background and perspectives. Plant Cell Physiol 55:237–240
Tounekti T, Joubert E, Hernández I, Munné-Bosch S (2013) Improving the polyphenol content of tea. Crit Rev Plant Sci 32:192–215
Unno K, Noda S, Kawasaki Y, Yamada H, Morita A, Iguchi K, Nakamura Y (2017) Reduced stress and improved sleep quality caused by Green Tea are associated with a reduced caffeine content. Nutrients 9(7):E777
Vary Z, Mullins E, McElwain JC, Doohan FM (2015) The severity of wheat diseases increases when plants and pathogens are acclimatized to elevated carbon dioxide. Global Chang Biol 21(7):2661–2669
Vuong QV, Bowyer MC, Roach PD (2011) L-Theanine: properties, synthesis and isolation from tea. J Sci Food Agric 91:1931–1939
Wijeratne MA, Anandacoomaraswamy A, Amarathunga MKSLD, Ratnasiri J, Basnayake BRSB, Kalra N (2007) Assessment of impact of climate change on productivity of tea (Camellia sinensis L.) plantations in Sri Lanka. J Nat Sci Found Sri Lanka 35(2):119–126
Yang SY, Zhang QG, Wei LI, Wang FW (2005) Effect of CO concentration increasing on the photosynthetic rate of *Camellia sinensis*. Chin J Agrometeorol 26(2):90–94
Yi C, Yao K, Cai S, Li H, Zhou J, Xia X, Shi K, Yu J, Foyer CH, Zhou Y (2015) High atmospheric carbon dioxide-dependent alleviation of salt stress is linked to RESPIRATORY BURST OXIDASE 1 (RBOH1)-dependent H_2O_2 production in tomato (Solanum lycopersicum). J Exp Bot 66:7391–7404
Zavala JA, Gog L, Giacometti R (2017) Anthropogenic increase in carbon dioxide modifies plant-insect interactions. Ann Appl Biol 170:68–77
Zavala JA, Nabity PD, DeLucia EH (2013) An emerging understanding of mechanisms governing insect herbivory under elevated CO_2. Annu Rev Entomol 58:79–97
Zhang G, Sakai H, Usui Y, Tokida T, Nakamura H, Zhu C, Fukuoka M, Kobayashi K, Hasegawa T (2015a) Grain growth of different rice cultivars under elevated CO_2 concentrations affects yield and quality. Field Crop Res 179:72–80
Zhang S, Li X, Sun Z, Shao S, Hu L, Ye M, Zhou Y, Xia X, Yu J, Shi K (2015b) Antagonism between phytohormone signalling underlies the variation in disease susceptibility of tomato plants under elevated CO_2. J Exp Bot 66:1951–1963
Zhao N, Wang G, Norris A, Chen X, Chen F (2013) Studying plant secondary metabolism in the age of genomics. Crit Rev Plant Sci 32:369–382

Chapter 8
Tea Plants and Air Pollutants

Lorenzo Cotrozzi, Cristina Nali, Elisa Pellegrini, and Giacomo Lorenzini

Abstract The major tea-growing regions of the world are located in Asia, where tea contributes substantially to their economy. It is known how the rapid development of the economy, twinned to global change, has created in many districts of industrialized countries critical levels of air pollution. Abiotic stresses may affect plant growth, quality, and distribution. This is particularly important for specialty crops such as tea, where functional quality is determined by phytonutrients, secondary metabolites, and bioactive components that play a pivotal role in plant defense and acclimation/adaptation/resilience to environmental stresses. Stress conditions such as drought, heat, light extremes, salinity, and toxic metals in the substrate have been the subject of intense researches, and the sensitivity of tea plants to these constraints has been tested by the scientific community through field and controlled experiments. Tea plants present high leaf surface areas, and exchange with atmosphere is elevated. However, little is known about the way air pollution affects tea responses and how this species is able to counteract this insult. In this chapter, the existing literature reporting the effects of air pollution on the tea plant is reviewed with the aim to examine physiological, biochemical, and molecular responses found in this species. To the best of our knowledge, only the impacts of few air pollutants have been somehow assessed on tea plants, and several responses are still poorly understood. Thus, more research on the impact of air pollution on tea plants is needed. This is of pivotal importance also because commercial tea samples may contain significant quantities of contaminants, which may be transferred to the consumer. No doubt that health national/international bodies should pay more attention to this issue and adopt safe standards of pollution content in the commodities of one of the world's most popular beverages, highly appreciated also by young people because of its pleasant aroma, flavor, and potential positive effect on mood.

L. Cotrozzi · C. Nali · E. Pellegrini (✉) · G. Lorenzini
Department of Agriculture, Food and Environment, University of Pisa, Pisa, Italy
e-mail: lorenzo.cotrozzi@for.unipi.it; cristina.nali@unipi.it; elisa.pellegrini@unipi.it; giacomo.lorenzini@unipi.it

W.-Y. Han et al. (eds.), *Stress Physiology of Tea in the Face of Climate Change*,
https://doi.org/10.1007/978-981-13-2140-5_8

Keywords Air pollutants · *Camellia sinensis* · Fluorides · Oxidative stress · Ozone · Sulfur dioxide

1 Introduction

Atmospheric pollution is one of the most demanding environmental problems currently facing society. The Intergovernmental Panel on Climate Change (IPCC 2013) highlighted that (1) the warming of the climate system is unequivocal (i.e., the last three decades have been successively warmer at the Earth's surface than any preceding decade since 1850), (2) the human influence has been the dominant cause of the observed warming since the mid-twentieth century, and (3) the continued emissions of greenhouse gases will cause further warming and changes in all the components of the climate system. Global change is now a scientific certainty as the effects of biotic and abiotic stresses are already detectable and will be more evident in the next years. This is particularly true in Southeast Asia, which is characterized by severe air pollution associated with a rapid urbanization and rushing economic development (Song et al. 2017). Air pollutants [i.e., tropospheric ozone (O_3), nitrogen oxides (NO_x), fluorides (F), sulfur dioxide (SO_2), carbon dioxide (CO_2), particulate matter, and heavy metals] can induce negative impacts on human health (WHO 2013). In addition, they may also affect ecosystem functioning and agricultural productivity, having adverse effects on growth, development, and longevity of plants.

Abiotic stresses associated with climate change (such as high temperature, drought, UV radiations, and air pollution) may cause considerable losses to agricultural production worldwide reducing average yields of several crops by more than 50% (Upadhyaya and Panda 2013). This is important for specialty crops such as tea [*Camellia sinensis* (L.) O. Kuntze], an evergreen shrub indigenous to Indochina cultivated in more than 50 countries around the world, with a huge economic value (tea is the world's most widely consumed beverage after water; around half of the world's population drinks tea). Indeed, tea functional quality is determined by phytonutrients, secondary metabolites, and bioactive food components that play a pivotal role in plant defense and acclimation/adaptation/resilience to environmental stresses (Zhou et al. 2014). Stress conditions such as drought, heat and light extremes, salinity, and toxic metals in the substrate have been the subject of intense researches, and the sensitivity of tea has been tested by the scientific community through field and controlled experiments (Upadhyaya and Panda 2013). However, little is still known about the way air pollution modulates plant responses and how it affects crop growth, distribution, and quality (Ahmed et al. 2014). Consequently, a priority for research on tea plants will be to elucidate this aspect in order to propose suitable measures to counteract any adverse impacts (Wijeratne et al. 2007).

In this chapter, therefore, the literature reporting the effects of air pollution on the tea plant is reviewed with the aim to examine physiological, biochemical, and molecular responses found in this species.

2 Ozone

Ground-level O_3 is a secondary pollutant generated when primary precursors, mainly nitrogen oxides (NO_x) and volatile organic compounds emitted by sources such as vehicle exhausts and industrial activities, are exposed to solar light ("photochemical smog mechanism"). This kind of environmental problem, first reported in the 1950s in California (Haagen-Smit et al. 1952), has been found in a mess of urban areas throughout the world. O_3 is also the third most harmful greenhouse (climate-altering) gas and is one of the most powerful oxidants known. The mixing ratios of tropospheric O_3 have increased considerably since pre-industrial times (The Royal Society 2008): the rate is approximately 0.5–2% per year over mid-latitudes of the Northern Hemisphere (IPCC 2013). Episodes of high concentrations of surface O_3 over large parts of Europe, North America, and Southeast Asia usually occur during the summer in dry and sunny weather conditions. Hourly averages of ozone concentrations in Europe, Japan, and the United States usually range between 20 and 60 ppb (for O_{-3}, 1 ppb is 1.96 $\mu g\ m^{-3}$ at 20 °C and 101.325 kPa), but peaks exceeding 200 ppb may be sometimes reached around metropolises (Hendriks et al. 2016) or during summer heat wave episodes (Pellegrini et al. 2007; Lorenzini et al. 2014a, b). Elevated concentrations in rural areas are caused by the transport of O_3 itself and its precursors from urban regions over distances of hundreds of kilometers, so to easily involve agricultural and remote areas (Stockwell et al. 1997). Moreover, models show that O_3 is expected to increase 20–25% by 2050 and as much as 40–60% by 2100, in quickly developing countries such as China and India (Wang et al. 2017). No surprise that O_3 has received continuous attention in the last decades from both the scientific and regulatory communities. In fact, O_3 can no longer be considered a mere local air quality issue, but it is a global problem requiring a global solution.

The overall impact of O_3 on natural plant communities and crop productivity (in terms of both quality and quantity) is relevant. Exposure to O_3 can result in leaf injury such as chlorosis and minute necrotic spots, but the appearance and the severity of these lesions differ among species, varieties, and genotypes (Tamaoki 2008). Final impact depends upon the quantity of the pollutant entering the leaf through open stomata and the duration of the exposure. Long-term exposures to medium-low levels of O_3 may result in decline in photosynthesis, growth inhibition, and premature senescence, normally even without/before the onset of/in addition to visible foliar symptoms (Long and Naidu 2002; Gottardini et al. 2014; Pellegrini et al. 2015).

The negative impact of O_3 on vegetation has been well documented in the North America and Europe for their locally important species, whereas very little is known so far about Asia (Zhang et al. 2012). Despite the economic importance of tea and the dramatically increasing concentrations of O_3 in tea-growing countries, only a few species of the Theaceae family have been investigated (Zhang et al. 2012; Li et al. 2016), and very little is known about the effects of O_3 on *C. sinensis*. To the best of our knowledge, only Liu et al. (2009) assessed the responses of derivatives of this species to O_3. According to Upadhyaya and Panda (2013), this is because although O_3 is considered a major abiotic stress that can affect growth and productivity of tea plants, other abiotic stresses such as high temperatures, heavy metals, and drought appear to have a pivotal significance. For this plant, coming back to Liu et al. (2009), the effects of (unrealistic) O_3 treatments (320 ppb for 15 to 45 minutes) on the aroma characteristic of Puer tea derivatives have been investigated: the amount of aroma components (acids and ester compounds) significantly decreased through the O_3 exposure. In addition, new aldehyde and ketone compounds were emitted only in O_3-treated tea derivatives, confirming that O_3 altered the composition of volatile organic compounds produced by fermented tea derivatives that have important functions in protecting plants (and/or derivatives) from abiotic insults, such as oxidative stress (Pellegrini et al. 2012).

3 Fluorides

Fluorides are naturally present in sediments, soils, water, and air, but F is not an essential element for plant life: the leaf background concentration is very low (1–10 ppm dry weight; Weinstein 1977). However, few plants, such as many tropical species included in the Theaceae family, have the unique capacity to accumulate well over 100 ppm of F even when the atmospheric and soil available F concentrations are at background levels (Weinstein and Laurence 1999). Fluorine (univalent poisonous gaseous halogen) is the most electronegative element in the periodic table and the most chemically reactive of elements (García and Borghino 2015).

F is an accumulative toxin in plant leaves and accumulation is gradual over time. F inhibits photosynthesis and many other physiological processes. Moving in the transpiration stream from roots or through stomata, it accumulates in leaf margins, so that peculiar F injury symptoms on broadleaf plants include marginal and tip necrosis that spread inward. Drought stress or salt toxicity can have very similar symptoms, and this makes sometimes complicate the diagnosis.

In the air, F is present in either gaseous or particulate form, and both are released from a large number of natural and anthropic sources, such as volcanoes, aluminum smelters, brick and ceramic factories, and phosphate fertilizer industries (Brougham et al. 2013) (Fig. 8.1). The common gaseous form is hydrogen fluoride (HF) that can be considered one of the most dangerous air pollutants in absolute, in relation to its high toxicity (1–3 orders of magnitude more than other major pollutants). HF in air is absorbed by plants through stomata and passes into the transpiration stream where it is carried to its final site of accumulation in the tips and margins of leaves (Jacobson et al. 1969).

The particulate forms (such as sodium fluoride and cryolite) adhere to the leaf surfaces and may be dissolved (if water soluble), enter leaves through cuticle, and/or are washed off by heavy rainfall (Davison and Blakemore 1976) (Fig. 8.2). For most of the twentieth century, environmental concern about F was focused on inorganic

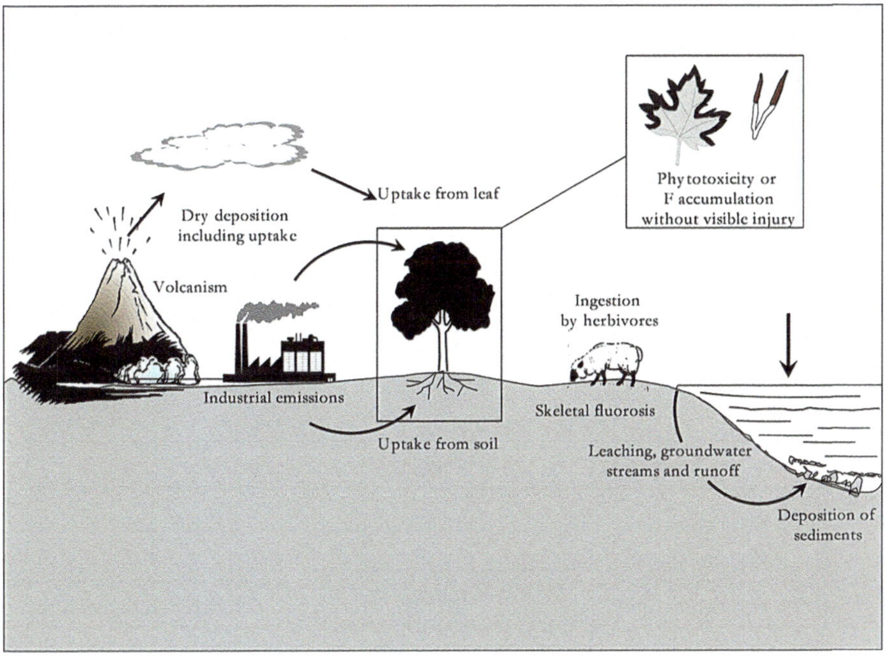

Fig. 8.1 Biogeochemical cycle of fluorides (modified from Weinstein and Davison 2004)

compounds, but there are about 30 naturally occurring and many manufactured organofluorides (e.g., monofluoroacetate) that occur in, or are released into, the environment and are of toxicological relevance (Weinstein and Davison 2004). There is a vast range of tolerance to F among plants. If the accumulation of F exceeds a threshold for a species, marginal and interveinal chloroses are usually the first symptoms observed. In some cases, before the appearance of these visual symptoms, changes in the physiological responses of plants have been observed, such as composition and function of cell membranes, photosynthesis (P_n), respiration, antioxidant enzyme activities, and metabolism of carbohydrates (Fornasiero 2001). Consequently, F can modify or disrupt metabolic processes and cause reductions in growth. Little is known about the physiological and structural effects of F in tolerant plants.

It is well known that the tea plant is a hyperaccumulator of F, being able to accumulate 871–1337 ppm fresh matter (fm) in mature leaves and more than 2000 ppm fm in old ones, even without exhibiting visible symptoms of toxicity (Ruan and Wong 2001; Shu et al. 2003). Several factors can influence the concentration of F in tea leaves, including (1) soil characteristics, (2) tea varieties, (3) environmental conditions, and (4) the maturity of harvested leaves (Fernandez-Caceres et al. 2001; Alvarez-Ayuso et al. 2011). Some reports led to a better understanding of F uptake in tea plants that is linearly correlated to the external F concentrations and readily transported, in particular to the leaves, leading to a considerably higher F concentration in mature or old leaves than in other organs (Ruan et al. 2003).

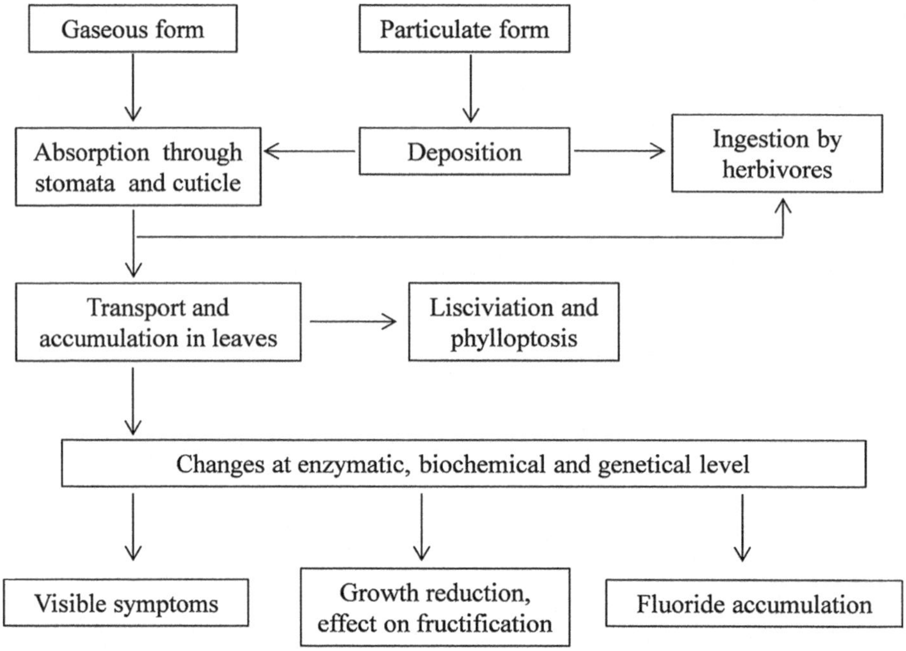

Fig. 8.2 Pathways of gaseous and particulate fluoride fate in the plant (Redrawn and modified from Weinstein and Davison 2004)

The tolerance of hyperaccumulator species can be related to the element forms, compartmentalization within cell vacuoles, and metabolism of reactive oxygen species (ROS) (Cai et al. 2016). The different forms of F and their distribution in organelles in a tea plant have been reported (Ruan and Wong 2001). Gao et al. (2014) found that most of the F was present in the cell walls and the soluble fractions of the leaves and vacuoles contained 98% of F. Oxidative stress triggered by high doses of F has not been studied in detail. Recently, Cai et al. (2016) elucidated the ability of tea plants to overcome F-induced stress by investigating (1) P_n, (2) photosystem II performance, (3) cellular ultrastructure, and (4) antioxidant enzyme activities. One-year-old plants of *C. sinensis* cv. Shu Cha Zao were exposed to six levels (0 to 50 ppm) of aerial F for 30 days. Aerial F concentrations above 5 ppm caused decreases in photosynthetic and chlorophyll fluorescence parameters. In addition, the activity of the antioxidant enzymes significantly decreased with increasing F levels. It was suggested that the enzymatic antioxidant defense system did not sufficiently protect the tissues under severe F stress, according to previous results reported by Li et al. (2011). F treatments increased lipid peroxidation, electrolyte leakage, and hydrogen peroxide content. However, the increase of proline content suggested that under F-induced stress, leaves may adjust the osmotic pressure and scavenge OH via proline accumulation in order to protect plants from oxidative damage.

Although not focused on *C. sinensis*, it is interesting to report that similar physiological responses were found by Yang et al. (2015a) in four other Theaceae tree species (*Camellia oleifera*, *Gordonia axillaris*, *Schima superba*, and *Tutcheria championi*) exposed to high levels of HF (500 ppm in the air, for 12 h) under controlled environmental conditions. The impact was different among species. Cell membranes of all the species were damaged, and a significant decrease in chlorophyll fluorescence parameters was observed. In *C. oleifera*, *G. axillaris*, and *T. championi* leaves, the electrolyte leakage values showed a significant increase, suggesting that an alteration of membrane permeability occurred. By contrast, in *S. superba* the electrolyte leakage values significantly declined. In addition, *C. oleifera* showed a slight reduction of the chlorophyll fluorescence parameters. Consequently, authors concluded that one of the most tolerant species to F was *C. oleifera*. These results were consistent with previous researches carried out on five species of *Camellia* (*C. nitidissima*, *C. hongkongensis*, *C. changii*, *C. oleifera*, and *C. caudata*) grown in a polluted site of the reservoir of Tianzi (China) near ceramic factories (Hu and Yin 2013). Authors estimated some biometric parameters and results showed that *C. nitidissima*, *C. hongkongensis*, *C. changii*, and *C. oleifera* may be considered more tolerant than *C. caudate*.

As stated above, the ability of accumulating F in tea plants is significantly different from variety to variety (Ruan and Wong 2001; Shu et al. 2003). Because of the long-term allogamy, tea plants might have a wide range of genetic backgrounds that affect the absorption and accumulation of F (Lai et al. 2001). Wang et al. (2013) selected two varieties ["Pingyangtezao" (PY) and "Fudingdabai" (FD)] and investigated their physiological response to F in the substrate. F treatment (0.1–0.3 mM of NH_4F added to a Hoagland solution) induced (1) decrease of P_n and ATPase activities, (2) changes in saturated fatty acid composition, and (3) anatomical structure injury. FD had higher net photosynthetic rate, stomatal conductance, water-use efficiency, and antioxidant enzymatic activities than PY. By contrast, higher increases in levels of saturated fatty acids (i.e., palmitic and stearic acids) were observed in PY. FD is more tolerant to F, as shown by a better maintenance of P_n and antioxidant enzyme activity, better control of membrane lipid degradation, and thus more efficient protection of metabolic processes and organic structure.

How F influences the secondary metabolism of plants is still unclear. Catechins are the main components of tea polyphenols and include ungallated and gallated catechins. These compounds are strictly related to the flavor and quality of tea and contribute to its astringency (Liu et al. 2017). The biosynthesis of catechins occurs through the main branch of the phenylpropanoid pathway and is related to the pentose phosphate, shikimate, and flavonoid pathway (Eungwanichayapant and Popluechai 2009). Catechins play a critical role in the normal physiological metabolism of tea plants. In particular, they act as a hydrogen carrier in the respiration chain and actively participate in plant morphology as a substrate for the production of lignin. In addition, they confer physiological functions that enable plants to survive and adapt to environmental stress (i.e., UV protection), and their carbon frame plays a role as a type of energy storage (Yang et al. 2003; Xia and Gao 2009). Recently Yang et al. (2015b) investigated the effects of F treatments [0–16 ppm, as

NH_4F added to a Hoagland solution] on the content of catechins and the activity of the major enzymes involved in the catechin biosynthetic pathway using two tea cultivars ["Fuyunliuhao" (FY) and "Wuniuzao" (WNZ)]. High aerial F concentrations ($\geq$ 8 ppm) induced (1) significant decreases of total polyphenol, catechin, and lignin contents, (2) inhibition of enzymatic activities (i.e., phenylalanine ammonia-lyase, cinnamate 4-hydroxylase, chalcone isomerase, and dihydroflavonol 4-reductase), and (3) growth reduction of both varieties. However, FY had a higher lignin content than WNZ. Consequently, FY may have poorer ability to compound lignin or chelate F, which can explain its higher sensitivity to F.

But the real issue when dealing with tea plants and F pollution is not the plant itself: it is the presence of F in the food chain when tea drinks invariably contain considerable levels of F. This is regarded as an essential micronutrient in animals, needed to prevent dental decay and promote healthy bone growth, although excess F can lead to a pathological condition called "fluorosis". Toxic F exposure from tea is not just a theoretical concern: the infusions may contain 24 to 83% of the total F contents of the original tea leaves (Fung et al. 1999). Fluorosis causes teeth discoloration and bone pain and stiffness; so (heavy) tea drinkers may be receiving F doses that put them at risk for skeletal fluorosis. Fluorosis usually only occurs in area where natural levels of F in drinking water are relevant (Hallanger Johnson et al. 2007). The older the leaves are, the more F they may contain: when it comes to harvesting, younger tea leaves are used to make higher-quality premium tea. Older, F-rich leaves are chosen to make the cheaper economy brands of tea (lower grade), such as brick tea, which therefore may contain high amounts of F (Lu et al. 2004). The F levels of tea bushes collected from five plantations from the Chinese Province of Guangdong were tested (Fung et al. 1999): brick tea released an extremely high F content (7.34 ppm), in comparison with black tea (1.89 ppm) or green tea (1.60 ppm). The cumulative F contents detected in the tea liquor obtained by repeated infusion steps were slightly higher than those prepared by continuous extraction. US experts recommend 4 mg of F daily for adults, with an "upper tolerable daily intake" of 10 mg. So, an adult consuming 1 liter of economy tea daily, containing 6 mg per liter of F, would be getting 75–120% of the recommended F allowance (Chan et al. 2013). However, only a few case reports document F toxicity from tea (Whyte et al. 2008; Izuora et al. 2011), but it is certain that some heavy tea drinkers suffering from fluorosis from tea drinking might not be correctly diagnosed (Yi and Cao 2008). So, health benefits of tea risk to be contrasted with potential detrimental effects from F extracted from leaves if plants grow in a polluted environment.

4 Sulfur Dioxide

SO_2, an integral part of the sulfur (S) biogeochemical cycle, is since long regarded as a pollutant of environmental significance. The largest source of SO_2 in ambient air is the burning of fossil fuels by thermal power plants and many other industrial facilities. Minor sources include ships and all engines/heavy equipment that burn (low-quality) fuels with a high S content, metal refinement from ores, and natural sources, such as volcanoes and wild forest fires (Fig. 8.3). SO_2 and other sulfur oxides contribute to rain acidification which can harm sensitive forest ecosystems (Bell and Treshow 2002). There has been a shift in the regional distribution of SO_2 emissions: the percentage contribution to global emissions from the United States, Canada, and Europe has declined substantially over recent decades, while the contribution from Asia has increased. Currently, the centrally planned Asia region, dominated by China, is believed to be the largest contributor to global SO_2 emissions (Smith and Pitcher 2001). So, many tea growers may be deeply involved.

Both the visual symptoms and biochemical effects of SO_2 on plants are well documented: at the pH values encountered within the chloroplast, SO_2 is mainly converted to sulfite, which is known to inhibit CO_2 assimilation, Rubisco activity, photophosphorylation, and the operation of the triose phosphate translocator (Bell and Treshow 2002).

The phytotoxicity of SO_2 depends on its concentration and on the duration of exposure and is influenced by the S status in plants (Long and Hullan 2012). Tea leaves contain many S compounds that play important roles for the quality of green tea. Suzuki et al. (2013) evaluated the effects of high SO_2 concentrations (range 0.25–10 ppm, for 1 to 24 h) on the chemical compositions of tea shoots (Experiment 1) and leaves (Experiment 2). In Experiment 1, the contents of organic acids in tea cells significantly are decreased by SO_2 exposure at 1 ppm. This suggests that cellular membranes were broken and intracellular components were leaked. In

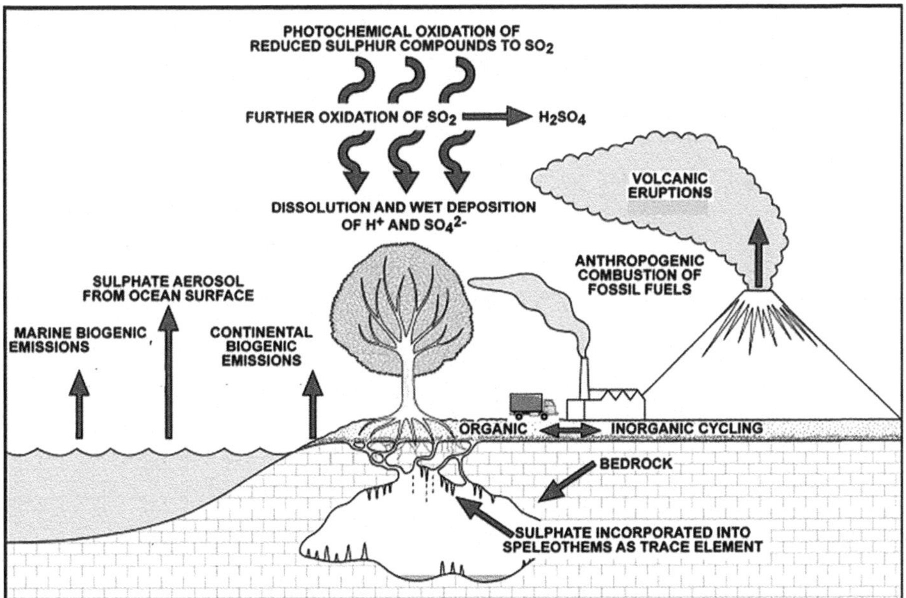

Fig. 8.3 Pathways of sulfur movement and transfer in the environment (Modified from Wynn et al. 2008)

Experiment 2, leaves exhibited foliar symptoms at the end of the treatment, and the photosynthetic process was significantly altered. However, the content of S-containing metabolites (such as alanine and γ-amino butyric acid) tended to increase in SO_2-treated young leaves, confirming that they absorbed and assimilated this gas as a S nutrient source. It is known that the resistant capability to SO_2 is significantly different from species to species. Hu and Yin (2013) reported that five species of *Camellia* grown in a highly SO_2-contaminated site showed a different degree of tolerance/resistance in terms of key plant growth parameters.

5 Carbon Dioxide

CO_2 is a basic part of the carbon (C) natural cycle and is the single most important greenhouse gas. CO_2 is derived from the combustion of fossil fuels. Sources of CO_2 include volcanic eruptions, wild fires and organic matter combustion, and respiratory process of living organisms (Abas and Khan 2014) (Fig. 8.4). In the last 200 years, global atmospheric CO_2 concentration has risen from the 280 ppm of the pre-industrial age to the present 400 (and even more) ppm, and some models predict a further rise (may be up to 700 ppm) within this century (IPCC 2013).

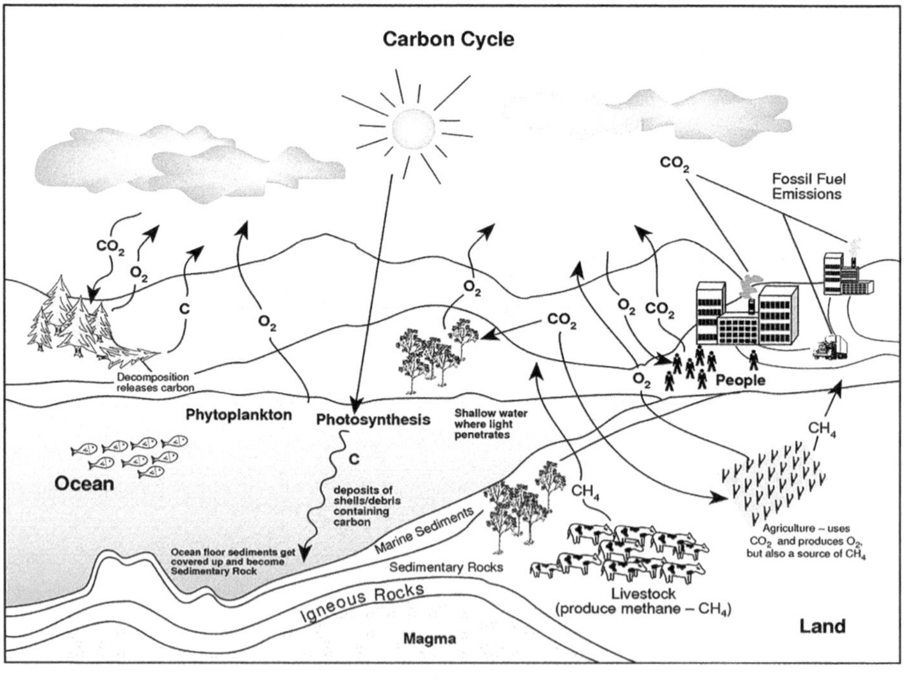

Fig. 8.4 Pathways of carbon movement and transfer in the environment (Modified from IPCC 2013)

Rising atmospheric CO_2 concentration is the primary factor enhancing land C uptake and influencing greater land-air C exchange. It may have a particular long-term impact on the terrestrial C balance at both the regional and global scales (Drake et al. 1997). Consequently, biogeochemical cycling and atmospheric budget of CO_2 play a key role in global climate change (Liu et al. 2014). No doubt that rising atmospheric CO_2 is a major concern for crop growth and production in the face of climate change (Porter and Semenov 2005; Lobell et al. 2011).

There has been general agreement on the beneficial effects of elevated (at least up to certain levels) CO_2 concentration on crop yield, probably due to increased P_n, C:N ratio, and water-use efficiency (Ainsworth and Long 2005). For an important economic crop such as tea, response of photosynthetic activity to variations of the atmospheric CO_2 is important not only to determine spatial and temporal variation of leaf P_n but also to determine how the productivity would respond to long-term climate change (De Costa et al. 2007). Smith et al. (1993) observed a positive and linear correlation between instantaneous P_n values of tea leaves and CO_2 concentration (351–490 ppm). De Costa et al. (2007) showed that the maximum values of photosynthetic process in young leaves of tea can be raised temporarily up to 40–60 μmol m^{-2} s^{-1} by artificial extreme (and unrealistic) CO_2 enrichment (1500 ppm). Similarly, maximum P_n around 30 μmol m^{-2} s^{-1} was reached around 1000–1200 ppm in mature field-grown tea plants (Mohotti 1998), confirming the beneficial effects of elevated CO_2 concentration on the photosynthetic process. However, a substantial genotypic variation in the response to increased CO_2 concentrations was observed in tea plants. Anandacoomaraswamy et al. (cited by Mohotti 1998) showed that while the whole plant dry weight total and the root dry weights of seedlings of *C. sinensis* "TRI 3019" increased significantly at elevated CO_2 concentrations (over a 3-month period of CO_2 enrichment), those of another genotype (TRI 3072) did not show a significant response under the same conditions. As reported before, such genotypic variation has been also shown for other air pollutants.

Strong relationships between environmental factors and tea yield are reported. For example, the temperature extremes affect quality yield of tea leaves and restrict their cultivation (Liu et al. 2016). However these studies do not include the direct influence of ambient CO_2 on crop yield and functional quality (Wijeratne and Fordham 1996; Ahmed et al. 2014). Wijeratne et al. (2007) investigated the effect of CO_2 air enrichment on the yield of tea plants under field conditions. After 18 months of treatment, the rise in CO_2 concentration from ambient level (370 ppm) to 600 ppm increased the yield of 33–37%. The concomitant increase of P_n and transpiration rates confirms the CO_2 "fertilization effect." Recently, Xin et al. (2016) evaluated the impact of elevated CO_2 concentration (800 ppm, 2 weeks) on endogenous caffeine content in tea seedlings. Caffeine content was significantly decreased in leaves under elevated CO_2 (−23% in comparison with ambient air). This suggests that elevated CO_2 concentrations not only have a direct effect on the photosynthetic process and on plant growth (as reported before) but also induce the reallocation of secondary metabolites that might contribute to the synthesis of signaling molecules (such as salicylic and jasmonic acids) altering plant defense.

6 Conclusive Remarks

Tea is cultivated in almost 40 countries over the world, but production is heavily concentrated in just a handful of regions, mainly located in Asia, where contributes substantially to their economy. It is known how the rapid development of the economy, twinned to global change, has created in many districts of these countries critical levels of air pollution. Tea plant roots and leaves absorb noxious chemical species from the soil and the air. This has an effect in reducing yield (quantity) and altering the distinctive flavor of tea liquor (quality). However, to the best of our knowledge, only the impacts of few air pollutants have been somehow assessed on tea plants, and several responses are still poorly understood. Thus, more research on the molecular and physiological adaptations in tea plants to air pollution with special regard to O_3 is needed.

Furthermore, this chapter highlights that commercial tea samples may contain significant quantities of contaminants, which may be transferred to the consumer. Many people drink tea solely for its healthy effects: they do not wonder that sometimes their cup could be as dangerous as a glass of dirty water. Translocation of chemicals from leaves to tea liquor depends on many variables, such as the cultivar, the age of leaves [as a general rule, older, more mature leaves have had more time for deposition/accumulation (Kavanagh and Renehan 1998) and may contain up to 20 times more F than younger tea leaves], and the way of home treatment (e.g., infusion vs decoction, time length of extraction; single infusion vs repeated infusion; Fung et al. 1999). Safety of tea consumers may also be challenged by other chemical residues not covered in the present review, such as pesticides, mycotoxins, polycyclic aromatic hydrocarbons, aluminum, lead, and other heavy metals (Chen and Wan 1997; Lin et al. 2006; Malir et al. 2014). No doubt that health national/international bodies should pay more attention to this issue and adopt safe standards of pollution content in the commodities of the world's most popular beverage, highly appreciated also by young people. A tight program of monitoring campaigns and analyses should be provided as well.

References

Abas N, Khan N (2014) Carbon conundrum, climate change, CO_2 capture and consumptions. J CO_2 Util 8:39–48

Ahmed S, Stepp JR, Orians C et al (2014) Effects of extreme climate events on tea (*Camellia sinensis*) functional quality validate indigenous farmer knowledge and sensory preferences in tropical China. PLoS One 9:e109126

Ainsworth EA, Long SP (2005) What have we learned from 15 years of free-air CO_2 enrichment (FACE)? A meta-analytic review of the responses of photosynthesis, canopy properties and plant production to rising CO_2. New Phytol 165:351–371

Alvarez-Ayuso E, Gimenes A, Ballesteros JC (2011) Fluoride accumulation by plants grown in acid soils amended with flue gas desulphurisation gypsum. J Hazard Mater 192:1659–1666

Bell JNB, Treshow M (2002) Air pollution and plant life. Wiley, London

Brougham KM, Roberts SR, Davison AW et al (2013) The impact of aluminium smelter shut-down on the concentration of fluoride in vegetation and soils. Environ Pollut 178:89–96

Cai H, Zhu X, Peng C et al (2016) Critical factors determining fluoride concentration in tea leaves produced from Anhui province, China. Ecotoxicol Environ Saf 131:14–21

Chan L, Mehra A, Saikat S et al (2013) Human exposure assessment of fluoride from tea (*Camellia sinensis* L.): a UK based issue? Food Res Int 51:564–570

Chen ZM, Wan HB (1997) Degradation of pesticides on plant surfaces and its prediction – a case study on tea plant. Environ Monit Assess 44:303–313

Davison A, Blakemore J (1976) Factors determining fluoride accumulation in forage. In: Mansfield T (ed) Effects of air pollution on plants. Cambridge University Press, Cambridge, pp 17–30

De Costa WAJM, Mohotti AJ, Wijeratne MA (2007) Ecophysiology of tea. Braz J Plant Physiol 19:299–332

Drake BG, Gonzalez-Meler MA, Long SP (1997) More efficient plants: a consequence of rising atmospheric CO_2? Annu Rev Plant Physiol Plant Mol Biol 48:609–639

Eungwanichayapant P, Popluechai S (2009) Accumulation of catechins in tea in relation to accumulation of mRNA genes involved in catechin biosynthesis. Plant Physiol Biochem 47:94–97

Fernandez-Caceres PL, Martín MJ, Pablos F et al (2001) Differentiation of tea (*Camellia sinensis*) varieties and their geographical origin according to their metal content. J Agric Food Chem 49:4775–4779

Fornasiero RB (2001) Phytotoxic effects of fluorides. Plant Sci 161:979–985

Fung KF, Zhang ZQ, Wong JWC et al (1999) Fluoride contents in tea and soil from tea plantations and the release of fluoride into tea liquor during infusion. Environ Pollut 104:197–205

Gao HJ, Zhao Q, Zhang XC et al (2014) Localization of fluoride and aluminum in subcellular fractions of tea leaves and roots. J Agric Food Chem 62:2313–2319

García MG, Borghino L (2015) Fluoride in the context of the environment. In: Preedy WR (ed) Fluorine: chemistry, analysis, function and effects. Royal Society of Chemistry, Cambridge, pp 3–21

Gottardini E, Cristofori A, Cristofolini F et al (2014) Chlorophyll-related indicators are linked to visible ozone symptoms: evidence from a field study on native *Viburnum lantana* L. plants in northern Italy. Ecol Indic 39:65–74

Haagen-Smit AJ, Darley EF, Zaitlin M et al (1952) Investigation on injury to plants from air pollution in the Los Angeles area. Plant Physiol 27:18–34

Hallanger Johnson JE, Kearns AE, Doran PM et al (2007) F-related bone disease associated with habitual tea consumption. Mayo Clin Proc 82:719–724

Hendriks C, Forsell N, Kiesewetter G et al (2016) Ozone concentrations and damage for realistic future European climate and air quality scenarios. Atmos Environ 144:208–219

Hu X-C, Yin A (2013) Growth responses of five *Camellia* species to air pollution from sulfur dioxide and fluoride. Int Camellia J 45:94–97

IPCC (2013) Climate change 2013: fifth assessment report of the intergovernmental panel on climate change. IPCC, Geneva

Izuora K, Twombly JG, Whitford GM et al (2011) Skeletal fluorosis from brewed tea. J Clin Endocrinol Metab 96:2318–2324

Jacobson JJ, Weinstein LH, McCune DC et al (1969) The accumulation of fluorine by plants. J Air Pollut Control Assoc 16:412–417

Kavanagh D, Renehan JJ (1998) Fluoride in tea - its dental significance: a review. Ir Dent Assoc 44:100–105

Lai JA, Yang WC, Hsia JY (2001) An assessment of genetic relationships in cultivated tea clones and native wild tea in Taiwan using RAPD and ISSR markers. Bot Bull Acad Sin 42:93–100

Li C, Zheng Y, Zhou J et al (2011) Changes of leaf antioxidant system, photosynthesis and ultrastructure in tea plant under the stress of fluorine. Biol Plantarum 55:563–566

Li P, Calatayud V, Gao F et al (2016) Differences in ozone sensitivity among woody species are related to leaf morphology and antioxidant levels. Tree Physiol 36:1105–1116

Lin D, Zhu L, He W et al (2006) Tea plant uptake and translocation of polycyclic aromatic hydrocarbons from water and around air. J Agric Food Chem 54:3658–3662
Liu T-X, Wang J, Y-n C (2009) Effects of ozone treatments on the aromatic characteristic of puer tea. Modern Food Sci Tech 25:944–948
Liu L, Zhou LX, Zhang XC et al (2014) Background variations of atmospheric CO_2 and carbon-stable isotopes at Waliguan and Shangdlanzi stations in China. J Geophys Res Atmos 119:5602–5612
Liu ZW, Wu ZJ, Li XH et al (2016) Identification, classification, and expression profiles of heat shock transcription factors in tea plant (*Camellia sinensis*) under temperature stress. Gene 576:52–59
Liu G-F, Han Z-X, Feng L et al (2017) Metabolic flux redirection and transcriptomic reprogramming in the albino tea cultivar 'Yu-Jin-Xiang' with emphasis on catechin production. Sci Rep 7:45062. https://doi.org/10.1038/srep45062
Lobell DB, Schlenker W, Costa-Roberts J (2011) Climate trends and global crop production since 1980. Science 333:616–620
Long L, Hullan Y (2012) Effect of sulfur dioxide on ROS production, gene expression and antioxidant enzyme activity in *Arabidopsis* plants. J Plant Physiol 58:46–53
Long SP, Naidu SL (2002) Effects of oxidants at the biochemical, cell and physiological levels, with particular reference to ozone. In: Bell JNB, Treshow M (eds) Air pollution and plant life. John Wiley and Sons, Chichester, pp 69–88
Lorenzini G, Nali C, Pellegrini E (2014a) Summer heat waves, agriculture, forestry and related issues: an introduction. Agrochimica 58(Special Issue):3–19
Lorenzini G, Pellegrini E, Campanella A et al (2014b) It's not just the heat and the drought: the role of ozone air pollution in the 2012 heat wave. Agrochimica 58(Special Issue):40–52
Lu Y, Guo WF, Yang XQ (2004) Fluoride content in tea and its relationship with tea quality. J Agric Food Chem 52:4472–4476
Malir F, Ostry V, Pfohl-Leszkowicz A et al (2014) Transfer of ochratoxin A into tea and coffee beverages. Toxins 6:3438–3453
Mohotti AJ (1998) Effect of irradiance and N nutrition on photosynthesis of tea (*Camellia sinensis* (L.) O. Kuntze) in comparison with sunflower (*Helianthus annuus* L.). PhD dissertation, University of Reading
Pellegrini E, Lorenzini G, Nali C (2007) The 2003 European heat wave: which role for ozone? Some data from Tuscany, Central Italy. Water Air Soil Pollut 181:401–408
Pellegrini E, Cioni PL, Francini A et al (2012) Volatiles emission patterns in poplar clones varying in response to ozone. J Chem Ecol 38:924–932
Pellegrini E, Campanella A, Lorenzini G et al (2015) Ecophysiological and antioxidant of *Salvia officinalis* under ozone stress. Environ Sci Pollut Res 22:13083–13093
Porter JR, Semenov MA (2005) Crop responses to climatic variations. Phil Trans R Soc B 360:2021–2035
Ruan J, Wong MH (2001) Accumulation of fluoride and aluminium related to different varieties of tea plant. Environ Geochem Health 23:53–63
Ruan J, Ma LF, Shi YZ et al (2003) Uptake of fluoride by tea plants (*Camellia sinensis* L.) and the impact of aluminium. J Sci Food Agric 83:1342–1348
Shu WS, Zhang ZQ, Lan CY et al (2003) Fluoride and aluminium concentrations of tea plants and tea products from Sichuan Province, PR China. Chemosphere 52:1474–1482
Smith RI, Harvey FJ, Cannell MGR (1993) Effects of light, temperature, irrigation and fertilizer on photosynthetic rate in tea (*Camellia sinensis*). Exp Agric 29:291–306
Smith SJH, Pitcher TML (2001) Global and regional anthropogenic sulfur dioxide emissions. Glob Planet Change 29:99–119
Song C, Wu L, Xie Y et al (2017) Air pollution in China: status and spatiotemporal variations. Environ Pollut 227:334–347

Stockwell WR, Kramm G, Scheel H-E et al (1997) Ozone formation, destruction and exposure in Europe and the United States. In: Sandermann H, Wellburn AR, Heath RL (eds) Ecological Studies, Vol. 127 – Forest decline and ozone. Springer-Verlag, Berlin, pp 1–38

Suzuki K, Tanaka Y, Kamimura R et al. (2013) The effects of various gas treatments on chemical compositions of tea (*Camellia sinensis* L.). In: abstracts 5th international conference on O-CHA(tea) culture and science, Shizuoka

Tamaoki M (2008) The role of phytohormone signaling in ozone-induced cell death in plants. Plant Signal Behav 3:166–174

The Royal Society (2008) Ground-level ozone in the 21st century: future, trends, impact and policy implications: royal society policy document 15/08, London, pp 1–132

Upadhyaya H, Panda SK (2013) Abiotic stress responses in tea (*Camellia sinensis* L. Kuntze): an overview. Rev Agri Sci 1:1–10

Wang LX, Tang JH, Xiao B et al (2013) Variation of photosynthesis, fatty acid composition, ATPase and acid phosphatase activities, and anatomical structure of two tea (*Camellia sinensis* L. Kuntze) cultivars in response to fluoride. Sci World J 2013:1–9

Wang T, Xue L, Brimblecombe P et al (2017) Ozone pollution in China: a review of concentrations, meteorological influences, chemical precursors, and effects. Sci Tot Environ 575:1582–1596

Weinstein LH (1977) Fluoride and plant life. J Occup Med 19:49–78

Weinstein LH, Davison AW (2004) Fluorides in the environment. CABI Publishing, Cambridge

Weinstein LH, Laurence JA (1999) Indigenous and cultivated plants as bioindicators. In: Biologic markers of air-pollution stress and damage in forests. National Academy Press, Washington, DC, pp 194–204

WHO (2013) Review of evidence in health aspects of air pollution - REVIHAAP project. Technical Report, Bonn

Whyte MP, Totty WG, Lim VT et al (2008) Skeletal fluorosis from instant tea. J Bone Miner Res 23:759–769

Wijeratne MA, Fordham R (1996) Effect of environmental factors on growth and yield of tea (*Camellia sinensis* L.) in the low country wet zone of Sri Lanka. Sri Lanka J Tea Sci 64:21–34

Wijeratne MA, Anandacoomarasqamy A, Amarathunga MK et al (2007) Assessment of impact of climate change on productivity of tea (*Camellia sinensis* L.) plantations in Sri Lanka. J Natn Sci Foundation Sri Lanka 35:119–126

Wynn PM, Fairchild IJ, Frisia S et al (2008) Isotopic archives of sulphate in speleothems. Geo Cosmo Acta 72:2465–2477

Xia T, Gao L (2009) Advances in biosynthesis pathways and regulation of flavonoids and catechins. Sci Agric Sinica 42:2899–2908

Xin L, Golam JA, Zhixin L et al (2016) Decreased biosynthesis of jasmonic acid via lipoxygenase pathway compromised caffeine-induced resistance to *Colletotrichum gloeosporioides* under elevated CO_2 in tea seedlings. Phytopathology 106:1270–1277

Yang X, Yu Z, Zhang B et al (2003) The physiological function of tea polyphenols in tea tree. Tea polyphenols chemistry. Shanghai Science and Technology Press, Shanghai, pp 66–71

Yang B, Sun T, Chen H et al (2015a) Effects of hydrogen fluoride-stress on physiological characteristics of theaceae tree seedlings. In: proceed 2nd International Conference on education, management and computing technology. Atlantis Press, Amsterdam

Yang X, Yu Z, Zhang B et al (2015b) Effect of fluoride on the biosynthesis of catechins in tea [*Camellia sinensis* (L.) O. Kuntze] leaves. Sci Hortic 184:78–84

Yi J, Cao J (2008) Tea and fluorosis. J Fluor Chem 129:73–138

Zhang W, Feng Z, Wang X et al (2012) Responses of native broadleaved woody species to elevated ozone in subtropical China. Environ Pollut 163:149–157

Zhou L, Xu H, Mischke S et al (2014) Exogenous abscisic acid significantly affects proteome in tea plant (*Camellia sinensis*) exposed to drought stress. Hortic Res 1:14029

Chapter 9
Nutrient Deficiency and Abundance in Tea Plants: Metabolism to Productivity

Roghieh Hajiboland

Abstract Soil fertility and nutrient availability is the most important component of the environment that determines plants' productivity. In agricultural ecosystems, due to continuous removal of nutrients from soil through harvesting, fertilizer application is necessary for maintenance of yield. Fertilizer application is a common management practice in tea production. However, heavy application of fertilizers does not always increase the yield and quality of tea products but instead can cause problems such as contamination of surrounding water systems, acidification of soil and emission of high levels of N_2O, one of the major greenhouse gasses. Thus, a balanced fertilization may not only reduce production cost but also help to protect the environment. In this chapter effect of each nutrient on plant growth and metabolism is shortly explained by an emphasis on the nutritional physiology of plants, its availability in soil and effect of fertilizers. Then the effect of deficiency or excess of each nutrient and its application on tea are discussed.

Keywords *Camellia sinensis* · Beneficial elements · Fertilizers · Macronutrients · Micronutrients · Tea quality

1 Introduction

Plants' productivity is highly dependent on the availability of nutrients in soil. In tea, continuous plucking of leaves removes the nutrients from soil, thus making it necessary to replenish the nutrients to plant (Tabu et al. 2015). Fertilizer application is a common management practice in tea production and has a significant effect on the yield and quality of the final product (Qiu et al. 2014; Venkatesan and Ganapathy 2004). The influence of an applied fertilizer on tea yield arises from its effects on the shoot extension rate and the rate of regeneration (Mokaya 2016). In addition, fertilizer application significantly influences green and black tea quality (Ruan et al. 2007b).

R. Hajiboland (✉)
Plant Science Department, University of Tabriz, Tabriz, Iran
e-mail: ehsan@tabrizu.ac.ir

W.-Y. Han et al. (eds.), *Stress Physiology of Tea in the Face of Climate Change*,
https://doi.org/10.1007/978-981-13-2140-5_9

1.1 Essential Nutrients and Their Role in Plants

A 'plant nutrient' is a chemical element that is essential for vegetative and reproductive growth of plants. An essential element (i) must be required for a plant to complete its life cycle; thus, the element has a function in plant metabolism and plays a direct role in plant growth and reproduction. In the absence of the essential element or with severe deficiency, the plant will die before it completes the cycle from seed to seed; (ii) no other element substitutes fully for the function of this element; a partial substitution might be possible; and (iii) all plants require the element and the essentiality is universal amongst plants. With short supply of the nutrient, abnormal growth or symptoms of deficiency will develop as a result of the disrupted metabolism; thus, plant may be able to complete its life cycle with restricted growth and abnormal appearance (Mengel and Kirkby 2001).

1.2 Macronutrients, Micronutrients and Beneficial Elements

Seventeen elements are considered to have met the criteria for designation as plant nutrients. Carbon, hydrogen and oxygen are derived from air or water. The other 14 elements are obtained from soil or nutrient solutions. Elements that are required in considerable quantities, i.e. 'macronutrients', generally accumulate to 1 mg g^{-1} dry weight and upward in plant tissues, e.g. nitrogen, phosphorus, sulphur, calcium, magnesium and potassium. Most of the elements required in small quantities in plants, i.e. 'micronutrients', generally accumulate to amounts less than 100 μg g^{-1} dry weight of plant tissues, e.g. iron, zinc, manganese, copper, molybdenum, nickel, boron and chlorine.

Elements can affect plant growth without being required for the plant to complete its life cycle. Elements that stimulate growth, but are not essential, or are essential only for certain plant species or under specific conditions, are termed beneficial elements. This definition applies in particular to sodium, silicon, selenium, cobalt and aluminium.

2 Nitrogen

2.1 Nitrogen in Plant and Soil

After carbon, nitrogen (N) is the element required in largest amounts by plants: about 1–5% (10–50 mg g^{-1} dry weight) of total plant dry matter consists of N, which is an integral constituent of proteins, nucleic acids, chlorophyll, coenzymes, phytohormones and secondary metabolites (Barker and Bryson 2007; Hawkesford et al. 2012). The availability of N to roots is a decisive factor for plant growth. The major sources of N taken up by the roots of higher plants are nitrate (NO_3^-) and ammonium (NH_4^+). Whether ammonium or nitrate as sole source of N supply is better for growth and yield formation of plants depends on many factors. Generally, plants adapted to soils which are acid or have a low redox potential (e.g. wetlands) have a preference for ammonium. In contrast, plants adapted to calcareous, high pH soils utilize nitrate preferentially. However, highest growth rates and plant yields are obtained by combined supply of both ammonium and nitrate (Hawkesford et al. 2012).

As ammonium or nitrate comprises about 80% of the total cations and anions taken up by plants, the form of N has a strong impact on the uptake of other cations and anions, on cellular pH regulation and on rhizosphere pH. Rhizosphere chemistry can be affected by the form of N taken up; ammonium supply may reduce rhizosphere pH through a net excretion of protons, whereas nitrate supply may increase rhizosphere pH through a net uptake of protons from the rhizosphere (Hinsinger et al. 2003).

Nitrogen-deficient plants are typically stunted, with narrow leaves. Chlorosis caused by N deficiency typically begins in the older leaves as N is remobilized to younger leaves. At the field scale, N-deficient crops appear pale green or even yellow (Hawkesford et al. 2012).

Nitrogen is usually the most deficient nutrient in cultivated soils of the world, and fertilization of these soils with N is required. Approximately 100 million tons of N fertilizers were applied globally in 2008 (FAO 2008). Nitrate is generally present in higher concentrations (1–5 mM) than ammonium (20–200 μM) in the solution of agricultural soils. Nitrate is more mobile in the soil than ammonium and therefore more available to plants and more susceptible to leaching. In unfertilized agricultural soils, ammonium can be present in higher concentrations than nitrate, and amino acids provide an additional source of N. Ammonium and amino acids are also the dominating plant-available N forms in acid forest soils (Miller and Cramer 2004; Rennenberg et al. 2009).

Anhydrous ammonia, aqua ammonia, urea (46% N), ammonium nitrate, ammonium sulphate, ammonium phosphates and many other N-containing fertilizers include double-salt mixtures such as ammonium nitrate sulphate (30% N), ammonium phosphate nitrate (25% N), urea-ammonium phosphate (25–34% N), nitric phosphate and ammoniated superphosphate (8% N) that are used as N fertilizers. Organic fertilizers are derived from plant and animal sources and are proteinaceous materials. Costs of handling, shipping and spreading of the bulky, low-analysis organic materials have led to their decline in usage with time. Also, many of the proteinaceous by-products of food processing have higher value as feeds for poultry and livestock than as fertilizers. Nevertheless, demand for organic fertilizers remains, as organic farmers require these products in the maintenance of soil fertility on their cropland (Barker and Bryson 2007).

Generally, only 40–50% of the applied N fertilizer is utilized by the crop. The N which is lost from the plant-soil system can result in environmental problems, including water and air pollution. Nutrient use efficiency of a crop is the yield obtained per unit of available nutrient in soil (Hirel et al. 2011). Nitrogen use efficiency (NUE) is the ratio between the total biomass of output, e.g. crop yield, and the N input, e.g. N supplied in fertilizers and/or residual N present in the soil. Nitrogen use efficiency is divided into two components: N uptake efficiency, i.e. the ability of the plant to remove N from the soil, and the utilization efficiency, i.e. the ability to use N to produce biomass or grain yield. In crops, large amounts of N fertilizer are required to attain maximum yield, and NUE is estimated to be less than 50% (Zhu 2000; Hawkesford et al. 2012).

In addition to soil and water pollution, incomplete capture and poor conversion of fertilizer-derived N causes global warming through emissions of nitrous oxide (N_2O). As a consequence, plant breeding aiming at development of new crop genotypes with better NUE has a high priority. Plant breeding for better NUE is focused on the different physiological processes which affect N uptake from the soil, N translocation, N assimilation and N redistribution. Identification of genotypes which grow and yield well under low N conditions is particularly needed for a successful outcome (Hirel et al. 2007; Masclaux-Daubresse et al. 2010).

2.2 *Nitrogen Nutrition in Tea*

Nitrogen nutrition enhances yield in tea under favourable conditions of temperature, relative humidity, rainfall and evaporation without any adverse effects of large amounts of N supply (Owuor 1997). Favourable yields ranging from 5800 to 6400 kg 'made tea' ha^{-1} per year in high yielding clones have been obtained with increasing N supply of up to 600 kg ha^{-1} per year. Green tea quality is closely correlated with the total N and free amino acid concentrations, and the incredible response of the tea plant to heavy application of N fertilizers is stimulated by frequent shoot harvest (Ruan and Wu 2004; Cheruiyot et al. 2009).

Under field conditions, N deficiency in tea leads to shortening of the internodes and lighter than normal green colour in young flush. The younger leaves may end up being yellow in colour, while the lower mature leaves might remain dark green, and they progressively turn yellow as the deficiency increases. Nitrogen deficiency may be also due to lower ability of feeder roots to take up N from soil, for example, during the drought season as the roots die out or during the cold season as their growth rate is greatly reduced (TRFK 2002; Hajiboland 2017).

In tea grown in sand culture and supplied with nutrient solution, leaves at early stage of deficiency show a generalized yellowing, while under extreme N-deficient conditions, leaves show more advanced symptoms. In these leaves, the brown areas of dead tissue commence as a large number of small dying patches which expand into larger areas. The leaf generally drops at this stage (Pethiyadoga and Krishnapillai 1970). The symptom of pinkish-yellow flush is observed only in some tea clones (Mulder and Visser 1960). The powers of recovery of tea plants with even acute N deficiency symptoms are striking. Within 2 weeks of return to a solution containing

N, the green colour of leaves is restored and vigorous growth commences (Pethiyadoga and Krishnapillai 1970).

Nitrogen is deficient in the tea plant if N leaf content is less than 30 mg g^{-1} dry weight, mildly deficient when it is between 30 and 35 mg g^{-1} dry weight and sufficiently supplied when it is more than 35 mg g^{-1} dry weight (Owuor 1997). Photochemical efficiency and CO_2 fixation rate of tea seedlings grown in nutrient solutions increased with increasing N supply up to 105 ppm N. The response of leaf photosynthesis rate is observed immediately after applying the N treatments (De Costa et al. 2007).

Tea plant is well-adapted to NH_4^+-rich environments and thus prefers NH_4^+ to NO_3^-(Ruan et al. 2000). Although it may be expected that uptake of NH_4^+ by tea plants may decline, whereas that of NO_3^- is unaffected or even increased at slightly acid pH levels (Hinsinger et al. 2003), it has been observed that absorption of NH_4^+ is 2- to 3.4-fold faster than NO_3^- irrespective the soil pH (Ruan et al. 2007b). The preference in tea for NH_4^+, and its poor utilization capacity for NO_3^-, may reflect its ecological position as a typical representative of climax vegetation species that is adapted to forest soils enriched with NH_4^+ as the predominant inorganic N form (Kronzucker et al. 2003). Interestingly, low root-zone pH reduces root growth only in NO_3^--fed plants; thus, NO_3^--supplied plants display higher sensitivity to low root-zone pH (Ruan et al. 2007b). In addition, NO_3^- absorption is inhibited by NH_4^+, an effect that has been attributed to repressive action on NO_3^- influx, which is likely accompanied by the downregulation of transporters for NO_3^- by NH_4^+ and/or their downstream metabolites at transcriptional and post-transcriptional levels (Glass et al. 2002). By contrast, by use of scanning ion-selective electrode technique, it has been recently shown that tea roots have a preference for NH_4^+, and presence of NO_3^- had a negative effect on NH_4^+ influx, while NH_4^+ had a positive effect on NO_3^-influx (Ruan et al. 2016).

Tea plants supplied with NO_3^- show yellowish leaves resembling N deficiency and grow much slower than those receiving NH_4^+ or NH_4^+-NO_3^-. The visual N deficiency-like symptoms and lower concentrations of free amino acids are also supporting indicators of inadequate N status of these plants and suggest that the reduced total N concentration in NO_3^--supplied plants is related to the low NO_3^- absorption rate (Ruan et al. 2007b). Tea plant is able to increase root glutamine synthetase activity substantially under conditions of high demand due to NH_4^+ supply providing essential capacity to assimilate the majority of NH_4^+ in the roots in order to avoid any excessive accumulation of lethal concentrations (Magalhaes and Huber 1989). In contrast, the glutamine synthetase activity of NO_3^--fed plants is downregulated due to N deficiency resulting from declining absorption. In addition, NH_4^+-treated plants develop a larger total leaf area than NO_3^--supplied plants, indicating a stronger source capacity, and improve carbohydrate status in the roots (Ruan et al. 2007b).

Nevertheless, uptake and utilization of NO_3^- by tea plants could not likely be neglected. While nitrification in strongly acidic tea soils fertilized with NH_4^+ is expected to be low (Wickramasinghe et al. 1985), substantial nitrification and a large pool of NO_3^- in tea soils of very low pH have been detected (Hayatsu and

Kosuge 1993). Increase in nitrate reductase activity was noted due to the increase in N or K fertilizer doses (De Costa et al. 2007). Interestingly, there is a significant correlation between green leaf yield and nitrate reductase activity determined at first and second harvesting rounds after fertilizer application (Venkatesan and Ganapathy 2004). Surprisingly, result of an experiment on tea plants supplied with nutrient solution in sand culture medium which provides a constant ratio of NH_4^+:NO_3^- with minimum nitrification for 14 months showed that a mixture of ammonium and nitrate is preferred by tea plants; the best performance resulted when 60–80% of the N is available in the form of nitrate (Pethiyadoga and Krishnapillai 1977).

In tea plantations, especially those destined for green tea production, large amounts of N fertilizers are applied in the form of NH_4 or urea (Ruan and Wu 2004). Foliar application of urea has been also practised in tea plantations (Ruan and Gerendás 2015). It has been observed that urea is rapidly absorbed by tea leaves and exported mainly to the young shoots, representing the major sink. Insufficient supply of K and S negatively affects the absorption and utilization of foliar urea fertilization by weakening sink N strength imposed by plant growth due to nutrient deficiency (Ruan and Gerendás 2015).

3 Phosphorus

3.1 Phosphorous in Plant and Soil

Phosphorus (P) is utilized in the fully oxidized and hydrated form as orthophosphate. Plants typically absorb either $H_2PO_4^-$ or HPO_4^{2-}, depending on the pH of the growing medium (Sanchez 2007). Phosphate is a structural component of phospholipids, nucleic acids, nucleotides, coenzymes and phosphoproteins. Phosphate in plant either remains as inorganic phosphate (Pi) or it is esterified through a hydroxyl group to a carbon chain (C-O-P) as a simple phosphate ester (e.g. sugar phosphate) or attached to another phosphate by the energy-rich pyrophosphate bond (P)~(P) (e.g. in ATP). Another type of phosphate bond is the relatively stable diester (C-P-C). In this association phosphate forms a bridging group connecting units to more complex or macromolecular structures (Hawkesford et al. 2012).

Total P in plant tissue ranges from about 1–10 mg g^{-1} dry weight. Phosphorus deficiency suppresses or delays growth and maturity. Plant stems or leaves are sometimes dark green, often developing red and purple colours. Sustained P deficiency will produce smaller-sized fruit and limited harvestable vegetable mass. Because P is mobile in plants, it is translocated readily from old to young leaves as deficiency occurs, and chlorosis and necrosis on older leaves are sometimes observed (Sanchez 2007; Hawkesford et al. 2012).

There is considerable variability in P efficiency amongst species of crops. Mechanisms of P-utilization efficiency have been classified into three broad classes including (i) secretion or exudation of chemical compounds into the rhizosphere,

(ii) variation in the geometry or architecture of the root system and (iii) association with microorganisms (Kochian et al. 2004).

Soils contain organic and inorganic P compounds. Organic P is much more mobile in soils than inorganic sources. Phospholipids and nucleic acids that enter the soil are degraded rapidly by soil microorganisms. Soils contain a wide range of microorganisms capable of releasing inorganic orthophosphate from organic phosphates catalysed by the enzymes collectively called phosphatases. Conditions that favour the activities of these organisms, such as warm temperatures and near-neutral pH values, also favour mineralization of organic P in soils (Sanchez 2007).

In almost all soils, P is often the most limiting mineral nutrient. Its bioavailability is low due to binding to soil mineral surfaces and fixation into organic forms; available phosphate (Pi) in the soil solution is commonly 1–2 μM (Sanchez 2007). Inorganic P entering the soil solution, by mineralization or fertilizer additions, is rapidly converted into less available forms through sorption and precipitation reactions. Phosphorus released to the soil solution from the mineralization of organic matter might be taken up by the microbial population, taken up by growing plants, transferred to the soil inorganic pool or less likely lost by leaching and runoff. The availability of P to plants is highest when there is a moderate pH of about 5.5–7 and becomes decreasingly available at pH below 5.5 or above 7 (Sanchez 2007). In very acidic soils, it combines with hydroxides of iron and aluminium to form compounds that are unavailable to plants (Fig. 9.1). Lack of mobility of the phosphate ion makes phosphate uptake dependent on root development (Neumann and Römheld 2012).

Most cropping systems show the best response to water-soluble P fertilizers. In contrast to N, a limited amount of fertilizer P can be absorbed by plant foliage, and only a modest portion of the crop's total P requirement can be met by foliar application (Sanchez 2007).

3.2 Phosphorous Nutrition in Tea

In tea P plays a major role in the formation of new wood and roots. It has been stated that P deficiency in tea appeared as the absence of brightness particularly in the mature leaves. There is also excessive dieback of young and old woody stem under P deficiency (TRFK 2002; Hajiboland 2017).

Phosphorus deficiency decreased photosynthetic electron transport capacity by impairing the whole electron transport chain from the PSII donor side up to the PSI, thus decreasing ATP content which limits ribulose bisphosphate (RuBP) regeneration and, hence, the rate of CO_2 assimilation. Energy dissipation is enhanced to protect P-deficient leaves from photo-oxidative damage in high light (Lin et al. 2009). Phosphorous deficiency influences green tea quality. It has been observed that the sensory and biochemical qualities of green tea were lowered by P deficiency (Lin et al. 2012). Phosphorous-deficient green tea displayed decreased concentrations of polyphenols, flavonoids, total free amino acids, theanine and asparaginic acid and glutamic acid and increased concentrations of water-soluble sugars, valine,

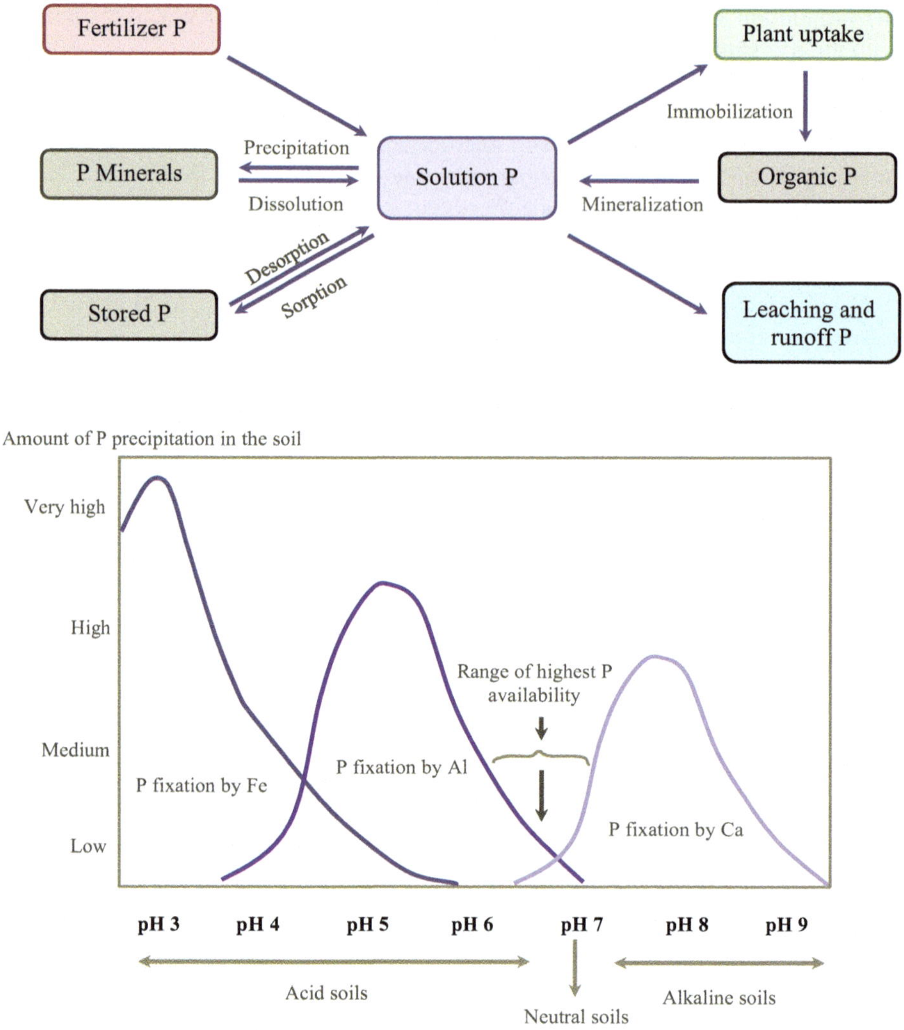

Fig. 9.1 Phosphorus cycle in agricultural soils (above) and P fixation as influenced by soil pH (below)

γ-aminobutyric acid, proline and cysteine, and ratio of total polyphenols to total free amino acids but unchanged concentrations of total catechins and epigallocatechin gallate (Lin et al. 2012).

Considering low pH and an expected low P availability in soil of tea plantations, several workers in different tea-growing countries have attempted to determine the optimum requirements of phosphate fertilizer. The majority of these works reported relatively low yield increase or even no response to P fertilizer in tea (Konishi et al. 1985; Ruan et al. 2000). Addition of P fertilizers at higher levels has been recommended by some authors to provide an excess over that lost by precipitation, and it has been stated that yield responses can occur only to heavy applications (Konishi et al. 1985). In a sand culture experiment with tea plants that were supplied with nutrient solution and grown for 14 months, omission of P has failed to produce any characteristic symptoms. Plants grown without P showed no adverse effects, and overall growth was as good as in the controls (Pethiyadoga and Krishnapillai 1970). In a greenhouse study on young tea plants grown from seed, it has been observed that the P requirement for an optimum growth is as low as 50 μM in the hydroponic medium that is much less than P requirement for optimum growth for many other crop species (Salehi and Hajiboland 2008). In the absence of any P in the nutrient solution made with double-distilled water, plants' height, leaf number and surface were reduced, but no leaf symptoms could be observed after 3-month growth (Salehi and Hajiboland 2008).

Mechanisms for the high P efficiency in tea plants have been investigated by some authors. A high internal use efficiency has been attributed to a high re-translocation of P from mature to the young growing leaves (Hajiboland and Salehi 2014). Considering that P-deficient plants may use PPi-dependent glycolytic reactions and PPi-dependent vacuolar H^+-transport preferentially, thus conserving limited pools of ATP, it has been observed that in P-deficient tea plants, the levels of expression of the genes for ATP-phosphofructokinase (PFK) and V-ATPase subunit A decreased, whereas the levels of expression of the genes for PPi-PFK and V-PPiase increased (Lin et al. 2010).This may contribute to the high internal utilization of P in tea plants under P starvation. Induction of bypass enzymes involved in tissue P economy was also observed for NADP-malic enzyme (ME), NAD-ME and pyruvate kinase (PK) that improved the ability to use P efficiently in the leaves (Lin et al. 2011).

As a mechanism for high external use efficiency, tea plant is also capable for release of malate and citrate into rhizosphere. Ten-month-old tea plants supplied for 17 weeks with 0, 40, 80, 160, 400 or 1000 μM P were studied for the effects of P supply on root citrate and malate release, the concentrations of malate and citrate and the activities of acid-metabolizing enzymes in roots. Root accumulation and release of malate and citrate was induced by 0 and 40 μM P accompanied by increased activities of phospho*enol*pyruvate carboxylase (PEPC), phospho*enol*pyruvate phosphatase (PEPP), citrate synthase (CS) and NAD-malic enzyme (NAD-ME) and decreased activities of pyruvate kinase (PK), NADP-ME and NADP-isocitrate dehydrogenase in roots (Lin et al. 2011). Collectively, these data indicated that tea plants use both external and internal mechanisms for increasing P efficiency and to lower P requirement for growth.

Association of roots with arbuscular mycorrhizal fungi (AMF) is a strategy for improvement of roots' ability for P capture in soils poor in available P. Reports on colonization of tea plants by AMF are very rare and may be questioned. Mridha (1999) stated that a large number of AMF spores from six genera were available in the soil of tea plantations in Bangladesh. However, occurrence of AMF spores in soil does not necessarily mean AMF association with tea plants (Aliasgharzad et al. 2001). Singh et al. (2008) studied the effect of soil-based AMF consortia on growth of tea plants and reported AMF colonization in the roots and growth promotion of the tea plants. The rhizosphere of tea plants has a very specific chemistry because of very low pH and phenolic-rich root exudates. In addition, microbial groups growing around tea roots show antagonistic behaviour and reduce populations of other microorganisms in the rhizosphere (Pandey and Palni 1996). These conditions may also adversely affect capability of AMF to colonize roots of tea plants. Tea belongs to the family Theaceae in the order Ericales. From about 27 families belonging to Ericales, 10 families consist of species without any type of mycorrhizal association (Wang and Qiu 2006). Records of mycorrhizal types in other families of Ericales are dominated by arbuscular mycorrhiza (at least 14 families), while ericoid mycorrhiza is found only in Ericaceae and Diapensiceae (Wang and Qiu 2006). The family Theaceae consists of two AMF species including *Camellia japonica* L. (Mejstrik 1974) and *Pyrenaria cheliensis* (Muthukumar et al. 2003) and no ericoid mycorrhizal species. Results of soil and plant analysis for AMF association in tea genotypes and different plantations in north Iran revealed that AMF spores are abundant in the rhizosphere soil of tea and accompanying (weed) species. Three species including two species of *Rhizophagus* (*R. fasciculatus* and *R. intraradices*) and one species of *Archaeospora* were identified in soil samples around roots or in the bulk soil. However, both microscopic observations of stained roots and the PLFA (phospholipid fatty acid) and NLFA (neutral lipid fatty acid) analyses as signature molecules showed no occurrence of AMF association in tea roots. Of the five weed plant species accompanying tea plants in these gardens, four species were mycorrhizal in different extent. This suggests that soils of these gardens have infective propagules of AMF. In pot culture experiment, tea plants inoculated with AMF (*Rhizophagus intraradices*) showed also no mycorrhizal colonization (Aliasgharzad et al. 2011).

4 Sulphur

4.1 Sulphur in Plant and Soil

The most important source of sulphur (S) is sulphate (SO_4^{2-}) that is taken up by the roots and then transported in the xylem and phloem. Sulphur is a constituent of the amino acids cysteine and methionine and hence of proteins. Both amino acids are precursors of other S-containing compounds (Haneklaus et al. 2007).

Sulphur is a structural constituent in the coenzymes and secondary plant products (e.g. R1-C-S-C-R2) or acts as a functional group (e.g. R-SH) directly involved in metabolic reactions. About 2% of the organically reduced S in plants is present in the water-soluble thiol (-SH) fraction, and under normal conditions, the tripeptide glutathione accounts for more than 90% of this fraction. Sulphur requirement for optimal growth varies between 1 and 5 mg g^{-1} dry weight of plants (Hawkesford et al. 2012).

Interruption of S supply decreases root hydraulic conductivity, stomatal aperture and net photosynthesis. The reduced leaf area in S-deficient plants is the result of both smaller size and particularly the number of leaf cells. Unlike N, S deficiency-induced leaf chlorosis is more uniformly distributed between old and new leaves and is affected by the N supply. Sulphur deficiency symptoms may occur either in young (in combination with sufficient N) or in old (in combination with low N) leaves, indicating that the extent of remobilization and re-translocation from older leaves depends on the rate of N deficiency-induced leaf senescence (Hawkesford et al. 2012).

Worldwide, S deficiency in crop production is quite common in rural areas, particularly in high rainfall areas, for example, in the humid tropics and temperate climates and in highly leached soils. Under these conditions, the application of N fertilizers is ineffective unless S is applied simultaneously (Malhi et al. 2005).

4.2 Sulphur Nutrition in Tea

Sulphur deficiency in tea generally results in reduction of plant growth rate with greater effect on shoots. In a sand culture experiment, symptoms of S deficiency appeared within 3–4 months of commencement of treatment (Pethiyadoga and Krishnapillai 1970). Plants deprived of S show a general yellowing of the leaves and a decrease of leaf size; the younger leaves are usually affected before older ones. The young shoots assume a prominent yellow colour and constitute an easily recognizable feature of S deficiency. If the deficiency remains uncorrected for a long time, the yellowed leaves finally show necrosis of tips and margins with subsequent defoliation. Sulphur deficiency is easily corrected by resupply of this element (Pethiyadoga and Krishnapillai 1970).

As a consequence of reduced S inputs through continuous application of urea instead of ammonium sulphate to tea soils, S deficiency is becoming a major factor that limits yield and reduces quality of crops under intensive tea cultivation (Ananthacumaraswamy et al. 2003). The plant parts above pruning cut need 1.0 kg S for every 100 kg crop, and out of 1.0 kg S assimilated 0.5 kg is recycled, while 0.5 kg S is removed from the field. Therefore, in order to produce 2000 kg of marketable tea ha^{-1} $year^{-1}$, 10 kg of S is consequently removed. However, the plants need three to four times more S in order to compensate for leaching losses and reduction of S to sulphides during monsoon period (Gunaratne et al. 2008).

Sulphur acts not only as a nutrient in tea plants but also as an acaricide and fungicide. Sulphur-based compounds of different formulations were found effective in controlling mite pests in tea. Thus, it has been suggested that S fertilizers can be incorporated in integrated nutrient management as well as integrated pest management programmes in tea cultivation (Ahmed et al. 2011).

Free amino acids in tea leaves are important chemical constituents that considerably influence the quality of tea, especially that of green tea (Ruan et al. 1999). A comparison between K_2CO_3 (or KCl) vs K_2SO_4 and MgO vs $MgSO_4$ revealed significantly higher amino acid contents in tea leaves when both potassium and magnesium were applied in the sulphate form, indicating that S plays an important role in the accumulation of amino acids in tea leaves (Ruan et al. 1998).

5 Calcium

5.1 Calcium in Plant and Soil

Calcium (Ca) is one of the most abundant elements in the lithosphere, and it is readily available in most soils. Calcium has several distinct functions in higher plants that can be divided into three main areas: (i) effects on membranes, (ii) effects on cell walls and (iii) effects on enzymes due to its function as second messenger and in signalling events (Hawkesford et al. 2012).

Calcium ions bridge phosphate and carboxylate groups of phospholipids and proteins at membrane surfaces, helping to maintain membrane structure. In the middle of the membrane, Ca also interacts with proteins as an integral part of membranes. Calcium is a key element in the structure of primary cell walls. Calcium bound as Ca-pectate in the middle lamella is essential for strengthening cell walls and plant tissues. The degradation of pectates mediated by polygalacturonase is strongly inhibited by high Ca concentrations. Hence, in Ca-deficient tissue, polygalacturonase activity is increased, and a typical symptom of Ca deficiency is the disintegration of cell walls and the collapse of the affected tissues, such as petioles, upper parts of stems and fruits (Pilbeam Morley 2007; Hawkesford et al. 2012).

The ability of Ca to function as second messenger is based on the very low cytosolic free Ca^{2+} concentrations in plant cells and the chemistry of Ca^{2+} which allows it to alter the conformation of proteins to which it binds. Environmental and developmental signals can activate Ca^{2+} channels in cell membranes that catalyse rapid Ca^{2+} influx to the cytosol and increase cytosolic free Ca^{2+} concentrations (Hepler 2005; Dodd et al. 2010).

Despite the abundance of Ca, plants suffer from a range of Ca deficiency disorders that affect tissues or organs that are naturally low in Ca. In these disorders, the shortage of calcium in the tissues causes a general collapse of membrane and cell wall structure, allowing leakage of phenolic precursors into the cytoplasm. Oxidation of polyphenols within the affected tissues gives rise to melanin compounds and necrosis. With the general breakdown of cell walls and membranes, microbial infection is frequently a secondary effect (Pilbeam Morley 2007).

The concentrations of Ca in plants are similar to the concentrations of potassium, in the range 1–50 mg Ca g^{-1} dry matter, and the critical concentrations of Ca are different depending on plant species. Plants supplied with N as ammonium tend to have much lower concentrations of cations, including Ca, than plants supplied with nitrate (Pilbeam Morley 2007).

Calcium is a major component of soils. One of the most important natural sources of Ca is underlying limestone or chalk, where it occurs as Ca carbonate (calcite).

Calcium in rocks also occurs as a mixture of Ca and magnesium carbonates (dolomite). Soils over such rocks often contain large amounts of Ca carbonate. Although Ca carbonate is sparingly water soluble, it can be removed by leaching so that the overlying soil may be depleted of Ca carbonate and be acidic (Pilbeam Morley 2007).

In very acid soils, there is a shortage of available Ca, and upon application of lime, Ca^{2+} from the Ca carbonate ($CaCO_3$) exchanges for $Al(OH)_2^+$ and H^+ ions on the cation-exchange sites. The $Al(OH)_2^+$ ions give rise to insoluble $Al(OH)_3$ that precipitates; the H^+ ions react with bicarbonate $(HCO_3)^-$ that arises during the dissolution of $CaCO_3$ in the soil water. This reaction leads to the formation of CO_2 and water as neutral products (Miller and Donahue 1990; Meriño-Gergichevich et al. 2010). In plants grown on acid soils, application of $CaCO_3$ results in displacement of Al^{3+} and H^+ ions by Ca^{2+} ions in the root plasma membrane and declines Al^{3+} toxicity. In addition, the more neutral or alkaline pH brought about by liming gives a more favourable environment for the microorganisms of the N cycle, enhancing the cycling of N from organic matter. It also restricts the uptake of heavy metals (Pilbeam Morley 2007).

5.2 *Ca Nutrition in Tea*

In tea plants, another action of lime is to decrease the concentration of fluorine (F) in tea plants (Ruan et al. 2004). This crop accumulates high concentrations of F from soils of normal F concentration (see Sect. 9.3). The action of liming in limiting F concentrations in tea plants is surprising given that the uptake of F is higher from more neutral soil than from acid soil and given that liming may increase the water-soluble F content of the soil. In this case, it appears that the Ca^{2+} in the lime either affects cell wall and plasma membrane permeability or changes the speciation of the F in the soil (Ruan et al. 2004).

The concentration of nitrous oxide (N_2O) in the atmosphere has increased over recent decades. The global warming potential of N_2O is approximately 300 times that of carbon dioxide. Nitrous oxide is involved in the destruction of the ozone layer. The majority of anthropogenic N_2O is emitted by agricultural activities, and agricultural soils are the major source of N_2O. Nitrous oxide is produced during nitrification and denitrification in the soil (Bouwman et al. 1995; Davidson and Kingerlee 1997). Akiyama et al. (2006) reported that the estimated mean N_2O emission and estimated annual fertilizer-induced emission factors for tea fields in Japan are 24.3 ± 16.3 kg N ha^{-1} and $2.82 \pm 1.80\%$, respectively. These emission factors were much higher than those of other upland fields ($0.62 \pm 0.48\%$) and paddy fields ($0.31 \pm 0.31\%$). Because of the high rate of N application and the high N_2O emission factor, reducing N_2O emissions from green tea fields is a significant step towards reducing N_2O emissions from agricultural soil. Some new methods of fertilizer application have been proposed to increase the efficiency of N uptake by tea plants and reduction of N_2O emission from tea fields. Amongst the new methods of fertilizer application, application of lime nitrogen (calcium cyanamide, $Ca(CN)_2$) has been recommended for tea soils. Calcium cyanamide fertilizers contain approximately 20% N and 50% lime in the form of calcium oxide (CaO). In the soil, $Ca(CN)_2$ is broken down into urea and dicyandiamide (DCD). Dicyandiamide inhibits ammonia oxidation by microbes and reduces nitrification, thus allowing ammonium-N to remain in the soil longer. This condition is favourable for the growth of tea plants that prefer ammonium to nitrate (see Sect. 2.2) and because the amount of N leached from the soil is reduced. In addition, N_2O emissions are also reduced by directly limiting the nitrification process by DCD and by indirectly limiting the denitrification process by suppressing rapid formation of nitrate (Bhatia et al. 2010). However, the effectiveness of DCD in reducing N_2O emissions from tea fields strongly depends on the timing of lime nitrogen application and on climate conditions, including rainfall and temperature (Hirono and Nonaka, 2014).

An adequate Ca nutrition improves tolerance of tea plant to abiotic stresses such as drought. It has been observed that application of $CaCl_2$ resulted in an increase in the capacity of antioxidative mechanism during post-drought recovery and also in decreased metabolism of reactive oxygen species resulting in an improved drought recovery potential of tea genotypes (Upadhyaya et al. 2011).

6 Magnesium

6.1 Magnesium in Plant and Soil

Magnesium (Mg) has major physiological and molecular roles in plants, such as being a component of the chlorophyll molecule; a cofactor for many enzymatic processes associated with phosphorylation, dephosphorylation and hydrolysis of various compounds; and a structural stabilizer for various nucleotides (Merhaut 2007; Hawkesford et al. 2012). About 15–30% of the total Mg in plants is associated with the chlorophyll molecule. The other 70–85% of the Mg in plants is associated with the role of Mg as a cofactor in various enzymatic processes, the regulation of membrane channels and receptor proteins and the structural role in stabilizing proteins and the configurations of DNA and RNA strands (Merhaut 2007; Hawkesford et al. 2012).

Since Mg is an integral component of the chlorophyll molecule and the enzymatic processes associated with photosynthesis and respiration, the assimilation of carbon and energy transformations will be affected directly by inadequate Mg. This effect was related to the decreased enzymatic activity as well as the decrease in photosynthetic capacity due to the loss in assimilating leaf area, occurring mainly as a consequence of leaf necrosis and defoliation (Hermans and Verbruggen 2005).

Early symptoms of Mg deficiency may be noted by fading and yellowing of the tips of old leaves, which progress interveinally towards the base and midrib of leaves, giving a mottled or herringbone appearance. In later stages of development, deficiency symptoms may be difficult to distinguish from those of potassium deficiency (Merhaut 2007).

Conditions of the soil and rhizosphere such as drought or irregular water availability, poor drainage or excessive leaching, low soil pH or cold temperatures will exaggerate Mg deficiency symptoms. Low soil pH is also associated with a low supply or depletion of Mg, possibly due to leaching; however, research suggests that impairment of root growth in acid soils (pH 4.3–4.7) also may hinder Mg absorption (Merhaut 2007).

Some studies have shown that the toxic effects of aluminium are reduced when Mg was introduced into the nutrient solution (Keltjens 1995) and subsequently increased the production and excretion of citrate from the root tips. Citrate binds with free aluminium, forming nontoxic aluminium-citrate complexes (Merhaut 2007). Considering surface soils, sandy soils typically have the lowest Mg concentrations, and clay soils typically have the highest Mg concentrations. Leached soils such as lateritic soils and podzols tend to be low on Mg (Merhaut 2007).

6.2 Magnesium Nutrition in Tea

Most tea fields in tea-growing countries were established on highly leached red soils located in the subtropical areas. Mg is highly prone to leaching loss, and Mg deficiency in tea is a widespread problem in tea-producing countries such as China (Ruan et al. 1998). The leaf concentration of Mg in tea in sufficient range has been reported to be in the range of 1.2–3.3 mg g^{-1} dry weight (Merhaut 2007). Field experiments demonstrated significant yield response and quality improvement following Mg fertilization. It has been reported that Mg deficiency results in a loss of yield and quality of tea in Nigerian plantations (Obatolu 1999). A spray of 30% MgO corrected magnesium deficiency within 14 days and increased growth from 16 to 134%. Two applications of a 20% solution were required to correct deficiencies. A second application of the 30% solution was toxic to the tea plants (Obatolu 1999).

A sufficient supply of Mg in nutrient solution increases biomass production and concentrations of free amino acids, notably theanine in young shoots and roots. Magnesium nutritional status is an important factor influencing the mobility of amino acids and sugars via xylem and phloem especially when N and carbon reserves are remobilized to support spring growth of young shoots (Ruan et al. 2012).

Under Mg-deficient tea plants, an interveinal or generalized chlorosis of the older leaves as the characteristic symptoms of Mg deficiency has been observed in some clones. The leaves were hard and brittle. However, in the other clones, there was a very obvious premature shedding of the lower leaves, and within a relatively short time, the plants were completely defoliated with the exception of small tufts of young leaves at the branch ends (Pethiyadoga and Krishnapillai 1970). Analyses of some of these shed leaves revealed Mg contents as low as 0.2–0.9 mg g^{-1} dry weight. Growth is very markedly reduced, and affected plants die relatively rapidly if not supplied with a nutrient solution including Mg (Pethiyadoga and Krishnapillai 1970).

7 Potassium

7.1 Potassium in Plant and Soil

Potassium (K) as a univalent cation is characterized by high mobility in plants, is the most abundant cation in the cytosol and contributes substantially to the osmotic potential of cells and tissues of glycophytic plant species (Mengel 2007; Hawkesford et al. 2012). Potassium ion (K^+) has an outstanding role in plant-water relations. Due to its high concentrations in the cytosol and chloroplasts, it balances the charge of soluble (e.g. organic acid anions and inorganic anions) and insoluble anions and thus facilitates stabilizing the pH between 7 and 8 in these compartments, which is the optimum for most enzyme reactions. The functions of K^+ in cell extension and other turgor-driven processes are related to the K^+ concentration in the vacuoles where it can be replaced to a varying degree by other cations (Na^+, Mg^{2+}, Ca^{2+}) or organic solutes (e.g. sugars) (Mengel 2007; Hawkesford et al. 2012).

Potassium deficiency alters assimilate partitioning and thus changes metabolite concentrations in vegetative plant organs. Potassium is required in higher concentrations for protein synthesis than for enzyme activation. Photosynthesis is strongly reduced in K-deficient leaves. Potassium affects photosynthesis at various levels, e.g. via its function in stomatal regulation, in the establishment of the transmembrane pH gradient necessary for photophosphorylation and in the scavenging of reactive oxygen species (Cakmak 2005; Hawkesford et al. 2012).

Cell extension not only in leaves but also in roots is positively correlated with their K concentration. Potassium deficiency significantly reduces turgor, cell size and leaf area in expanding leaves. Reduced leaf extension rate is a most sensitive indicator of K deficiency in the field and under controlled conditions in hydroponics (Hawkesford et al. 2012). A high osmotic potential in the stele of roots is a prerequisite for turgor pressure-driven solute transport in the xylem and for the water balance of plants. Potassium has important functions in both the loading of sucrose and the rate of the mass flow-driven solute transport in the sieve tubes of the phloem (Cakmak 2005; Ahanger et al. 2016).

The frequently observed positive effects of K fertilization on crop yields under adverse conditions have been interpreted as evidence that K increases the resistance of plants against biotic and abiotic stress. It is well established that K-deficient plants are more susceptible to abiotic and biotic stresses. Examples are the enhanced plant injury of K-deficient plants under high-light intensity, drought, low temperature, iron toxicity and pest and disease pressure. Thus, when exposed to such stresses, an optimum K nutritional status is critical for stress resistance of plants (Cakmak 2005; Hawkesford et al. 2012; Wang et al. 2013).

After N, K is the nutrient required in the largest amount by plants. The K requirement for optimal plant growth is 20–50 mg g^{-1} in vegetative parts, fleshy fruits and tubers. When K is deficient, growth is retarded, and net transport of K^+ from mature leaves and stems is enhanced. Under severe deficiency these organs become chlorotic and necrotic, depending on the light intensity to which the leaves are exposed.

Also lignification of vascular bundles is impaired, a factor which may contribute to the higher susceptibility of K-deficient plants to lodging (Cakmak 2005; Hawkesford et al. 2012; Wang et al. 2013).

7.2 Potassium Nutrition in Tea

Similar with N, large amounts of K are removed from the soil through harvesting, and tea has a K requirement of high to moderate (Hamid 2006; Sultana et al. 2014). The amount of K removed by harvesting in tea has been estimated in the range of 23 mg K g^{-1} plant matter (Mengel and Kirkby 2001). Potassium concentration in tea (in the youngest fully developed leaf) at the mid of the vegetation season was reported 16–23 mg g^{-1} DW and considered as sufficient (Mengel 2007).

Depletion of starch reserve in the roots, degeneration of feeder roots and dieback and accumulation of nitrates occur under K deficiency in tea. Although the overall size of the plant is not evidently affected by K deficiency, it is very apparent that development of wood is markedly reduced and the stems remain sparse and slender (Pethiyadoga and Krishnapillai 1970; TRFK 2002). In K-deficient tea, bush branches are thin and weak and the recovery from pruning takes more time. The size of leaves is greatly reduced in K-deficient plants and leaf deformation is also common. There is also tip and margin burn of mature leaves that become loosely attached to the stem, and shaking of the deficient bush leads to fall of leaves (TRFK 2002; Jessy 2010). In a sand culture experiment, tea plants supplied with nutrient solution without K, deficiency symptoms become evident within about 4–4.5 months from commencement of treatment. There are interesting clonal differences in the manifestation of deficiency symptoms. Some of the leaves show a pronounced purple colour (bronzing) over the entire leaf blade particularly on the underside of the leaf. Saprophytic fungi are often seen to have colonized these dead patches (Pethiyadoga and Krishnapillai 1970).

Return of deficient plants with advanced symptoms to K-containing medium restores readily normal growth. Although the already affected mature leaves show no signs of improvement, new leaves produced are normal. In the complete absence of K, even relatively small amount of sodium (supplied as sodium dihydrogen phosphate) aggravates these leaf symptoms (Pethiyadoga and Krishnapillai 1970).

8 Micronutrients

Current fertilizer recommendations for tea have always emphasized on macronutrients, and despite continuous removal through harvesting similar with macronutrients, micronutrients have not been taken into consideration. Availability of micronutrients is affected by various factors amongst them soil properties and root uptake efficiency, interactions of roots with microorganisms in the rhizosphere and interactions amongst the micronutrients themselves. High soil pH results in retention of micronutrients in soil (Rengel 2015), thus limiting their uptake by tea plants (Hajiboland 2017).

8.1 Iron

Iron in Plant and Soil

The ability of iron (Fe) to undergo a valence change between Fe^{3+} and Fe^{2+} is important in its functions. Iron is a component of two major groups of proteins, namely, the haem proteins and the Fe-S proteins. A characteristic haem Fe-porphyrin complex acts as a prosthetic group of the cytochromes, catalase, peroxidase and leghemoglobin. Iron is essential for chlorophyll biosynthesis and chloroplast biogenesis and function. Along with the Fe requirement in some haem enzymes, Fe has a function in Fe-S proteins including ferredoxin, which has a strong involvement with the light-dependent reactions of photosynthesis (Hänsch and Mendel 2009; Broadley et al. 2012).

The typical symptoms of Fe deficiency in plants are chlorotic leaves. Often the veins remain green, whereas the laminae are yellow, and a fine reticulate pattern develops with the darker green veins contrasting markedly with a lighter green or yellow background. Iron deficiency causes marked changes in the ultrastructure of chloroplasts, with thylakoid grana being absent under extreme deficiency and the chloroplasts being smaller. Iron is not easily re-translocated as long as the leaves are not senescent; thus, the younger leaves tend to be more affected than the older leaves (Römheld and Nicolic 2007; Broadley et al. 2012).

The major factor affecting acquisition of Fe by plants is soil pH, with high pH making Fe less available and giving rise to chlorosis. In lime-induced chlorosis, the soil bicarbonate reduces Fe solubility due to the high pH in the rhizosphere and at the root uptake site (Römheld and Nicolic 2007).

Iron toxicity is not a common problem in the field, except in acid soils and under flooded conditions. Fe toxicity symptom is known as 'bronzing' and includes small reddish-brown spots on the leaves, which gradually extend to the older leaves. The whole leaf may turn brown, and the older leaves may die prematurely. In some species, leaves may become darker in colour, and roots may turn brown (Römheld and Nicolic 2007).

Iron Nutrition in Tea

Considering the low pH in tea soils, Fe deficiency does not seem to be a problem for tea growth. There is no report from a field experiment, but in an experiment with tea grown in sand irrigated with nutrient solution without Fe, no Fe deficiency symptoms were observed in the experimental plants after about 4 months (Pethiyadoga and Krishnapillai 1970). It is likely that tea has a high capacity for Fe uptake, and a high Fe availability in acid soil may rather induce Fe toxicity in this species. It has been observed that one of the mechanisms for Al-induced growth stimulation might be the alleviation of a latent Fe toxicity occurring in tea plants without Al supply (Hajiboland et al. 2013) (see also Sect. 9.2).

8.2 Zinc

Zinc in Plant and Soil

Zinc (Zn) is the second most abundant transition metal in living organisms after Fe, and its average total concentration in cultivated soils is around 65 mg kg^{-1}. Zinc is taken up by roots predominantly as a divalent cation (Zn^{2+}), but at high pH, it is probably absorbed as a monovalent cation ($ZnOH^{+}$) (Alloway 2009).

The metabolic functions of Zn are based on its strong tendency to form tetrahedral complexes with N-, O- and particularly S-ligands, and thus it plays a catalytic and structural role in enzyme reactions. Zinc is an integral component of enzyme structures and has catalytic, coactive or structural functions (Hänsch and Mendel 2009; Broadley et al. 2012). Zinc is a structural component of the ribosomes, is required for maintenance of integrity of biomembranes and is important in cell division. In addition, a class of Zn-dependent proteins (Zn-metalloproteins) have been identified which are involved in DNA replication, transcription and, thus, regulation of gene expression (Hänsch and Mendel 2009; Broadley et al. 2012).

Many of the most obvious symptoms of Zn deficiency such as leaf chlorosis and necrosis inhibited shoot elongation and increased membrane permeability which are expressions of oxidative stress brought about by higher generation of reactive oxygen species and an impaired detoxification system in Zn-deficient plant (Cakmak 2000).

Zinc deficiency is common in plants growing in highly weathered acid or calcareous soils. Zinc deficiency initially appears in all plants as interveinal chlorosis (mottling) in which lighter green to pale yellow colour appears between the midrib and secondary veins. Developing leaves are smaller than normal, and the internodes are short (‘little leaf’ and ‘rosette’ disease) (Storey 2007). Supply of plants with high P intensifies Zn deficiency due to mechanisms acting both within plant and in soil (Broadley et al. 2012).

Zinc is unavailable in alkaline soils because of formation of insoluble $ZnCO_3$ and in acid soil where Zn is in competition with nickel. Zinc may be associated with soil organic matter, hydrous oxides and carbonates via adsorption, surface complex formations, ion exchange, incorporation into the crystal lattice and co-precipitation (Alloway 2009). Along with root Zn application, foliar application is practised for many crop plants. Mixing urea-ammonium nitrate fertilizer with $ZnSO_4$ increases leaf Zn concentration compared to using $ZnSO_4$ alone. Zinc nitrate is more efficient than $ZnSO_4$ in increasing leaflet concentration; its hydrophilic nature makes it more appropriate for arid climates (Storey 2007).

Zn Nutrition in Tea

The leaf analyses at plucking indicated a concentration of <3 μg g^{-1} Zn as Zn status in the mature leaves of tea (Fahmy 1977). Zinc deficiency in tea plant leads to high mortality and stunted growth when severely affected; the young leaves appear narrow, erect and form a rosette at the apex of the stem (Nelson 2006). Zinc deficiency in tea plants grown in greenhouse had appeared after 2-month growth under low Zn conditions as chlorosis, growth impairment and dieback of shoots emerged (Mukhopadhyay et al. 2013). Ultrastructural study of leaves revealed disorganization of cellular organelles in Zn-deficient tea plants. Reduction of net photosynthetic rate and content of chlorophyll and activation of antioxidative defence concomitant with an upregulation of respective genes have been reported for Zn-deficient tea plants. Nevertheless, antioxidant system as a whole did not afford sufficient protection against oxidative damage (Mukhopadhyay et al. 2013). Zinc is not easily absorbed by tea plant, and its deficiency can be corrected by foliar application, while soil application is not effective. Application of Zn improves chlorophyll content, net photosynthesis rate and water use efficiency and consequently increases made tea yield (Kumar et al 2014; Hajiboland 2017).

8.3 *Manganese, Copper, Molybdenum and Nickel*

Manganese in Plant and Soil

Manganese (Mn) can exist in different oxidation states; in plants, Mn (II) is the dominant form, but it can readily be oxidized to Mn (III) and Mn (IV) and therefore plays an important role in redox processes (Broadley et al. 2012). Although a relatively large number of enzymes are activated by Mn, there are only a small number of Mn-containing enzymes, namely, the Mn-protein in PS II (Mn-containing enzyme is the 33 kDa polypeptide of the water-splitting system in PS II), the Mn-containing superoxide dismutase (MnSOD) and oxalate oxidase, i.e. a secreted multimeric glycosylated Mn-containing enzyme and Mn-containing purple acid phosphatase (Broadley et al. 2012). Manganese is important for photosynthesis and oxygen evolution, synthesis of proteins, carbohydrates and lipids, cell division and extension (Hänsch and Mendel 2009; Broadley et al. 2012).

In Mn-deficient plants, dry matter production, net photosynthesis and chlorophyll content decline rapidly. Manganese-deficient plants are more susceptible to damage by freezing temperatures and a range of soilborne root-rotting fungal diseases (Millaleo et al. 2010; Broadley et al. 2012; Hajiboland 2012).

Manganese deficiency is abundant in plants growing in soils derived from parent material inherently low in Mn and in highly leached tropical soils. It is also common in soils of high pH containing free carbonates, particularly when combined with high organic matter content. In acid soils, in contrast, the amount of exchangeable Mn (mainly Mn^{2+}) increases in the soil solution. Lime application is a key factor in decreasing soluble Mn in acid soils. Interestingly, the highest concentrations of soluble and exchangeable Mn are found after hot, dry summers and under warm waterlogged conditions in acid soils. This is probably due to the inhibition of Mn-oxidizing organisms, thereby allowing the chemical reduction of Mn oxides in these soils (Millaleo et al. 2010). Manganese deficiency can be corrected by soil or foliar application of $MnSO_4$.

Unfortunately, information is lacking on the Mn availability and probable deficiency in tea. The only data on tea Mn concentration reported <50 μg g^{-1} in the mature leaves and explained it as deficient (Fahmy 1977).

Copper in Plant and Soil

Copper (Cu) is a redox-active transition element with roles in photosynthesis, respiration, carbon and N metabolism and protection against oxidative stress. Like Fe, it participates in electron transfer reactions (Broadley et al. 2012). Most of the functions of Cu as a plant nutrient are based on enzymatically bound Cu which catalyses redox reactions.

There are more than 100 different Cu-containing proteins in plants (Yruela 2009). About 50% of Cu found in plants is present in chloroplasts, bound to plastocyanin, where it participates in photosynthetic reactions (Hänsch and Mendel 2009). Other major forms include Cu-binding chaperones and numerous enzymes, particularly single and multi-Cu-containing oxidase enzymes (Burkhead et al. 2009). Copper is also part of the ethylene receptor and is involved in Mo-cofactor biosynthesis. Copper is a structural component of superoxide dismutase, cytochrome c oxidase, ascorbate oxidase, diamine oxidases and polyphenol oxidases and plays important roles in carbohydrate, lipid and N metabolism, lignification, pollen formation and fertilization (Broadley et al. 2012).

Copper deficiency is often observed in plants growing on soils either low in total Cu, e.g. ferrallitic and ferruginous coarse-textured soils or calcareous soils derived from chalk, and on soils high in organic matter where Cu is complexed with organic substances. High N availability can also lead to Cu deficiency (Kopsell and Kopsell 2007).

The availability of Cu can be low in many soils and this can be corrected by soil or foliar applications. Soil applications of inorganic Cu as $CuSO_4$ or oxide forms, or slow-release metal compounds, sewage sludges or manures are often appropriate for long-term effects. Foliar applications of Cu in the form of inorganic salts, oxides or chelates can be used to rapidly correct Cu deficiency in soil-grown plants (Kopsell and Kopsell 2007).

Considering the role of Cu in phenolic metabolism in general and its structural role in polyphenol oxidase, i.e. a phenolic-metabolizing enzyme, in particular, it is expected that Cu nutritional status has a profound influence on the quality of green and black tea. However, information is lacking on the Cu availability in tea soils and its probable deficiency in tea plants.

Molybdenum in Plant and Soil

The functions of molybdenum (Mo) as a plant nutrient are related to the valency changes it undergoes as a metal component of enzymes. In higher plants, only few enzymes have been found to contain Mo as a cofactor, including nitrate reductase, xanthine dehydrogenase, sulphite reductase and aldehyde oxidase; the latter enzyme play important roles in the response and resistance to various stresses (Kaiser et al. 2005; Hänsch and Mendel 2009).

In Mo-deficient plants, symptoms of N deficiency and stunted growth and chlorosis in young leaves are common. In dicotyledonous species, a strong reduction in size and irregularities in leaf blade formation ('whiptail') are the most typical visual symptoms, caused by local necrosis in the tissue and insufficient differentiation of vascular bundles in the early stages of leaf development (Broadley et al. 2012).

Because of its function in nitrate reductase, Mo requirement of nitrate-fed plants is higher than ammonium-fed plants. Without Mo supply, however, ammonium-fed plants also develop deficiency symptoms likely because of nitrate formation in the rooting medium through nitrification. In many species, deficiency symptom is observed independent from N supply form as the consequence of metabolic disturbances in the absence of Mo (Broadley et al. 2012).

Molybdenum deficiency is not common in most agricultural cropping areas (Kaiser et al. 2005). Molybdenum fertilization through foliar sprays can effectively supplement internal Mo deficiencies and rescue the activity of molybdoenzymes. Sulphate and molybdate are strongly competing anions during uptake by the roots. Therefore, sulphate-containing soil amendments such as gypsum and single superphosphate which contains sulphate reduce Mo uptake (Broadley et al. 2012).

Molybdenum is very rare in soils, and is usually present in a highly soluble form, making it susceptible to leaching. In acid soils, in contrast, Mo attaches to mineral surfaces that would prevent its escape into the groundwater but would also impede uptake by plants. In addition, in the leaf litter layer, most of the Mo forms strong complexes with plant-derived tannins and tannin-like compounds; Mo binds to these organic ligands across a wide pH range (Wichard et al. 2009). All factors affecting Mo availability could be of high importance in tea soils, but information is lacking on Mo leaching, fixation or availability in tea soils, and there is no data on Mo nutritional status in this species.

Nickel in Plant and Soil

Nickel (Ni) is involved in the function of at least nine proteins of which urease and the Ni-urease accessory protein (Eu3) have roles in plants. Symptoms of Ni deficiency suggest additional essential roles are likely (Hänsch and Mendel 2009; Broadley et al. 2012).

When supplied with urea as sole N source and in the absence of Ni, growth of different plant species is reduced, urea accumulates and toxicity symptoms occurs as severe necrosis of the leaf tips. Root and shoot growth is significantly lower in the

Ni-deficient plants, which are less green and develop interveinal chlorosis and necrosis, and the terminal two centimetres of the leaves fail to unfold. Nickel deficiency also results in a marked enhancement in plant senescence and a reduction in tissue Fe concentrations (Hänsch and Mendel 2009; Tejada-Jiménez et al. 2009). Considering that urea is the most common N source in tea plantations, it is necessary to provide plants with adequate Ni. However, information is lacking on the Ni availability and probable deficiency in tea.

8.4 Chlorine and Boron

Chlorine in Plant and Soil

Chlorine (Cl) is classified as a micronutrient, but it is often taken up by plants at levels comparable to a macronutrient. Chlorine is readily taken up by plants in the electrically charged form as chloride ion (Cl^-). As an essential element, Cl has several biochemical and physiological functions within plants and is required for optimal enzyme activity of asparagine synthethase, amylase and ATPase. In photosynthesis, Cl is an essential cofactor for the activation of the oxygen-evolving enzyme associated with photosystem II (Tejada-Jiménez et al. 2009; Broadley et al. 2012). The osmoregulatory function in specific tissues requires concentrations of Cl^- that are typical of a macronutrient. Along with K, Cl^- participates in stomatal opening, serves to balance charges from cations and is an alternative to the formation of malate in its charge-balancing role. This role of Cl^- may be of greater importance when cation uptake exceeds anion uptake, as often occurs with plants provided with ammonium nutrition (Broadley et al. 2012).

Supplies of Cl in nature are often plentiful, and obvious symptoms of deficiency are seldom observed. The most commonly described symptom of Cl deficiency is wilting of leaves, especially at the margins. Addition of Cl has been reported to reduce the severity of several foliar and root diseases on different crops (Heckman 2007). In acid soils, Cl inhibits nitrification, and application of N in the ammonium form can lower rhizosphere pH and influence microbial populations and nutrient availability in the rhizosphere. Chlorine can also enhance Mn availability by promoting Mn-reducing microorganisms in soil. Chloride, like nitrate, is susceptible to loss from soil by leaching in areas of high rainfall (Heckman 2007).

How a crop is harvested influences the amount of Cl^- in soil. When harvested only as seed, the amount of Cl^- removed is limited, but when harvested as green biomass, the amount of Cl^- removal may be substantial. Because Cl^- leaches from ageing leaves, harvest of mature biomass may remove only about half as much Cl^- as does harvest before the onset of senescence (Heckman 2007).

A high quality of green tea is associated with a high amino acid-to-catechin ratio (Ruan et al. 1999). It has been shown that excessive Cl supply is detrimental for amino acid accumulation. Soil-grown tea plants supplied with different levels of K as K_2SO_4 or KCl exhibited increased concentrations of free amino acid in young shoots only when supplied with K_2SO_4, and the negative effect of KCl supply was mainly due to a reduced concentration of theanine. Chloride application reduced yield of young shoots, and severity of leaf damage was related to the concentration of Cl^- in leaves (Ruan et al. 2007a).

Boron in Plant and Soil

Boron (B) is a member of the metalloid group of elements. Boric acid ($B(OH)_3$) and borate ($B(OH)_4^-$) are the two main forms of B in plant cells. Within the cell, $B(OH)_3$ form complexes with diols and polyols, particularly with *cis*-diols, either as monoester or diester. Boron is neither an enzyme constituent nor is there convincing evidence that it directly affects enzyme activities. There are many postulated roles of B including sugar transport, cell wall synthesis, lignification, cell wall structure,carbohydrate metabolism, RNA metabolism, respiration, auxin metabolism, phenol metabolism and membranes integrity. This long list might indicate that B is involved in a number of metabolic pathways and/or deficiency results in a 'cascade effect' due to disruption of a critical and central cellular process (Broadley et al. 2012).

A high proportion of total plant B is associated with cell wall pectins. Boron in the cell wall predominantly cross-links the apiosyl residue in the A side chain of each of two neighbouring monomeric RGII molecules to form a dimeric B-dRGII pectin complex (O'Neill et al. 2001).

Boron deficiency causes a wide range of anatomical, physiological and biochemical symptoms. These include inhibition of apical growth, necrosis of terminal buds, reduction in leaf expansion, breaking of tissues due to brittleness and fragility, abortion of flower initials and shedding of fruits (Brown et al. 2002). One of the most rapid responses to B deficiency is inhibition or cessation of root elongation and shoot meristematic growth. In roots, B deficiency results in the roots with a stubby and bushy appearance. In shoots, complete inhibition of meristematic growth occurs in many species, whereas meristem death occurs in others (Cakmak and Römheld 1997).

The changes in cell wall formation and composition result in serious physiological disturbances in plants grown under low B supply. For example, B deficiency enhances the number of Al-binding sites in cell walls, possibly due to increasing amount of unmethylated pectin in the root tips, resulting in higher Al concentrations and greater Al damage in roots (Yu et al. 2009). Impairments in development and organization of primary cell walls associated with B deficiency have adverse impacts on form, wood quality and cold tolerance of trees (Wang et al. 2015).

Accumulation of phenols is a typical feature of B-deficient plants. It has been suggested that the formation of borate complexes with certain phenols may be involved in the regulation of the concentration of free phenols and the rate of synthesis of phenol alcohols as precursors of lignin biosynthesis. Accordingly, under B deficiency, phenols accumulate and polyphenol oxidase activity is increased (Cakmak and Römheld 1997).

Soil reaction or soil pH is an important factor affecting availability of B in soils. Generally, B becomes less available to plants with increasing soil pH. Liming of soil decreases the plant B accumulation when soil B reserves are high. Generally, soils that have developed in humid regions have low amounts of plant-available B because of leaching. Thus, plants growing in regosols, sandy podzols, alluvial soils, organic soils and low humic gleys tend to develop B deficiencies because of low soil B reserves (Shorrocks 1997).

Boron Nutrition in Tea

Boron deficiency and effect of B fertilizers on growth and yield of tea have been extensively studied (Njoloma 2012; Kumar et al. 2014). Boron content in soil was found to decrease with the increase in the age of tea. High incidences of Cu, B and Zn deficiencies in leaf analyses of these elements were shown from approximately 170 mature clonal tea fields in Malawi (Njoloma 2012). Soil analysis from 11 districts of Assam and 4 districts of west Bengal reported B deficiency in 17% and 64% of the samples, respectively (Baruah et al. 2011). These authors showed that although some of surface soil samples in Assam have B below detection limit, it was associated with pollution of drinking water sources in this area due to leaching of B during monsoon rains from surface soils beyond the root zone (Baruah et al. 2011). Since the range of soil B concentration between deficiency and toxicity symptoms is very narrow (Broadley et al. 2012) and the majority of data from field studies did not show considerable improvement of tea yield by B application either to the soil or leaves, application of B fertilizers to tea gardens is not routinely recommended for many of tea-growing areas. Excessive B can cause off colour and stunting of tea plant growth. As B concentrations in plant tissue increase to toxic levels, older foliage may exhibit leaf edge burning or necrosis. Some plants will also develop black spots on older foliage (Baruah et al. 2011). In addition, adverse effects of high B content in tea infusion for human health (Zioła-Frankowska et al. 2014) raise concerns on application of B fertilizers for tea fields.

In tea plant grown hydroponically under greenhouse conditions, visual B deficiency symptoms was observed 30–35 days after treatment as reduced shoot height. Leaf symptoms which appeared only in the young leaves include reduced surface area, dark green colour and rarely curling of leaf lamina. Roots of B-deficient plants were poorly branched and tended to turn light purple. Boron-deficient plants have up to 40% lower shoot and root dry weight. The number of leaves and height of shoot were also reduced by about 42% and 17%, respectively (Hajiboland et al. 2011) (Fig. 9.2). Data of greenhouse study on tea revealed that a considerable re-translocation of B occurs from mature to growing leaves. As related to potential B mobilizing and/or complexing compounds, it has been observed that sucrose, glucose and fructose exist in the leaf exudate and phloem sap, while xylitol, sorbitol, mannitol, maltose, galactose, cellobiose or raffinose are not found indicating a considerable re-translocation of B in tea plant despite lacking polyol compounds (Hajiboland et al. 2013b).

Fig. 9.2 Boron deficnecy in tea plants grown from seed and cultivated hydroponicaly for 6 weeks under green house conditions

Boron Nutrition and Stress Tolerance in Tea

In general, B deficiency is intensified under drought conditions due to lower B mobility in soil (Broadley et al. 2012). In turn, low B plants are susceptible to drought stress because of lower shoot-root translocation and re-translocation from mature to young leaves (Hajiboland and Bastani 2012). Although B-deficient plants are more conservative in relation to water loss following elevated stomatal limitation, their water potential is lower than in B-sufficient plants irrespective of the watering regime because of lower water uptake (Hajiboland and Bastani 2012). Boron deficiency reduces root hydraulic conductivity (Apostol and Zwiazek 2004) likely due to disturbance in the formation of functional xylem vessels (Broadley et al. 2012). A decline in root hydraulic conductance may also be caused by a reduction in new root growth in B-deficient plants. Under the combined effects of B deficiency and water stress, the reduction in the CO_2 assimilation was mainly related to non-stomatal factors (Hajiboland and Bastani 2012).

9 Aluminium and Fluorine

9.1 Aluminium in Plant and Soil

Aluminium (Al) is the third most abundant metallic element in the earth's crust and exists in soils in many mineral forms including hydrous oxides, aluminosilicates, sulphates and phosphates. In acid soils, an appreciable portion of the

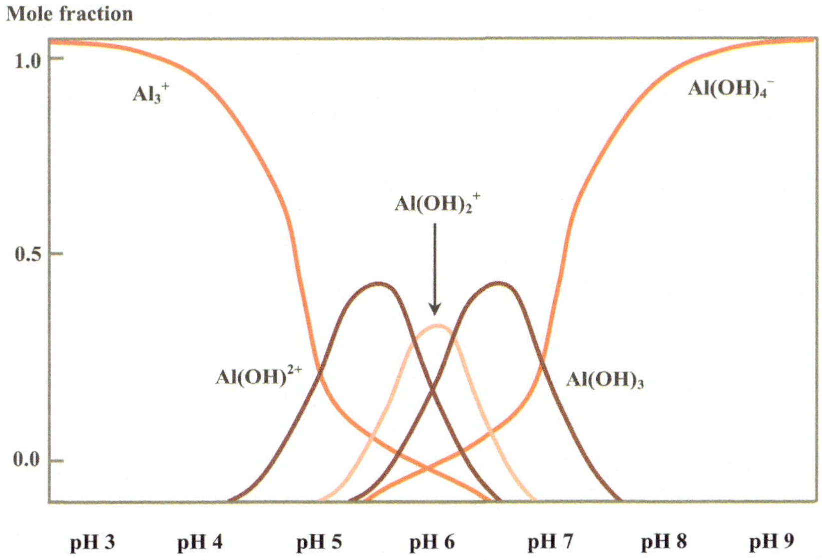

Fig. 9.3 Speciation of aluminum as affected by soil pH

cation-exchange capacity is satisfied by Al ions, i.e. the soils have a high Al saturation. Aluminium released from soil minerals into the soil solution under acid conditions, or the Al in nutrient solutions of pH < 4.0, is mainly $Al(H_2O)_6^{3+}$ (referred to as Al_3+). As the pH increases, the total Al concentration of the solution decreases, but mononuclear hydrolysis products such as $Al(OH)^{2+}$ and $Al(OH)_2^+$ are formed as intermediates in the precipitation of solid $Al(OH)^0{}_3$. Above pH 7, the solution Al concentration increases again due to the formation of the aluminate ion $Al(OH)_4^-$. At elevated OH^-/Al ratios in solution, polynuclear hydroxyl Al species may form (Miyasaka et al. 2007) (Fig. 9.3).

Aluminium toxicity is usually the major growth-limiting factor for crop production in acid soils. The initial and most obvious symptom is inhibition of root growth, and injured roots are characteristically stubby with reduced growth of the main axis and inhibited lateral root formation. Aluminium toxicity also interferes directly with active ion uptake processes functioning across the root cell plasma membrane (Poschenrieder et al. 2008, 2015).

Relative to Al accumulation, there appears to be two groups of plant species: Al excluders and Al accumulators (Miyasaka et al. 2007). Most plant species, particularly crop plants, are Al excluders with an Al content of about 200 μg g^{-1} in leaves. Al accumulators are defined as those plants with 1000 μg Al g^{-1} or greater in leaves. Aluminium accumulation is found frequently amongst perennial, woody species in tropical rain forests (Miyasaka et al. 2007). Tea is the only crop species with Al accumulation ability.

9.2 Aluminium Functions in Tea

Aluminium promotes growth of tea (Ghanati et al. 2005; Hajiboland et al. 2013a; Hajiboland 2017). A more than threefold increase of root biomass was observed for tea plants supplied with 300 μM Al (125 μM Al^{3+} activity). Total root length was positively related to root Al concentrations ($r = 0.98$). Greater root surface area may improve water and nutrient uptake by the plants. Al-induced growth stimulation in tea was mediated by higher photosynthesis rate and nitrate reductase activity and increased antioxidant defence (Hajiboland et al. 2013a). Recently, an RNA-Seq transcriptome analysis for study of mechanisms for Al tolerance and accumulation in tea plant implied that there are some common and distinct Al-tolerant mechanism between tea plants and rice, *Arabidopsis* and buckwheat (Li et al. 2017).

Tea accumulates Al in leaves at concentrations as high as 30,000 and 600 μg g^{-1} dry weight in old and young leaves, respectively (Matsumoto et al. 1976). Investigations using energy-dispersive microanalysis (EDXMA) have shown that in old tea leaves, Al is mainly accumulated in the cell walls of epidermal cells. However, in younger tea leaves, the EDXMA was not sufficiently sensitive to obtain Al localization maps or to detect Al in symplast compartments (Carr et al. 2003). Using low-energy X-ray fluorescence spectromicroscopy (LEXRF), it was shown that Al is mainly localized in the cell walls of the leaf epidermal cells, while almost no Al signal was obtained from the leaf symplast. The results suggest that the retention of Al in the epidermal leaf apoplast represents the main tolerance mechanism to Al in tea plants (Tolra et al. 2011). Conventional and confocal laser scanning microscopy images of morin-stained roots indicated a high fluorescence signal in the caps and adjacent meristematic cells (Hajiboland and Poschenrieder 2016). Towards basal parts, however, Al tended to accumulate mainly in the root hairs, rhizodermal and endodermal cell layers and slightly in the cortex, while it was clearly excluded from the central cylinder. A high Al-morin signal was detected from the cell wall compared with other parts of the cells. Relatively high green fluorescence signal was emitted from the epidermal cell layer, trichomes, vascular bundle region and stomatal cells of particularly young leaves (Hajiboland and Poschenrieder 2016) (Fig. 9.4). By means of ^{27}Al-NMR, it was shown that Al is chelated by phenolic compounds (mainly catechin) in the leaves of tea (Nagata et al. 1992). In the apical parts of tea roots (0–20 mm), however, Al presents mainly as complexes with oxalate (Morita et al. 2008).

It has been observed that Al stimulates uptake of Ca, Mg, K and Mn, whereas the uptake of Fe, Cu and Zn is retarded (Fung et al. 2008). In the absence of Al, tea roots accumulate Fe up to toxic levels, and its concentration in the young leaves reached up to 1000 μg Fe g^{-1} dry weight that is above the Fe toxicity threshold in most species. Supply of Al reduced Fe uptake and transport. These results indicate that Al-induced growth stimulation might be due to alleviation of a latent Fe toxicity occurring in tea plants without Al supply (Hajiboland et al. 2013).

In acid soils, in addition to higher Al availability, there is higher probability of B deficiency due to higher leaching losses (Shorrocks 1997). Within plants, the main

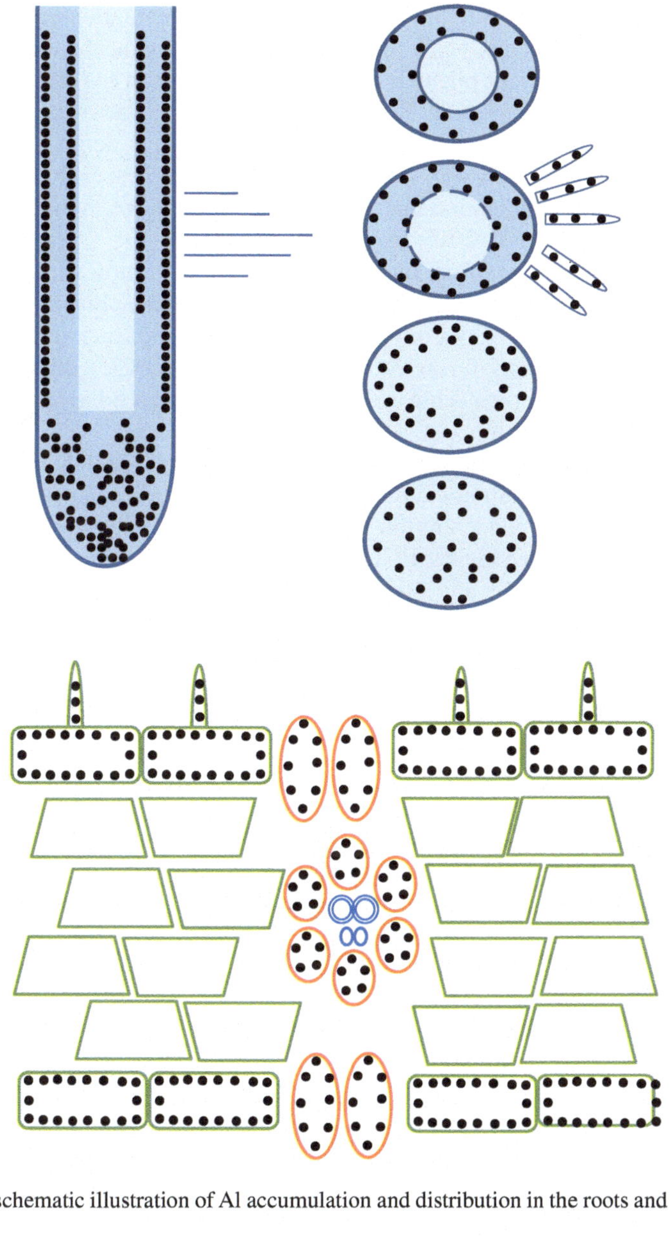

Fig. 9.4 A schematic illustration of Al accumulation and distribution in the roots and leaves of tea plants

Al species is probably $Al(OH)_3$ (Kochian 1995) which has structural similarity to B(OH). Considering the above mentioned factors and the fact that both Al and B primarily target plant cell wall, interactions between these elements are to be expected. It has been observed that growth of B-deficient plants is completely resumed by Al supplementation or even surpassed control plants regarding shoot biomass (Hajiboland et al. 2015; Hajiboland 2017). Upregulation of carbon and N metabolism and enhancement of B uptake and transport are mechanisms for growth amelioration of B-deficient plants by Al supplementation in tea (Hajiboland et al. 2015). The positive Al effect in B-deficient plants was related to an Al-induced increase of B contents in the root cell walls. Moreover, in B-deficient plants, more Al was partitioned into cell wall-bound fractions in both leaves and roots than in B-sufficient plants, an indication that B deficiency reduced the mobility of Al in the tea plants (Hajiboland et al. 2015; Hajiboland 2017).

Aluminium has a great effect on tea phenolic concentration and profile. Green tea leaves contain polyphenols, particularly flavan-3-ols (catechins). The five major variants of catechins are trihydroxy-flavan-3-ol gallic acid ester (epigallocatechin gallate, EGCG), trihydroxy-flavan-3-ol (epigallocatechin, EGC), dihydroxy-flavan-3-ol gallic acid ester (epicatechin gallate, ECG), dihydroxy-flavan-3-ol (epicatechin, EC) and gallocatechin (GC) (Harbowy and Balentine 1997, Anesini et al. 2008). In addition to an increase in the activity of phenolic-metabolizing enzymes including phenylalanine ammonia lyase, polyphenol oxidase and peroxidases in the leaves, total phenolic concentration increases with Al treatment. Amongst the important phenolics in tea leaves, EGC decreased, while ECG increased in the presence of Al in both young and old leaves (Hajiboland et al. unpublished data). Considering the chelating capacity of phenolic compounds in the following order EC < EGC < ECG < EGCG (Inoue et al. 2002), reduction of EGC concentration accompanied by increase of ECG concentration suggests an improvement of chelating capacity of leaf phenolics upon Al treatment. Moreover, ECG and EGCG have a high antioxidant capacity and therefore have a large role in protection against oxidative stress. A high correlation coefficient was obtained between Al and ECG and between Al and total phenols in tea (Chen et al. 2011).

Interestingly, Al treatment decreased lignin synthesis in both leaves and roots (Hajiboland et al. 2015) that may contribute in an extended growth period and is responsible for higher growth rate in tea plant in the presence of Al (Fig. 9.5).

9.3 *Fluorine Accumulation and Function in Tea*

Fluorine (F) exists mainly as fluoride in the environment. Fluoride content in soil ranges from under 20 to several thousand mg kg^{-1} (Weinstein and Davison 2004) and is derived from the parent materials, atmospheric deposition from geological and anthropogenic activities and applications of fertilizers (e.g. phosphate fertilizers) (Weinstein and Davison 2004).

To most plants, F is phytotoxic by influencing metabolism; disturbing the cellular homeostasis of important cations, such as calcium and magnesium; and inhibiting the activities of several enzymes, such as phosphoryl transfer enzymes and enolases (Barbier et al. 2010). Less than 10 mg kg^{-1} F is usually found in plants from F-uncontaminated areas. Tea plants, nevertheless, can accumulate large amounts of F in mature leaves from soils of normal F availabilities without toxicity

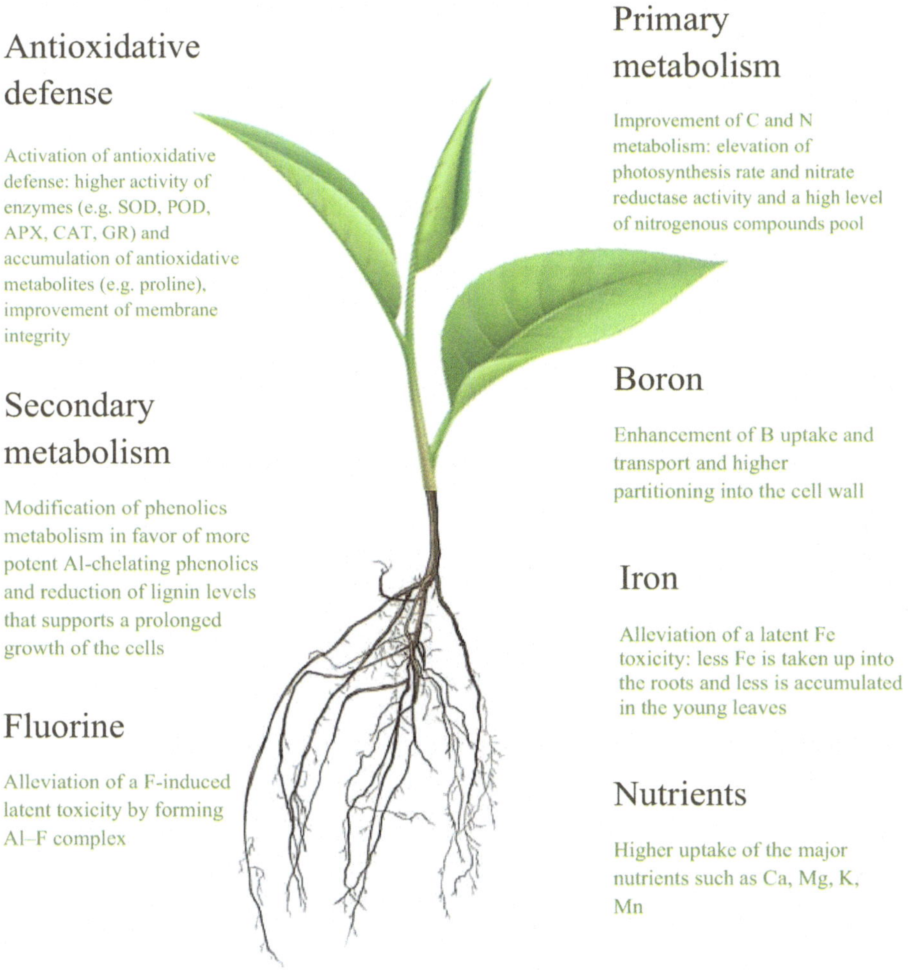

Fig. 9.5 A summary of mechanisms for Al-induced growth stimulation in tea plants

symptoms. The reasons for high F accumulation by tea plants are not clear. It was previously thought that plant roots take up F passively (Weinstein and Davison 2004). This view has been challenged by a recent study demonstrating that F is taken up by tea roots through active processes (Zhang et al. 2013).

The significant characteristic of F uptake by tea plants is that most of the F taken up is readily transported to the above-ground organs, in particular to the leaves, even within a relative short uptake time. In other species this process is usually limited, leading to a higher F concentration in roots than in shoots (Ruan et al. 2003). The ready translocation of F from root to shoot in tea plants is also reflected in field investigations, which showed that most of the F in tea plants was found in the mature or old leaves (Yumada and Hattori 1977; Fung et al. 1999).

It has been shown that F concentrations lower than 4 mg L^{-1} promote the biosynthesis of catechins in tea plants by upregulating the activity of related biosynthetic enzymes, whereas F concentrations more than 8 mg L^{-1} restrain tea plant growth and the synthesis of catechins (Yang et al. 2015).

F uptake by tea plants is affected both by pH and by Ca levels in the medium (Ruan et al. 2004). The reduced F uptake following Ca application appeared not to be due simply to the precipitation of CaF_2 in solution and soil or to the complexing of Ca and F in roots, although these factors cannot be dismissed. It was more likely due to the effect of Ca on the properties of cell wall or membrane permeability in the solution experiments and to alteration of F speciation and their quantities in soil solutions following Ca application (Ruan et al. 2004).

It has been found that F contents in tea plant are more strongly correlated with Al than with other elements (Yumada and Hattori 1977; Takmaz-Nisancioglu and Davison 1988). It appears that the increased uptake of F in tea plants by Al could not be explained by their interaction in the soil, since F availability was actually decreased with Al application and labile Al levels in the soil (Ruan et al. 2003). The presence of Al in solution or in soil increases the uptake of F and translocation to leaves of tea plants (Ruan et al. 2003). The positive relationship between F and Al concentrations in tea plants, and also in other plant species, led to the hypothesis that the uptake of F and Al is somewhat linked. It has been proposed that tea plants take up Al and F in the forms of Al-F complexes (Nagata et al. 1993). However, other experimental evidences showed that Al-mediated facilitation of F accumulation in tea plants might have occurred mainly in the processes of uptake and been less likely during the translocation from roots to shoots (Ruan et al. 2003).

Interestingly, Al may hinder F toxicity in tea plants. It is highly probable that the simultaneous accumulation of large amounts of Al and F is the reason for the lack of F toxicity symptoms in tea plants grown in typically acidic soils. Yang et al. (2016) observed that Al greatly alleviated the inhibitory effect of F on root and shoot growth of tea plants by forming Al-F complexes.

10 Effect of Nutrient Supply Level on the Quality of Tea

Both green and black tea qualities are influenced by the level and form of N supply to plants (Ruan 2005). Quality of green tea progressively increases with N supply, and producers of green tea, e.g. in Japan, apply up to 1200 kg N ha^{-1} per year to maximize its quality (Owuor 1997). Green tea quality and its flavour components are greatly influenced by concentrations of free amino acids, fatty acids, polyphenols (mainly catechins) and caffeine. There is a positive relationship between N concentration and accumulation of these quality-related components of tea plants (Ruan 2005). In tea plant grown hydroponically, concentrations of free amino acids were considerably higher in NH_4^+ than in NO_3^--fed plants. Furthermore, the relative allocation of absorbed N to free amino acids, particularly theanine and glutamine, was substantially increased by NH_4^+ nutrition. It has also been reported that raising the N supply leads to higher arginine concentration and consequences less favourable taste of produced green tea (Ruan et al. 2007b; Hajiboland 2017).

In contrast to green tea, black tea quality significantly declines with increase in N supply in some aspects including flavour quality (Owuor et al. 2000). Accordingly, the optimum N supply for maximum tea yield produces poor quality of black tea. Increased N supply is thought to reduce catechin levels in tea shoots, leading to low formation of the aflavins and the arubigins which are critical parameters of black tea quality. High N has also been demonstrated to increase the levels of unsaturated fatty acids and also to reduce the ratio of Group II: Group I volatile favour compounds, all of which depress black tea quality. The recommended optimum N rate which compromises for better yield and quality of black tea is within the range of 100–200 kg N ha^{-1} $year^{-1}$ (Owuor et al. 2000, 2006).

As related to the micronutrients, leaf Cu and Zn are important factors in determining the fermenting characteristics of leaf and therefore quality of the final product (Ogunmoyela et al. 1994). Application of fertilizers as a combination of N, K and micronutrients improves caffeine content of tea leaves. In other report (Njoloma 2012), it has been observed that thearubigin concentration was the only biochemical quality parameter that was affected by the micronutrient (Cu, Zn and B) foliar sprays. An increase in the thearubigin/theaflavin ratio by foliar spray of Cu, Zn and B indicated that more catechins were being converted to thearubigin than theaflavin during fermentation (Njoloma 2012).

In greenhouse-grown tea plants, modification in the metabolism of phenolic compounds under B deficiency conditions has been observed as rising of phenylalanine ammonia lyase activity in the young leaves. However, activity of polyphenol oxidase which catalyses the oxidation of catechins to theaflavins and thearubigins was only marginally affected in the young leaves of B-deficient tea plant. Soluble phenolics fraction was increased up to 3.4-fold in the young leaves while did not influence by B nutrition in the old leaves and roots. Cell wall-bound phenolic and lignin content were lower in B-deficient plants compared with B-sufficient ones (Hajiboland et al. 2013b; Hajiboland 2017).

Many studies have been conducted to investigate the F concentrations in tea plants or infused teas. Consumption of certain elements such as F, Al, Cu and Zn, through drinking tea on human health, has attracted a lot of attention (Ruan and Wong 2001). Though not essential to plants, F is an essential element of bone and tooth health of mammals. In many countries, F is used as an essential additive of toothpaste or supplemented in drinking water. However, overexposure to F could be detrimental to human health. The guideline values of 0.5–1.5 mg F L^{-1} in water have been imposed in many countries. Tea plant takes up F from soil and accumulates it in its leaves. Concentration of F in the young spring shoots including a bud with three leaves ranged from 104 to 158, while in mature leaves, the F concentrations varied from 871 to 1337 mg kg^{-1}, which were eight to ten times higher than in young shoots (Ruan and Wong 2001). A substantial amount of F is released during tea infusion (Fung et al. 1999), and nearly all (about 94.9%) of the F released is available to consumers, making tea one of the main F sources (WHO 1984).

The concentration of Al in tea and its implication on human health have been a topic of concern (Natesan and Ranganathan 1990; Owuor et al. 1990; Wong et al. 1998). It is generally believed that excessive intake of Al is harmful to health, which could cause kidney weakness (Jackson and Huang 1983). More recently, Alzheimer's disease has been claimed to be linked with a high Al content in the human brain, derived from alum-treated drinking water over a long period of time (McLachlan 1995).

Many factors influence the concentrations of elements in tea plant and its product. These included the maturity of leaves used for tea manufacturing, fertilizers added to the soil, soil conditions and the variety of tea plants (Owuor et al. 1990). It is therefore possible to reduce Al and F concentrations in tea through selection of clones and using young leaves for making tea products (Ruan and Wong 2001).

11 Environmental Consequences of Intensive Use of Fertilizers in Tea Production

Heavy application of fertilizers does not always increase the yield and quality of tea products. Some environmental problems such as an increased level of nitrate in the surrounding water systems (Hirono et al. 2009), acidification of soil and emission of high levels of N_2O, one of the major greenhouse gasses (Tokuda and Hayatsu 2004), are the consequences of heavy application of fertilizers particularly N.

11.1 Environmental Effects of N Fertilizers

The current application of N fertilizers in tea fields is usually beyond 800 Kg N ha^{-1} $year^{-1}$ and in some cases even reaches 2500 kg ha^{-1} $year^{-1}$ that is much higher than those in other crop fields (Hirono and Nonaka 2014). It has been stated that out of inorganic NPK applied annually, only 25–50% could be taken up by plants and the rest is lost through leaching, erosion, denitrification or volatilization (in case of N), or they could be immobilized in soil organic matter (Roberts 2008).

The heavy fertilizer input influences negatively water quality in tea-producing regions (Hajiboland 2017). The increasing input of fertilizers in tea fields may contribute to the deterioration of surface water quality. It was found out that a high concentration of nitrate is drained back to ponds around tea fields, and excessive leaching leads to underground water pollution (Hirono et al. 2009). In addition, the nitrous oxide emission potential is significantly high in tea fields with an N application rate of over 1000 Kg ha^{-1} $year^{-1}$ (Hirono and Nonaka 2012).

In a study on subtropical tea fields in southeast China, organic, slow-release and conventional chemical fertilizers were applied to different plots. Although equivalent N and P were applied in each fertilization treatment, the organic and slow-release fertilizer treatments had the lowest annual concentration of total N and P in runoff. This suggests that replacement of conventional chemical fertilizers with organic or slow-release fertilizers in tea fields could reduce N and P losses while maintaining tea yields (Yang et al. 2012).

It was found that the nutrient use efficiency of NPK is significantly enhanced by the application of tablet form of NPK fertilizer as a slow-release form (Jagadeeswaran et al. 2005). In order to reduce nitrate leaching, it has been recommended to use coated N fertilizers in tea fields. Calcium cyanamide ($CaCN_2$) can give a slow release of N and improve soil fertility and reduce the infestation of the soil with soilborne pathogens and weeds (Oh et al. 2006). Compared with the conventional high N application treatment (1100 kg N ha^{-1}) without lime N, the low N application (400 kg N ha^{-1}) with $CaCN_2$ effectively stopped soil acidification as well as achieved the same or slightly higher levels in tea yield and in total N and amino acid content of tea shoots (Oh et al. 2006). Higher efficiency of N uptake by tea plants and reduction of N_2O emission from tea fields was also reported for $CaCN_2$ as N fertilizer (see Sect. 2.2).

11.2 Soil pH and Microbial Ecology of Soils

Continuous use of ammonium N fertilizer in tea field leads to lower pH, affects adversely soil microorganisms (Thenmozhi et al. 2012), causes degradation of some base elements in the soil (K, Mg and Ca) and increases the levels of Mn and Al causing nutrient imbalance in the soil (TRFK 2012). Soil acidity increases with the increase of tea cultivation period. Soil pH of cultivated soils decreased by 1.37, 1.62 and 1.85, respectively, after 13, 34 and 54 years of tea plantation compared to the soil of the unused land (Wang et al. 2010). In addition, soil acidification rate at early stages of tea cultivation is higher than those at the later stages. The acidification rate for a period of 0–13 years was found to be 4.40 kmol H^+ ha^{-1} $year^{-1}$ (Wang et al. 2010). It has been demonstrated that tea cultivation intensity and duration have a strong impact on the microbial community structure, biomass and its functioning most probably through soil acidification and fertilizer addition (Han et al. 2007).

The availability and concentration of macronutrients increase with an increase in the soil pH (Nath 2013). Inorganic fertilizer requirement for young tea plants can be reduced by use of poultry manure, while the mature tea chemical fertilizer requirement can be minimized or avoided by use of cow dung as the mature tea plants require more N and P compared to the younger ones (Sultana et al. 2014).

11.3 Effect of Fertilizer on Induction of Drought Stress

Tea plants respond to high N application only under optimal irrigation or rainfall that is well distributed in the year (Carr 1972; Hajiboland 2017). It was found that high rates of N fertilizer in tea raise the optimal soil water content required for tea and subsequently lower tea productivity and survival during drought (Cheruiyot et al. 2009). Application rate of N fertilizer above 200 kg ha^{-1} limits growth and yield of tea during drought season (Ng'etich 1999). The lack of tolerance to water stress in well-fertilized tea plant is mainly due to modification of assimilate partitioning leading to an increased leaf to root mass ratios as well as reduction of soil-plant hydraulic conductivity (Cheruiyot et al. 2009).

In a study on the effect of nutrients on the post-drought stress recovery, four nutrients including K, Ca, Mn and B were applied as foliar spray at relatively low concentrations (50–100 μM). Considering biomass, water relations, antioxidative defence and phytochemical parameters, application of these nutrients, in particular K and Ca, improved post-drought recovery potential in tea plant (Upadhyaya et al. 2012).

Acknowledgement This work was modified from the paper published by the author in *Folia Horticulturae* journal (Hajiboland 2017, 29/2: 199–220). The related contents are reused with the permission.

References

Ahanger MA, Morad-Talab N, Abd-Allah EF, Ahmad P, Hajiboland R (2016) Plant growth under drought stress: significance of mineral nutrients. In: Ahmad P (ed) Water stress and crop plants: a sustainable approach. Wiley, New York, pp 649–668

Ahmed M, Hoque AR, Mamun MS (2011) Efficacy of different Sulphur formulations against red spider mite (Oligonychus coffeae Nietner) of tea in Bangladesh. Int J Sustain Agric Technol 7:29–33

Akiyama H, Yan XY, Yagi K (2006) Estimations of emission factors for fertilizer-induced direct N_2O emissions from agricultural soils in Japan: summary of available data. Soil Sci Plant Nutr 52:774–787

Aliasgharzad N, Hajiboland R, Olsson PA (2011) Lack of arbuscular mycorrhizal colonisation in tea (*Camellia sinensis* L.) plants cultivated in northern Iran. Symbiosis 5:91–95

Aliasgharzad N, Saleh Rastin N, Towfighi H, Alizadeh A (2001) Occurrence of arbuscular mycorrhizal fungi in saline soils of the Tabriz plain in relation to some physical and chemical properties of soil. Mycorrhiza 11:119–122

Alloway BJ (2009) Soil factors associated with zinc deficiency in crops and humans. Environ Geochem Health 31:537–548

Ananthacumaraswamy S, Hettiarachchi LSK, Dissanayake SM (2003) Soil and foliar sulfur status in some tea plantations of Sri Lanka. Commun Soil Sci Plant Anal 34:11–12

Anesini C, Ge F, Filip R (2008) Total polyphenol content and antioxidant capacity of commercially available tea (*Camellia sinensis*) in Argentina. J Agric Food Chem 56:9225–9229

Apostol KG, Zwiazek JJ (2004) Boron and water uptake in jack pine (*Pinus banksiana*) seedlings. Environ Exp Bot 51:145–153

Barbier O, Arreola-Mendoza L, Del Razo LM (2010) Molecular mechanisms of fluoride toxicity. Chem Biol Interact 188:319–333

Barker AV, Bryson GM (2007) Nitrogen. In: Allen V, Barker AV, Pilbeam DJ (eds) Handbook of plant nutrition. CRC Press/Taylor & Francis Group, Boca Raton, pp 21–50

Baruah BK, Haque A, Das B, Medhi C, Misra AK (2011) Boron in soil and water samples in some tea garden belt of Golaghat district. Assam Adv Appl Sci Res 2:298–305

Bhatia A, Sasmal S, Jain N, Pathak H, Kumar R, Singh A (2010) Mitigating nitrous oxide emission from soil under conventional and no-tillage in wheat using nitrification inhibitors. Agric Ecosyst Environ 136:247–253

Bouwman A, Van der Hoek K, Olivier J (1995) Uncertainties in the global source distribution of nitrous oxide. J Geophys Res 100:2785–2800

Broadley M, Brown P, Cakmak I, Rengel Z, Zhao F (2012) Function of nutrients: micronutrients. In: Marschner P (ed) Marschner's mineral nutrition of higher plants. Academic Press, London, pp 191–248

Brown PH, Bellaloui N, Bassil ES, Ruiz J, Hu H, Pfeffer H, Dannel F, Römheld V (2002) Boron in plant biology. Plant Biol 4:205–223

Burkhead JL, Gogolin Reynolds KA, Abdel-Ghany SE, Cohu CM, Pilon M (2009) Copper homeostasis. New Phytol 182:799–816

Cakmak I (2000) Possible roles of zinc in protecting plant cells from damage by reactive oxygen species. New Phytol 146:185–205

Cakmak I (2005) The role of potassium in alleviating detrimental effects of abiotic stresses in plants. J Plant Nutr Soil Sci 168:521–530

Cakmak I, Römheld V (1997) Boron deficiency-induced impairments of cellular functions in plants. Plant Soil 193:71–83

Carr HP, Lombi E, Kupper H, Mcgrath SP, Wong MH (2003) Accumulation and distribution of aluminium and other elements in tea (*Camellia sinensis*) leaves. Agronomie 23:705–710

Carr MKV (1972) The climatic requirements of the tea plant: a review. Exp Agric 8:1–14

Chen YM, Tsao TM, Liu CC, Lin KC, Wang MK (2011) Aluminium and nutrients induce changes in the profiles of phenolic substances in tea plants (*Camellia sinensis* CV TTES, no. 12 (TTE)). J Sci Food Agric 91:1111–1117

Cheruiyot EK, Mumera LM, Ng'etich WK, Hassanali A, Wachira FN (2009) High fertilizer rates increase susceptibility of tea to water stress. J Plant Nutr 33:115–129

Davidson EA, Kingerlee W (1997) A global inventory of nitric oxide emissions from soils. Nutr Cycl Agroecosyst 48:37–50

De Costa WJ, Mohotti A, Wijeratne M (2007) Ecophysiology of tea. Braz J Plant Physiol 19:299–332

Dodd AN, Kudla J, Sanders D (2010) The language of calcium signaling. Annu Rev Plant Biol 61:593–620

Fahmy FN (1977) Soil and leaf analyses in relation to tree crop nutrition in Papua New Guinea. In: Abstracts of the conference on classification and management of tropical soils. Malaysian Society of Soil Science, Kuala Lumpur, Malaysia, August 1977

FAO (Food and Agriculture Organization) (2008) Current world fertilizer trends and outlook to 2011/12. Food and Agriculture Organization of the United Nation, Rome, p 44

Fung KF, Carr HP, Zhang J, Wong MH (2008) Growth and nutrient uptake of tea under different aluminium concentrations. J Sci Food Agric 88:1582–1591

Fung KF, Zhang ZQ, Wong JWC, Wong MH (1999) Fluoride contents in tea and soil from tea plantations and the release of fluoride into tea liquor during infusion. Environ Pollut 104:197–205

Ghanati F, Morita A, Yokota H (2005) Effects of aluminum on the growth of tea plant and activation of antioxidant system. Plant Soil 276:133–141

Glass ADM, Britto DT, Kaiser BN, Kinghorn JR, Kronzucker HJ, Kumar A, Okamoto M, Rawat S, Siddiqi MY, Unkles SE, Vidmar JJ (2002) The regulation of nitrate and ammonium transport systems in plants. J Exp Bot 53:855–864

Gunaratne GP, Hettiarachchi LSK, Jayakody AN (2008) An investigation on factors responsible for sulphur nutrition of tea in Sri Lanka. Sri Lanka J Tea Sci 73:49–65

Hajiboland R, Barceló J, Poschenrieder C, Tolrà R (2013) Amelioration of iron toxicity: a mechanism for aluminum-induced growth stimulation in tea plants. J Inorg Biochem 128:183–187

Hajiboland R, Bastani S, Bahrami-Rad S, Poschenrieder C (2015) Interactions between aluminum and boron in tea (*Camellia sinensis*) plants. Acta Physiol Plant 37:54

Hajiboland R, Salehi SY (2014) Remobilization of phosphorus in tea plants. J Plant Nutr 37:1522–1533
Hajiboland R (2012) Effect of micronutrient deficiencies on plants stress responses. In: Ahmad P, Prasad MNV (eds) Abiotic stress responses in plants. Springer, New York, pp 283–329
Hajiboland R, Bahrami-Rad S, Barceló J, Poschenrieder C (2013a) Mechanisms of aluminum-induced growth stimulation in tea (*Camellia sinensis*). J Plant Nutr Soil Sci 176:616–625
Hajiboland R, Bahrami-Rad S, Bastani S, Tolrà R, Poschenrieder C (2013b) Boron re-translocation in tea (*Camellia sinensis* (L.) O. Kuntze) plants. Acta Physiol Plant 35:2373–2381
Hajiboland R, Bastani S (2012) Tolerance to water stress in boron-deficient tea (*Camellia sinensis*) plants. Folia Hort 24:41–51
Hajiboland R, Bastani S, Bahrami-Rad S (2011) Effect of light intensity on photosynthesis and antioxidant defense in boron deficient tea plants. Acta Biol Szeged 55:265–272
Hajiboland R, Poschenrieder C (2016) Localization and compartmentation of Al in the leaves and roots of tea plants. Phyton (Argentina) 84:86–100
Hajiboland R (2017) Environmental and nutritional requirements for tea cultivation. Folia Hort 29:199–220
Hamid FS (2006) Yield and quality of tea under varying conditions of soil and nitrogen availability. PhD dissertation, University Islamabad, Pakistan
Han W, Kemmitt SJ, Brookes PC (2007) Soil microbial biomass and activity in Chinese tea gardens of varying stand age and productivity. Soil Biol Biochem 39:1468–1478
Haneklaus S, Bloem E, Schnug E, de Kok LJ, Stulen I (2007) Sulphur. In: Allen V, Barker AV, Pilbeam DJ (eds) Handbook of plant nutrition. CRC Press/Taylor & Francis Group, Boca Raton, pp 183–238
Hänsch R, Mendel RR (2009) Physiological functions of mineral micronutrients (Cu, Zn, Mn, Fe, Ni, Mo, B, Cl). Curr Opin Plant Biol 12:259–266
Harbowy ME, Balentine DA (1997) Tea chemistry. Crit Rev Plant Sci 16:415–480
Hawkesford M, Horst W, Kichey T, Lambers H, Schjoerring J, Skrumsager I, White P (2012) Function of macronutrients. In: Marschner P (ed) Marschner's mineral nutrition of higher plants. Academic, London, pp 135–189
Hayatsu M, Kosuge N (1993) Autotrophic nitrification in acid tea soils. Soil Sci Plant Nutr 39:209–217
Heckman JR (2007) Chlorine. In: Allen V, Barker AV, Pilbeam DJ (eds) Handbook of plant nutrition. CRC Press/Taylor & Francis Group, Boca Raton, pp 79–291
Hepler PK (2005) Calcium: a central regulator of plant growth and development. Plant Cell 17:2142–2155
Hermans C, Verbruggen N (2005) Physiological characterization of Mg deficiency in *Arabidopsis thaliana*. J Exp Bot 56:2153–2161
Hinsinger P, Plassard C, Tang C, Jaillard B (2003) Origins of root-mediated pH changes in the rhizosphere and their responses to environmental constraints: a review. Plant Soil 248:43–59
Hirel B, Tétu T, Lea PJ, Dubois F (2011) Improving nitrogen use efficiency in crops for sustainable agriculture. Sustainability 3:1452–1485
Hirel B, Le Gouis J, Ney B, Gallais A (2007) The challenge of improving nitrogen use efficiency in crop plants: towards a more central role for genetic variability and quantitative genetics within integrated approaches. J Exp Bot 58:2369–2387
Hirono Y, Nonaka K (2014) Effects of application of lime nitrogen and dicyandiamide on nitrous oxide emissions from green tea fields. Soil Sci Plant Nutr 60:276–285
Hirono Y, Nonaka K (2012) Nitrous oxide emissions from green tea fields in Japan: contribution of emissions from soil between rows and soil under the canopy of tea plants. Soil Sci Plant Nutr 58:384–392
Hirono Y, Watanabe I, Nonaka K (2009) Trends in water quality around an intensive tea-growing area in Shizuoka, Japan. Soil Sci Plant Nutr 55:783–792
Inoue MB, Inoue M, Fernando Q, Valcic S, Timmermann BN (2002) Potentiometric and ^{1}H NMR studies of complexation of Al^{3+} with (−)-epigallocatechin gallate, a major active constituent of green tea. J Inorg Biochem 88:7–13

Jackson ML, Huang PM (1983) Aluminum of acid soils in the food chain and senility. Sci Total Environ 28:269–276

Jagadeeswaran R, Murugappan V, Govindaswamy M (2005) Effect of slow release NPK fertilizer sources on the nutrient use efficiency in turmeric (*Curcuma longa* L.). World J Agric Sci 1:65–69

Jessy MD (2010) Potassium management in plantation crops with special reference to tea, coffee and rubber. Karnataka J Agric Sci 24:67–74

Kaiser BN, Gridley KL, Ngaire Brady J, Phillips T, Tyerman SD (2005) The role of molybdenum in agricultural plant production. Ann Bot 96:745–754

Keltjens WG (1995) Magnesium uptake by Al-stressed maize plants with special emphasis on cation interactions at root exchange sites. Plant Soil 171:141–146

Kochian LV, Hoekenga OA, Piñeros MA (2004) How do crop plants tolerate acid soils? Mechanisms of aluminum tolerance and phosphorous efficiency. Annu Rev Plant Biol 55:459–493

Kochian LV (1995) Cellular mechanisms of aluminum toxicity and resistance in plants. Annu Rev Plant Biol 46:237–260

Konishi S, Miyamoto S, Taki T (1985) Stimulatory effects of aluminum on tea plants grown under low and high phosphorus supply. Soil Sci Plant Nutr 31:361–368

Kopsell DE, Kopsell DA (2007) Copper. In: Allen V, Barker AV, Pilbeam DJ (eds) Handbook of plant nutrition. CRC Press/Taylor & Francis Group, Boca Raton, pp 293–328

Kronzucker HJ, Siddiqi MY, Glass AD, Britto DT (2003) Root ammonium transport efficiency as a determinant in forest colonization patterns: an hypothesis. Physiol Plant 117:164–170

Kumar R, Singh AK, Bisen JS, Choubey M, Singh M, Bera B (2014) Influence of foliar application of micronutrients on physiological characteristics and yield of Darjeeling tea (*Camellia sinensis* L). In: Abstracts of the 3rd international conference on agriculture and horticulture, Hyderabad, India, 27–29 October 2014

Li Y, Huang J, Song X, Zhang Z, Jiang Y, Zhu Y, Zhao H, Ni D (2017) An RNA-Seq transcriptome analysis revealing novel insights into aluminum tolerance and accumulation in tea plant. Planta 1:1–3

Lin ZH, Chen LS, Chen RB, Zhang FZ, Jiang HX, Tang N, Smith BR (2011) Root release and metabolism of organic acids in tea plants in response to phosphorus supply. J Plant Physiol 168:644–652

Lin ZH, Chen LS, Chen RB, Zhang FZ, Jiang HX, Tang N (2009) CO_2 assimilation, ribulose-1,5-bisphosphate carboxylase/oxygenase, carbohydrates and photosynthetic electron transport probed by the JIP-test, of tea leaves in response to phosphorus supply. BMC Plant Biol 9:43

Lin ZH, Chen LS, Chen RB, Zhang FZ, Yang LT, Tang N (2010) Expression of genes for two phosphofructokinases, tonoplast ATPase subunit A, and pyrophosphatase of tea roots in response to phosphorus-deficiency. J Hortic Sci Biotechnol 85:449–453

Lin ZH, Qi YP, Chen RB, Zhang FZ, Chen LS (2012) Effects of phosphorus supply on the quality of green tea. Food Chem 130:908–914

Magalhaes JR, Huber DM (1989) Ammonium assimilation in different plant species as affected by nitrogen form and pH control in solution culture. Fert Res 21:1–6

Malhi SS, Schoenau JJ, Grant CA (2005) A review of Sulphur fertilizer management for optimum yield and quality of canola in the Canadian Great Plains. Can J Plant Sci 85:297–307

Masclaux-Daubresse C, Daniel-Vedele F, Dechorgnat J, Chardon F, Gaufichon L, Suzuki A (2010) Nitrogen uptake, assimilation and remobilization in plants: challenges for sustainable and productive agriculture. Ann Bot 105:1141–1157

Matsumoto H, Hirasawa E, Morimura S, Takahashi E (1976) Localization of aluminium in tea leaves. Plant Cell Physiol 17:627–631

McLachlan DR (1995) Aluminium and the risk for Alzheimer's disease. Environmetrics 6:233–275

Mejstrik L (1974) The frequency of vesicular-arbuscular mycorrhizae in the roots of *Camellia japonica* L. from different sites in New Zealand. Pac Sci 28:73–77

Mengel K, Kirkby EA (2001) Principles of plant nutrition, 5th edn. *Kluwer Academic Publisher*, Dordrecht

Mengel K (2007) Potassium. In: Allen V, Barker AV, Pilbeam DJ (eds) Handbook of plant nutrition. CRC Press/Taylor & Francis Group, Boca Raton, pp 91–120

Merhaut DJ (2007) Magnesium. In: Allen V, Barker AV, Pilbeam DJ (eds) Handbook of plant nutrition. CRC Press/Taylor & Francis Group, Boca Raton, pp 145–181

Meriño-Gergichevich C, Alberdi M, Ivanov AG, Reyes-Díaz M (2010) Al^{3+}/Ca^{2+} interaction in plants growing in acid soils: Al-phytotoxicity response to calcareous amendments. J Soil Sci Plant Nutr 10:217–243

Millaleo R, Reyes-Díaz M, Ivanov AG, Mora ML, Alberdi M (2010) Manganese as essential and toxic element for plants: transport, accumulation and resistance mechanisms. J Soil Sci Plant Nutr 10:470–481

Miller RW, Donahue RL (1990) Soils. An introduction to soils and plant growth, 6th edn. Prentice-Hall, Englewood Cliffs

Miller AJ, Cramer MD (2004) Root nitrogen acquisition and assimilation. Plant Soil 274:1–36

Miyasaka SC, Hue NV, Dunn MA (2007) Alluminum. In: Allen V, Barker AV, Pilbeam DJ (eds) Handbook of plant nutrition. CRC Press/Taylor & Francis Group, Boca Raton, pp 439–497

Mokaya BN (2016) Effect of varying rates of organic and inorganic fertilizers on growth, yield and nutrient use efficiency of clonal tea (*Camellia sinensis* [L.] O. Kuntze). Dissetation, University of Nairobi, Kenya

Morita A, Yanagisawa O, Takatsu S, Maeda S, Hiradate S (2008) Mechanism for the detoxification of aluminum in roots of tea plant (*Camellia sinensis* (L.) Kuntze). Phytochemistry 69:147–153

Mridha MAU (1999) Vesicular-arbuscular mycorrhizal fungi of soils of tea plantations. Jpn Soc Soil Sci Plant Nutr 45:397

Mukhopadhyay M, Das A, Subba P, Bantawa P, Sarkar B, Ghosh P, Mondal TK (2013) Structural, physiological, and biochemical profiling of tea plants under zinc stress. Biol Plant 57:474–480

Mulder D, Visser T (1960) Nitrogen deficiency in clonal plants. Tea Q 31:153–155

Muthukumar T, Sha LQ, Yang XD, Cao M, Tang JW, Zheng Z (2003) Mycorrhiza of plants in different vegetation types in tropical ecosystems of Xishuangbanna, Southwest China. Mycorrhiza 13:289–297

Nagata T, Hayatsu M, Kosuge N (1993) Aluminium kinetics in the tea plant using ^{27}Al and ^{19}F NMR. Phytochemistry 32:771–775

Nagata T, Hayatsu M, Kosuge N (1992) Identification of aluminium forms in tea leaves by ^{27}Al NMR. Phytochemistry 31:1215–1218

Natesan S, Ranganathan V (1990) Content of various elements in different parts of the tea plant and in infusions of black tea from southern India. J Sci Food Agric 5:125–139

Nath TN (2013) Status of macronutrients in some selected tea growing soils in Dibrugarh Sivasagar districts of Assam. India Int J Sci Res 2:273–275

Nelson S (2006) Zinc deficiency in tea (*Camellia Sinensis*). Plant disease. https://www.ctahr.hawaii.edu/oc/freepubs/pdf/PD-34.pdf. Accessed 30 Aug 2017

Neumann G, Römheld V (2012) Rhizosphere chemistry in relation to plant nutrition. In: Marschner P (ed) Marschner's mineral nutrition of higher plants. Academic Press, London, pp 347–368

Ng'etich WK (1999) Effects of different applied nitrogen rates on yield and plant survival during periods of water stress. Tea 20:61–65

Njoloma C (2012) Application of foliar sprays containing copper, zinc and boron to mature clonal tea (*Camellia sinensis*): effect on yield and quality. Dissetation, University of Pretoria, South Africa

Obatolu CR (1999) Correction of magnesium deficiency in tea plants through foliar applications. Commun Soil Sci Plant Anal 30:1649–1655

Ogunmoyela OA, Obatolu CR, Adetunji MT (1994) Effect of soil micronutrient status on the fermentation characteristics and organoleptic quality of Nigerian tea. Afr Crop Sci J 2:87–92

Oh K, Kato T, Li ZP, Li FY (2006) Environmental problems from tea cultivation in Japan and a control measure using calcium. Pedosphere 16:770–777

O'Neill MA, Eberhard S, Albersheim P, Darvill AG (2001) Requirement of borate cross-linking of cell wall rhamnogalacturonan II for *Arabidopsis* growth. Science 294:846–849

Owuor PO (1997) Fertilizer use in tea: the case of nitrogen. Tea 18:132–143
Owuor PO, Ng'etich WK, Obanda M (2000) Quality response of clonal black tea to nitrogen fertilizer, plucking interval and plucking standard. J Sci Food Agric 80:438–446
Owuor PP, Obanda M, Nyirenda HE, Mphangwe NIK, Wright LP, Apostolides Z (2006) The relationship between some chemical parameters and sensory evaluations for plain black tea (*Camellia sinensis*) produced in Kenya and comparison with similar teas from Malawi and South Africa. Food Chem 97:644–653
Owuor PO, Gone FO, Onchiri DB, Jumba IO (1990) Levels of aluminium in green leaf of clonal teas, black tea and black tea liquors, and effects of rates of nitrogen fertilizers on the aluminium black tea contents. Food Chem 35:59–68
Pandey A, Palni LMS (1996) The rhizosphere effect of tea on soil microbes in a Himalayan monsoonal location. Biol Fertil Soils 21:131–137
Pethiyadoga U, Krishnapillai S (1970) Studies on the mineral nutrition of tea 2. Experimentally-induced major nutrient deficiency symptoms. Tea Q 41:107–120
Pethiyagoda U, Krishnapillai S (1977) Studies on the mineral nutrition of tea 4-ammonium and nitrate as forms of nitrogen for tea plants in sand culture. Tea Q 42:109–117
Pilbeam DJ, Morley PS (2007) Calcium. In: Allen V, Barker AV, Pilbeam DJ (eds) Handbook of plant nutrition. CRC Press/Taylor & Francis Group, Boca Raton, pp 121–144
Poschenrieder C, Gunsé B, Corrales I, Barceló JA (2008) Glance into aluminum toxicity and resistance in plants. Sci Total Environ 400:356–368
Poschenrieder C, Tolra R, Hajiboland R, Arroyave C, Barceló J (2015) Mechanisms of hyper-resistance and hyper-tolerance to aluminum in plants. In: Panda SK, Baluška F (eds) Aluminum stress adaptation in plants. Springer, Cham, pp 81–98
Qiu SI, Wang LM, Huang DF, Lin XJ (2014) Effects of fertilization regimes on tea yields, soil fertility, and soil microbial diversity. Chilean J Agric Res 74:333–339
Rengel Z (2015) Availability of Mn, Zn and Fe in the rhizosphere. J Soil Sci Plant Nutr 15:397–409
Rennenberg H, Dannenmann M, Gessler A, Kreuzwieser J, Simon J, Papen H (2009) Nitrogen balance in forest soils: nutritional limitations of plants under climate change stresses. Plant Biol 14:4–23
Roberts TL (2008) Improving nutrient use efficiency. Turk J Agric For 32:177–182
Römheld V, Nicolic M (2007) Iron. In: Allen V, Barker AV, Pilbeam DJ (eds) Handbook of plant nutrition. CRC Press/Taylor & Francis Group, Boca Raton, pp 329–350
Ruan J, Gerendás J, Haerdter R, Sattelmacher B (2007a) Effect of alternative anions (cl$^-$ vs. SO_4^{2-}) on concentrations of free amino acids in young tea plants. J Plant Nutr Soil Sci 170:49–58
Ruan J, Gerendás J, Härdter R, Sattelmacher B (2007b) Effect of nitrogen form and root-zone pH on growth and nitrogen uptake of tea (*Camellia sinensis*) plants. AnnBot 99:301–310
Ruan J, Gerendás J (2015) Absorption of foliar-applied urea-^{15}N and the impact of low nitrogen, potassium, magnesium and sulfur nutritional status in tea (*Camellia sinensis* L.) plants. Soil Sci Plant Nutr 6:653–663
Ruan J, Lifeng M, Shi Y, Han W (2004) The impact of pH and calcium on the uptake of fluoride by tea plants (*Camellia sinensis* L.). Ann Bot 93:97–105
Ruan J, Ma L, Shi Y, Han W (2003) Uptake of uoride by tea plants (Camellia sinensis l.) and the impact of aluminium. J Sci Food Agric 83:1342–1348
Ruan J, Ma L, Yang Y (2012) Magnesium nutrition on accumulation and transport of amino acids in tea plants. J Sci Food Agric 92:1375–1383
Ruan J, Wong MH (2001) Accumulation of fluoride and aluminium related to different varieties of tea plant. Environ Geochem Health 23:53–63
Ruan J (2005) Quality-related constituents in tea (*Camellia sinensis* (L.) O. Kuntze) as affected by the form and concentration of nitrogen and the supply of chloride. Dissetation, Christian-Albrechts Universität Kiel
Ruan J, Wei K, Wand L, Cheng H, Zhang F, Wu L, Bai P, Zhang C (2016) Characteristics of NH_4^+ and NO_3^- fluxes in tea (*Camellia sinensis*) roots measured by scanning ion-selective electrode technique. Sci Rep 6:38370

Ruan J, Zhang F, Wong MH (2000) Effect of nitrogen form and phosphorus source on the growth, nutrient uptake and rhizosphere soil property of *Camellia sinensis* L. Plant Soil 223:65–73
Ruan JY, Wu X (2004) Nutrient input and evaluation of fertilization efficiency in typical tea areas of China. In: Härdter R, Xie J, Zhou J, Fan Q (eds) Nutrient management in China. Part I. Nutrient balances and nutrient cycling in agro-ecosystems. International Potash Institute, Basel, pp 367–375
Ruan JY, Wu X, Härdter R (1999) Effects of potassium and magnesium nutrition on quality components of different types of tea. J Sci Food Agric 79:47–52
Ruan JY, Wu X, Ye Y, Härdter R (1998) Effect of potassium, magnesium and Sulphur applied in different forms of fertilisers on free amino acid content in leaves of tea (*Camellia sinensis* L.). J Sci Food Agric 76:389–396
Salehi SY, Hajiboland R (2008) A high internal phosphorus use efficiency in tea (*Camellia sinensis L.)* plants. Asian J Plant Sci 7:30–36
Sanchez CA (2007) Phosphorus. In: Allen V, Barker AV, Pilbeam DJ (eds) Handbook of plant nutrition. CRC Press/Taylor & Francis Group, Boca Raton, pp 51–90
Shorrocks VM (1997) The occurrence and correction of boron deficiency. Plant Soil 193:121–148
Singh S, Pandey A, Chaurasia B, Palni LMS (2008) Diversity of arbuscular mycorrhizal fungi associated with the rhizosphere of tea growing in 'natural' and 'cultivated' ecosites. Biol Fertil Soils 44:491–500
Storey JB (2007) Zinc. In: Allen V, Barker AV, Pilbeam DJ (eds) Handbook of plant nutrition. CRC Press/Taylor & Francis Group, Boca Raton, pp 411–435
Sultana J, Siddique MNA, Kamaruzzaman M, Halim MA (2014) Conventional to ecological: tea plantation soil management in Panchagarh District of Bangladesh. J Sci Technol Environ Inform 1:27–35
Tabu IM, Kekana VM, Kamau DM (2015) Effects of varying ratios and rates of enriched cattle manure on leaf nitrogen content, yield and quality of tea (*Camellia sinensis*). J Agric Sci 7:175–181
Takmaz-Nisancioglu S, Davison AW (1988) Effects of aluminium on fluoride uptake by plants. New Phytol 109:149–155
Tejada-Jiménez M, Galván A, Fernández E, Llamas Á (2009) Homeostasis of the micronutrients Ni, Mo and cl with specific biochemical functions. Curr Opin Plant Biol 12:358–363
Thenmozhi K, Manian S, Paulsamy S (2012) Influence of long term nitrogen and potassium fertilization on the biochemistry of tea soil. J Res Agric 1:124–135
Tokuda SI, Hayatsu M (2004) Nitrous oxide flux from a tea field amended with a large amount of nitrogen fertilizer and soil environmental factors controlling the flux. Soil Sci Plant Nutr 50:365–374
Tolra R, Vogel-Mikuš K, Hajiboland R, Kump P, Pongrac P, Kaulich B, Gianoncelli A, Babin V, Barceló J, Regvar M, Poschenrieder C (2011) Localization of aluminium in tea (Camellia sinensis) leaves using low energy X-ray fluorescence spectro-microscopy. J Plant Res 124:165–172
TRFK (Tea Research Foundation of Kenya) (2002) Tea growers' handbook, 5th edn. Kijabe Printing Press, Kijabe
TRFK (Tea Research Foundation of Kenya) (2012) Tea cultivation manual for good agricultural practices, 1st edn. TRFK (Tea Research Foundation of Kenya), Nairobi
Upadhyaya H, Dutta BK, Sahoo L, Panda SK (2012) Comparative effect of Ca, K, Mn and B on post-drought stress recovery in tea [*Camellia sinensis* (L.) O Kuntze]. Am J Plant Sci 3:443–460
Upadhyaya H, Panda SK, Dutta BK (2011) $CaCl_2$ improves post-drought recovery potential in *Camellia sinensis* (L) O. Kuntze. Plant Cell Rep 30:495–503
Venkatesan S, Ganapathy MN (2004) Nitrate reductase activity in tea as influenced by various levels of nitrogen and potassium fertilizers. Commun Soil Sci Plant Anal 35:1283–1291
Wang B, Qiu YL (2006) Phylogenetic distribution and evolution of mycorrhizas in land plants. Mycorrhiza 16:299–363
Wang H, Xu RK, Wang N, Li XH (2010) Soil acidification of alfisols as influenced by tea cultivation in eastern China. Pedosphere 20:799–806

Wang M, Zheng Q, Shen Q, Guo S (2013) The critical role of potassium in plant stress response. Int J Mol Sci 14:7370–7390
Wang N, Yang C, Pan Z, Liu Y, Peng SA (2015) Boron deficiency in woody plants: various responses and tolerance mechanisms. Front Plant Sci 6:916
Weinstein LH, Davison AW (2004) Fluorides in the environment. CABI Publishing, Wallingford
WHO (World Health Organization) (1984) Fluorine and fluoride. Environmental health criteria 36. WHO (World Health Organization), Geneva, p 136
Wichard T, Mishra B, Myneni SC, Bellenger JP, Kraepiel AM (2009) Storage and bioavailability of molybdenum in soils increased by organic matter complexation. Nat Geosci 2:625
Wickramasinghe KN, Rodgers GA, Jenkinson DS (1985) Nitrification in acid tea soils and a neutral grassland soil: effects of nitrification inhibitors and inorganic salts. Soil Biol Biochem 17:249–252
Wong MH, Zhang ZQ, Wong JW, Lan CY (1998) Trace metal contents (Al, Cu and Zn) of tea: tea and soil from two tea plantations, and tea products from different provinces of China. Environ Geochem Health 20:87–94
Yang J, Yang Z, Zou J (2012) Effects of rainfall and fertilizer types on nitrogen and phosphorus concentrations in surface runoff from subtropical tea fields in Zhejiang, China. Nutr Cycl Agroecosyst 93:297–307
Yang X, Yu Z, Zhang B, Huang J, Zhang Y, Fang F, Li C, Zhu H, Chen Y (2015) Effect of fluoride on the biosynthesis of catechins in tea [*Camellia sinensis* (L.) O. Kuntze] leaves. Sci Hortic 184:78–84
Yang Y, Liu Y, Huang CF, de Silva J, Zhao FJ (2016) Aluminium alleviates fluoride toxicity in tea (*Camellia sinensis*). Plant Soil 402:179–190
Yruela I (2009) Copper in plants: acquisition, transport and interactions. Funct Plant Biol 36:409–430
Yu M, Shen R, Xiao H, Xu M, Wang H, Wang H, Zeng Q, Bian J (2009) Boron alleviates aluminum toxicity in pea (*Pisum sativum*). Plant Soil 314:87–98
Yumada H, Hattori T (1977) Investigation of the relationship between aluminium and fluorine in plants (part I). Relationship between aluminium and fluorine in tea leaves. Jpn J Soil Sci Plant Nutr 48:253–261
Zhang L, Li Q, Ma L, Ruan J (2013) Characterization of fluoride uptake by roots of tea plants (*Camellia sinensis* (L.) O. Kuntze). Plant Soil 366:659–669
Zhu Z (2000) Loss of fertilizer N from the plant–soil system and the strategies and techniques for its reduction in China. Soil Environ Sci 9:1–6
Zioła-frankowska A, Frankowski M, Novotny K, Kanicky V (2014) Comparison of the level of boron concentrations in black teas with fruit teas available on the polish market. Sci World J 2014:1. https://doi.org/10.1155/2014/898425

Chapter 10
Differential Changes in Tea Quality as Influenced by Insect Herbivory

Eric R. Scott and Colin M. Orians

Abstract Tea quality depends on plant metabolites that impact flavor, aroma, and health-beneficial properties. Plants respond to insect herbivory by altering the concentration and blend of these metabolites, and many secondary metabolites are produced only after insect attack. Research in tea and other plants shows that insect herbivores affect the concentrations of metabolites important to tea quality such as volatiles, polyphenols, methylxanthines, and amino acids. Plants, including tea, respond differently to different insect herbivores by producing different blends of metabolites. Tea plant metabolites also vary in their responses to increasing herbivore density which results in a change in metabolite blends as herbivore density changes. Because climate change is predicted to impact the density and species composition of insect herbivores in tea-growing regions of the world, induction of metabolic changes by insect herbivores represents a potentially important indirect effect of climate change on tea quality. Although it is often assumed that insect attack is detrimental to tea quality, there are some cases where tea quality is improved by herbivore-induced changes in tea metabolites. It is therefore possible that allowing some insect herbivory could be an important strategy for mitigating detrimental effects of climate on tea quality.

Keywords Secondary metabolites · Insect herbivory · Induced metabolites · Eastern Beauty oolong · Tea quality · Polyphenols · Methylxanthines · Catechins · Amino acids · Pests

---

E. R. Scott · C. M. Orians (✉)
Department of Biology, Tufts University, Medford, MA, USA
e-mail: colin.orians@tufts.edu

W.-Y. Han et al. (eds.), *Stress Physiology of Tea in the Face of Climate Change*,
https://doi.org/10.1007/978-981-13-2140-5_10

1 Introduction

Insect herbivores are the primary consumers of plants and are responsible for an estimated 13% crop loss globally (Pimentel 1991). Tea (*Camellia sinensis*) is no exception—all parts of the tea plant, including the leaves, are known to be fed on by at least one species of insect herbivore, and insect herbivory can result in 11%–55% loss in tea yield (Hazarika et al. 2009).

However, plants are not defenseless. Insect herbivores are important selective agents on plants, and this has led to the evolution of secondary metabolites that function in plant defense against herbivores (Ehrlich and Raven 1964; Berenbaum and Zangerl 1998). Plant secondary metabolites are organic compounds produced by plants that are not directly involved in primary metabolism (i.e., not carbohydrates, lipids, or amino acids). Secondary metabolites as a class have a diverse range of functions including protecting from abiotic stress such as UVB radiation, high temperatures, and drought and also mediating plant–plant and plant–enemy interactions (Raguso et al. 2015; Kessler 2015). In defense against insect herbivores, plant secondary metabolites can act as toxins, as feeding deterrents, by reducing digestibility of plant tissue or by attracting natural enemies (Bostock 2005; Cipollini and Heil 2010; Dicke and Baldwin 2010). These secondary metabolites can be produced by plants constitutively or induced by damage from herbivores or exposure to volatiles released from neighboring damaged plants (Hare 2011) and are ubiquitous among plants including those consumed by humans.

Plant secondary metabolites are consumed by humans for their flavor, aroma, psychoactive properties, antimicrobial and preservative qualities, and medicinal and health benefits. Secondary metabolites are therefore important for the quality of many crops. For example, the flavor of hops depends largely on blends of terpenes and alpha acids (Kishimoto et al. 2005; Sharp et al. 2014), the antioxidant capacity and astringent taste of wine often depend on the concentration and blend of polyphenolics (Sáenz-Navajas et al. 2010; Granato et al. 2011), and the flavor of coffee is influenced by the concentration of chlorogenic acid and trigonelline in coffee beans (Vaast et al. 2006). It therefore follows that insect herbivory should be important for crop quality, since production of many secondary metabolites is influenced by herbivory.

In addition to secondary metabolites, some primary metabolites important for crop quality are influenced by herbivory. Free amino acids are primary metabolites responsible for the umami taste of food and are an important contributor to quality of crops including tea (Kaneko et al. 2006; Feng et al. 2014; Narukawa et al. 2014). Insect herbivory has been shown to alter nitrogen dynamics in plants and cause alterations in free amino acid content which could have an impact on the quality of crop plants (Gómez et al. 2010; Ángeles-López et al. 2016).

For tea, quality can be a more important determinant of farmer profits than yield, leading some tea farmers to forgo harvesting in times of increased production but reduced quality (Ahmed et al. 2014). In addition to decreasing yields, insect herbivores also impact the quality of crop plants by altering leaf appearance and inducing changes in plant metabolite production and distribution among tissues. Changes in

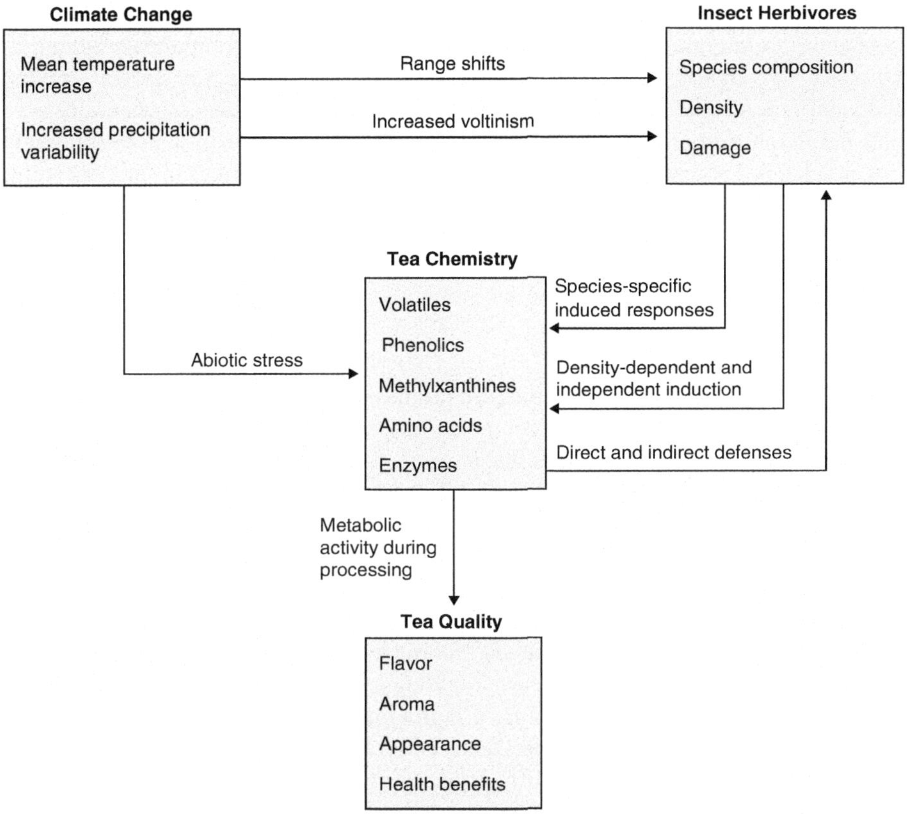

Fig. 10.1 Conceptual diagram of direct and indirect effects of climate change on tea chemistry and tea quality

tea plant metabolites can impact aspects of processed tea quality including flavor, aroma, appearance, and concentrations of health-beneficial compounds.

In this chapter, we explore the metabolites that most strongly influence tea quality, how insect herbivores affect the concentrations of those metabolites, and how climate change may have indirect impacts on tea quality mediated by effects on herbivore population dynamics and feeding. These indirect impacts of climate change on quality are likely to depend strongly on the insect herbivore species, insect herbivore density, tea cultivar, and processing methods (Fig. 10.1).

2 Chemistry of Tea Quality

Aspects of tea quality include appearance, aroma, flavor, and health-beneficial properties. All of these aspects of quality are influenced by tea plant metabolites. The main secondary metabolites in tea that contribute to quality for humans include methylxanthines, phenolics, and volatiles. Amino acids, especially L-theanine, also contribute to tea quality. All of these important tea quality components belong to classes of metabolites that are known to be influenced by insect herbivory, and the secondary metabolites (methylxanthines, phenolics, and volatiles) all generally have anti-herbivore properties. All of these metabolites are present in tea constitutively but are also influenced by herbivory. We will discuss how changes in the concentrations and blends of these metabolites are affected by herbivory in Section 3.

2.1 Tea Volatiles

Although a great diversity of volatiles are produced by tea leaves during tea processing, many volatiles and volatile precursors important for tea quality are produced by live tea plants, especially by plants that have been damaged by herbivores (Cho et al. 2007; Gohain et al. 2012; Cai et al. 2014). Volatiles are typically the earliest induced response from damaged plants, especially the so-called green leaf volatiles (GLVs). GLVs are a group of mostly C_6 compounds that are produced by constitutive enzymes from membrane lipid precursors that are released upon cell damage (ul Hassan et al. 2015). Other classes of volatiles include terpenes, aromatic compounds, and nitrogen- and sulfur-containing compounds. Volatiles can have many roles in plant–herbivore interactions, but generally they are thought to have evolved as damage signals that attract predators and parasitoids of herbivores or signal neighboring plants of impending attack (Takabayashi and Dicke 1996; Karban et al. 2013).

Volatiles comprise a very small fraction of tea leaves by weight, but because of low odor thresholds of many volatiles, they contribute strongly to tea aroma, flavor, and, therefore, quality. The volatile fraction of brewed tea can contain hundreds of unique compounds, many of which have aromas detectible by humans (Kowalsick et al. 2014). The aroma of brewed tea is due to volatile organic compounds in tea leaves released upon brewing. These volatiles can be present in live tea plants at the time of harvest or can be produced by tea leaves during processing, as the leaves remain metabolically active until the fixation step which heats leaves to denature enzymes and halt enzymatic browning as well as other metabolic processes (Cho et al. 2007; Gohain et al. 2012).

Aroma is an important aspect of tea quality, and many famous teas are described by their aromas. For example, Darjeeling black tea is often described as having a "muscatel grape" aroma (Kawakami et al. 1995; Gohain et al. 2012), and different cultivars of the tea plant processed as Phoenix oolongs (凤凰单丛 *fenghuang dancong*) are named by the aroma they impart on the finished tea (e.g., 蜜兰香 *mi lan xiang* or "honey orchid fragrance"). Besides being important for the aroma of brewed tea (via orthonasal olfaction, i.e., "smelling"), volatiles are an extremely important contributor to overall flavor and can interact with basic tastes in complex ways (via retronasal olfaction). For example, in strawberries, overall flavor intensity is best explained by total and specific volatile concentration (Schwieterman et al. 2014), and in tomatoes, specific volatiles can increase the perception of sweetness without a change in sugar content (Tieman et al. 2012). Similarly, volatiles can modulate the perception of other basic tastes important to tea quality such as bitterness (Drewnowski 2001).

2.2 *Tea Methylxanthines*

The methylxanthines are purine alkaloid compounds found in tea and other plant species produced from the precursor xanthosine by methylation and removal of ribose (Wang et al. 2016). Tea methylxanthines include caffeine, theobromine, and theophylline. In contrast to volatiles and phenolics, caffeine (the major methylxanthine in tea) is less affected by processing, resulting in similar concentrations in finished tea regardless of the type of tea (green, black, oolong, etc.) (Yung-Sheng Lin et al. 2003; Chin et al. 2008). However, the caffeine content of fresh tea leaves has been shown to vary greatly among cultivars and leaf age (Yung-Sheng Lin et al. 2003; Fang et al. 2016).

Caffeine is thought to have evolved as an anti-herbivore defense and acts as a repellant and toxicant to insects and other herbivorous pests (Hollingsworth et al. 2003; Kim et al. 2011). Variation in caffeine content has been associated with pest resistance in tea (Hewavitharanage et al. 1999). Additionally, caffeine may have antimicrobial activity in tea. Li et al. (2016a) found that a reduction in caffeine concentration in tea leaves grown under elevated CO2 was associated with a sharp increase in susceptibility to brown blight disease. In tea, caffeine is generally more concentrated in younger leaves compared to older leaves, possibly because younger leaves are more susceptible to herbivores and disease and/or more valuable to the tea plant (Yung-Sheng Lin et al. 2003).

Caffeine is the most widely used psychoactive substance by humans and occurs naturally in other plants used to produce beverages including coffee, cacao, kola (*Cola* sp.), yerba mate (*Ilex paraguariensis*), and guayusa (*Ilex guayusa*). Consumption of caffeine affects mood, cognitive function, alertness, and motor function with few negative health impacts in humans (Fredholm et al. 1999). Caffeine, one of the most abundant compounds in brewed tea, is also a major contributor to the bitter taste of tea and an important determinant of tea quality (Yu et al. 2014; Zhang et al. 2017). Additionally, caffeine may impact tea flavor in more complex ways since bitterness can affect the sensory perception of some volatiles (Drewnowski 2001).

2.3 *Tea Polyphenols*

Polyphenols are secondary metabolites found in all plant families with a wide variety of structures characterized by multiple phenol units (Levin 1971). They serve a wide variety of functions in plants including protection against light damage

(Close and McArthur 2002), microbial infection (Bansal et al. 2013), and herbivory (Levin 1971).

Tea is especially rich in one category of phenolics known as flavanols. The primary flavanols in tea are known as tea catechins and include catechin (C), epicatechin (EC), gallocatechin (GC), epigallocatechin (EGC), catechin gallate (CG), and epigallocatechin gallate (EGCG) which comprise up to 30% of tea by dry weight (Yung-Sheng Lin et al. 2003). Catechins are synthesized via the phenylpropanoid pathway with the first step catalyzed by phenylalanine ammonia lyase (PAL). Polyphenol profiles can change dramatically during tea processing, especially for black and oolong teas that undergo enzymatic browning (known as "oxidation" or "fermentation" in the tea industry). During the processing of black and oolong teas, tea leaves are bruised by shaking or rolling which combines tea catechins with polyphenol oxidase (PPO), initiating polymerization reactions that form thearubigins and other oligomers (Tanaka et al. 2009).

Catechins have been associated with disease and herbivore resistance in a variety of plants (Treutter 2006). Upon damage to plant cells, catechins, which are stored in the vacuole, are mixed with PPO, which is stored in plastids. This results in conversion of catechins to quinones through an oxidation reaction. Quinones are then able to polymerize with other quinones or polyphenols and can also bind to proteins (Holderbaum et al. 2010). These quinones can be toxic to insects and can reduce protein digestibility for vertebrates (Barbehenn and Constabel 2011). Production of polyphenols can also be induced by herbivory which activates the jasmonic acid signaling pathway which, in turn, has been shown to upregulate PAL and PPO (Nafie et al. 2011; Wang et al. 2016; Qingshan et al. 2017).

Many of the health-beneficial properties of tea, including anticancer, cardiovascular protective, and antimicrobial properties, have been ascribed to catechins (Dufresne and Farnworth 2001; McKay and Blumberg 2002; Crespy and Williamson 2004; Velayutham et al. 2008; Dias et al. 2013). Catechins also impart sweet, bitter, or astringent flavor to tea depending on concentration and catechin identity (Zhang et al. 2016). Catechins are also important precursor molecules during processing of black and oolong teas for the formation of thearubigins and other oligomers that contribute to the flavor and characteristic red color of brewed black and oolong teas (Shao et al. 1995; Tanaka et al. 2009).

2.4 Tea Amino Acids

Free amino acids are important contributors to flavor and other aspects of crop quality, and although they are primary metabolites not directly involved in plant defense, their concentrations can be influenced by herbivory. Additionally, free amino acids are typically used by plants for nitrogen transport and may be important indicators of nutritional quality for nitrogen-limited herbivores (Brodbeck et al. 1990; Fischer et al. 1998). Glutamate, glutamine, aspartate, and asparagine are typically the primary transport amino acids found in plant phloem and xylem, although the tea plant (and other *Camellia* species) uses L-theanine, a glutamic acid derivative, as its primary transport amino acid (Fischer et al. 1998; Vuong et al. 2011). Free amino acids can taste sweet, bitter, or umami depending on their identity. For example, L theanine and glutamate taste umami, aspartate tastes sweet and umami, and arginine tastes bitter (Zhang et al. 2017).

Insect herbivory has been shown to alter amino acid content of plant tissues in several insect–plant systems including whiteflies on *Capsicum annuum* and gall forming aphids on the Japanese rowan tree (*Sorbus commixta*) (Koyama et al. 2004; Ángeles-López et al. 2016). Additionally, application of exogenous methyl jasmonate, a plant hormone commonly used to mimic the effects of insect herbivory, caused a reduction of amino acid content in tomato leaves (Gómez et al. 2010). Therefore, changes in amino acid concentration represent an important mechanism by which herbivory might influence crop quality.

The main free amino acid in tea is L-theanine, a glutamate derivative, which accounts for 1%–2% of the dry weight of tea (Nobre et al. 2008). L-Theanine is synthesized in tea roots from glutamate and transported to leaves via phloem where, among other functions, it serves as a precursor for catechins (Kito et al. 1968). While L-theanine itself has no known anti-herbivore properties, its concentration in tea leaves is known to be affected by environmental factors including elevation, shade, and temperature (Saijo 1980; Li et al. 2015; Han et al. 2017). To our knowledge, there are no studies of the effects of insect herbivores on the amino acid content of tea leaves or on L-theanine concentration specifically, but given the findings from other plant–insect systems, an effect of herbivory on L-theanine concentration is likely.

L-Theanine is the primary contributor to umami flavor of tea which is important for tea quality, especially that of green teas (Kaneko et al. 2006; Narukawa et al. 2014; Han et al. 2017). Japanese shade-grown green teas such as gyokuro or tencha (used in the production of matcha) are especially high in L-theanine giving them a characteristic umami taste (Saijo 1980). In addition to an umami taste, L-theanine has been shown to have numerous effects on human health. Both in vitro and in vivo studies have shown effects on cancer prevention, cardiovascular health, immune system function, and obesity (Vuong et al. 2011). L-Theanine also crosses the blood–brain barrier and has psychoactive effects in humans. Consumption of L theanine has been shown to induce a relaxed state without drowsiness (Nobre et al. 2008). However, it is unclear whether the L-theanine content of a typical cup of tea is sufficient to produce any mood-altering effects on humans (Camfield et al. 2014).

While there are other classes of compounds that directly and indirectly impact the quality of tea, volatiles, polyphenols, methylxanthines, and amino acids are primary contributors to the flavor, aroma, and health benefits of tea. Because of the importance of these classes of compounds, they are the most well studied in the tea literature. Based on ecological studies of other plant–insect interactions, these classes of compounds are also very likely to be influenced by insect herbivory on tea plants. However, it is important to note that these metabolites will also vary in concentration due to plant genotype, environment, and management practices unrelated to herbivory. Therefore, the relative importance of herbivory compared to other factors on tea quality will likely depend on a variety of factors including the metabolite class, the identity of the insect herbivore, and a variety of environmental factors. Despite this, in at least some cases, insect herbivory is an important determinant of tea quality.

3 Herbivore-Induced Changes in Tea Metabolites

Some plant secondary metabolites are produced by plants constitutively, while the production of other metabolites is only induced by damage. Constitutive secondary metabolites are present in their active form in plant tissues even before damage and are known to differ among plant genotypes and specifically among tea cultivars (Orians et al. 1996; Kallenbach et al. 2012; Fang et al. 2016). In contrast, induced metabolites are produced or increase in concentration only after damage (Hare 2011). Damage-induced changes in metabolites can be caused by insect herbivory, mechanical damage, and plant hormones such as methyl jasmonate and volatiles released by neighboring plants (Lortzing and Steppuhn 2016). Many induced changes are almost immediate such as when secondary metabolites are stored as glycosides and mixed with glycosidases upon damage to become activated and released (Bones and Rossiter 1996). Other induced metabolites are released hours or even days after the initial damage as a result of de novo synthesis. For example, hybrid poplar leaves attacked by chrysomelid beetles synthesized and released monoterpenes with emission peaking 4 days after the initial attack (Brilli et al. 2009).

In addition, differences in herbivore species and plant genotype can result in the induction of different suites of secondary metabolites. Plant responses to chewing herbivores are different from responses to phloem feeders (Li et al. 2016b), and the responses to specialists are different from responses to generalist herbivores (Lankau 2007). Even herbivore species with similar feeding modes may elicit different responses from plants (de Oliveira et al. 2015; Unsicker et al. 2015). Plant genotypes may also vary in their induced responses to herbivory. That is, some genotypes may respond to herbivory with a greater production of induced metabolites or the production of a different blend of metabolites (Rasmann et al. 2009).

Tea plants are attacked by a variety of herbivores from 12 insect orders with different feeding modes (Hazarika et al. 2009; Ye et al. 2014). Major pests include chewing herbivores like the tea looper, *Ectropis obliqua* (Lepidoptera: Geometridae), the tea tortricids (Lepidoptera: Tortricidae), and the tea weevil, *Myllocerinus aurolineatus* (Coleoptera: Curculionidae); cell-rupture feeders like the tea leafhopper, *Empoasca onukii* (Hemiptera: Cicadellidae), various mites (Acarina), and the yellow tea thrip, *Scirtothrips dorsalis* (Thysanoptera: Thripidae); and phloem feeders like the tea mosquito bug, *Helopeltis theivora* (Hemiptera: Miridae), and the tea aphid, *Toxoptera aurantii* (Hemiptera: Aphidoidea). Induced responses of tea plants to herbivores have been studied by analyzing volatiles, nonvolatiles like catechins and methylxanthines, transcriptomes, and enzyme activity, and results show that tea plants respond differently to different herbivores and that response may vary by tea plant genotype. A few studies have additionally attempted to link insect herbivore induced metabolites to changes in tea quality, although most work is focused on understanding the response of the tea plant, not the effect on finished tea metabolites.

3.1 Chewing Herbivores on Tea

Important tea pests that feed by chewing include caterpillars of several species and tea weevils. Chewing herbivores cause more mechanical damage and remove more leaf area compared to other feeding modes. Damage from chewing feeders generally results in the immediate release of GLVs and activation of the jasmonic acid pathway (Lortzing and Steppuhn 2016). In addition to GLVs, other volatiles such as terpenoids and aromatic compounds are commonly released by tea plants in response to chewing herbivores.

Feeding by chewing herbivores on tea typically results in rapid release of a blend of volatiles. Damage by the small tea tortrix, *Adoxophyes honmai*, resulted in the emission of (Z)-3-hexen-1-ol, linalool, α-farnesene, benzyl nitrile, indole, ocimenes, and nerolidol by tea plants (Dong et al. 2011). However, this had no effect on the volatiles stored in tea leaves, indicating that they were released immediately upon attack. Therefore, it is unclear what effect tea tortrix damage might have upon processed tea quality. A similar, but not identical, blend of GLVs, aromatic volatiles, and terpenoids was released in response to the tea geometrid, *Ectropis obliqua*, and the tea weevil, *Myllocerinus aurolineatus,* which are also chewing feeders (Sun et al. 2010; Sun et al. 2014). In addition to volatile production, feeding by *E. obliqua* resulted in an upregulation of genes involved in catechin and caffeine synthesis (Wang et al. 2015; Wang et al. 2016).

The oral secretions of chewing herbivores may allow plants to recognize them and mount appropriate responses (Alborn et al. 1997). For example, despite similarity in the type of damage created, different species of lepidopteran larvae induce different plant hormone responses in *Nicotiana attenuata* (Diezel et al. 2009). Oral secretions from herbivores may also function to suppress plant defenses. For example, Yang et al. (2013) found that exposing mechanically damaged tea leaves to *E. obliqua* regurgitant suppressed PPO activity. Reduced PPO activity could reduce the effectiveness of catechins as anti-herbivore defense compounds and could have an impact on enzymatic browning during tea processing. However, the upregulation of catechin synthesis in response to *E. obliqua* attack found by Wang et al. (Wang et al. 2016) could compensate for this.

It may be tempting to think that chewing herbivory would not affect quality since leaves with extensive damage or missing leaf area may be avoided during harvest or harvest could be delayed until after pest control measures were put into place. However, Dong et al. (2011) showed that exposing undamaged tea leaves to volatiles released by damaged tea leaves resulted in metabolomic changes in the undamaged leaves. Additionally, Cai et al. (2012) observed continued production of volatiles by weevil-damaged leaves after herbivore removal. Therefore, chewing herbivore damage may result in systemic induction and the induction of defenses that lasts through time that impacts the quality of non-damaged leaves.

3.2 Tea-Feeding Aphids

Insects in the family Hemiptera typically have piercing-sucking mouthparts, and herbivorous species often use these mouthparts to penetrate plant tissues and feed on phloem, xylem, or cell contents. Aphids feed on phloem with minimal physical damage to plant tissues by moving their stylet through intercellular spaces and secreting a salivary sheath (Tjallingii and Esch 1993). Feeding by aphids often activates the salicylic acid pathway in plants in addition to the JA pathway (Walling 2000).

Han and Chen (2002) showed that attack by tea aphids resulted in higher emission of benzaldehyde and E-hexenal from tea shoots compared to control treatment. Benzaldehyde is stored in tea leaves as a cyanogenic glycoside and released by a ß-glucosidase upon damage. However, mechanical damage didn't produce high levels of benzaldehyde in this study, implying that aphid attack doesn't simply damage plant cells and might induce de novo production of benzaldehyde or its cyanogenic glycoside precursor. This has the potential to impact tea quality since benzaldehyde has an almond, cherry, or bitter aroma depending on its concentration (The Good Scents Company). Aphid damage can sometimes be concentrated and patchy, so aphid impacts on quality through induced metabolites might depend on whether induction is systemic or more localized.

3.3 Tea Mosquito Bug

Unlike aphid damage, attack by the tea mosquito bug, *Helopeltis theivora* (Hemiptera: Miridae), causes extensive necrosis around the feeding site as a result of extraoral digestion of plant tissue (Sarker and Mukhopadhyay 2006). Upon attack by the tea mosquito bug, tea plants showed increased activity of PPO but decreased PAL activity resulting in the decreased total phenolic concentrations and concentrations of some catechins (Chakraborty and Chakraborty 2005). The literature shows contrasting results on the downstream effect of this reduction in catechins on theaflavin and thearubigin production during black tea processing (Chakraborty and Chakraborty 2005; Chowdhury et al. 2016). However, in another study, researchers found flavanol biosynthesis genes upregulated by *H. theivora* feeding, although the response was cultivar specific with a resistant cultivar showing greater upregulation than a susceptible cultivar (Bandyopadhyay et al. 2014). In contrast to phenolics, feeding by *H. theivora* had little to no impact on caffeine concentration or biosynthesis. Contrasting results in the literature on the effects of *H. theivora* on tea metabolites and a lack of studies on the effects of *H. theivora* on volatile metabolites demonstrate that more research is needed to fully understand the effects of important tea pests on tea quality.

3.4 Cell-Rupture Feeding Tea Herbivores

Cell-rupture feeders are typically small (< 3 mm) and use their mouthparts to break plant cell walls and then feed on cell contents while secreting a watery saliva that may prevent some plant defense responses (Jin et al. 2012). Cell-rupture feeding results in mechanical damage to leaves, but unlike chewing feeders, leaf area isn't reduced. Typical cell-rupture feeders include leafhoppers, thrips, and mites and are an important group of pests on tea and other crops globally (Hazarika et al. 2009; Ye et al. 2014). Heavy damage by cell-rupture feeders can result in a reduction in photosynthesis, stunting of growth, and even leaf abscission.

One of the major cell-rupture feeders on tea is the tea green leafhopper. Although it was previously believed that tea farms in mainland China, Taiwan, and Japan were attacked by different species of leafhoppers, recent molecular and morphological work has shown that they are all the same species, *Empoasca onukii* (Fu et al. 2014; Qin et al. 2015; Shi et al. 2015). Electrical penetration graph (EPG) studies of *Empoasca* species have shown that they use their stylets in a rasping motion to rupture cells and ingest their contents rather than feeding on phloem (Jin et al. 2012; Miao et al. 2014). Defenses induced by leafhopper damage are thought to be important to the development of the unique aroma of some teas (Box 10.1).

Box 10.1: Eastern Beauty Tea: A Case Study in the Benefits of Herbivory for Quality

History and Geography

Eastern Beauty oolong tea (东方美人茶 *dongfang meiren*, also translated as "Oriental Beauty oolong") is a style of tea produced mainly in Hsinchu, Miaoli, and Taoyuan counties, Taiwan. The earliest records of this style of tea date to 1933 and indicate that it originated in Beipu, Hsinchu county, and at the time was called *peng feng cha* or "braggart's tea" (Lawrence Zhang, personal communication. Earliest record found in (Taiwan Bureau of Productive Industries 1933)). This style of tea is only produced using tea leaves that have been attacked by tea green leafhoppers (*E. onukii*) because of the unique flavor produced as a result of induction of certain volatiles (Qin et al. 2015).

Other so-called "bug-bitten" teas are also produced from leafhopper attacked tea plants including Precious Concubine oolong (贵妃美人茶 *guifei meiren*) and black teas often referred to as "honey fragrance" (蜜香红茶 *mixiang hong cha*). The production of Eastern Beauty oolong has also spread beyond Taiwan. Today, tea farms in Fujian and Guangdong Provinces in Mainland China (Wen-Yan Han, personal communication), and in tea-growing regions of Thailand and Vietnam, also produce Eastern Beauty oolong (Tom 2015; Hatvala Tea and Coffee 2016). The benefits of leafhopper attack may not be limited to East Asian teas either. There is some evidence that the unique flavors of second flush Darjeeling tea may be due to attack by leafhoppers and thrips (Gohain et al. 2012).

(continued)

Box 10.1 (continued)

Interaction Between Induction by Herbivores and by Processing
Tea farmers producing Eastern Beauty tea recognize interactions between the leafhopper damage and development of aroma during tea processing (Writer 2017). Attack by *E. onukii* induces the production of a suite of volatiles in live tea plants (linalool, hotrienol, methyl salicylate, phenylacetaldehyde, 2-phenylethanol, E-ß-ocimene, DMNT, and (E,E)-α-farnesene) that may serve to attract predators such as jumping spiders and contribute to tea aroma (Gao et al. 2003; Gohain et al. 2012; Cai et al. 2014; Yang et al. 2017). However, some of these induced volatiles, especially linalool, may also serve as precursors to additional aroma compounds developed during the oxidation step of processing when leaves are mechanically damaged by shaking and metabolites are exposed to enzymes and oxygen. In addition, Gohain et al. (2012) found that tea attacked by leafhoppers and thrips showed upregulation of alcohol dehydrogenase, geraniol synthase, several linalool synthases, and other enzymes related to volatile secondary metabolite production during the enzymatic browning step of Darjeeling black tea processing relative to undamaged leaves. This resulted in a significant increase in volatile concentration of finished tea. Therefore, cell-rupture feeders may improve tea quality through a combination of induction of precursor metabolites (such as linalool) and priming (upregulation of metabolic pathways), which, when combined with mechanical damage during tea processing, results in a greater concentration and unique blend of volatiles in the finished tea (Cho et al. 2007).

Adoption by Tea Farmers
Leafhopper feeding reduces photosynthesis and slows shoot growth, thus reducing yields (Fu et al. 2014). However, leafhopper damage does not directly reduce leaf area, and improvements in quality due to induced changes in metabolites may outweigh reductions in yield in terms of farmer profits. Additionally, Eastern Beauty oolong is typically produced from June through August during a time when tea quality may already be reduced due to climatic conditions (Ahmed et al. 2014). Therefore, allowing leafhopper damage may be a way for farmers to ameliorate decreases in tea quality during summer months and improve profits. In order to produce Eastern Beauty oolong, farmers may also reduce or eliminate insecticide spraying which reduces costs and environmental impact. Improvement in tea quality, increase in farmer profits, and the reduced need for insecticides may contribute to the spread of "bug bitten" tea production beyond Taiwan.

In comparison to chewing feeders, cell-rupture feeders still cause some mechanical damage to leaves, but may result in lower production of GLVs. In comparison with the chewing tea geometrid *E. obliqua*, damage by the *E. onukii* resulted in a lower induction of GLVs relative to terpenoids and a greater induction of methyl salicylate (Cai et al. 2014). Similarly, in comparison with the blend of GLVs, aromatic volatiles, and terpenoids released by tea leaves in response to chewing damage by the small tea tortrix, the cell-piercing damage of Kanzawa spider mites (*Tetranychus kanzawai*) induced a much simpler blend of only ocimenes and α-farnesene (Dong et al. 2011). Gohain et al. (2012) compared the volatile profile and gene expression of leaves exposed to damage by tea thrips (*Scirtothrips dorsalis*) or tea green leafhoppers to undamaged plants and found that damage from both thrips and leafhoppers increased volatile production and induced the upregulation of key genes involved in volatile metabolism. Species-specific effects were also apparent as thrip-infested leaves produced more linalool and linalool oxides than leafhopper-infested leaves. Thrip-infested leaves also produced some unique volatiles not present in the volatile blend from leafhopper-damaged leaves, namely, isoamyl alcohol, 2-methylbutanol, and 2-phenylethanol.

In addition to inducing volatile production, cell-rupture feeders also induce changes in nonvolatile metabolites. Mild infestation by tea green leafhoppers resulted in upregulation of caffeine synthesis genes and genes involved in cell lignification as well as a downregulation of genes involved in photosynthesis (Yang et al. 2011). A more recent study on the effects of leafhopper feeding showed decreased chlorophyll and soluble carbohydrate content of attacked leaves as well as increased total polyphenol content (Li et al. 2017). Lignification and reduction in photosynthesis and chlorophyll content are consistent with the typical damage leafhopper attack produces—slightly curled yellow leaves with stunted growth (Cho et al. 2007). These metabolic changes have potential impacts on tea quality. Lignification causes leaves to become tougher and unsuitable for processing certain types of tea that require pliable young leaves. Changes in caffeine content as a result of leafhopper damage have yet to be verified by metabolomic studies but may be another pathway by which leafhopper damage impacts the flavor profile of finished tea.

4 Importance of Herbivore Density to Secondary Metabolite Induction

The phenomenon of secondary metabolite induction by herbivores has existed for decades (Karban 1989). However, most studies on herbivore-induced metabolites compare control, undamaged plants to plants with some constant number or density of herbivores. That is, few studies explicitly test the effects of herbivore density on plant induction. Studies that do explicitly test the effects of herbivore density on induction find that plants sometimes have density-dependent responses and other times do not (all or nothing type induction). For example, one cultivar of cabbage (*Brassica oleracea* var. *capitata* cv Shikidori) was found to have a density dependent volatile production response to feeding by cabbage white larvae (*Pieris rapae*) but a strong density-independent response to diamondback moth larvae (*Plutella xylostella*) (Shiojiri et al. 2010). Another cultivar, kale (*B. oleracea* var. *acephala*), on the other hand, produced volatiles in a density-dependent manner in response to diamondback moth larval density, indicating that the difference in induced responses was particular to the host plant genotype, not the herbivore. It is therefore unsurprising when studies on induced metabolites use the same plant–insect system but obtain differing results due to inconsistent or arbitrary herbivore densities (Horiuchi et al. 2003).

Additionally, individual secondary metabolites, even those in the same class of compounds, may not all respond the same way to changes in herbivore density. For example, Horiuchi et al. (2003) identified ten volatile compounds induced by spider mites (*Tetranychus urticae*) and caterpillars (*Spodoptera exigua*) feeding on lima bean leaves, but only five and six volatiles, respectively, varied in concentration in response to spider mite and caterpillar density. This non-collinear response (i.e., not all metabolites increasing in a density-dependent manner) to increasing spider mite density also had important implications to the information contained in the induced volatile blend. Conspecifics were attracted to the volatile blend induced by low densities of herbivorous spider mites but repelled by volatiles induced by high densities.

The response of tea plant metabolites to increasing herbivore density may also be complex and non-collinear with different classes of compounds or even individual compounds responding in different ways—some may show density-dependent induction, and others may be induced, but not density-dependent. These potentially complex responses to herbivore density may have important implications for tea quality (Fig. 10.2). However, few studies on the effects of herbivore density on tea plant secondary metabolites exist, to our knowledge. Cai et al. (2012) reported differences in the number and concentration of volatiles detected in the headspace of tea plants after attack by 40 or 100 tea weevils. Tea plants with a greater weevil density emitted more volatiles at higher concentrations compared to those infested with a low density, and the emission of volatiles lasted for a longer duration after herbivore removal. The same research group showed that tea plant volatiles also vary in response to the density of tea green leafhoppers, *E. onukii* (Cai et al. 2014). They detected 31 volatiles from tea plants attacked by a high density of *E. onukii* and only 25 compounds from a low-density treatment. However, the differences in volatile profile due to herbivore density were much less than differences due to herbivore identity or between damaged and undamaged plants.

Li et al. (2017) found no significant linear relationship between leafhopper damage index and total polyphenol content of tea plants exposed to 0, 5, 10, or 20 leafhoppers. However, there may be a nonlinear relationship between leafhopper damage and polyphenol content (this is not tested by the authors). The authors did present results showing differences between the polyphenol contents of plants exposed to different numbers of leafhoppers, but these differences were only significant between some density treatments and only at 1, 4, and 10 (not 7) days of exposure to herbivory. At all sampling dates, however, the highest mean polyphenol content is in the medium density treatment (10 leafhoppers), not the highest density

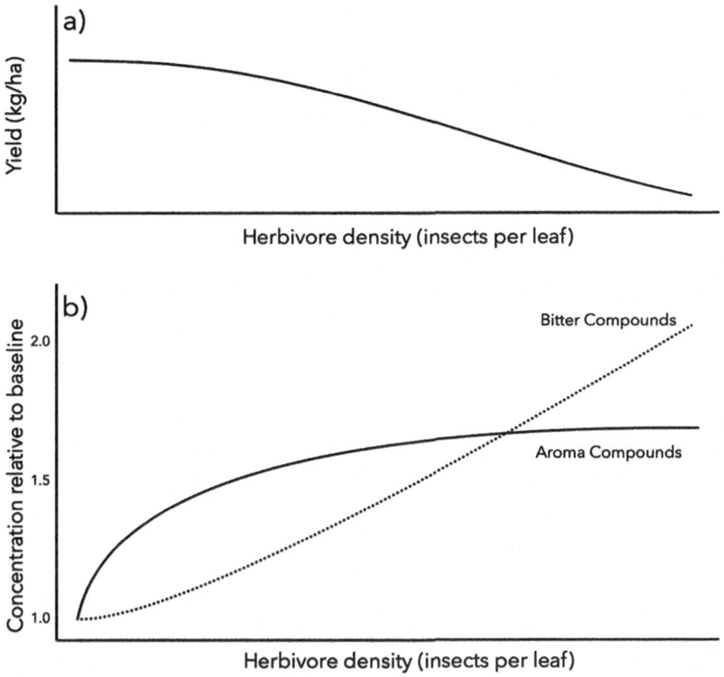

Fig. 10.2 Hypothetical response of metabolites and yield to increasing herbivore density. (**a**) Yield decreases as insect herbivore density increases. (**b**) Some metabolites are released rapidly at even low herbivore densities, while others are generally slower to be induced and/or require greater damage to trigger induction. In this hypothetical example, compounds that contribute positively to aroma and quality are induced rapidly at low densities and reach maximum induction at medium densities, while bitter-tasting compounds increase linearly with herbivore density. In this hypothetical example, under low herbivore density, yield isn't reduced dramatically, and quality may be improved due to a relatively greater induction of aroma compounds compared to bitter compounds. Therefore, a farmer might choose to allow some damage in order to improve crop quality (see Box 10.1), while at higher densities, the induction of bitter compounds and reduction in yield might be unacceptable

(20 leafhoppers). This demonstrates the need for studies that treat insect herbivore density on tea (or damage index) as a continuous variable that can be used to model the (potentially nonlinear) effects of herbivory on tea chemistry.

More studies using different herbivore species that elucidate the relationship between herbivore density or damage and tea metabolites are needed to better understand the effects of insect herbivory on quality under natural conditions. Economic thresholds are routinely established in regard to insect pest densities and yield loss (Fig. 10.2a). However, in order to establish similar economic thresholds for tea quality, it will be necessary to integrate the responses of the tea plant metabolome and yield to continuous variation in pest density (Fig. 10.2b) and to better understand the interactions between herbivory and other environmental factors. This will be especially important in the face of a changing climate that could impact herbivore densities across time.

5 Effects of Climate Change on Insect Herbivory on Tea

Climate change is expected to have a strong impact on subtropical climates which include the majority of tea-growing regions (IPCC 2007). The direct impacts of climate change on tea include alteration in tea quality through changes in temperature (Chaps. 2, 3), precipitation (Chap. 4), atmospheric CO_2 (Chap. 7), and ozone (Chap. 6). However, the effects of climate change on insect herbivores are likely to have strong indirect impacts on tea quality as well. The increases in mean annual temperature, extreme temperature events, and precipitation variability under global climate change are expected to impact insect herbivore abundances, range, seasonality, and behavior (Rosenzweig et al. 2001). These changes in herbivore populations will have impacts on tea secondary metabolites through induced defenses and affect tea quality.

Warming, but not extreme temperature events, is predicted to directly benefit many insect species. As temperatures increase, insect herbivores generally show increases in vital rates, such as development rate or egg hatching rates. Additionally, winter mortality may decrease, and emergence times of insect herbivores may advance due to a disproportionate increase in nighttime and winter temperatures (Rosenzweig et al. 2001; Bale et al. 2002). Therefore, an increase in mean temperature is likely to have the strongest effect on population densities of multivoltine insect herbivores by increasing the number of generations per season (Bale et al. 2002). However, increasing temperatures may also increase population density and activity of insect predators which may abate the increase in herbivore density (Berggren et al. 2009) (Fig. 10.3). Additionally, extreme high temperatures may increase insect mortality (Hogg 1985; Reineke and Hauck 2012). In addition to increasing insect vital rates, warming is also expected to increase insect activity and potentially feeding rates, since insect herbivores are poikilotherms, and it may also result in range increases for some insects (Berggren et al. 2009).

The effects of increased precipitation variability on insect herbivore population density are less predictable, although long wet periods may increase insect mortal-

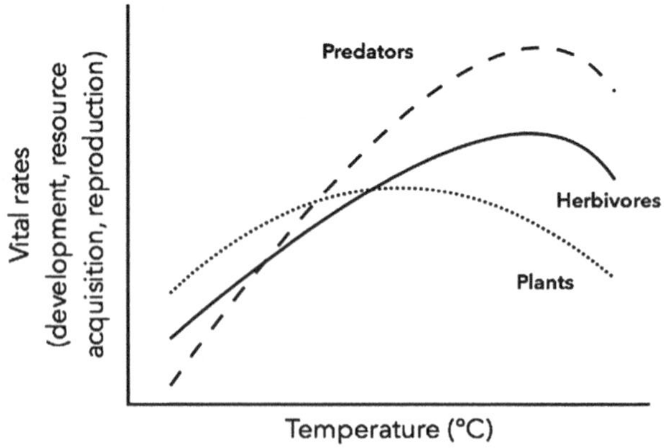

Fig. 10.3 A conceptual model showing the effects of increased mean temperatures on plants, herbivores, and predators. At high temperatures, plant growth decreases due to photorespiration. Predator vital rates increase more as temperature increases relative to herbivores because of higher temperature sensitivity of movement and prey capture. (Adapted from Berggren et al. 2009)

ity by encouraging the spread of entomopathogens through increased relative humidity, while prolonged dry periods may result in mortality due to desiccation (Thacker et al. 1997; Luz and Fargues 1999). Additionally, storms may contribute to direct mortality or hinder their dispersal of very small insect herbivores (Morsello et al. 2008).

Warming due to climate change is predicted to increase growth rate of plants less than insects, despite an increase in atmospheric CO_2. This is a result of a steep increase in respiration and photorespiration as temperature increases, while photosynthesis has a temperature optimum and decreases at high temperatures (Ayres 1993; Berggren et al. 2009) (Fig. 10.3). As temperatures increase, eventually photosynthesis slows down, and photorespiration increases, thus reducing net carbon assimilation and slowing plant growth. The differential increase in insect population growth and plant shoot growth due to climate warming could lead to an effective increase in herbivore density on tea (number of insects per young leaf). Slow tea shoot growth could also reduce the ability of tea plants to "outgrow" herbivore damage which has consequences for tea quality since only the youngest leaves are harvested. Additionally, increases in atmospheric CO_2 concentrations may lead to increased carbon to nitrogen ratios in plant tissue, making it less nutritious and increasing consumption of leaves by insect herbivores (DeLucia et al. 2012). Therefore, increases in effective herbivore density and damage (e.g., insects per young leaf, % leaf area damaged) may result from increased herbivore population growth relative to plant growth.

Although it is likely that climate change will result in increased herbivory on tea, tea farmers may be able to adapt by changing management practices to reduce herbivore numbers, increase tea shoot growth, or take advantage of insect-induced changes in tea plant chemistry. For example, Ahmed et al. (2013) found that simulated herbivory (methyl jasmonate treatment) partially offsets the impacts of drought on total methylxanthine and total phenolic concentrations of greenhouse-grown tea plants. Additionally, for some, teas slow tea shoot growth relative to insect densities may be beneficial since damage by some herbivores may actually increase tea quality through induced secondary metabolites without great losses in yield (Box 10.1).

6 Conclusion

For tea, quality can be very important for consumers and for farmer livelihood—sometimes more so than yield. All of the components of tea quality including color, flavor, aroma, and health benefits result from metabolites whose concentrations and blends are known to respond to insect herbivory. In fact, many secondary metabolites in tea may have evolved specifically to respond to herbivores. Therefore, it is essential that we understand the impacts of herbivory on tea chemistry and the impacts of tea chemistry on quality.

The impact of herbivory on plant metabolites is known to depend on herbivore density, herbivore identity, and plant genotype. However, current research on the effects of herbivores on tea metabolites is limited, especially in terms of the effects of herbivore density and interactions between herbivore identity and tea cultivar. Other plant–insect systems have shown great variation in inducibility of metabolite classes among plant genotypes and herbivore-specific induced responses including variation in the density dependency of induction (Shiojiri et al. 2010). In order to better understand the impact of herbivory on tea quality, we need to understand interactive effects of herbivore identity, herbivore density, cultivar, and other environmental factors on a wide variety of metabolite classes. High-throughput metabolomic techniques combined with powerful multivariate analysis tools will be essential in understanding the effects of insect herbivory on tea quality.

Much of the past research on the effects of herbivory on tea plants focuses on ecological interactions and does not follow the impacts of herbivory on tea plants through to the finished product in order to directly assess changes in quality. For example, it is unclear what role tea processing methods may play in terms of interaction with induced metabolite changes due to herbivory. Tea leaves remain metabolically active during tea processing until the fixation step. Therefore, induction of precursor metabolites or upregulation of enzymes in response to herbivory may result in interactions between herbivory and processing method that are largely unexplored.

It is also important to study the response of tea plant metabolites to herbivory in a context of climate change. A warming climate is likely to result in changes in herbivore densities and timing of herbivore emergence relative to tea plant growth as well as changes in herbivore ranges in both altitude and latitude. Because impacts of herbivory on tea quality may depend on herbivore density and identity, climate change is likely to affect the impacts of herbivory on tea quality.

However, there is some good news—effects of herbivory on tea quality are not necessarily negative. Bug-bitten teas like Eastern Beauty oolong may present a model for tea farmers to adapt to climate change by taking advantage of insect herbivory to ameliorate quality impacts of adverse weather conditions (Box 10.1). Although currently only a limited set of tea processing styles explicitly take advantage of insect herbivory to improve flavor quality, the potential for beneficial effects of herbivory may not be limited to these types of tea or to improvements in flavor. Besides flavor, health benefits are of major interest to tea consumers, and many health-beneficial compounds in tea may be increased in concentration in response to herbivores.

Another complicating factor not discussed here is that tea cultivars differ in their susceptibility to herbivory (Shaha et al. 2014). Understanding interactions between susceptibility to herbivory and the metabolic changes that happen as a result of damage will likely be important to finding suitable cultivars to take advantage of secondary metabolite induction in order to improve quality.

Tea may provide a model for other quality-centric crop systems around the world where secondary metabolites are particularly important relative to yield. Because the classes of metabolites important for tea quality are shared by many other crops, an improved understanding of how herbivory impacts tea quality could inform other crop systems. And because there are cases where herbivory actually improves tea quality, there may also be the potential to improve the quality of other crops by taking advantage of the natural response of plants to their insect herbivores in order to affect changes in flavor, aroma, or health benefits to consumers.

Acknowledgments The authors would like to thank Professor Lawrence Zhang and David Campbell for information about the origins of Eastern Beauty oolong tea, as well as Clarissa Wei, Lew Perin, and Michael Coffee for advice on the translation of tea names.

References

Ahmed S, Orians CM, Griffin T, et al (2013) Effects of water availability and Pest pressures on tea (Camellia sinensis) growth and functional quality. AoB plants 6:plt054

Ahmed S, Stepp JR, Orians CM et al (2014) Effects of extreme climate events on tea (*Camellia sinensis*) functional quality validate indigenous farmer knowledge and sensory preferences in tropical China. PLoS One 9:e109126

Alborn HT, Turlings TCJ, Jones TH et al (1997) An elicitor of plant volatiles from beet armyworm oral secretion. Science 276:945–949

Ángeles-López YI, Rivera-Bustamante RF, Heil M (2016) Colonization by phloem-feeding herbivore overrides effects of plant virus on amino acid composition in phloem of chili plants. J Chem Ecol:1–4

Ayres MP (1993) Plant defense, herbivory, and climate change. In: Kareiva PM, Kingsolver JG, Huey RB (eds) Biotic interactions and global change. Sinauer Associates Incorporated, Sunderland, MA

Bale JS, Masters GJ, Hodkinson ID et al (2002) Herbivory in global climate change research: direct effects of rising temperature on insect herbivores. Glob Change Biol 8:1–16

Bandyopadhyay T, Gohain B, Bharalee R et al (2014) Molecular landscape of *Helopeltis theivora* induced transcriptome and defense gene expression in tea. Plant Mol Biol Rep 33:1042–1057

Bansal S, Choudhary S, Sharma M et al (2013) Tea: a native source of antimicrobial agents. Food Res Int 53:568–584

Barbehenn RV, Constabel CP (2011) Tannins in plant-herbivore interactions. Phytochemistry 72:1551–1565

Berenbaum MR, Zangerl AR (1998) Chemical phenotype matching between a plant and its insect herbivore. PNAS 95:13743–13748

Berggren Å, Björkman C, Bylund H, Ayres MP (2009) The distribution and abundance of animal populations in a climate of uncertainty. Oikos 118:1121–1126

Bones AM, Rossiter JT (1996) The myrosinase-glucosinolate system, its organisation and biochemistry. Physiol Plantarum 97:194–208

Bostock RM (2005) Signal crosstalk and induced resistance: straddling the line between cost and benefit. Annu Rev Phytopathol 43:545–580

Brilli F, Ciccioli P, Frattoni M et al (2009) Constitutive and herbivore-induced monoterpenes emitted by *Populus* x *euroamericana* leaves are key volatiles that orient *Chrysomela populi* beetles. Plant Cell Environ 32:542–552

Brodbeck BV, Mizell RF, French WJ et al (1990) Amino acids as determinants of host preference for the xylem feeding leafhopper, *Homalodisca coagulata* (Homoptera: Cicadellidae). Oecologia 83:338–345

Cai X-M, Sun X-L, Dong W-X et al (2012) Variability and stability of tea weevil-induced volatile emissions from tea plants with different weevil densities, photoperiod and infestation duration. Insect Science 19:507–517

Cai X-M, Sun XL, Dong WX et al (2014) Herbivore species, infestation time, and herbivore density affect induced volatiles in tea plants. Chemoecology 24:1–14

Camfield DA, Stough C, Farrimond J, Scholey AB (2014) Acute effects of tea constituents L-theanine, caffeine, and epigallocatechin gallate on cognitive function and mood: a systematic review and meta-analysis. Nutr Rev 72:507–522

Chakraborty U, Chakraborty N (2005) Impact of environmental factors on infestation of tea leaves by *Helopeltis theivora*, and associated changes in flavonoid flavor components and enzyme activities. Phytoparasitica 33:88–96

Chin JM, Merves ML, Goldberger BA et al (2008) Caffeine content of brewed teas. J Anal Toxicol 32:702–704

Cho J-Y, Mizutani M, Shimizu B-I et al (2007) Chemical profiling and gene expression profiling during the manufacturing process of Taiwan oolong tea "Oriental Beauty". Biosci Biotechnol Biochem 71:1476–1486

Chowdhury RS, Moly IS, Ahmed M et al (2016) Impact of the mosquito bug (*Helopeltis theivora*) infestation on the quality of tea (*Camellia sinensis*). Bangladesh J Zool 44:197–207

Cipollini D, Heil M (2010) Costs and benefits of induced resistance to herbivores and pathogens in plants. Plant Sci Rev 5:1–25

Close DC, McArthur C (2002) Rethinking the role of many plant phenolics—protection from photodamage not herbivores? Oikos 99:166–172

Crespy V, Williamson G (2004) A review of the health effects of green tea catechins in in vivo animal models. J Nutr 134:3431S–3440S

de Oliveira EF, Pallini A, Janssen A (2015) Herbivores with similar feeding modes interact through the induction of different plant responses. Oecologia 180:1–10

DeLucia EH, Nabity PD, Zavala JA, Berenbaum MR (2012) Climate change: resetting plant-insect interactions. Plant Physiol 160:1677–1685

Dias TR, Tomás G, Teixeira NF, Alves MG (2013) White tea (Camellia Sinensis (L.)): antioxidant properties and beneficial health effects. Int J Food Sci Nutr Diet 2:19–26

Dicke M, Baldwin IT (2010) The evolutionary context for herbivore-induced plant volatiles: beyond the "cry for help". Trends Plant Sci 15:167–175

Diezel C, von CC D, Gaquerel E, Baldwin IT (2009) Different lepidopteran elicitors account for cross-talk in herbivory-induced phytohormone signaling. Plant Physiol 150:1576–1586

Dong F, Yang Z, Baldermann S et al (2011) Herbivore-induced volatiles from tea (*Camellia sinensis*) plants and their involvement in Intraplant communication and changes in endogenous nonvolatile metabolites. J Agric Food Chem 59:13131–13135

Drewnowski A (2001) The science and complexity of bitter taste. Nutr Rev 59:163–169

Dufresne CJ, Farnworth ER (2001) A review of latest research findings on the health promotion properties of tea. J Nutr Biochem 12:404–421

Ehrlich PR, Raven PH (1964) Butterflies and plants: a study in coevolution. Evolution 18:586–608

Fang R, Redfern SP, Kirkup D et al (2016) Variation of theanine, phenolic, and methylxanthine compounds in 21 cultivars of Camellia sinensis harvested in different seasons. Food Chem 220:571–526

Feng L, Gao MJ, Hou RY et al (2014) Determination of quality constituents in the young leaves of albino tea cultivars. Food Chem 155:98–104

Fischer W-N, André B, Rentsch D et al (1998) Amino acid transport in plants. Trends Plant Sci 3:188–195

Fredholm BB, Bättig K, Holmén J, Nehlig A (1999) Actions of caffeine in the brain with special reference to factors that contribute to its widespread use. Pharmacol Rev 51:83–133

Fu J-Y, Han B-Y, Xiao Q (2014) Mitochondrial *COI* and 16sRNA evidence for a single species hypothesis of *E. vitis*, *J. formosana* and *E. onukii* in East Asia. PLoS One 9:e115259

Gao J, Zhao D, Chen Z (2003) Predatory function of *Evarcha albaria* upon *Empoasca vitis*. Chin J Trop Crop 25:72–74

Gohain B, Borchetia S, Bhorali P et al (2012) Understanding Darjeeling tea flavour on a molecular basis. Plant Mol Biol 78:577–597

Gómez S, Ferrieri RA, Schueller M, Orians CM (2010) Methyl jasmonate elicits rapid changes in carbon and nitrogen dynamics in tomato. New Phytol 188:835–844

Granato D, Katayama FCU, de Castro IA (2011) Phenolic composition of South American red wines classified according to their antioxidant activity, retail price and sensory quality. Food Chem 129:366–373

Han BY, Chen ZM (2002) Composition of the volatiles from intact and mechanically pierced tea aphid–tea shoot complexes and their attraction to natural enemies of the tea aphid. J Agric Food Chem 50:2571–2575

Han W-Y, Huang J-G, Li X et al (2017) Altitudinal effects on the quality of green tea in East China: a climate change perspective. Eur Food Res Technol 243:323–330

Hare JD (2011) Ecological role of volatiles produced by plants in response to damage by herbivorous insects. Annu Rev Entomol 56:161–180

Hatvala Tea and Coffee (2016) Oriental beauty: estate oolong tea, son La, Vietnam. In: Hatvala Tea and Coffee. http://drinksbeansandleaves.com/Vietnamese-Tea/Vietnam-Oolong-Tea/Oriental Beauty Oolong-Tea-Vietnam. Accessed 4 Oct 2017

Hazarika LK, Bhuyan M, Hazarika BN (2009) Insect pests of tea and their management. Annu Rev Entomol 54:267–284

Hewavitharanage P, Karunaratne S, Kumar NS (1999) Effect of caffeine on shot-hole borer beetle (*Xyleborusfornicatus*) of tea (*Camellia sinensis*). Phytochemistry 51:35–41

Hogg DB (1985) Potato leafhopper (Homoptera: Cicadellidae) immature development, life tables, and population dynamics under fluctuating temperature regimes. Environ Entomol 14:349–355

Holderbaum DF, Kon T, Kudo T, Guerra MP (2010) Enzymatic browning, polyphenol oxidase activity, and polyphenols in four apple cultivars: dynamics during fruit development. Hortscience 45:1150–1154

Hollingsworth RG, Armstrong JW, Campbell E (2003) Caffeine as a novel toxicant for slugs and snails. Ann Appl Biol 142:91–97

Horiuchi J-I, Arimura G-I, Ozawa R et al (2003) A comparison of the responses of *Tetranychus urticae* (Acari: Tetranychidae) and *Phytoseiulus persimilis* (Acari: Phytoseiidae) to volatiles emitted from lima bean leaves with different levels of damage made by *T. urticae* or *Spodoptera exigua* (Lepidoptera: Noctuidae). Appl Entomol Zool 38:109–116

IPCC (2007) Contribution of working group I to the fourth assessment report of the intergovernmental panel on climate change, 2007. Cambridge, UK

Jin S, Chen ZM, Backus EA et al (2012) Characterization of EPG waveforms for the tea green leafhopper, *Empoasca vitis* Göthe (Hemiptera: Cicadellidae), on tea plants and their correlation with stylet activities. J Insect Physiol 58:1235–1244

Kallenbach M, Bonaventure G, Gilardoni PA et al (2012) *Empoasca* leafhoppers attack wild tobacco plants in a jasmonate-dependent manner and identify jasmonate mutants in natural populations. PNAS 109:E1548–E1557

Kaneko S, Kumazawa K, Masuda H et al (2006) Molecular and sensory studies on the umami taste of Japanese green tea. J Agric Food Chem 54:2688–2694

Karban R (1989) Induced plant responses to herbivory. Annu Rev Ecol Syst 20:331–348

Karban R, Shiojiri K, Ishizaki S et al (2013) Kin recognition affects plant communication and defence. Proc R Soc 280:20123062

Kawakami M, Ganguly SN, Banerjee J, Kobayashi A (1995) Aroma composition of oolong tea and black tea by brewed extraction method and characterizing compounds of Darjeeling tea aroma. J Agric Food Chem 43:200–207

Kessler A (2015) The information landscape of plant constitutive and induced secondary metabolite production. Curr Opin Insect Sci 8:47–53

Kim Y-S, Lim S, Kang K-K et al (2011) Resistance against beet armyworms and cotton aphids in caffeine-producing transgenic chrysanthemum. Plant Biotechnol 28:393–395

Kishimoto T, Wanikawa A, Kagami N, Kawatsura K (2005) Analysis of hop-derived Terpenoids in beer and evaluation of their behavior using the stir bar–Sorptive extraction method with GC-MS. J Agric Food Chem 53:4701–4707

Kito M, Kokura H, Izaki J, Sasaoka K (1968) Theanine, a precursor of the phloroglucinol nucleus of catechins in tea plants. Phytochemistry 7:599–603

Kowalsick A, Kfoury N, Robbat A Jr et al (2014) Metabolite profiling of *Camellia sinensis* by automated sequential, multidimensional gas chromatography/mass spectrometry reveals strong monsoon effects on tea constituents. J Chromatogr A 1370:230–239

Koyama Y, Yao I, Akimoto S-I (2004) Aphid galls accumulate high concentrations of amino acids: a support for the nutrition hypothesis for gall formation. Entomol Exp Appl 113:35–44

Lankau RA (2007) Specialist and generalist herbivores exert opposing selection on a chemical defense. New Phytol 175:176–184

Levin DA (1971) Plant Phenolics: an ecological perspective. Am Nat 105:157–181

Li C-F, Yao M-Z, Ma C-L et al (2015) Differential metabolic profiles during the albescent stages of "Anji Baicha" (*Camellia sinensis*). PLoS One 10:e0139996

Li X, Ahammed GJ, Li Z et al (2016a) Decreased biosynthesis of Jasmonic acid via lipoxygenase pathway compromised caffeine-induced resistance to Colletotrichum gloeosporioides under elevated CO2 in tea seedlings. Phytopathology 106:1270–1277

Li Y, Dicke M, Kroes A et al (2016b) Interactive effects of cabbage aphid and Caterpillar herbivory on transcription of plant genes associated with Phytohormonal Signalling in wild cabbage. J Chem Ecol 42:1–13

Li J-Y, Shi M-Z, Fu J-W et al (2017) Physiological and biochemical responses of *Camellia sinensis* to stress associated with *Empoasca vitis* feeding. Arthropod Plant Interact 24:1–11

Lin Y-S, Tsai Y-J, Jyh-Shyan Tsay A, Lin J-K (2003) Factors affecting the levels of tea polyphenols and caffeine in tea leaves. J Agric Food Chem 51:1864–1873

Lortzing T, Steppuhn A (2016) Jasmonate signalling in plants shapes plant–insect interaction ecology. Curr Opin Insect Sci 14:32–39

Luz C, Fargues J (1999) Dependence of the entomopathogenic fungus, *Beauveria bassiana*, on high humidity for infection of *Rhodnius prolixus*. Mycopathologia 146:33–41

McKay DL, Blumberg JB (2002) The role of tea in human health: an update. J Am Coll Nutr 21:1–13

Miao J, Han BY, Zhang QH (2014) Probing behavior of *Empoasca vitis* (Homoptera: Cicadellidae) on resistant and susceptible cultivars of tea plants. J Insect Sci 14:223–223

Morsello SC, Groves RL, Nault BA, Kennedy GG (2008) Temperature and precipitation affect seasonal patterns of dispersing tobacco Thrips, *Frankliniella fusca*, and onion Thrips, *Thrips tabaci* (Thysanoptera: Thripidae) caught on sticky traps. Environ Entomol 37:79–86

Nafie E, Hathout T, Mokadem Al AS (2011) Jasmonic acid elicits oxidative defense and detoxification systems in *Cucumis melo* L. cells. Braz J Plant Physiol 23:161–174

Narukawa M, Toda Y, Nakagita T et al (2014) L-Theanine elicits umami taste via the T1R1 + T1R3 umami taste receptor. Amino Acids 46:1583–1587

Nobre AC, Rao A, Owen GN (2008) L-theanine, a natural constituent in tea, and its effect on mental state. Asia Pac J Clin Nutr 17:167–168

Orians C, Roche BM, Fritz RS (1996) The genetic basis for variation in the concentration of phenolic glycosides in *Salix sericea*: an analysis of heritability. Biochem Syst Ecol 24:719–724

Pimentel D (1991) Diversification of biological control strategies in agriculture. Crop Prot 10:243–253

Qin D, Zhang L, Xiao Q et al (2015) Clarification of the identity of the tea green leafhopper based on morphological comparison between Chinese and Japanese specimens. PLoS One 10:e0139202

Qingshan X, Junyan Z, Shiqi Z et al (2017) Transcriptome profiling using single-molecule direct RNA sequencing approach for in-depth understanding of genes in secondary metabolism pathways of *Camellia sinensis*. Front Plant Sci 8:1205

Raguso RA, Agrawal AA, Douglas AE et al (2015) The raison d'être of chemical ecology. Ecology 96:617–630

Rasmann S, Johnson MD, Agrawal AA (2009) Induced responses to herbivory and jasmonate in three milkweed species. J Chem Ecol 35:1326–1334

Reineke A, Hauck M (2012) Larval development of *Empoasca vitis* and *Edwardsiana rosae* (Homoptera: Cicadellidae) at different temperatures on grapevine leaves. J Appl Entomol 136:656–664
Rosenzweig C, Iglesias A, Yang XB et al (2001) Climate change and extreme weather events; implications for food production, plant diseases, and pests. Glob Chang Hum Health 2:90–104
Sáenz-Navajas M-P, Campo E, Fernández-Zurbano P et al (2010) An assessment of the effects of wine volatiles on the perception of taste and astringency in wine. Food Chem 121:1139–1149
Saijo R (1980) Effect of shade treatment on biosynthesis of catechins in tea plants. Plant Cell Physiol 21:989–998
Sarker M, Mukhopadhyay A (2006) Studies on salivary and midgut enzymes of a major sucking pest of tea, *Helopeltis theivora* (Heteroptera: Miridae) from Darjeeling Plains, India. J Ent Res Soc 8:27–36
Schwieterman ML, Colquhoun TA, Jaworski EA et al (2014) Strawberry flavor: diverse chemical compositions, a seasonal influence, and effects on sensory perception. PLoS One 9:e88446–e88412
Shaha S, Yadava R, Boruab PK (2014) Biochemical Defence mechanism in *Camellia sinensis* against *Helopeltis theivora*. Int J Plant Anim Environ Sci 4:246–253
Shao W, Powell C, Clifford MN (1995) The analysis by HPLC of green, black and Pu'er teas produced in Yunnan. J Sci Food Agric 69:535–540
Sharp DC, Townsend MS, Qian Y, Shellhammer TH (2014) Effect of harvest maturity on the chemical composition of Cascade and Willamette hops. J Am Soc Brew Chem 72:231–238
Shi LQ, Zeng ZH, Huang HS et al (2015) Identification of *Empoasca onukii* (Hemiptera: Cicadellidae) and monitoring of its populations in the tea plantations of South China. J Econ Entomol 108:1025–1033
Shiojiri K, Ozawa R, Kugimiya S et al (2010) Herbivore-specific, density-dependent induction of plant volatiles: honest or "cry wolf" signals? PLoS One 5:e12161
Sun X-L, Wang G-C, Cai X-M et al (2010) The tea weevil, *Myllocerinus aurolineatus*, is attracted to volatiles induced by conspecifics. J Chem Ecol 36:388–395
Sun X-L, Wang G-C, Gao Y et al (2014) Volatiles emitted from tea plants infested by *Ectropis obliqua* larvae are attractive to conspecific moths. J Chem Ecol 40:1080–1089
Taiwan Bureau of Productive Industries (1933) Issen en no kōkyū cha. Taiwan no chagyō 16(1):30
Takabayashi J, Dicke M (1996) Plant–carnivore mutualism through herbivore-induced carnivore attractants. Trends Plant Sci 1:109–113
Tanaka T, Mine C, Watarumi S et al (2009) Production of Theaflavins and Theasinensins during tea fermentation. In: Biologically active natural products. American Chemical Society, Washington, DC, pp 188–196
Thacker JI, Thieme T, Dixon AFG (1997) Forecasting of periodic fluctuations in annual abundance of the bean aphid: the role of density dependence and weather. J Appl Entomol 121:137–145
The Good Scents Company Flavor and Fragrance Information Catalog. In: http://www.thegoodscentscompany.com/allprod.html. Accessed 7 Sept 2017
Tieman D, Bliss P, McIntyre LM et al (2012) The chemical interactions underlying tomato flavor preferences. Curr Biol 22:1035–1039
Tjallingii WF, Esch TH (1993) Fine structure of aphid stylet routes in plant tissues in correlation with EPG signals. Physiol Entomol 18:317–328
Tom (2015) DMS "Cha Nang Ngam" Oriental Beauty Oolong Tea - North Thailand's "Dong Fang Mei Ren." In: Siam Tea Blog. http://siamteas.com/2015/05/28/dms-cha-nang-ngam-oriental beauty-oolong-tea-north-thailands-dong-fang-mei-ren/. Accessed 4 Oct 2017
Treutter D (2006) Significance of flavonoids in plant resistance: a review. Environ Chem Lett 4:147–157
Ul Hassan MN, Zainal Z, Ismail I (2015) Green leaf volatiles: biosynthesis, biological functions and their applications in biotechnology. Plant Biotechnol J 13:727–739
Unsicker SB, Gershenzon J, Köllner TG (2015) Beetle feeding induces a different volatile emission pattern from black poplar foliage than caterpillar herbivory. Plant Signal Behav 10:e987522

Vaast P, Bertrand B, Perriot JJ et al (2006) Fruit thinning and shade improve bean characteristics and beverage quality of coffee (*Coffea arabica* L.) under optimal conditions. J Sci Food Agric 86:197–204

Velayutham P, Babu A, Liu D (2008) Green tea catechins and cardiovascular health: an update. Curr Med Chem 15:1840–1850

Vuong QV, Bowyer MC, Roach PD (2011) L-Theanine: properties, synthesis and isolation from tea. J Sci Food Agric 91:1931–1939

Walling LL (2000) The myriad plant responses to herbivores. J Plant Growth Regul 19:195–216

Wang D, Li C-F, Ma C-L, Chen L (2015) Novel insights into the molecular mechanisms underlying the resistance of *Camellia sinensis* to *Ectropis oblique* provided by strategic transcriptomic comparisons. Sci Hortic 192:429–440

Wang Y-N, Tang L, Hou Y et al (2016) Differential transcriptome analysis of leaves of tea plant (*Camellia sinensis*) provides comprehensive insights into the defense responses to *Ectropis oblique* attack using RNA-Seq. Funct Integr Genomics 16:383–398

Writer S (2017) "Relying on Heaven": natural farming and "Eco-tea" in Taiwan. In: Lewis T (ed) Green Asia. Taylor & Francis, New York, pp 51–65

Yang H, Xie S, Wang L et al (2011) Identification of up-regulated genes in tea leaves under mild infestation of green leafhopper. Sci Hortic 130:476–481

Yang Z-W, Duan X-N, Jin S et al (2013) Regurgitant derived from the tea geometrid *Ectropis obliqua* suppresses wound-induced polyphenol oxidases activity in tea plants. J Chem Ecol 39:744–751

Yang T-B, Liu J, Yuan L-Y et al (2017) Molecular identification of spiders preying on *Empoasca vitis* in a tea plantation. Sci Rep 7:7784

Ye G-Y, Xiao Q, Chen M et al (2014) Tea: biological control of insect and mite pests in China. Biol Control 68:73–91

Yu P, Yeo AS-L, Low M-Y, Zhou W (2014) Identifying key non-volatile compounds in ready-to drink green tea and their impact on taste profile. Food Chem 155:9–16

Zhang Y-N, Yin J-F, Chen J-X et al (2016) Improving the sweet aftertaste of green tea infusion with tannase. Food Chem 192:470–476

Zhang H, Li Y, Lv Y et al (2017) Influence of brewing conditions on taste components in Fuding white tea infusions. J Sci Food Agric 97:2826–2833

Chapter 11
Biochemical, Physiological and Molecular Defence Mechanisms of Tea Plants Against Pathogenic Agents Under Changing Climate Conditions

Aziz Karakaya and Murat Dikilitas

Abstract Tea is one of the most widely consumed beverages after water in the world. It has delicate requirements in terms of agricultural and factorial processing. Tea is a popular drink; however, its biotic and abiotic stresses are lowering the quality of brewed tea. Researchers have concentrated mostly on its cultivation and processing or its sustainability and breeding. One of the main reasons behind this, tea needs specific places such as tropic or subtropic areas with high altitude and irrigation regimes. However, its diseases are as important as other diseases of crop plants. Moreover, diseases on tea could have more impact due to extensive climate changes that could have potential to reduce crop production through increased temperature, reduced humidity and prolonged drought conditions. The microorganisms adapted to live in these harsh conditions would also create more drastic consequences via increased levels of toxins and pathogenic enzymes that would increase the pathogenicity and the virulence of the microorganisms as well as reducing crop productivity and quality. In this chapter, we evaluated diseases occurring on tea plants and biochemical, physiological and molecular defence mechanisms of tea plants against pathogenic agents. Possible behaviours of pathogenic agents under projected climate change issues are also discussed.

Keywords Tea diseases · *Camellia sinensis* · Tea defence mechanisms · Climate change

A. Karakaya (✉)
Department of Plant Protection, Faculty of Agriculture, Ankara University, Ankara, Turkey
e-mail: karakaya@agri.ankara.edu.tr

M. Dikilitas
Department of Plant Protection, Faculty of Agriculture, Harran University, Şanlıurfa, Turkey

W.-Y. Han et al. (eds.), *Stress Physiology of Tea in the Face of Climate Change*,
https://doi.org/10.1007/978-981-13-2140-5_11

1 Introduction

Tea plant (*Camellia sinensis* (L.), O. Kuntze) is an evergreen wild shrub. Tea, a common drink in the world, is produced from the leaves of the tea plant. It was discovered around 2700 BC in South-East Asia. When it is cultivated, it is kept at a low level to enable the young shoots to produce young leaves. Today, tea is cultivated in more than 50 countries including China, Japan, Sri Lanka, India, Turkey, Indonesia, Australia, Russia and high altitudes of South American countries and Africa (Hajiboland 2017). Depending on the procedure of harvested leaves, fermented or non-fermented, tea becomes black or green. However, climatic conditions such as hot or cold weather, drought, water stress, the high temperature in most of the seasons as well as abiotic and biotic stress agents could limit the crop production with the loss of quality. Climate and soil characteristics are the most important ecological factors for tea cultivation. Tea plants, in general, require a rainfall between 1500 and 4000 mm; however, consecutive dry seasons more than 3 months could limit the performance of tea plants (Ahmed et al. 2014). The optimum temperature, on the other hand, is between 18 and 20 °C. Temperature over 30 and below 12 °C could limit the growth of tea plants; however, temperatures below 5 °C could not be tolerated, and therefore, tea plants are not frost-tolerant (Lu et al. 2017). Humidity between 70% and 90% is required; therefore, dry conditions that reduce the humidity in leaves could interfere with the production of leaves.

For the soil conditions, as in all plants, tea plants thrive better in healthy and rich soils in terms of organic components. They prefer little acidic conditions; pH between 4.5 and 6.5 is good for tea growing (Ruan 2004).

Biotic stress agents, on the other hand, could vary as in the case of abiotic stress factors. To overcome these limitations, use of resistant varieties and chemical control could be effective. Development of resistant cultivars is an important measure of disease control, and plant breeding is economically important. However, diseases remain a major cause of quality and yield losses in tea production. New races of pathogenic microorganisms, as well as the introduction of new pathogens in tea cultivation areas along with the cultivars differing in resistance to diseases, create more disease incidences. Combined effects of pathogenic microorganisms and abiotic stress factors could be able to create more stressful areas for tea cultivation. The combined stress could either be additive or in worst cases, it could be synergistic to one another. In most cases, abiotic stress could trigger and increase the pathogenicity of microorganisms, and in some cases, it increases the virulence of pathogens (Dikilitas et al. 2016). No one should expect that the combination of one or more stress agents would nullify each other since the pathogenic agents somehow would find a way to adapt themselves to the conditions of abiotic stress agents.

Although tea is consumed commonly, it has not yet received proper attention in terms of diseases and abiotic stresses under changing climate conditions. In this chapter, beneficial effects of tea were summarized, and its diseases and common abiotic stress agents were discussed, and the possible combined effects of disease and abiotic stress interactions were evaluated. It is important to note that the use of resistant cultivars or application of pesticides for stress agents should be used on time before the critical level is passed; otherwise, generation of multiple stress-resistant tea cultivars for stressful conditions would be ineffective under combined stress conditions.

2 Characteristics of Tea

Tea plant is a member of Theaceae family. The cultivated taxa consist of three main natural hybrids. They are *Camellia sinensis* (L.) O. Kuntze or China type, *C. assamica* or Assam (Indian) type and *C. assamica* subsp. *lasiocalyx* (Planchon ex Watt.) or Cambod type (Mondal 2016). Freshly harvested tea leaf is processed and consumed differently in different parts of the world, mostly black; the rest is green and oolong (Kuroda and Hara 1999). Whether it is black or green, tea contains biochemical components with antioxidant characteristics. For example, the black tea contains lower catechins and higher theaflavins than those of green tea (Subramanian et al. 1999). Tea is also rich in theanine, a unique amino acid, proteins, caffeine, vitamin C, carbohydrates, polysaccharides and lipids (Kaur et al. 2014). High oxidant levels such as reactive nitrogen species (RNS) and reactive oxygen species (ROS) generated by stress agents in human body could be reduced or prevented via the supply of antioxidant diet system. As in the other antioxidant substances, tea is rich in terms of antioxidant contents and could be used to remediate the toxic effects of oxidants. For example, tea polyphenols have strong scavenging properties for free oxygen radicals; therefore, they reduce the oxidant status and contribute the increase of antioxidants in cell (Bruno et al. 2014). It has many health benefits, like anticancer, antioxidative, astringent activity, and contains few important vitamins like B, C, E and K and minerals like manganese, zinc, etc. Tea polyphenols are also able to induce enzymes that detoxify carcinogens, thus inhibiting cancer initiation or carcinogenesis. Catechins, another chemical component with high antioxidant activity, interact with an enzyme in human intestines to suppress glucose uptake and prevent the occurrence of diabetes (Kim and Kim 2013). Kidney diseases and osteoarthritis could be relieved via intake of tea. For example, tea catechins relieve high oxidative stress and improve renal blood circulation and ease the pain caused by the pressure in the renal blood vessels (Khalesi et al. 2014). Tea polyphenols are also effective in relieving the arthritis. Other components such as theanine as well as catechin have neuro-protective effects; therefore, they are quite effective in preventing cerebral stroke. Components in tea also effectively control the flu-like symptoms. Due to antibacterial properties, influenza and infections of the respiratory tract and lungs could be significantly reduced via consuming tea (Law et al. 2003). Tea also raises basic metabolic rate, and therefore, it helps to control the body weight with a daily exercise. It is well known that Japanese people who take 8–10 cups of tea per day are good at keeping their body weight at constant even if their ages increase. It contains caffeine with no calorie thus keeping people alert when needed without gaining extra calorie. The tea flavour comprises both aroma and taste which is the most important factor in determining the tea quality. Catechins contribute up to 30% of dry weight, and they are responsible for taste; on the other hand, volatile flavour compounds consist 0.01–0.02% of the total dry weight (Millin 1987; Sharma et al. 2011). Catechins possess antioxidant activity and play important roles for inhibition of carcinogenesis and mutagenesis through their medicinal and antimicrobial properties (Yang 1997). Catechins have been reported to increase in stress and play a significant role in plant defence (Sharma et al. 2011). Flavour component such as 2-phenylethanol also possess antimicrobial properties and play a significant role for protecting flowers and fruits (Zhu et al. 2005). In different countries more than 500 tea cultivars have been bred and released (Chen et al. 2012).

Tea is one of the most important beverages consumed in the world. Ramkumar et al. (2016) stated that total polyphenols, polyphenol oxidase (PPO) activity, amino acid content, enzymatic antioxidants and catechins are considered to be potential quality parameters in characterization of tea. Gene expression pattern and biochemical characterization of *Camellia sinensis* clone P/11/15 were analysed. They found that the selected clone possessed potent catechin and polyphenol as well as sufficient peroxidase (POD), polyphenol oxidase (PPO) and catalase (CAT) activities. Therefore, biochemical constituents determine the quality of tea. However, tea plants are pretty much affected by abiotic and biotic stress factors that reduce the crop production and quality.

3 Pathogenic Agents in Tea Cultivation Areas and Biochemical, Physiological and Molecular Defence Mechanisms of Tea Plants

During infection by the pathogen, the defence mechanisms of plants will be activated which result in various biochemical and molecular changes. Phytoanticipins, phytoalexins, and other postinfectionally accumulated antimicrobial substances play a crucial role in the plant defence mechanisms (González-Lamothe et al. 2009). The biochemical contents of tea determine the level of resistance or tolerance to diseases or abiotic stress factors. The biochemical changes such as phenolics and flavonoid compounds also determine the quality of tea (Pereira et al. 2014). For example, infestation degree of tea insect pests and diseases is closely related to the amino acid content in tea leaves. Takeda (2003) reported that tea cultivars with high arginine content were usually a target for infestation by *Tetranychus kanzawai.* Similarly, the contents of soluble sugars in tea shoots can be correlated with the tea plant susceptibility (Ponmurugan and Baby 2007). On the contrary, the caffeine contents in tea shoots have been found to be related to the pink mite resistance (Yao and Chen 2012). In general, the average contents of caffeine and total amino acids have been characterized with resistant tea cultivars. Here, it is important to note that virus diseases are less common in tea plant compared with other plants. It is thought that the high content of polyphenols in tea leaves is believed to play important roles for the resistance (Cheruiyot et al. 2007). For example, Lang'at et al. (1998) reported that the leaf content of total polyphenols inversely correlated to the size of lesions developed following inoculation with the fungal pathogen *Pestalotiopsis theae*. Phenol and orthodihydroxy phenol contents were also increased in resistant tea cultivars following inoculation with *Glomerella cingulata*, *Pestalotiopsis theae* and *Corticium theae* (Chakraborty et al. 2005a). They noticed that the phenolic substances were oxidized to highly reactive quinones which play important roles for preventing the metabolic activities of parasite and host cells. In some studies, increases of antimicrobial substances such as pyrocatechol were also noticed following inoculation with *Pestalotiopsis theae* (Chakraborty et al. 2005a). Some enzyme activities in tea plants were reported to be closely related to the tea plant resistance. In general, POD and PPO activities increased in resistant cultivars following inoculation. Similarly, phenylalanine ammonia-lyase (PAL) enzyme increase in tea leaves was characterized with resistance upon infection of fungal pathogens. For example, Chakraborty et al. (2005a) stated that the highest activity was found in a resistant tea cultivar, while the susceptible cultivar showed a very low increase in activity following inoculation. It is also important to note that the accumulation of chitinase and β-1,3-glucanase after inoculation are characterized with resistant cultivars (Chakraborty et al. 2005b).

Greater accumulation of phenolics during infection makes the plants more resistant and reduces pathogen attack, and phenolics are naturally antifungal compounds (Chandra et al. 2014). Higher accumulation of phenolic compounds results in higher rate of esterification of hydroxycinnamic acid derivatives which yields lignin. Lignin is associated with resistance of plants (Matern and Kneusel 1988). The biochemical changes and metabolites produced at infection stage also act as signalling molecules for the expression of defence-related genes. These changes include the synthesis of pathogenesis-related (PR) proteins, phytoalexins, rapid accumulation of reactive oxygen species (ROS) and defence-related enzymes, non-enzymatic metabolites and structural changes (Senthilkumar et al. 2012). The production of ROS by the host is toxic to infecting pathogens; however, ROS can also disrupt the normal metabolism of the host through oxidative damage to proteins, nucleic acids and lipids. Imbalance towards the oxidative status with the increase of ROS could initiate the stress and break the defence barriers. The production of ROS in a rapid and controlled manner results in early signalling that could play important roles in the expression of defence-related genes and biochemical pathways. Expression of genes and the production of antioxidant enzymes (POD, APX, SOD, CAT) and non-enzymatic metabolites such as glutathione, proline, ascorbate etc. can take roles in protecting cells from oxidative damage through scavenging ROS (Bhattacharjee 2010; Palanisamy and Mandal 2014). However, this could also affect the quality parameters of crop plants.

Tea diseases, in general, cause serious damage to leaves, shoots and roots. However, leaf and shoot diseases are of more concern to tea growers than to the growers of other plants due to the economic importance of leaves. There are various disease agents; here, we introduced some of these disease agents with their symptoms and some infection mechanisms on tea plants. Defence mechanisms of tea plants for particular disease agents were also evaluated. From the disease agents, *Colletotrichum* and *Pestalotiopsis* species cause significant diseases and crop losses on tea plants. They have great potentials on a wide range of crops in terms of disease progression. The disease caused by *Colletotrichum* species is called anthracnose which is one of the most damaging diseases of tea and other crops. More than one *Colletotrichum* species cause disease on tea plants (Ali et al. 1993; Keith et al. 2006; Karakaya et al. 2011). Quite a few *Colletotrichum* species such as *C. camelliae*, *C. fioriniae*, *C. cliviae*, *C. karstii*, *C. fructicola*, *C. siamense*, *C. endophytica*, *C. aenigma*, *C. wuxiense* and *C. truncatum* were isolated from *Camellia sinensis*. *C. fructicola* and *C. camelliae* were found as the dominant species causing anthracnose in *C. sinensis* (Wang et al. 2016a; Liu et al. 2015). *C. gloeosporioides*, *C. boninense* and an unidentified species of *Glomerella* sp. were reported in tea plants in Turkey (Karakaya et al. 2011) (Fig. 11.1). *Glomerella cingulata* which is the teleomorph of *Colletotrichum gloeosporioides* was also observed in tea gardens causing diseases (Karakaya and Çelik Oğuz 2013). The disease is also called as brown blight. High humidity, prolonged periods of leaf wetness, poor air circulation

Fig. 11.1 Symptoms of tea diseases. (**a**) Leaf symptom caused by *Pestalotiopsis theae*, (**b**) twig symptom caused by *Pestalotiopsis theae,* (**c**) leaf symptom caused by a *Pestalotiopsis* sp., (**d**,**e**) leaf symptoms caused by *Colletotrichum gloeosporioides*, (**f**) shoot and bud blight incited by *Botrytis cinerea*, (**g**) leaf symptom caused by *Phoma* sp

or high temperature favours the disease. It often causes flower and fruit drop in tea oil plants leading to 50% or more crop loss (Xinzhang et al. 2012). For example, Cao et al. (2014) assessed *C. gloeosporioides* for the resistance mechanisms in oil tree varieties by monitoring biochemical and physiological parameters of differentially resistant cultivars. The fruit coats were analysed for anthocyanins, tannins, pH, the activity of three enzymes (PAL, PPO and POD), soluble sugar content, free radical scavenging capacity and buffer capacity. Soluble contents, anthocyanins and free radical scavenging capacity of the resistant varieties were all related to anthracnose resistance. They found that soluble contents of the metabolites and anthocyanins in resistant varieties were nearly twofold higher than those of susceptible cultivars of *C. oleifera*. Therefore, this fungus is able to reduce the crop yield and able to change the biochemical characteristics of the host plant. Since tea plants are rich in secondary metabolites, their roles in anthracnose resistance are of great importance. Xinzhang et al. (2012) stated that the polyphenols, catechol and salicylic acid content of the resistant varieties were five to ten times higher than those of the more susceptible varieties. In resistant varieties flavonoid content was also significantly higher. No difference was observed between different cultivars in terms of activities of the defence-related enzymes PAL and PPO. However, after pathogen inoculation, enzyme activity of resistant cultivars improved markedly. No change was observed in enzyme activity of susceptible cultivars. Ponmurugan and Baby (2007) reported that the growth characteristics, physiological and biochemical parameters such as nitrogen, total sugar, amino acids, polyphenol and catechins were reduced significantly in infected plants. They stated that the reduction was more pronounced in susceptible cultivar. They concluded that as a result of pathogen infection, all physiological, biochemical and morphological parameters were depleted to some extent in the plants which ultimately affected the tea quality. Cao et al. (2014) stated that extracts from highly resistant varieties of *C. oleifera* fruit coats performed more efficiently in the scavenging of free radicals than those from susceptible varieties. The enzyme activities such as PPO, PAL and POD in highly resistant varieties rose rapidly, and continuously, on the other hand, enzyme activities of medium-resistant and highly susceptible varieties initially increased, and then a decrease occurred. No significant differences between different cultivars were observed in terms of buffer capacity, tannins and pH. Wang et al. (2016b) compared the contents of catechins, caffeine and total phenolics in two cultivars with different resistance levels to anthracnose during *C. fructicola* infection. They stated that critical regulatory genes, (+)-catechin (C), (−)-epigallocatechin-3-gallate (EGCG) and caffeine were induced in *C. fructicola*-resistant tissues. In vitro antifungal tests and electron microscopy studies revealed that caffeine more strongly inhibited the mycelial growth than tea polyphenols and catechins. Wang et al. (2016c) also compared the resistance of tea cultivars to elucidate the molecular and biochemical pathways between resistant and susceptible ones. They tested the response of anthracnose-resistant cultivar ZC108 and its parent cultivar LJ43 to *C. camelliae* infection. The cultivar ZC108 showed greater resistance than LJ43. Tea cultivar ZC108 also exhibited earlier sprouting in the spring and produced different plant architecture and leaf shape. The genes were also differentially expressed between LJ43 and ZC108 cultivars. The genes encoding secondary metabolites such as phenylpropanoids, phenylalanines and flavonoids and involving in signalling and plant hormone biosynthesis were changed. Quantitative real-time PCR analysis showed that 27 selected genes in infected and uninfected tea leaves were expressed. Genes encoding a MADS-box transcription factor, phenylpropanoid metabolism pathway components (CCR, CAD, beta-glucosidase, POD, PAL and ALDH) and NBS-LRR disease-resistant protein were among those differentially expressed in cultivar ZC108. Chen et al. (2017) stated that *C. acutatum*, a new causal agent associated with *C. sinensis*, showed high phenotypic and genotypic diversity compared with the more commonly reported *C. gloeosporioides*. Pathogenicity tests validated that both species were causal agents of tea brown blight disease and were highly pathogenic to tea leaves. However, the two groups of *C. gloeosporioides* with low levels of variability within their ITS and GAPDH regions differed in their virulence.

Another important tea disease is grey blight caused by *Pestalotiopsis theae*. Diseases caused by *Pestalotiopsis* species are common in tea-growing regions of the world (Karakaya and Çelik Oğuz 2013; Ertaş et al. 2016; Ertaş and Karakaya 2018). *Pestalotiopsis* species commonly cause a leaf disease called grey blight; however, the fungus can also cause disease in tea twigs (Karakaya and Çelik Oğuz 2013) (Fig. 11.1). *Pestalotiopsis theae*, *P. longiseta*, *P. guepinii*, *P. adusta*, *P. furcata*, *P. natrassii*, *P. camelliae* and *P. palmarum* have been reported from tea plants (Ertaş and Karakaya 2018). High temperature, poor air circulation, prolonged periods of leaf wetness or high humidity favours the disease (Keith et al. 2006). Senthilkumar et al. (2012) stated that the infection with *P. theae* in leaves resulted in a big loss to the tea industry. They used suppressive subtractive hybridization (SSH) technique to unravel the differential gene expression pattern during grey blight disease development. Complementary DNA from *P. theae*-infected and *P. theae*-uninfected leaves of disease-tolerant tea cultivar UPASI-10 was used as tester and driver populations, respectively. They demonstrated the efficiency of SSH as a tool in enriching grey blight disease related up- and down-regulated genes in tea. They showed that many genes related to disease resistance were suppressed during *P. theae* infection. Sanjay and Baby (2007) stated that the biochemical changes such as stomatal conductance, photosynthetic rate, chlorophyll fluorescence and water use efficiency were also decreased in the infected leaves, while stomatal resistance was increased in tea leaves infected by *Pestalotiopsis.* Biochemical constituents were higher in healthy leaves compared to those in infected leaves, except for polyphenol and catechin contents. Chlorophyll and carotenoid contents were decreased significantly. Activity of PAL and tyrosine ammonia-lyase (TAL) enzymes was found higher in the healthy zone of infected leaves than those of uninfected leaves while decreased in the lesion region. On the other hand, the activity of PPO and POD enzymes was higher in the lesion region and healthy portion of the infected leaves. Koh et al. (2001) stated that grey blight caused by *P. longiseta* had more disease impact on Japanese tea cv. Yabukita than the Korean local cultivar. Although this indicated that the varieties responded differently and the pathogen had a different pattern on resistant and susceptible varieties, however, both varieties were severely damaged during the third harvest period. High temperature and humidity favoured the disease. Sometimes, more than one disease could be found on tea leaves. For example, the presence of *Glomerella cingulata* on the margin of grey blight lesion caused by *Pestalotiopsis longiseta* at the late stage suggested that the pathogenic fungi of tea grey blight were replaced by the brown blight fungus *Glomerella cingulata* (Ando 1992). Ando (1992) also stated that *G. cingulata* invaded mycelial colonies of *P. longiseta* to a moderate extent in paired cultures on potato dextrose agar and caused abnormalities in *P. longiseta* hyphae. Ando (1992) suggested that *G. cingulata* was more aggressive than *P. longiseta* in vivo as well. Under in vitro conditions, *P. longiseta* phytotoxins ((+)-epiepoxydon and oxysporone) were slightly inhibitory to mycelial growth of *G. cingulata*. The author suggested that *P. longiseta* invaded intercellular spaces in the host tissues and secreted the phytotoxins that cause necrosis, but did not invade the cells. On the other hand, *G. cingulata* readily invaded tissues damaged by these toxins; thus resistance level of tea to this fungus was reduced. *P. longiseta* growth was gradually suppressed as the food supply was depleted, and *G. cingulata* became dominant, eventually replacing *P. longiseta*. Horikawa (1986) stated that *P. longiseta* caused shoot blight on susceptible cultivars and were isolated from young and fresh lesions on the twigs; however, *Glomerella cingulata* was more frequently isolated from older and well-developed lesions. Horikawa (1986) revealed that *P. longiseta* primarily infected the young shoots through wounds made by the abscission of scale leaves or incomplete leaves and then *G. cingulata* secondarily colonized the tissue pre-infected by *P. longiseta* and expanded the lesion areas. Ando et al. (1985) stated that the responses of varieties against the attack by *P. longiseta* following wound inoculation of 18 varieties in the field with *P. longiseta* showed variation. Yabukita, Sayamidori and Natsumidori always formed large lesions, while Benihomare, Kanayamidori, Makinoharawase and Yaeho always formed small lesions. Other varieties had an intermediate response, tending towards the less susceptible. Generally, detached leaf tests also confirmed these results. Inoculation of 6 varieties with *P. theae* produced smaller lesions. Takeda (2002) stated that tea plant resistance to *P. longiseta* was controlled by two independent dominant resistance genes Pl_1 and Pl_2, based on the parent-offspring genetic analysis. Palanisamy and Mandal (2014) selected three tolerant and three susceptible tea clones. One tea bush from each cultivar was inoculated with *P. theae*. Catalase (CAT), glutathione reductase (GR), peroxidase (POD), polyphenol oxidase (PPO), superoxide dismutase (SOD) and ascorbate peroxidase (APO) enzymes and their activities were compared between susceptible and tolerant tea cultivars. The infected leaves of the susceptible cultivars showed a higher amount of enzyme activity when compared with the tolerant cultivars. They claimed that the more antioxidant enzymes resulted in more susceptible cultivars. However, this was not the case in Xinzhang et al. (2012) study. Although Palanisamy and Mandal (2014) showed that the antioxidant enzymes had increased in susceptible cultivars, however, they did not provide any results regarding the duration of this enzyme activity. Their results were possibly early defence response and hypersensitivity regarding with the susceptible cultivars in general. Therefore, general idea is that the high antioxidant activity with prolonged period makes the cultivars resistant. Minimal activity in tolerant cultivars at early stages could be related to lower expression of defence-related genes that the stress level was not regarded as high as those of susceptible cultivars. For example Senthilkumar et al. (2012) stated that majority of the upregulated genes are related to hypersensitive response and ROS production. Takeda and Nagatomi (1998) subjected to tea (*Camellia sinensis*) cv. Yabukita to γ-irradiation. Three lines obtained through backcrosses with parent cv. Yakubita showed higher level of resistance to *Pestalotiopsis longiseta* than the parent cv. Yakubita. Tea fragrance, water colour and leaf colour scores were higher in the resistant lines than in the parent cv. Yabukita. Takeda (2003) determined variations in tea cultivars in terms of phenotypic and genotypic characteristics to *Pestalotiopsis longiseta.* The majority of the *Camellia sinensis* var. *assamica* plants showed a high level of resistance and very low variation. *C. sinensis* var. *sinensis* exhibited a wider genetic diversity in terms of resistance to *P. longiseta*. It is stated that highly resistant tea genotypes could be used in breeding studies.

Another important disease in tea plants is caused by *Phomopsis theae*. The fungus causes stem, collar and branch canker disease of tea, and it is common in important tea-growing regions of the world (Anonymous 1993). This disease limits tea production in Kenya and causes damage frequently (Mutai and Cheramgoi 2009). The disease agent can also cause leaf blights (Karakaya 2009). *Phomopsis* can attack tea plants at any stage of growth; however, younger plants are more susceptible (Rattan 1993). Rattan (1993) observed more Phomopsis disease in water-stressed areas. The drought conditions as a result of climate change can induce more disease incidences. This will affect tea quality and quantity as well. The reduction in quality parameters such as total sugar, protein, amino acids, nitrogen, catechins and polyphenols was more noticeable in susceptible cultivar compared to tolerant cultivar. Mphangwe et al. (2013) determined that six RAPD primers generated nine specific bands that were associated with *Phomopsis theae*, tolerance to drought and black tea quality, low temperature, high temperature and high-yield traits. They concluded that the RAPD markers identified could improve the precision and efficiency of tea selection and breeding. Ponmurugan and Baby (2001) carried out a greenhouse study to evaluate the effect of predisposing factors on Phomopsis canker incidence. Highest disease incidence was observed in the gravelly soil where the soil was mixed with gravel at a 1:3 ratio (56%), followed by mulching with coir pith (40%) close to the collar and artificially induced drought (40%). Also, deep planting and fertilizer application close to the collar resulted in significant disease level. Optimal watering and planting and careful application of fertilizers and as well as mulching away from the collar did not lead to disease incidence. Canker size was the largest on plants which were mulched close to the collar. Under field conditions, the growth characteristics such as height, dry mass, stomatal conductance, photosynthetic and transpiration rates, etc. were reduced significantly more in susceptible tea cultivar than in tolerant tea cultivar. Ponmurugan and Baby (2007) conducted a study to observe biochemical, morphological and physiological changes in tea plants following *Phomopsis theae* infection. Transpiration rate, photosynthetic rate, water use efficiency and stomatal conductance were studied in tolerant and susceptible cultivars. Growth characteristics (plant strength, height and dry weight) and biochemical parameters (amino acids, total sugar, protein, nitrogen, polyphenols and catechins) were evaluated using diseased and healthy tea plants. All growth characteristics, biochemical and physiological parameters were reduced in diseased tea plants compared to healthy ones. The reduction was more noticeable in susceptible cultivar compared to tolerant cultivar. Following inoculation, a clear variation in the canker size in the susceptible and tolerant cultivars was observed. Also, healthy cultivars had significant bark moisture content and levels of root saccharides, which were more noticeable in the tolerant cultivar. Under field conditions, Ponmurugan et al. (2007) found a significant reduction in the biochemical and physiological constituents in tea leaves infected with blister blight pathogen *Exobasidium vexans*, grey blight pathogen *Pestalotiopsis theae*, brown blight pathogen *Colletotrichum camelliae*, sooty mould pathogen *Capnodium theae* and red rust *Cephaleuros parasiticus*. The growth characteristics such as height, dry mass, stomatal conductance, water use efficiency and photosynthetic and transpiration rates were reduced significantly more in susceptible tea cultivar than in tolerant tea cultivar. Total sugars, amino acids, nitrogen, polyphenols, proteins and catechin contents were reduced in diseased plant leaves. The reduction was more noticeable in susceptible than tolerant cultivar.

Botrytis cinerea, another important disease, causes blight of young leaves, buds, shoots and flowers, shoot canker and leaf spots of tea (Karakaya and Bayraktar 2010) (Fig. 11.1). *B. cinerea* produces large, reddish brown, irregular lesions with zoned patterns on the fully opened, overwintered soft tea leaves (Hamaya 1981; Khodaparast and Hedjaroude 1996). However, detailed biochemical, physiological and molecular studies are still needed related to this pathogen and its effects on tea quality.

Blister blight disease is an important disease of tea in China, Bangladesh, India, Cambodia, Indonesia, Japan, Nepal, Malaysia, Thailand, Vietnam and Sri Lanka. It is caused by the obligate parasite fungus *Exobasidium vexans*. It is among the most serious leaf diseases. No alternate host of the fungus is known. This pathogen mainly attacks succulent, young and tender leaves. It spreads by wind-borne basidiospores and infection occurs most likely through stomata. After germination of mycelia, the mycelium grows intercellularly, and the basidia fruiting bodies form below the lower epidermis. As the mycelia develop, they rupture the epidermis and form the blister. *E. vexans* causes initial discolouration in infected leaves and form circular blisters. Moderate temperatures, prolonged leaf wetness, shade and high relative humidity favour the disease. Quality of tea prepared from diseased leaves becomes poor (Lehmann-Danzinger 2000; Mouli 2003). Climatic factors can influence the severity of blister blight. Shade, cultural practices, elevation and land aspects influence the severity of the disease. It causes at least 40% total yield loss as well as reducing the tea quality significantly through a reduction in total phenols and catechin content (Jayaswall et al. 2016). Temperatures between 10 and 23 °C, mean daily rainfall ranging from 20 to 100 mm, relative humidity >60%, leaf wetness of 11 hours and average weekly sunshine of 2–10 hours favour the disease. Several generations can occur within a single crop season. For this reason, the management of the disease is very difficult and repeated applications of fungicides are needed. From mature blisters, basidiospores discharge and spores are wind-borne. Discharge occurs fast during humid conditions (Booth 1983). Molecular mechanisms of defence against blister blight disease were investigated. Using RNA-seq using Illumina GAIIx at 4 different stages during ~20-day disease cycle transcriptome of blister blight interaction with susceptible and resistant tea genotypes showed that 149 defence-related transcripts/genes, namely, resistance genes, multidrug-resistant transporters, defence-related enzymes, retrotransposons, transcription factors, chaperons and metacaspases, were observed in resistant tea genotype (Jayaswall et al. 2016). They stated that jasmonic acid and salicylic acid possibly induced synthesis of antimicrobial compounds required to overcome the virulence of *E. vexans*. On the other hand, Mur et al. (2015) determined that jasmonic acid and salicylic acid were significantly reduced in blister blight disease and this was found directly linked to the disease progression. As a result of that, the fungus perturbed the cell wall and reduced many potentially antimicrobial compounds such as caffeine and phenols to aid disease progression. Therefore, antimicrobial compounds and metabolites have important characteristics in preventing disease occurrence.

Although, fungicide applications had shown encouraging results for controlling blister blight agent, however, a serious problem of fungicide residues and phytotoxicity created environmental and health issues. Since the pathogen enters leaf cells through stomata and proliferates in intercellular spaces and obtain nourishment through specialized haustoria and suppress the host defence by delivering effector proteins into plant cells and reduces the plant immunity, the use of chemicals have not been found very effective. Biocontrol agents such as *Gliocladium virens, Trichoderma harzianum, Pseudomonas fluorescens, Bacillus subtilis* and *Serratia marcescens* were not found to be very effective either. Tea accessions resistant to *E. vexans* are also limited (Jayaswall et al. 2016). Punyasiri et al. (2017) investigated the metabolites epicatechin gallate (ECg), epicatechin (EC), epigallocatechin gallate (EGCg), epigallocatechin (EGC), theobromine, gallic acid and caffeine in 6 non-beverage-type and 87 beverage-type tea accessions from Sri Lankan tea germplasm. It appears that high EGC and EC contents in green leaves seem to be a reliable marker for identifying high-quality black tea producing accessions. Molecular approaches have some promising sites. For example, Singh et al. (2015) inserted a plant pathogenesis-related gene, *Solanum tuberosum* class I chitinase gene (AF153195), into tea genome via *Agrobacterium*-mediated transformation. They succeeded in increasing resistance to fungal pathogen and formed hypersensitivity via transgenic plantlets.

Wu et al. (2013) found that different tea cultivars showed different resistance levels to blister blight pathogen *E. vexans* in China. Disease occurrence was closely related to rainfall. Shade status of tea gardens was also related to disease severity. Kumar and Shalini (2010) studied the enzymes APX, GR, SOD and CAT using *E. vexans* resistant and susceptible tea clones. Activities of the defence-related enzymes were significantly higher in blister blight-resistant clones. Chakraborty and Sharma (2007) stated that cross-reactive antigens shared by *Camellia sinensis* and *Exobasidium vexans* determined the disease pattern. They reported that cross-reactive antigens were concentrated mainly around mesophyll and epidermal cells in susceptible tea varieties and *E. vexans* isolates but not in resistant tea varieties, nonhosts or nonpathogens. The authors concluded that pathogenicity of *E. vexans* to different tea varieties was related to the level of antigenic similarity between host and pathogen. Immunological methods could be used in screening commercially cultivated tea varieties against blister blight pathogen.

Jeyaramraja et al. (2010) investigated the role of chitinase and physical barriers in resistance status of tea plants to *E. vexans*. Increased thickness of cuticle/epidermal layer and higher amounts of epicuticular wax conferred resistance to blister blight. Chakraborty et al. (2005b) evaluated the β-1,3-glucanase (βGLU), peroxidase (POX) and chitinase (CHT) in tea plants against blister blight disease-causing agent *E. vexans*. Greater accumulation of CHT and βGLU were detected in disease-free leaf samples. POX activity increase was seen in resistant tea plants. A general increase in CHT, βGLU and POD was observed in within 24–48 h of inoculation. On the other hand, Punyasiri et al. (2005) showed a decline in proteins at very early stages of infection. Several proteins were lost following infection with *E. vexans* infection. Gulati et al. (1999) also reported similar findings. They found that aromatic compounds particularly linalool, phenyl acetaldehyde, hexene-I-ol, geraniol, methyl salicylate, indole, nerolidol, β-ionone and several unassigned components were lower in diseased tea plants. Baby et al. (1998) reported that blister blight disease decreased the levels of polyphenols, catechins and enzymes present in green tea leaves and difficulties during withering and fermentation were observed. Similar findings were also reported that total colour and brightness of the tea liquor were decreased due to blister blight disease (Sugha et al. 1991). Pius et al. (1998) investigated the changes in the saccharine content of tea leaves following infection with *E. vexans*. Decreases in glucose and saccharose contents in diseased leaf areas were evident at the sporulation stage, and this lasted throughout the sporulation period. During the initiation of sporulation, fructose content increased sharply. This increase remained constant in blistered and non-blistered areas until the end of sporulation. Continuous decrease in the starch content of blistered leaf areas positively correlated with the leaf senescence in which the POD activity was high. Activity of acid invertase was also inversely proportional to the starch content, and this activity was closely related to the changes in glucose and saccharose. In blistered areas, a decrease in the protein and chlorophyll levels was observed. It could be concluded that important compounds such as protein, aromatic compounds, starch, etc. which would determine the quality of tea would be lost for defence against pathogen attacks. Therefore, under stress tea plants might lose their characteristics.

In field trials in Sri Lanka, Gunasekera et al. (1997) investigated the effects of ultraviolet-B (UV-B, 290–320 nm) radiation on blister blight disease. Exclusion of UV-B radiation increased the number of immature sites of infection and the number of sporulating mature blisters. The increased survival and germination of basidiospores were also evident when UV-B wavelengths were removed in vitro. The authors suggested that UV-B component of solar radiation might play an important role in the natural regulation of disease in the field.

Another important tea disease is caused by *Phyllosticta* spp. A few *Phyllosticta* species have been reported in tea plants (Chen and Chen 1982; Khodaparast and Hedjaroude 1996; Karakaya et al. 2011). In China, bud blight caused by a

Phyllosticta sp., white scab caused by *P. theaefolia*, red leaf spot incited by *P. theicola*, Phyllosticta leaf spot incited by *P. erratica* and grey spot caused by *P. dusana* are common. Phyllosticta leaf spot, grey spot and red leaf spot occur mainly on more mature leaves; on the other hand, bud blight and white scab occur on buds and young tea leaves. Gao and Guo (1999) and Li et al. (2016a) reported that *Phyllosticta theicola*, the red leaf spot pathogen, grew well at 22–28 °C under in vitro conditions and was one of the most damaging diseases of tea plants. Gao (1997) reported that successful mature leaf penetrations resulted in slow growth in leaf tissues and branched hyphae. Wounded leaves and horizontal tender leaves were easily infected. Leaf cuticle thickness, the palisade layer and the spongy cell arrangement variations among the tea cultivars significantly affected the disease incidence. Tea cultivars possessing more palisade layer and thick leaf cuticle showed a relatively strong resistance. Yie (1987) reported that the tea disease incidence increased with the increase in altitude. Disease incidence was 5% below 400 m and 43% above 800–1000 m.

Phoma spp. can cause leaf spots, bud blight and bud browning of tea plants (Mouli 1996; Karakaya et al. 2011; Yang et al. 2016) (Fig. 11.1). There are limited studies related to this fungus. Similarly, bird's eye spot disease caused by *Cercospora theae* in tea plants were reported (Gnanamangai et al. 2011).

In Taiwan, Chen et al. (1998) reported *Macrophoma theicola* as the causal organism of tea twig dieback. The fungus tolerated low water potentials and developed well at relatively high temperatures. The fungus was able to grow 0.3 cm linearly in 2 d at 38 °C and in 5 d at −71 bars. *Macrophoma theicola* was widespread in Georgia and caused premature withering of tea branches and stems. Spore germination and colony growth occurred at 3–28 °C, optimum 18 °C (Dzhalagoniya 1975). Sabanayagam et al. (1974) evaluated the resistance status of 24 tea clones for resistance to *M. theicola*. Five clones were found resistant to this fungus.

A number of root diseases affect tea quality and quantity. Primary root disease of tea in southern India was red root caused by *Poria hypolateritia*. Black root caused by *Rosellinia arcuata* and brown root caused by *Fomes* (*Phellinus*) *noxius* followed this disease. Diseases caused by *Xylaria* sp. and *Armillaria mellea* infrequently observed. Frequently encountered secondary root diseases were root rot caused by *Botryodiplodia theobromae* and charcoal stump disease caused by *Ustulina zonata* [*U. deusta*]). In waterlogged soils, violet root rot incited by *Sphaerostilbe repens* was also observed (Mouli 1988). Pot experiments revealed that root infection process of *Xylaria* sp. was very slow, increasing from 3% after 4 months to 42.1% after 32 months (Anonymous 1979).

Root diseases caused by *Ganoderma pseudoferreum* [*philippii*] (below 1000 m) and *Poria hypolateritia* (at altitudes above 900 m) were the most important diseases of tea in Indonesia (Mulder 1976). Root rot caused by *Pythium* spp., *Cylindrocladium* spp. or *Fusarium* spp. was also present in tea nurseries in India (Mouli 1997).

Armillaria root rot is an important disease of tea plants. In Kenya, up to 50% yield losses have been reported (Mwenje et al. 2006; Otieno et al. 2003). The fungus was the most challenging production factor in tea-growing areas. *Armillaria heimii* and *A. mellea* were the most common species in Africa. The main mode of spread among tea bushes was through the inoculum contact. Drought, insect attack and other diseases were among the predisposing factors. Distinguishing characteristics of the disease included sheets of creamy white mycelia inside the bark with mushroom like-smell, resin exudation and longitudinal cracks at the collar region (Cheruiyot 2014). Onsando et al. (1997) surveyed Armillaria root rot severity in the 12 tea-growing districts of Kenya between 1986 and 1990. The disease was present in all areas. Infected tea roots and residual tree roots caused the infection of tea plants most of the time. Rhizomorphs were rarely involved in infection. More rhizomorphs were observed in higher elevation plantations. In these plantations, rhizomorphs were present only on residual tree roots in the cooler, moister and deeper areas of the soil. In India, *A. mellea* causes extensive damage to tea at high altitudes (Satyanarayana et al. 1982).

Rosellinia species are soilborne disease-causing agents of many crops and woody plants in different parts of the world. *Rosellinia necatrix*, *Rosellinia arcuata* and *Rosellinia bunodes* cause diseases on tea plants (Sivanesan and Holliday 1972a; Sivanesan and Holliday 1972b; Sivanesan and Holliday 1972c). Wilt and death of diseased plants may occur. Detailed biochemical, physiological and molecular studies are needed on effects of soilborne diseases on tea plants and tea quality.

Pseudomonas syringae pv. *theae* is an important pathogen of tea plants in Japan and incites bacterial shoot blight disease of *Camellia sinensis* L. The disease was common between late autumn and early spring. Yields of the first crop were affected heavily (Yoshida 2016). The response of tea cultivars and lines were determined under field conditions. Seventeen tea cultivars (8 black tea and 9 green tea), and 13 tea lines exhibited highly resistant or resistant reactions. The remaining 135 accessions showed moderately resistant or lower reactions. Ikeda et al. (2012) studied the effects of inoculation date, leaf position and total nitrogen content of tea leaves on the development of bacterial shoot blight lesions incited by *Pseudomonas syringae* pv. *theae*. Shoots of 'Yabukita' (moderately resistant) and 'Shizu-Inzatsu 131' (susceptible) were collected and injected with the bacterium. The lesions on susceptible cultivar 'Shizu-Inzatsu 131' were larger than those on 'Yabukita' after each inoculation. Larger lesions on the upper leaves were found in each cultivar. A decrease in the nitrogen content of leaves from the upper leaves to the lower ones was observed. A significant positive correlation between leaf nitrogen content and lesion size in each cultivar was present. The authors concluded that total nitrogen content and maturity of tea leaves were among the environmental factors determining susceptibility to *Pseudomonas syringae* pv. *theae*.

4 Effects of Climate Change in Tea-Growing Areas

As biotic stress affects the crop production and quality, agroclimatic conditions also have potential to change the quality traits and predispose the plants to biotic stress factors. Despite improvements in crop yield technology, crop production is dependent on climate because of the effects of radiation, temperature and humidity on crop plants. Not only crop yield is affected but also pests and pathogens are affected. For example, Murugan et al. (2012) stated that the warming trend accompanying frequent dry and wet spells during the summer would have a favourable effect on diseases and insect pests. They stated that many pathogens and pest insects increased in recent years together with the increase of temperature in India. Although it is known that changes in climate affect the conditions of crop plants, however, we have to understand how these changes will affect the pests and pathogens inhabiting with the crop plants. Their physiological behaviours, biochemical and molecular responses should be evaluated under changing climate conditions. Therefore, future studies must focus on the combined interactions such as drought and pathogens, salinity and pathogens, temperature and pathogens, etc. on tea plants. Murugan et al. (2012) stated that during the last 30 years, the global temperature increased significantly and the largest significant upward trend and the greatest increase were observed in the daily temperature. Rainfall, on the other hand, during the rainy seasons showed a downward trend. Vyas et al. (2007) stated that low temperatures during winter dormancy alone or in combination with high solar irradiance might induce oxidative stress when ROS exceeds the scavenging capacity of the cell. Gill and Tuteja (2010) stated that under normal conditions, the ROS molecules are scavenged by various antioxidative defence mechanisms. The equilibrium between antioxidant and oxidant status could be perturbed with an increase of stress factors such as irradiation, drought, salinity, heavy metals, high or cold temperature, nutrient deficiency, air pollution, pesticide toxicity, pathogen or insect attacks in favour of oxidative stress. Imbalance towards the oxidative status led by ROS can cause significant damage to cell walls and cell components. The ROS molecules are highly reactive substances and quite toxic and cause damage to lipids, carbohydrates, nucleic acids and proteins, and they are capable of changing the gene expression pattern (Zhang et al. 2017). Since they affect gene expression and causes DNA damages in severe cases, breeding for plant tolerance under stress conditions is very difficult. Stress-induced ROS accumulation is only counteracted by enzymatic antioxidant systems including antioxidant enzymes such as CAT, POD, SOD and non-enzymatic molecules such as ascorbate, glutathione, α-tocopherol (vitamin E), carotenoids, vitamin C and flavonoids (Cai et al. 2016). In addition to them, proline, an amino acid, could be added to that list since it plays a significant role during osmotic and pathogenic stress (Bolat et al. 2016). Upadhyaya and Panda (2013) stated that the tea plant is commonly grown in rain-fed ecosystems; however, it experiences drought from time to time, and nowadays, drought is more prevalent globally than those of the last century. Drought stress causes water loss in plants, and eventually, dry mass, carotenoid, relative water content, chlorophyll and total phenolic contents of leaf and antioxidants like glutathione and ascorbate in tea plants are prone to decrease. Upadhyaya and Panda (2013) also stated that drought stress altered antioxidative response with an apparent decrease in mineral nutrient (Ca, Zn, Fe, Na, K and Mg) contents of tea leaves. It is suggested that mineral deficiency mediated drought stress-induced oxidative damage in tea plants. They suggested that foliar spray of K-, Ca- and Zn-containing nutrients played important roles in post-drought recovery in tea plants by improving antioxidant potential. Although the climate has been in changing trend, the current situation in climate change is unnatural due to human activities. This change is altering the pattern of rainfall and temperature.

5 What to Be Expected in Defence Responses of Tea Plants When Climate Changes Threaten the Crop Production of Tea-Cultivated Areas

The resistance of the tea plant to diseases largely depends on the genetic make-up of the plant, but the environment and variation induced by crossing between the different cultivars also influence this trait (Chen et al. 2012). Accelerated climate change has now been affecting the plants, behaviour of plant pathogens and their secondary metabolites. It could change the distribution and progress of diseases and affects the virulence of the pathogenic microorganisms. It may either suppress the pathogenicity of the pathogens via a decrease in sporulation and reduction in host cultivation areas or it may increase the occurrence of the disease via increased secretion of toxins and other metabolites and generate new pathotypes. These new pathotypes would negatively affect the quality of crops. Therefore, it is difficult to predict the occurrence, distribution and virulence of the disease agents in advance. For example, Ahmed et al. (2014) stated that tea growth during the monsoon season was up to 50% higher than that of spring drought. On the contrary, concentrations of methylxanthine and catechin secondary metabolites and other major compounds that determine tea functional quality were up to 50% lower during the monsoon period, on the other hand, antioxidant activity and total phenolic concentrations increased. They suggested that this inverse relationship between the growth of tea and concentrations of individual secondary metabolites results in a dilution effect of precipitation on the quality of tea. Therefore, generating a model system for the microorganisms thriving under modified atmosphere, drought or high-temperature conditions is very difficult due to poor understanding of the mechanisms of biotic and abiotic agents under changing climate conditions.

Not only virulence of the pathogens are important to tackle but new pathogenic races, pathotypes or strains may appear that must be prevented from the unintroduced areas. It is well known that crop yield and quality are heavily affected by climate change, directly and indirectly, due to the occurrence of disease agents since disease agents would adapt themselves to the conditions of crop plants to survive

(Dikilitas and Karakas 2012). After initial acclimation through secretion and synthesis of stress-related proteins, pathogenic agents have tended to adapt themselves to the conditions of crop plants and may act still as pathogens. This time their virulence may increase as in the case of *Fusarium* toxins (Kos et al. 2014) that the synthesis of its toxins may significantly increase upon adaptation to drought conditions. On the other hand, climate change is impacting crops, agroecosystems and our future worldwide. While it is evident that frequent climate events in many places are causing a decline in crop yields, however, the crop quality, as well as its resistance to additional stress, has not been acknowledged clearly. The combined stress could affect the breeding programmes to reach the desired levels of taste, aroma and yield.

In this section, possible changes and responses of tea pathogens were evaluated when they have to face up to changing climate conditions. This will help prioritize breeding objectives. The increase in levels of CO_2 and temperatures could have a direct effect on pests and diseases as well as in crops. As stated above these effects could be negative or positive. There is also an indirect effect of climate change on pathogens. Negative effects of climate on pathogens may increase or decrease the aggressiveness of the pathogen microorganisms. If climate changes on the pathogen are positive then the occurrence of the disease would become more prevalent and severity of the disease would possibly increase due to a reduction in defence mechanisms of crop plants (Dikilitas and Karakas 2014). If climate change is negative on the disease progress, the impact of pathogen would be minimal; however, recent publications have reported that the pathogens have been able to produce more toxins and secondary metabolites under harsh conditions that directly affect the quality and defence responses of the crop plants (Dikilitas et al. 2017). Therefore, these issues should be seriously questioned before the predicted climate changes take place. The increased levels of CO_2 lead to increased levels of simple sugars and lower nitrogen contents in leaves. The reduced contents of nitrogen in leaves would result in more damage due to an increased need for the attacking pathogen or insects. Under these circumstances, attacking organisms would need more nitrogen for their metabolic activities, and therefore, they would destroy more organic compounds to complete their life cycle. Higher temperatures resulted from global warming will also preserve the occurrence of microorganisms; therefore, more life cycles for the pests would be completed. Although we assume that elevated CO_2 and high temperatures may act synergistically on the pest or pathogens, however, we cannot foresee the changes of consequences except that we have to wait for the visual appearance of the symptoms whether they are going to be severe or less. Modelling of disease progress under climate changes is also difficult to assess if biotic stress agents would become aggressive. Surveying all fields or areas under projected climate changes would be costly and time-consuming, and in fact, it is almost impossible. However, from the results of previous works and in vitro studies performed under stress conditions, we can estimate what kinds of behaviours of pathogens would exhibit and what would be the consequences of pathogenicity in terms of crop production and its quality. Under these circumstances, spore production and enzymes or toxins produced by mycelia can assist in anticipating possible incidences of plant diseases. For the increase of air temperatures, the summation of the mean temperatures above a threshold level was accepted as a base for the pathogenicity that could be useful for anticipating the severity of plant diseases. Although high temperature at night reduces the leaf wetness thus causing lesser conidia germination, however, in the daytime it may cause more mycelial growth in case of sprinkle irrigation. Sharma et al. (2011) stated that decrease in catechin content, 2-phenylethanol and prephenate dehydratase (PDT) activity was evident in shoots of tea infected with blister blight pathogen when compared to healthy shoots. Water withheld for 8 days and drought stress was induced. An initial increase in the contents of the catechins, 2-phenoxyethanol and PDT activity and decrease with 3-day onwards with an increase in the severity of water stress was observed. Sharma et al. (2011) also showed that PDT enzyme from fresh tea shoots had a strong positive correlation with catechins suggesting that PDT activity played an important role in the biosynthesis of catechins. It is unfortunate that they did not produce any results regarding PDT and catechins in tea leaves under both *E. vexans* and drought conditions.

Not only temperature affects the distribution and aggressiveness of the diseases, but also fertilization or nutrition could be able to affect the disease progress. For example, Mkervali (1972) stated that leaf infection by *Colletotrichum camelliae* and *Pestalozzia theae* rose significantly with increasing nitrogen rates compared with the unfertilized control or phosphorus-potassium treatment but dropped with increasing potassium rates compared with nitrogen fertilization. Achary et al. (2017) stated that the application of phosphite might increase the synthesis of organic compounds, which could act as phytoalexins that inhibit growth and spread of pathogens in the host tissue.

Li et al. (2016b) stated that elevated CO_2 reduced endogenous caffeine (it has been known for its role in plant defence) content in tea leaves. This resulted in significant increases in susceptibility to *C. gloeosporioides*. Foliar application of caffeine increased the resistance of tea plants to some extent.

Climate changes have specific effects on microorganisms and plants. Both host susceptibility and pathogen development and survival rates could be affected. Climatic change effects will differ by geographical region and pathosystem (Elad and Pertot 2014). Interactions among environment, host, pathogen and potential vectors will affect plant diseases and their wide range of responses to factors influencing climate change (Garrett et al. 2016).

Different abiotic and biotic stress factors interact with each other over the plant defence system, and this leads to altered plant responses (Senthil-Kumar 2017). For example, drought conditions can influence pathogen spread and survival of the pathogens, and this can affect the disease incidence. Goto (1985) reported that drought conditions favoured multiplication of *Streptomyces scabies* in the rhizosphere region. Ramegowda et al. (2013) also reported that drought conditions affected response of plant defence responses and resulted in an increase in the ABA accumulation and increased susceptibility of plants to *Pseudomonas syringae*. The outline of biochemical and molecular mechanisms of tea plants infected with pathogens under climate change was summarized in Fig. 11.2.

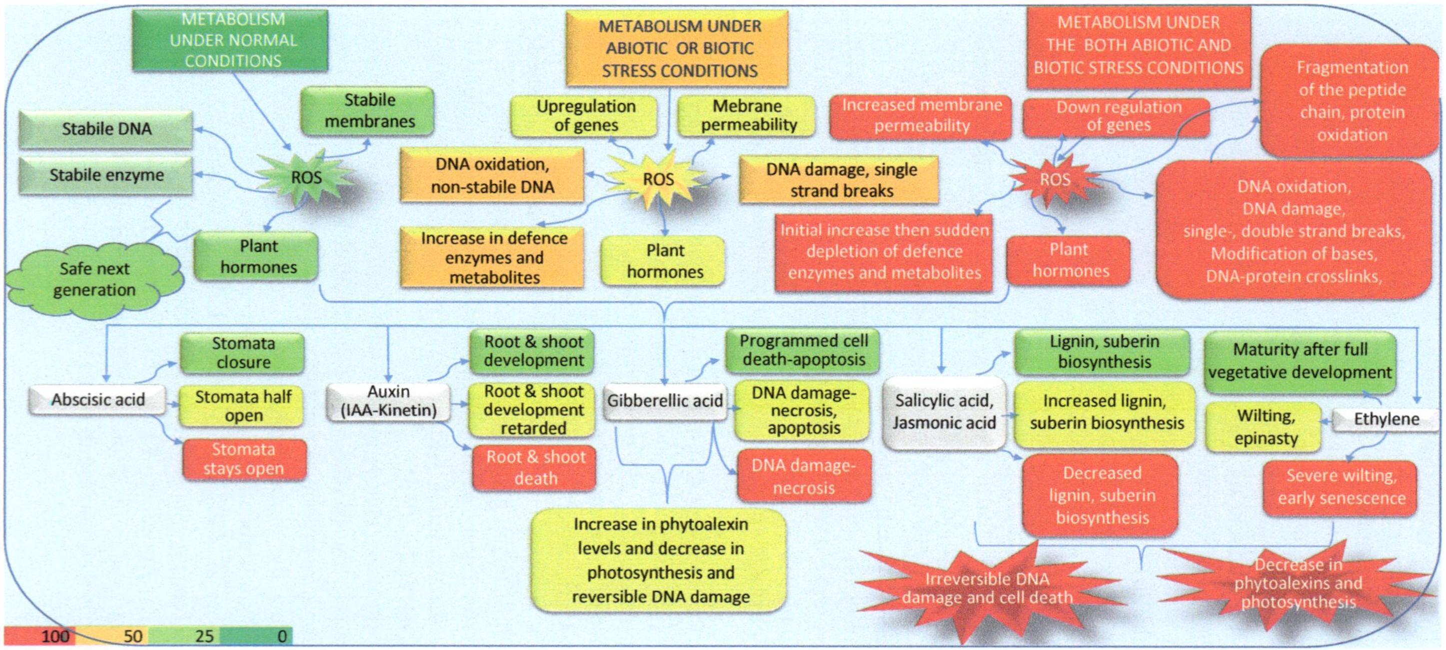

Fig. 11.2 Biochemical and molecular mechanisms of tea plants infected with pathogens under elevated temperature and CO_2 levels. At low concentrations, reactive oxygen species (ROS) acts as the secondary messenger. At moderate levels, ROS can induce various stress symptoms; however, these disorders could be repaired via activation of defence-related genes, enzymes and modulation of hormones and osmolytes. At high levels of stress (high temperature or CO_2 or both), ROS reaches the highest level and cannot be reduced with antioxidant defence mechanisms. ROS at very high level could damage DNA structure, hormones and enzymes. Although at this level the defence mechanisms are active, most of the defence-related genes are either downregulated or not expressed and defence-related enzymes could not modulate the defence response. DNA oxidation and DNA methylation are inevitable. Finally, DNA damages could occur via single or double strand breaks in which the repair is mostly impossible; therefore, cell death occurs at the final stage. Colour from green to red corresponds to the level of stress and dysfunctionality of the mechanism

Cell metabolism under normal conditions. At low concentrations, ROS act as the secondary messenger

Cell metabolism under abiotic or biotic stress conditions. Moderate ROS level

Cell metabolism under the combined abiotic and biotic stress conditions. High ROS level

6 Conclusions and Future Perspectives

Climate change is going to be one of the most devastating abiotic stress factors in this century. It is estimated that it costs the world more than 1.2 trillion US dollars (Gautam et al. 2013). The average global temperature in the last century increased as 0.74 degrees, and the atmospheric CO_2 concentration increased from 280 ppm in 1750 to 400 ppm in 2013 (Gautam et al. 2013). Such changes will have a drastic effect on the growth and cultivation of crops as well as microorganisms thriving in those habitats. Elevated temperature or CO_2 levels will cause more drastic results when compared to those of other abiotic and biotic interactions. It was reported that the impact of late blight (*Phytophthora infestans*) disease of potato and blast (*Pyricularia oryzae*) and sheath blight (*Rhizoctonia solani*) of rice increased under climate change. This case is also valid for other biotic stress agents under changing climate conditions for those which would not have been aggressive pathogens under normal climatic conditions. Therefore, disease management strategies should be reoriented in changing conditions for sustainable crop production. Changes in climate, in terms of temperature, would cause more handicaps in agriculture. It is estimated that temperature would increase by 3.4 degrees and CO_2 concentration would be 1250 ppm by the end of this century (Gill and Tuteja 2010; Gautam et al. 2013). Plant diseases are one of the most important biotic stress factors which have a direct impact on crop production. Under climate change, the case is going to be much worse. Theoretically, increase in CO_2 concentration may encourage the production of plant biomass; however, increase in temperature through CO_2 increase will increase the evapotranspiration, thus decreasing the water content of the soil and available water for plants. High temperature also results in accumulation of amino acids and sugar molecules in stressed leaves. However, these metabolites would be able to promote the development of biotrophic fungi and other pathogens. Also, increased plant density will lead to increase in wetness in leaf surfaces and thus leads to the easier infection by foliar plant pathogens. Elevated CO_2 concentration is able to affect the virulence of plant pathogens by altering leaf chemistry through the closure of stomata. With the increase of global temperature, dormant pathogens that are not able to induce pathogenicity could cause pathogenicity and even result in epidemics. This is also true for virus diseases, because long transport of plant material or vector would have a chance to survive where they are transported. Pathogens could easily find new suitable areas (Manici et al. 2014). Therefore, the occurrence of combined and abiotic stress is likely to be higher in future.

Although resistant cultivars have been developed, however, the use of pesticides is still in use due to the effects of severe fungal pathogens. The use of some synthetic fungicides for controlling fungal diseases has been banned or restricted due to the concern of environmental health and negative effects on non-target organisms as well as cost and possible development of pathogen resistance. Another alternative source could be the use of botanical extracts as fungicides to control fungal infection and detection of new antifungal compounds which have no side effects on the environment or on animal health. However, it should be remembered that the residue of pesticides on leaves could be extracted into the water when the tea is prepared with hot water. Therefore, use of pesticides should be minimized whatever their sources are. Although the development of resistant cultivars is the practical and least expensive solution to obtain high-yield crops with good quality, however, climate change would impact tea plants and create new races of pathogens after adaptation period. Therefore, new cultivars should be developed for considering these issues because pests, as well as pathogens, could thrive in hot and wetter climates, which is exactly the expected sort of climate change scenario in the future. Biotechnological methods can expedite the tea breeding process. Excessive applications of nitrogen may favour the pest growth and development of and decrease in the resistance of plants may occur.

Another approach could be the use of biologically active compounds. Saravanakumar et al. (2007) stated that foliar applications of plant growth-promoting rhizobacteria bio-formulations reduced the tea blister blight disease incidence. Also yield increase following the application of these bacteria was observed. In some studies, plant activators (PAs) have been used to induce plant disease resistance. For example, probenazole, a yeast extract preparation, tiadinil and prohydrojasmonas PAs were sprayed on field-grown 'Yabukita' tea leaves. Each PA suppressed lesion development of anthracnose and grey blight in PA-treated leaves (Yoshida et al. 2010). Ajay and Baby (2010) also stated that salicylic acid and acibenzolar-S-methyl benzo-[1,2,3]-thiadiazole-7-carboxylic acid S-methyl ester were successfully used against blister blight disease of tea. Chandra et al. (2014) used abiotic elicitor calcium chloride for inducing resistance in tea plants to *Exobasidium vexans*. Around 80% inhibition of blister blight was observed with foliar application of $CaCl_2$. Also, higher accumulation of total phenolics and the significant increase in the levels of PAL, POD, PPO and glucanase were observed. Elevated transcript levels of catalase, thaumatin, cinnamate 4-hydroxylase, flavonoid 30-hydroxylase and PAL genes were also noticed. Similarly, chitosan application induced systemic and durable tea plant resistance against blister blight disease agent (Chandra et al. 2017).

References

Achary VMM, Ram B, Manna M, Datta D, Bhatt A, Reddy MK, Agrawal PK (2017) Phosphite: a novel P fertilizer for weed management and pathogen control. Plant Biotechnol J 15(12):1493–1508

Ahmed S, Stepp JR, Orians C, Griffin T, Matyas C, Robbat A, Cash S, Xue D, Long C, Unachukwu U, Buckley S, Small D, Kennely E (2014) Effects of extreme climate events on tea (*Camellia sinensis*) functional quality validate indigenous farmer knowledge and sensory preferences in tropical China. PLoS One 9(10):e109126

Ajay D, Baby UI (2010) Induction of systemic resistance to *Exobasidium vexans* in tea through SAR elicitors. Phytoparasitica 38(1):53–60

Ali MA, Ali M, Huq M, Ahmed M (1993) In vitro studies on fungicides against *Colletotrichum gloeosporioides* (Penz.) Sac. The die back of tea. Sri Lanka J Tea Sci 62(1):25–31

Ando Y (1992) Mechanism of replacement of the tea gray blight fungus *Pestalotiopsis longiseta* by the tea brown blight fungus *Glomerella cingulata* on the border of tea gray blight lesions. Bull Natl Res Inst Veg Ornam Plants Tea Ser B 5:29–37

Ando Y, Hamaya E, Suzuki H (1985) Varietal differences in susceptibility to tea grey blight. Stud Tea 67:21–25

Anonymous (1979) Annual report Tea Scientific Department, United Planters' Association of Southern India for the period 1st January 1977 to 31st December 1977, 166 pp

Anonymous (1993) Distribution maps of plant diseases. *Phomopsis theae*. International Mycological Institute. CABI. Map No. 493

Baby UI, Ravichandran R, Ganesan V, Parthiban R, Sukumar S (1998) Effect of blister blight disease on the biochemical and quality constituents of green leaf and CTC tea. Trop Agric 75(4):452–456

Bhattacharjee S (2010) Sites of generation and physicochemical basis of formation of reactive oxygen species in plant cell. In: Gupta SD (ed) Reactive oxygen species and antioxidants in higher plants. Science Publishers, CRC Press, Taylor and Francis, Boca Raton, pp 1–30

Bolat I, Dikilitas M, Ikinci A, Ercisli S, Tonkaz T (2016) Morphological, physiological, biochemical characteristics and bud success responses of Myrobolan 29 C plum rootstock subjected to water stress. Can J Plant Sci 96:485–493

Booth C (1983) *Exobasidium vexans*. [Descriptions of fungi and bacteria]. IMI descriptions of fungi and bacteria. Sheet 779

Bruno RS, Bomser JA, Ferruzzi MG (2014) Antioxidant capacity of green tea (*Camellia sinensis*). In: Preedy V (ed) Processing and impact on antioxidants in beverages. Academic, Amsterdam, pp 33–39

Cai HM, Dong YY, Li YY, Li DX, Peng CY, Zhang ZZ, Wan XC (2016) Physiological and cellular responses to fluoride stress in tea (*Camellia sinensis*) leaves. Acta Physiol Plant 38(6):144

Cao Z, Shu Q, Dong C, Zhang X (2014) Variation in physiological characteristics at developmental stage in different disease-resistant varieties of *Camellia oleifera*. Pak J Bot 46(1):207–212

Chakraborty BN, Sharma M (2007) Serological detection and immunogold localization of cross-reactive antigens shared by *Camellia sinensis* and *Exobasidium vexans*. J Appl Microbiol 103(5):1669–1680

Chakraborty BN, Rana S, Das S, Das G, Som R, Datta S, Chakraborty U (2005a) Defense strategies of tea towards foliar pathogens. In: Chakraborty U, Chakraborty B (eds) Stress biology. Narosa Publishing House, New Delhi, pp 208–215

Chakraborty BN, Sharma M, Das RB (2005b) Defense responses in tea plants triggered by *Exobasidium vexans*. In: Chakraborty U, Chakraborty B (eds) Stress biology. Narosa Publishing House, New Delhi, pp 226–232

Chandra S, Chakraborty N, Chakraborty A, Rai R, Bera B, Acharya K (2014) Induction of defence response against blister blight by calcium chloride in tea. Arch Phytopathol Plant Protect 47(19):2400–2409

Chandra S, Chakraborty N, Panda K, Acharya K (2017) Chitosan-induced immunity in *Camellia sinensis* (L.) O. Kuntze against blister blight disease is mediated by nitric-oxide. Plant Physiol Biochem 115:298–307

Chen T-M, Chen S-F (1982) Diseases of tea and their control in the People's Republic of China. Plant Dis 66(10):961–965

Chen JS, Thseng FM, Ko WH (1998) *Macrophoma theicola*: the causal organism of tea twig die-back in Taiwan. Recent development in tea production. Proceedings of the International Symposium. Taiwan Tea Experiment Station (Chiu TF, Wang CH eds). February 29–March 5, 1988, pp 197–203

Chen L, Apostolides Z, Chen Z (2012) Global tea breeding: achievements, challenges and perspectives. Springer, Heidelberg, 384 pp

Chen Y, Qiao W, Zeng L, Shen D, Liu Z, Wang X, Tong H (2017) Characterization, pathogenicity, and phylogenetic analyses of *Colletotrichum* species associated with brown blight disease on *Camellia sinensis* in China. Plant Dis 101(6):1022–1028

Cheruiyot H (2014) Armillaria root rot (*Armillaria mellea*) of tea: biology, pathogenicity, symptoms and control. Tea 35(1):3–7

Cheruiyot EK, Mumera LM, Ng'etich WK, Hassanali A, Wachira F (2007) Polyphenols as potential indicators for drought tolerance in tea (*Camellia sinensis* L). Biosci Biotech Biochem 71(9):2190–2197

Dikilitas M, Karakas S (2012) Behaviour of plant pathogens for crops under stress during the determination of physiological, biochemical, and molecular approaches for salt stress tolerance. In: Ashraf M (ed) Crop production for agricultural improvement. Springer, Cham, pp 417–441

Dikilitas M, Karakas S (2014) Crop plants under saline-adapted fungal pathogens: an overview. In: Ahmad P, Rasool S (eds) Emerging technologies and management of crop stress tolerance, Volume II A sustainable approach, Elsevier/Academic, London/Sydney/San Francisco, pp 173–185

Dikilitas M, Karakas S, Hashem A, Abd Allah EF, Ahmad P (2016) Oxidative stress and plant responses to pathogens under drought conditions. In: Ahmad P (ed) Water stress and crop plants: a sustainable approach. Wiley, Chichester, pp 102–123

Dikilitas M, Yucel N, Dervis S (2017) Production of antioxidant and oxidant metabolites in tomato plants infected with *Verticillium dahliae* under saline conditions. In: Khan MIR, Khan NA (eds) Reactive oxygen species and antioxidant systems in plants: role and regulation under abiotic stress. Springer, Singapore, pp 315–329

Dzhalagoniya KT (1975) *Macrophoma theicola* on tea. Mikologiya i Fitopatologiya 9(2):133–135

Elad Y, Pertot I (2014) Climate change impacts on plant pathogens and plant diseases. J Crop Improve 28(1):99–139

Ertaş MN, Karakaya A (2018) Çay ve kivi bitkilerinde hastalık oluşturan *Pestalotiopsis* türleri. Harran Tarım ve Gıda Bilimleri Dergisi 22(1):152–168

Ertaş MN, Karakaya A, Çelik Oğuz A, Baştaş KK (2016) *Pestalotiopsis* species isolated from tea and kiwifruit plants in Turkey. Radovi Poljoprivrednog Fakulteta Univerziteta u Sarajevu (Works of the Faculty of Agriculture University of Sarajevo) 61, 66(1):264–268

Gao XH (1997) Correlation between the occurrence of tea red leaf spot and the structure and spatial position of leaves. J Tea Sci 17(1):21–26

Gao XH, Guo SH (1999) Studies on the epidemiology of tea red leaf spot (*Phyllosticta theicola*). Acta Phytophylacica Sin 26(2):133–136

Garrett KA, Nita M, De Wolf ED, Esker PD, Gomez-Montano L, Sparks AH (2016) Plant pathogens as indicators of climate change. In: Letcher T (ed) Climate change, 2nd edn. Elsevier, Amsterdam

Gautam HR, Bhardwaj ML, Kumar R (2013) Climate change and its impact on plant diseases. Curr Sci 105(12):1685–1691

Gill SS, Tuteja N (2010) Reactive oxygen species and antioxidant machinery in abiotic stress tolerance in crop plants. Plant Physiol Biochem 48:909–930

Gnanamangai BM, Ponmurugan P, Yazhini R, Pragadeesh SK (2011) PR enzyme activities of *Cercospora theae* causing bird's eye spot disease in tea plants (*Camellia sinensis* (L.) O. Kuntze). Plant Pathol J 10(1):13–21

González-Lamothe R, Mitchell G, Gattuso M, Diarra M, Malouin F, Bouarab K (2009) Plant antimicrobial agents and their effects on plant and human pathogens. Int J Mol Sci 10(12):3400–3419

Goto K (1985) The relative importance of precipitation and sugar content in potato peel for the detection of the incidence of common scab (*Streptomyces scabies*). Soil Sci Plant Nutr 31:419–425

Gulati A, Gulati A, Ravindranath SD, Gupta AK (1999) Variation in chemical composition and quality of tea (*Camellia sinensis*) with increasing blister blight (*Exobasidium vexans*) severity. Mycol Res 103(11):1380–1384

Gunasekera TS, Paul ND, Ayres PG (1997) The effects of ultraviolet-B (UV-B: 290–320 nm) radiation on blister blight disease of tea (*Camellia sinensis*). Plant Pathol 46(2):179–185

Hajiboland R (2017) Environmental and nutritional requirements for tea cultivation. Folia Hortic 29(2):199–220

Hamaya E (1981) Diseases of tea plant in Japan and their control. Rev Plant Protect Res 14:96–111

Horikawa T (1986) Involvement of *Pestalotia longiseta* Spegazzini in outbreak of tea shoot blight and its infection period and sites. Ann Phytopathol Soc Japn 52(5):766–771

Ikeda N, Hirono Y, Yoshida K (2012) Relationship between lesion development and nitrogen content and maturity of tea leaves after artificial inoculation with the causal agent of tea bacterial shoot blight, *Pseudomonas syringae* pv. *theae*. Bull Nat Inst Veg Tea Sci 11:99–106

Jayaswall K, Mahajan P, Singh G, Parmar R, Seth R, Raina A, Swarnkar M, Singh A, Shankar R, Sharma R (2016) Transcriptome analysis reveals candidate genes involved in blister blight defense in tea (*Camellia sinensis* (L) Kuntze). Sci Rep 6:30412

Jeyaramraja PR, Pius PK, Manian S, Meenakshi SN (2010) Role of physical barriers and chitinase in conferring blister blight resistance to *Camellia sinensis* (L.) O. Kuntze. Res J Parasitol 5(3):166–173

Karakaya A (2009) *Phomopsis theae* on *Camellia sinensis* in Turkey. J Plant Pathol 91:S4.105

Karakaya A, Bayraktar H (2010) Botrytis disease of tea in Turkey. J Phytopathol 158(10):705–707

Karakaya A, Çelik Oğuz A (2013) Rize ilinde ve Doğu Karadeniz bölgesinde çay bitkilerinde görülen hastalıklar. II. Rize Kalkınma Sempozyumu, 3–4 Mayıs 2013, Rize, Türkiye

Karakaya A, Moriwaki J, Sato T (2011) Tea diseases observed in Turkey. Asian Mycological Congress 2011 & the 12th International Marine and Freshwater Mycology Symposium. 7–11 August, 2011. Incheon, Korea

Kaur L, Jayasekera S, Moughan PJ (2014) Antioxidant quality of tea (*Camellia sinensis*) as affected by environmental factors. In: Preedy V (ed) Processing and impact on antioxidants in beverages. Elsevier, Amsterdam, pp 121–129

Keith L, Ko W-H, Sato DW (2006) Identification guide for diseases of tea (*Camellia sinensis*). Cooperative Extension Service. College of Tropical Agriculture and Human Resources, University of Hawai'i at Manoa, PD-33

Khalesi S, Sun J, Buys N, Jamshidi A, Nikbakht-Nasrabadi E, Khosravi-Boroujeni H (2014) Green tea catechins and blood pressure: a systematic review and meta-analysis of randomised controlled trials. Eur J Nutr 53(6):1299–1311

Khodaparast AS, Hedjaroude GA (1996) Fungal pathogens of tea plant in Northern Iran. Iran J Plant Path 32:233–243 (Pe), 168–170 (En)

Kim HM, Kim J (2013) The effects of green tea on obesity and type 2 diabetes. Diabetes Metab J 37(3):173–175

Koh YJ, Shin G-H, Hur J-S (2001) Seasonal occurrence and development of gray blight of tea plants in Korea. Plant Pathol J 17(1):40–44

Kos J, Hajnal EJ, Skrinjar M, Misan A, Mandi A, Jovanov P, Milovanovic I (2014) Presence of Fusarium toxins in maize from autonomous province of Vojvodina, Serbia. Food Control 46:98–101

Kumar RR, Shalini D (2010) Defense mechanism of antioxidative enzymes against blister blight pathogen. Newsletter UPASI Tea Research Foundation 20(2):2

Kuroda Y, Hara Y (1999) Antimutagenic and anticarcinogenic activity of tea polyphenols. Mutat Res 436:69–97

Lang'at JK, Otieno W, Musau JM (1998) Evaluation of some Kenyan tea (*Camellia sinensis*) clones for resistance/susceptibility to *Pestalotiopsis theae* (Sawada) as influenced by some chemical attributes of mature green leaf. Tea 19(1):6–10

Law M, Wald N, Morris J (2003) Lowering blood pressure to prevent myocardial infarction and stroke: a new preventive strategy. Health Technol Assess 7(31):1–94

Lehmann-Danzinger H (2000) Diseases and pests of tea: overview and possibilities of integrated pest and disease management. Tropenlandwirt 101(1):13–38

Li X, Wu J, Jin L, Chen Z, Zhang R, Zhang X (2016a) Pest population survey of the main tea garden in Guizhou. Guizhou Agric Sci 44(3):73–75

Li X, Ahammed GJ, Li Z, Tang M, Yan P, Han W (2016b) Decreased biosynthesis of jasmonic acid via lipoxygenase pathway compromised caffeine-induced resistance to *Colletotrichum gloeosporioides* under elevated CO in tea seedlings. Phytopathology 106:1270–1277

Liu F, Weir BS, Damm U, Crous PW, Wang Y, Liu B, Wang M, Zhang M, Cai L (2015) Unravelling *Colletotrichum* species associated with *Camellia*: employing ApMat and GS loci to resolve species in the *C. gloeosporioides* complex. Persoonia 35:63–86
Lu Y, Hu Y, Li P (2017) Consistency of electrical and physiological properties of tea leaves on indicating critical cold temperature. Biosyst Eng 159:89–96
Manici LM, Bregaglio S, Fumagalli D, Donatelli M (2014) Modelling soil borne fungal pathogens of arable crops under climate change. Int J Biometeorol 58:2071–2083
Matern U, Kneusel RE (1988) Phenolic compounds in plant disease resistance. Phytoparasitica 16:153–170
Millin DJ (1987) Factors affecting quality of tea. In: Herschdoerfer S (ed) Quality control in the food industry. Academic, London, pp 127–160
Mkervali VG (1972) The effect of nutrition on plant susceptibility to diseases. Subtropicheskie Kul'tury 2:167–170
Mondal TK (2016) Breeding and biotechnology of tea and its wild species. Springer, Cham
Mouli BC (1988) Root diseases of tea and their control. J Coffee Res 18(Suppl 1):29–37
Mouli C (primary collector) (1996) Diseases of Tea (*Camellia sinensis* (L.) O. Kuntze). https://www.apsnet.org/publications/commonnames/Pages/Tea.aspx
Mouli BC (1997) Diseases in tea nursery and their management. Planters' Chronicle 92(5):221–223
Mouli BC (2003) Blister blight of tea: biology, epidemiology and management. Ann Rev Plant Pathol 2:145–162
Mphangwe NIK, Vorster J, Steyn JM, Nyirenda EH, Taylor NJ, Apostolides Z (2013) Screening of tea (*Camellia sinensis*) for trait-associated molecular markers. Appl Biochem Biotechnol 171:437–449
Mulder D (1976) Diseases and disorders of tea (*Camellia sinensis* (L.) O. Kuntze) in Indonesia. Mededelingen van de Faculteit Landbouwwetenschappen, Rijksuniversiteit Gent 41(2(II)):1331–1335
Mur L, Hauck B, Winters A, Heald J, Lloyd A, Chakraborty U, Chakraborty B (2015) The development of tea blister caused by *Exobasidium vexans* in tea (*Camellia sinensis*) correlates with the reduced accumulation of some antimicrobial metabolites and the defence signals salicylic and jasmonic acids. Plant Pathol 64(6):1471–1483
Murugan M, Shetty PK, Ravi R, Anandhi A, Rajkumar AJ (2012) Climate change and crop yields in the Indian Cardamom Hills, 1978–2007 CE. Climatic Change 110:737–753
Mutai R, Cheramgoi E (2009) Topical issues: integrated management of stem canker – *Phomopsis theae*.TRFK. Q Bull 14(1):3–5
Mwenje E, Wingfield BD, Coetzee MPA, Nemato H, Wingfield MJ (2006) *Armillaria* species on tea in Kenya identified using isozyme and DNA sequence comparisons. Plant Pathol 55(3):343–350
Onsando JM, Wargo PM, Waudo SW (1997) Distribution, severity, and spread of *Armillaria* root disease in Kenya tea plantations. Plant Dis 81(2):133–137
Otieno W, Sierra AP, Termorshuizen A (2003) Characterization of *Armillaria* isolates from tea (*Camellia sinensis*) in Kenya. Mycologia 95(1):160–175
Palanisamy S, Mandal AK (2014) Susceptibility against grey blight disease-causing fungus *Pestalotiopsis* sp. in tea (*Camellia sinensis* (L.) O. Kuntze) cultivars is influenced by anti-oxidative enzymes. Appl Biochem Biotechnol 172(1):216–223
Pereira V, Knor F, Vellosa J, Beltrame F (2014) Determination of phenolic compounds and antioxidant activity of green, black and white teas of *Camellia sinensis* (L.) Kuntze, Theaceae. Revista Brasileira de Plantas Medicinais 16(3):490–498
Pius PK, Krishnamurthy KV, Nelson R (1998) Changes in saccharide metabolism induced by infection of *Camellia sinensis* by *Exobasidium vexans*. Biologia Plantarum 41(1):127–132
Ponmurugan P, Baby UI (2001) Effect of pre-disposing factors on phomopsis canker incidence. Newsletter – UPASI Tea Research Foundation 11(2):4
Ponmurugan P, Baby UI (2007) Morphological, physiological and biochemical changes in resistant and susceptible cultivars of tea in relation to *Phomopsis* disease. Plant Pathol J 6(1):91–94

Ponmurugan P, Baby UI, Rajkumar R (2007) Growth, photosynthetic and biochemical responses of tea cultivars infected with various diseases. Photosynthetica 45(1):143–146

Punyasiri PAN, Abeysinghe ISB, Kumar V (2005) Preformed and induced chemical resistance of tea leaf against *Exobasidium vexans* infection. J Chem Ecol 31(6):1315–1324

Punyasiri PAN, Jeganathan B, Kottawa-Arachchi JD, Ranatunga MAB, Abeysinghe ISB, Gunasekare MTK, Bandara BMR (2017) Genotypic variation in biochemical compounds of the Sri Lankan tea (*Camellia sinensis* L.) accessions and their relationships to quality and biotic stresses. J Horticult Sci Biotechnol 92(5):502–512

Ramegowda V, Senthil-Kumar M, Ishiga Y, Kaundal A, Udayakumar M, Mysore KS (2013) Drought stress acclimation imparts tolerance to *Sclerotinia sclerotiorum* and *Pseudomonas syringae* in *Nicotiana benthamiana*. Int J Mol Sci 14:9497–9513

Ramkumar S, Kumar PS, Sudhakar G, Anitha J, Geetha S, Mohankumar P, Gopalakrishnan VK (2016) Biochemical and molecular analysis of *Camellia sinensis* (L.) O. Kuntze tea from the selected P/11/15 clone. J Gen Eng Biotechnol 14:69–75

Rattan PS (1993) Incidence of Phomopsis stem and branch canker in Zimbabwe. Tea Research Foundation (Central Africa), Q Newsl October 1993 MW #49, 112: 23–24

Ruan J (2004) The impact of pH and calcium on the uptake of fluoride by tea plants (*Camellia sinensis* L.). Ann Bot 93(1):97–105

Sabanayagam JV, Samarakoon HH, Shanmuganathan N (1974) Susceptibility of some tea clones to stem canker caused by *Macrophoma theicola* Petch in the Low Country. Tea Q 44(1):74–78

Sanjay R, Baby UI (2007) Physiological and biochemical changes in tea leaves due to *Pestalotiopsis* infection. J Plant Crop 35(1):15–18

Saravanakumar D, Vijayakumar C, Kumar N, Samiyappan R (2007) PGPR-induced defense responses in the tea plant against blister blight disease. Crop Prot 26:556–565

Satyanarayana G, Padmanaban R, Barua GCS, Barua KC, Devnath SK (1982) *Armillaria mellea* – a primary root rot disease of tea (*Camellia sinensis*) in NE India. Two Bud 29(1):18–21

Senthil-Kumar M (ed) (2017) Plant tolerance to individual and concurrent stresses. Springer, New Delhi

Senthilkumar P, Thirugnanasambantham K, Mandal AKA (2012) Suppressive subtractive hybridization approach revealed differential expression of hypersensitive response and reactive oxygen species production genes in tea (*Camellia sinensis* (L.) O. Kuntze) leaves during *Pestalotiopsis theae* Infection. Appl Biochem Biotech 168:1917–1927

Sharma V, Joshi R, Gulati A (2011) Seasonal clonal variations and effects of stresses on quality chemicals and prephenate dehydratase enzyme activity in tea (*Camellia sinensis*). Eur Food Res Technol 232:307–317

Singh HR, Deka M, Das S (2015) Enhanced resistance to blister blight in transgenic tea (*Camellia sinensis* [L.] O. Kuntze) by overexpression of class I chitinase gene from potato (*Solanum tuberosum*). Funct Integr Genom 15(4):461–480

Sivanesan A, Holliday P (1972a) *Rosellinia necatrix*. [Descriptions of fungi and bacteria]. IMI Descriptions of Fungi and Bacteria. Sheet 352

Sivanesan A, Holliday P (1972b) *Rosellinia arcuata*. [Descriptions of fungi and bacteria]. IMI Descriptions of fungi and bacteria. Sheet 353

Sivanesan A, Holliday P (1972c) *Rosellinia bunodes*. [Descriptions of fungi and bacteria]. IMI descriptions of fungi and bacteria. Sheet 351

Subramanian N, Venkatesh P, Ganguli S, Sinkar VP (1999) Role of polyphenol oxidase and peroxidase in the generation of black tea theaflavins. J Agric Food Chem 47(7):2571–2578

Sugha SK, Singh BM, Sharma DK, Sharma KL (1991) Effect of blister blight on tea quality. J Plant Crop 19(1):58–60

Takeda Y (2002) Genetic analysis of tea gray blight resistance in tea plants. JARQ, Jpn Agric Res Q 36(3):143–150

Takeda Y (2003) Phenotypes and genotypes related to tea gray blight disease resistance in the genetic resources of tea in Japan. JARQ, Jpn Agric Res Q 37(1):31–35

Takeda Y, Nagatomi S (1998) Tea clones highly resistant to grey tea blight selected from γ-irradiated tea cultivar Yabukita. Report of the Kyushu Branch of the Crop Science Society of Japan 64:44–46

Upadhyaya H, Panda SK (2013) Abiotic stress responses in tea [*Camellia sinensis* L (O) Kuntze]: An overview. Rev Agric Sci 1:1–10

Vyas D, Kumar S, Ahuja PS (2007) Tea (*Camellia sinensis*) clones with shorter periods of winter dormancy exhibit lower accumulation of reactive oxygen species. Tree Physiol 27:1253–1259

Wang Y-C, Hao X-Y, Wang L, Xiao B, Wang X-C, Yang Y-J (2016a) Diverse *Colletotrichum* species cause anthracnose of tea plants (*Camellia sinensis* (L.) O. Kuntze) in China. Sci Rep 6:35287

Wang Y-C, Qian W-J, Li N-N, Hao X-Y, Wang L, Xiao B, Wang X-C, Yang Y-Z (2016b) Metabolic changes of caffeine in tea plant (*Camellia sinensis* (L.) O. Kuntze) as defense response to *Colletotrichum fructicola*. J Agric Food Chem 64(35):6685–6693

Wang L, Wang Y, Cao H, Hao X, Zeng J, Yang Y, Wang X (2016c) Transcriptome analysis of an anthracnose-resistant tea plant cultivar reveals genes associated with resistance to *Colletotrichum camelliae*. PLoS One 11(2):e0148535

Wu Q, Chen F, Lei Y, Zhang J, Zhou G, Zhou H (2013) Analysis on the occurrence and its influencing factors of tea blister blight in Lishui city. J Tea Sci 33(2):131–139

Xinzhang GY, Yang J, Shu Q (2012) Physiological mechanism of resistance to anthracnose of different *Camellia* varieties. Afr J Biotechnol 11(8):2026–2031

Yang CS (1997) Inhibition of carcinogenesis by tea. Nature 389:134–135

Yang W, Chen Y, Chen X, Yao Y, Zhou Y (2016) A new disease of tea plant caused by *Phoma adianticola*. J Tea Sci 36(1):59–67

Yao M, Chen L (2012) Tea Germplasm and breeding in China. In: Global tea breeding, Advanced topics in science and technology in China. Springer, Berlin/Heidelberg

Yie DM (1987) Occurrence and control of *Phyllosticta theaefolia* on tea plants in southwest Zhejiang mountainous areas. Plant Prot 13(3):13–14

Yoshida K (2016) Assay of bacterial shoot blight resistance among tea cultivars and breeding lines using a field inoculation test. Bull Nat Inst Veg Tea Sci 15:35–47

Yoshida K, Ogino A, Yamada K, Sonoda K (2010) Induction of disease resistance in tea (*Camellia sinensis* L.) by plant activators. Jpn Agric Res Q JARQ 44(4):391–398

Zhang QF, Liu MY, Ruan JY (2017) Metabolomics analysis reveals the metabolic and functional roles of flavonoids in light-sensitive tea leaves. BMC Plant Biol 17:64

Zhu J, Obrycki JJ, Ochieng SA, Baker TC, Pickett JA, Smiley D (2005) Attraction of two lacewing species to volatiles produced by host plants and aphid prey. Naturwissenschaften 92:277–281

Chapter 12
Plant Hormones as Mediators of Stress Response in Tea Plants

Liping Zhang, Golam Jalal Ahammed, Xin Li, Peng Yan, Lan Zhang, and Wen-Yan Han

Abstract The roles of hormones in plant growth, development and responses to stress have extensively been studied in model plants as well as some crop species. Some important genes, molecular players, and transcription factors (TFs) have been shown to play a vital role in mediating multiple hormone-regulated processes and stress responses. Recent transcriptomic analysis in tea plants identified some interesting candidates that are potentially involved in the hormonal control based on the knowledge of model plants. Identification of these genes suggests that modulation of hormone biosynthesis and signaling pathways can be effective toward manipulation of tea composition (quality) and stress tolerance; however, the issue has never been thoroughly reviewed. In this chapter, we aim to discuss those putative genes and TFs and link them with the response of tea plants to different environmental stimuli. In addition, the underlying physiological and molecular mechanisms of hormone-mediated responses to environmental stress are discussed.

Keywords Abiotic stress · Crosstalk · Plant hormone · Tea plant (*Camellia sinensis*) · Transcription factor (TF)

L. Zhang · P. Yan · L. Zhang
Key Laboratory of Tea Quality and Safety Control, Ministry of Agriculture, Tea Research Institute, Chinese Academy of Agricultural Sciences, Hangzhou, People's Republic of China

G. J. Ahammed
Department of Horticulture, College of Forestry, Henan University of Science and Technology, Luoyang, PR China
e-mail: ahammed@haust.edu.cn

X. Li · W.-Y. Han (✉)
Chinese Academy of Agricultural Sciences, Tea Research Institute, Hangzhou, PR China
e-mail: lixin@tricaas.com; hanwy@tricaas.com

W.-Y. Han et al. (eds.), *Stress Physiology of Tea in the Face of Climate Change*,
https://doi.org/10.1007/978-981-13-2140-5_12

1 Introduction

Tea is the most widely consumed nonalcoholic health drink in the world (Han et al. 2018; Wu et al. 2015). Tea plant (*Camellia sinensis*) is a perennial evergreen woody dicot plant species, mostly cultivated in the tropical and subtropical regions of Asia and Africa as a commercial beverage crop (Han et al. 2018; Jeyaraj et al. 2014; Wang et al. 2012). Just like other plants, tea plants are continuously challenged by a wide variety of abiotic stresses, such as high temperature, low temperature, drought, salinity, heavy metals, ozone, UV radiation, and so on (Zheng et al. 2016). For instance, low temperatures are one of the most critical environmental factors that limit the growth, survival, and geographical distribution of tea plants (Wang et al. 2013). Exposure of tea plants to low temperatures, including chilling and frost, drastically reduces the yield and quality of tea, leading to significant economic losses (Wang et al. 2017). Early spring frost can adversely impact the development of tea sprouts and often causes loss of tea production in many temperate regions (Wang et al. 2012). Since yield of tea is largely dependent on the production of leaf biomass, abiotic stress-induced leaf senescence causes severe diminutions of the yield and tea quality every year (Wu et al. 2015; Zhou et al. 2014; Wang et al. 2016). Due to the unprecedented climate change, the frequency and the magnitude of the extreme weather events that often cause abiotic stress have increased tremendously in the last several decades. Recent studies show that climate change not only affects tea yield and quality but also impacts overall tea system in China (Ahmed et al. 2014).

Plants perceive and respond to abiotic stresses in a highly coordinated and interactive manner involving a wide range of local and long-distance signals (Skubacz et al. 2016). For example, chilling initiates an acclimation response that prepares the plant for the more severe stress associated with freezing temperatures (Zwack and Rashotte 2015). Due to the local climate changes, tea plants undergo a cold acclimation process to enhance its freezing tolerance in winter. During the chilling acclimation process, the cold tolerance of tea plants is enhanced with the decrease in temperature but attenuated with the increase in temperature. The ability of tea plants to resist the cold is regulated by a series of genes involved in a complex regulatory network (Wang et al. 2013).

Plant hormones are small molecules that regulate plant growth, development, and responses to stress (Colebrook et al. 2014). Under abiotic stresses, hormonal signaling pathways act as key regulatory mediators to integrate adverse external cues within internal plant developmental programs (Golldack et al. 2013). Plant hormones such as auxin, gibberellins (GAs), cytokinins (CKs), abscisic acid (ABA), ethylene (ET), brassinosteroids (BRs), salicylic acid (SA), and jasmonates/jasmonic acids (JAs) have been implicated in the aforementioned integration. Recent advances in phytohormone research indicate that these hormones fine-tune plant growth and stress responses through crosstalk or interaction (Ahammed et al. 2015). In addition, transcription factors (TFs), the end-point molecular player of a signaling cascade (Šamajová et al. 2013), play central roles in gene expression by regulating expression of downstream genes as *trans*-acting elements via specific binding to *cis*-acting elements in the promoters of target genes (Mizoi et al. 2012).

TFs are key signal transmitters that regulate growth, development, and response to adverse abiotic stresses in plants through various signal transduction pathways (Wang et al. 2017; Wu et al. 2015). For instance, the ethylene-responsive factor (ERF) TFs are key downstream components of ET/JA signaling pathway (Müller and Munné-Bosch 2015). ERFs are involved in various processes of plant development and stress responses (Chakravarthy et al. 2003). Furthermore, the WRKY57 TF acts as a negative regulator of JA-induced leaf senescence (Skubacz et al. 2016). In all eukaryotes, mitogen-activated protein kinase (MAPK) signaling pathways are highly conserved modules that are most commonly composed of a number of protein kinases that phosphorylate and thereby change the activity of their respective target proteins (Šamajová et al. 2013). MAPK pathways are common and versatile signaling components that lie downstream of second messengers and hormones and play essential roles in abiotic stress, hormonal responses, the regulation of cell division, growth, differentiation, and programmed cell death (PCD) (Smékalová et al. 2014).

In this chapter, our aim is to explore putative hormone-related genes, TFs, and other modules and link them with the response of tea plants to different environmental stimuli based on the knowledge of model plants. In addition, particular attention has been paid to explore the changes in the tea quality and the underlying physiological and molecular mechanisms of hormone-mediated response to environmental stress.

2 Role of Hormones and Growth Regulators in Tea Plant Growth, Development, and Quality

Although plants require phytohormones in small quantities, they play essential roles in plant growth and development (Smékalová et al. 2014). Notably, plant growth cycle is linked to variations of endogenous levels of plant hormones, and there are specific balances for physiological processes (Kakkar and Nagar 1997). ABA is important in seed dormancy and germination process in plants (Wang et al. 2017). Ghanati and Ishkaa (2009) reported that ABA application causes inhibition of conversion of globular and heart embryos to shoot in the tissue culture of tea plant. Although ABA alone in the medium inhibits somatic embryogenesis, co-supplementation of ABA and glycine betaine can bring the induction stimulus in the tea explants (Akula et al. 2000). In tissue culture of tea, the positive effect of ABA on the induction process is dependent on the presence of glycine betaine in the medium (Table 12.1). Auxin positively regulates plant growth and development via the control of cell division, elongation, and differentiation (Skubacz et al. 2016). In tea, auxin stimulates adventitious root formation in the cuttings of tea plants. For example, indole-3-butyric acid (IBA) treatment can increase the rooting ability of micro cuttings of tea plant and is widely used in tea propagation (Wei et al. 2014). Application of 1 mg L^{-1} IBA causes the highest root length and root number in micro cutting of tea (Bidarigh et al. 2012) (Table 12.1). GA regulates the growth and

Table 12.1 Role of plant hormone and relevant molecular players in tea plants

Hormone and other modules	Growth, development and stress response	Role	Description	References
Abscisic acid (ABA)	Winter dormancy	Positive	The level of free abscisic acid (ABA) in tea remains high during dormancy periods, and it decreases very low during dormancy release with concomitant increase in bound ABA	Kakkar and Nagar (1997)
ABA	Shoot development	Negative	ABA application causes inhibition of conversion of globular and heart embryos to shoot in the tissue culture of tea plant	Ghanati and Ishkaa (2009)
ABA+ glycine betaine	Somatic embryogenesis	Positive	Co-supplementation of ABA and glycine betaine can bring the induction stimulus in the tea explants	Akula et al. (2000)
ABA	Drought tolerance	Positive	ABA plays an important role in the tea plant during drought stress	Zhou et al. (2014)
Auxin (IAA)	Dormancy release	Positive	The levels of free IAA reach highest value after dormancy release. Conjugated IAA levels increases with onset of dormancy	Nagar and Sood (2006)
Auxin (IBA)	Root formation	Positive	IBA stimulates adventitious root formation in the cuttings of tea plants by regulating related gene expression	Wei et al. (2014)
Auxin (IBA)	Root development	Positive	Application of 1 mg L^{-1} IBA causes the highest root length and root number in micro cutting of tea	Bidarigh et al. (2012)
Cytokinins (CKs)	Dormancy release	Positive	Endogenous CKs increase during dormancy release	Kakkar and Nagar (1997)
Gibberellins (GAs)	Dormancy release	Positive	The *della* mutants show reduced survival under freezing; GA_3 is effective in inducing early bud break in tea	Colebrook et al. (2014), Nagar and Kumar (2000)
GAs	Embryogenesis	Positive	In the developing liquid endosperm of tea, endogenous GAs levels remain at a maximum in the globular embryo stage	Koshioka et al. (1993)

(continued)

Table 12.1 (continued)

Hormone and other modules	Growth, development and stress response	Role	Description	References
Brassinosteroids (BRs)	Dormancy release	Positive	miRNA (cs-miR414)-mediated endogenous concentration of BR and its homeostasis might play a critical role in bud dormancy in tea plants	Jeyaraj et al. (2014)
BRs	Secondary metabolism	Positive	BRs improve secondary metabolism (flavonoid biosynthesis) in tea plants	Li et al. (2016)
Melatonin	Cold tolerance	Positive	Melatonin, a biostimulator, can also improve cold tolerance in tea plants	Li et al. (2018b)
Glycine betaine	Somatic embryogenesis	Positive	The rate of induction of somatic embryogenesis increases linearly with external betaine concentration.	Akula et al. (2000)
Polyamines (putrescine)	Winter dormancy	Positive	Polyamines play main roles in winter dormancy and dormancy release of tea. Putrescine remains high, while spermidine and spermine levels remain low during imposition of dormancy	Kakkar and Nagar (1997), Nagar and Kumar (2000)
Spermidine, spermine	Dormancy release	Positive	Spermidine and spermine levels elevated during dormancy release	Kakkar and Nagar (1997)
CsCBF1	Cold tolerance	Positive	Low temperature induces the expression of C-repeat/ dehydration-responsive element binding factors (CBFs) gene, *CsCBF1*	Wang et al. (2012)
CsCBF3	Cold and drought	Positive	*CsCBF3* regulates the expressions of downstream genes of cold responsive pathway (COR15a and COR78)	Yin et al. (2016)
CsAP2/ERF	Temperature stress response	Positive	Five *CsAP2/ERF* [APETALA 2 (AP2)/ethylene-responsive factor (ERF)] genes are related to temperature stresses	Wu et al. (2015)
CsbZIP6	Cold stress	Negative	CsbZIP6 negatively regulates cold stress response in tea plant	Wang et al. (2017)
CsHsf	Temperature stress	Positive/ negative	Temperature stress positively or negatively regulates *CsHsf* genes in tea plant	Liu et al. (2016)

development of plants including seed germination, the conversion between the vegetative and generative stages, leaf expansion, stem elongation, and flowering (Skubacz et al. 2016; Golldack et al. 2013). In the developing liquid endosperm of tea, endogenous GAs levels remain at a maximum in the globular embryo stage and then decrease rapidly during embryo maturation (Koshioka et al. 1993).

In addition to effect of hormones on cell division, differentiation and overall plant growth and development, plant hormones have been implicated in the regulation of tea quality (Li et al. 2016, 2017). Tea quality basically depends on several groups of bioactive compounds broadly classified as tea polyphenols (TP), amino acids (AA), and the ratio between them (Han et al. 2016). In addition, individual constituents of the aforementioned compounds such as catechins, theanine, and caffeine are the key determinants of tea quality. Previous studies showed that exogenous GA application could not only increase AA concentration but also decrease TP concentration, leading to a decreased TP to AA ratio in green tea (Liang et al. 1996). Exogenous indole-acetic acid (IAA) treatment increases polyphenol oxidase activity and the caffeine content (Chandra and Pandey 1998), while exogenous application of methyl jasmonate (MeJA) improves the aroma quality of black tea (Shi et al. 2014). Furthermore, individual treatment with GA and ABA alters the concentration of catechins (flavan-3-ols, major bioactive TP) as well as transcript levels of its biosynthetic genes such as *PHENYLALANINE AMMONIA-LYASE* (*CsPAL*), *CINNAMATE 4-HYDROXYLASE* (*CsC4H*), *P-COUMAROYL-COA LIGASE* (*Cs4CL*), *FLAVONOID 3-HYDROXYLASE* (*CsF3H*), and *ANTHOCYANIDIN REDUCTASE* (*CsANR*) (Singh et al. 2009; Rani et al. 2012).

Previously, our group has revealed that 24-epibrassinolide (EBL), a bioactive BRs, can promote photosynthesis in tea plants in a concentration-dependent manner. In addition, EBR increases concentrations of TP and AA and decreases the ratio of TP to AA. BR application enhances concentrations of catechins and theanine, but does not affect caffeine content. BR also improves activity of PAL and glutamine: 2-oxoglutarate aminotransferase (GOGAT) enzymes involved in catechins and theanine biosynthesis, respectively. Transcript analysis revealed that exogenous EBR upregulates the transcript levels of *CsPAL*, *glutamine synthetase* (*CsGS*), and *CsGOGAT* in tea leaves, implying that EBR improves the concentration of TP and AA by stimulating their biosynthesis (Li et al. 2016). This study advocates a high potential of exogenous BR in improving quality of summer tea. Furthermore, we unveiled the mechanism of BR-induced flavonoid biosynthesis in tea plants. We showed that BR enhances flavonoid concentration in tea leaves by inducing an increase in the endogenous concentration of nitric oxide (Li et al. 2017).

3 Role of Hormones and Other Molecular Players in Various Abiotic Stresses

Plants have evolved elaborate defense mechanisms for rapid sensing of signals from a changing environment as well as generating specific defense responses (Šamajová et al. 2013; Yoshida et al. 2014; Zelicourt et al. 2016; Zwack and Rashotte 2015).

Each signaling molecule, including various plant hormones and other modules, participates in distinct signaling pathways forming a broad crosstalking network to coordinate and fine-tune specific responses to different stresses (Smékalová et al. 2014).

Plant hormones play important roles in plant responses to abiotic stresses through activating multiple hormonal signaling pathways in plants (Fariduddin et al. 2014; Riemann et al. 2015; Colebrook et al. 2014). Several hormones function as key regulators of stress tolerance, connecting local stimuli to systemic responses (Zwack and Rashotte 2015). Hormones, in particular ABA, CK, SA, ET, and JA, play substantial direct or indirect roles in the response of plants to abiotic stress (Peleg and Blumwald 2011). In the following paragraphs, the roles of different hormones and other molecular players are discussed under a specific abiotic stress.

3.1 Drought Stress

Drought is a major abiotic stress that not only limits productivity but also affects tea quality. Prolonged drought stress results in defoliation and even plant death in *C. sinensis* (Upadhyaya et al. 2013). In tea plants, simulated drought stress (polyethylene glycol, PEG treatment) reduces chlorophyll content but increases lipid peroxidation and subsequent oxidative stress. Drought tolerance is a complex trait that can be mediated through a wide range of mechanisms (Kazan 2015). In particular, ABA plays a key role in drought tolerance in tea plants by improving protein transport, carbon metabolism, and expression of resistance proteins (Table 12.1). Exogenous ABA pretreatment increases chlorophyll and proline content in tea leaves but reduces lipid peroxidation, thereby improving drought tolerance in tea plants (Zhou et al. 2014). Moreover, proteomics analysis reveals that exogenous ABA-induced H_2O_2 production potentially enhances the expression of heat shock protein 70 (Hsp70), leading to enhanced drought tolerance in tea plants (Zhou et al. 2014). Drought stress induces ABA biosynthesis, which is the signal that triggers a number of molecular and cellular responses, ultimately resulting in stomatal closure (Verslues 2016). ABA stimulates second messengers such as reactive oxygen species (ROS), NO, and Ca^{2+}, followed by activation and inactivation of protein kinases/phosphatases that target the ion channels to prevent water loss (Zhou et al. 2014). ABI5 seems to be an ABA signaling regulator that integrates many different signals and influences the expression of stress-responsive genes. ABI5 activity is modulated by multiple protein interactions and posttranslational modifications such as phosphorylation, ubiquitination, sumoylation, and *S*-nitrosylation at the protein level. The complicated network of ABI5 often results in a both positive and negative ABA signaling feedback loops (Skubacz et al. 2016).

Plant hormone CK is also involved in drought tolerance in plants (Rasool et al. 2018). In general, drought stress decreases CK levels (Peleg and Blumwald 2011). Enhanced CK levels by direct manipulation in transgenic plants improves drought tolerance as reflected by drought phenotypes (Verslues 2016). Increased CK levels enhance plant survival under water-stress conditions, delay leaf senescence, and induce increased proline levels (Alvarez et al. 2008). The constitutive overexpression of *isopentenyl transferase* (*IPT*), a gene encoding a key step in CK biosynthesis, enhances endogenous CK level but decreases root growth under water stress (Smigocki and Owens 1989).

In addition to ABA and CKs, JAs and ET are involved in the regulation of stomatal movements. The conversion of JA precursor 12-oxophytodienoic acid (12-OPDA) to JA is inhibited under drought stress. OPDA either functions independently or together with ABA to trigger stomatal closure, leading to increased drought tolerance (Kazan 2015). Moreover, lipoxygenase 6 (LOX6)-mediated 12-OPDA may also function as a modulator of stomatal movements (Hasegawa et al. 2011). In rice, interaction of OsJAZ1 and OsbHLH148 activates the expression of *OsDREB1*, resulting in an enhanced drought (Seo et al. 2011). ET is involved in both stomatal opening and closure. Under stress conditions, ET-mediated production of antioxidant flavanols and subsequent accumulation in the guard cells negatively regulates stomatal closure by inhibiting ROS production (Kazan 2015). ET inhibits ABA-induced stomatal closure, while crosstalk between ABA and ET in regulating stomatal conductance suggests existence of convergent signaling pathways of these hormones in response to drought and contextual regulation of plant growth (Golldack et al. 2013). GA is also known to mediate the drought response in plants (Naser and Shani 2016), whereas a higher levels of endogenous auxin improve plant tolerance to water deficit (Skubacz et al. 2016).

Exogenous BR application was reported to induce drought tolerance in diverse plant species, including *Arabidopsis*, *Brassica napus*, and mustard (Fariduddin et al. 2014; Peleg and Blumwald 2011). In addition, BRs induce changes in antioxidative enzyme activities and contents of antioxidants, resulting in improved plant growth under drought stress (Fariduddin et al. 2014). In our own research, we found that exogenous BR increases flavonoids biosynthesis by inducing NO production in tea plants (Li et al. 2017). However, blocking BR biosynthesis by pharmacological approach abolishes BR-induced NO production and subsequent flavonoid accumulation. Since flavonoids are the key antioxidants in tea plants, BR-induced enhanced flavonoids may improve drought tolerance in tea plants. However, further studies are required to confirm such assumptions.

The role of MAPK signaling in plant responses to drought stress is known to act downstream of the secondary signaling mechanisms. Stomatal movements induced by drought stress are predominantly regulated by MPK9 and MPK12, which also respond to changes in ABA levels (Jammes et al. 2009). Long-term responses of plants to drought stress generally induce change of gene expression, which are mediated by transcriptionally and posttranslationally regulated TFs including WRKYs and basic leucine zipper factors (Smékalová et al. 2014). Overexpression studies and reduced resistance of *areb1 areb2 abf3* triple mutant to drought have jointly shown that three ABA-responsive bZIP TFs *AREB1/ABF2*, *AREB2/ABF4*, and *ABF3* have a pivotal regulatory function during response to drought stress, which is consistent with increased expression of these three AREB/ABFs induced by drought, high salinity, and ABA in vegetative tissues (Yoshida et al. 2014; Golldack et al. 2013). *AtERF7* inhibits the expression of ABA-responsive genes and thereby decreasing tolerance to drought stress in *Arabidopsis* (Müller and Munné-Bosch 2015).

From the above discussion, it is well conceived that ABA-induced drought tolerance involves participation of other molecular players such as DREB, WRKY, and AREB in the model plants. Similar mechanisms may also present in the tea plants, however, currently remain elusive. Therefore, further studies are to be carried out in tea to unveil the hormone-regulated drought tolerance mechanism in order to develop drought-tolerant tea genotypes and breed tea varieties with increased drought resistance, along with crop management practices that can mitigate drought stress (Zhou et al. 2014).

3.2 Salt Stress

Several lines of evidence suggest that ET pathway positively or negatively affects salt stress tolerance (Kazan 2015). The ET receptors ETR1 and ETR2 are two-component histidine protein kinases that function upstream of constitutive triple response (CTR1), which is a Raf-like MAPK3 and has therefore long been recognized as a negative regulator of the ET response pathway (Smékalová et al. 2014). CTR1 transmits the ET signal and phosphorylates the C-terminal domain of ethylene-insensitive2 (EIN2). The *ctr1* mutant shows enhanced survival of severe salt stress, because of the constitutive activation of ET responses (Ju et al. 2012). In *Arabidopsis*, salinity-induced EIN3/EIL1 conferred enhanced tolerance to salinity by promoting the ROS scavenging in an EIN2-independent manner (Tao et al. 2015). In addition to ET, JA pathway positively regulates salt tolerance in plants (Kazan 2015). GAs also improve seed germination under salinity, as well as alleviate salt stress-induced pigment loss (reviewed by Bali et al. 2017). There exists a link between DELLA function and plant survival under salt stress (Colebrook et al. 2014). For example, in *Arabidopsis* seedlings, salinity causes a reduction in endogenous bioactive GAs, which coincides with DELLA accumulation (Magome et al. 2008). Consistent with this, the depressed growth in quadruple-*della* mutant induced by salt stress is attenuated compared with wild-type *Arabidopsis* seedlings (Colebrook et al. 2014). DELLAs regulate salt-induced accumulation of ROS in *Arabidopsis*, and DELLAs control the expression of ROS scavenging enzymes as superoxide dismutase (SOD) and catalase (CAT) (Achard et al. 2008a). The GA-deficient biosynthetic mutants *ga1–3* show enhanced survival under salt stress (Magome et al. 2008).

Moreover, auxin especially IAA plays an essential role in plant adaptation to salt stress. IAA increases root and shoot growth of plants under salinity (Bali et al. 2017). Salt stress induces an increase in total auxin in developing xylem of *P. euphratica* (Popko et al. 2010) and disturbs the expression of the *PIN* genes and stabilizes AXR3/IAA17, which in turn reduces the size of the root meristem because of the lower auxin level (Skubacz et al. 2016). Studies showed that auxin receptors (TIR1 and AFB2) in *Arabidopsis* are involved in salinity stress tolerance by improving germination rate, chlorophyll contents, and root elongation (reviewed by Bali et al. 2017). Some studies have revealed that BRs significantly increased the dry mass accumulation and activities of antioxidant enzymes under salinity (reviewed by Rasool et al. 2018). Exogenous EBL treatment partly improves the tolerance of salt-sensitive rice seedlings to short-term salt stress. Furthermore, EBL treatment ameliorates NaCl-induced inhibition in seed germination of *Arabidopsis* and *Brassica napus* (Fariduddin et al. 2014). Higher levels of ABA have been reported in salt-stressed tobacco cells and alfalfa seedlings. It is also observed that ABA can increase salt stress in common bean and potato. Transgenic tomato rootstocks expressing *IPT* gene have enhanced root CK synthesis that can modify shoot hormonal balance under salinity stress (reviewed by Rasool et al. 2018).

Pretreatment of tomato with SA in hydroponic culture triggers ABA accumulation, leading to an improved acclimation to salt stress (Miura and Tada 2014). Systemin, a plant hormone that promotes JA production in tomato, can promote salt tolerance most likely in an ABA-dependent manner (Kazan 2015). ET regulates salinity response interacting with many stress signaling molecules, including plant hormones ABA, JA, SA, and BR. ET biosynthesis and signal transduction pathway interacted with ABA in regulating salinity response (Tao et al. 2015). ET signaling is at least partly integrated with GA signaling at the level of DELLA function under salt stress (Colebrook et al. 2014). SA signaling enhances salt and oxidative stress tolerance in *Arabidopsis* by the induction of the *NONEXPRESSER OF PATHOGENESIS RELATED1* (*NPR1*) gene (Müller and Munné-Bosch 2015). SA activates the accumulation of ABA in tomato plants under salinity stress, which ultimately facilitates the osmotic adjustment (reviewed by Bali et al. 2017). SA-induced inhibition of 1-aminocyclopropane-l-carboxylic acid (ACC) in *Vigna radiata* plants under salinity stress leads to the repression of ET synthesis and steps up plant growth and photosynthesis (Khan and Khan 2014). In maize plants, SA increases the level of IAA and decreases the level of ABA under salinity (reviewed by Rasool et al. 2018).

Overexpression of many *ERF* genes could enhance salt tolerance, most of which seem to be independent of ET signaling (Tao et al. 2015). The ERF subfamily transcription factor ERF6 has been implicated in control of *GA2ox6* in response to mild osmotic stress (Colebrook et al. 2014). Previous studies showed that salt stress induced the expression of a NAC-type transcription factor gene *AtNAC2* (Tao et al. 2015). At the MAP2K level, it was shown that salt stress has different effects on a range of proteins in rice, including OsMKK1, 3, 4, 6, and 10–2 (Smékalová et al. 2014). In *Arabidopsis*, where salinity induces the activation of MPK4 and MPK6, it was shown that MKK2 is the major activator of these MAPKs and that null *mkk2* mutants with impaired MKK2 function are hypersensitive to salt (Teige et al. 2004). On the other hand, MKK9 activates MPK3 and MPK6 and appears to have a negative impact on salinity tolerance since *mkk9* knockout mutants exhibit reduced salt. Interestingly, *Arabidopsis mekk1* mutants exhibit improved growth under high salinity conditions (Smékalová et al. 2014).

The AOC-dependent JA biosynthesis pathway confers salt tolerance in a MYC2-dependent but an ABA-independent manner (Kazan 2015). In rice, salinity triggers

MAPK cascades to stabilize ACSs, leading to enhanced ET production and ET signaling, which then promote ROS accumulation and growth inhibition (Tao et al. 2015). The GA-deficient phenotype of the activation-tagged *dwarf and delayed flowering-1* (*ddf1*) line leads to the elucidation of the role of the DDF1 transcription factor in directly promoting the expression of *AtGA2ox7* on exposure to salt stress (Magome et al. 2008; Colebrook et al. 2014).

Salt- and hyperosmosis-induced secondary messengers activate diverse signaling proteins of the MAPK family members. In *Arabidopsis*, salt concentrations that are high enough to induce a hyperosmotic response in addition to sodium toxicity cause the activation of MKKK20, which lies upstream of MPK6 (Smékalová et al. 2014). Hyperosmotic stress and salinity both cause differential activation of the three major stress-related *Arabidopsis* MPKs, namely, MPK3, MPK4, and MPK6 (Šamajová et al. 2013). The *cis*-acting element, ABA-responsive element (ABRE), and a group of transcription factors, ABRE-binding protein/ABRE-binding factors (AREB/ABFs), have pivotal functions in ABA-dependent gene expression. Similarly, a *cis*-element, dehydration-responsive element/C-repeat (DRE/CRT), and DRE-/CRT-binding protein 2 (DREB2) transcription factors play key roles in ABA-independent gene expression in response to osmotic stress (Yoshida et al. 2014).

3.3 Temperature Stress

Temperature (low and high) stress affects a number of plant physiological and biochemical processes to modulate plant responses through altered transcription and posttranslational modifications (Khan et al. 2015). In general, low temperatures during the growing season have negative effect on tea quality through altered accumulation of metabolites in shoots (Zheng et al. 2016). Extreme low temperatures can even damage tea shoots and reduces tea yield (Li et al. 2018a). However, tea plants have evolved sophisticated mechanisms to survive under extreme low temperature during winter. Winter dormancy is a crucial adaptation mechanism used by tea plants to survive adverse low-temperature conditions. During dormancy period, no fresh leaves are produced in tea plants (Zheng et al. 2016).

However, the yield and quality of tea in the subsequent season depend on bud dormancy. Previous studies showed that the balance of naturally occurring growth-promoting hormones and inhibiting substances might determine the state of growth or dormancy of tea shoots (Kakkar and Nagar 1997; Nagar and Kumar 2000). For example, the level of free ABA in tea is high during dormancy periods, and it declines to a very low level during dormancy release with concomitant increase in bound ABA, while endogenous CKs increase during dormancy release (Kakkar and Nagar 1997). Free GA-like activity is extremely low during the dormancy period of tea plant, while GA_3 is quite effective in inducing early bud break in tea (Nagar and Kumar 2000) (Table 12.1). In *Arabidopsis* seedlings, cold stress triggers a reduction in bioactive GA, promotes DELLA accumulation, and results in DELLA-mediated growth restriction (Achard et al. 2008a). The *della* mutants show reduced survival under freezing (Colebrook et al. 2014), indicating potential role of GA-DELLA module in winter dormancy in tea plants. The levels of free IAA are very low in deep dormant tea shoots, and they reach the highest value after dormancy release. Conjugated IAA levels increase with onset of dormancy and reach the maximum value when free IAA levels remain the lowest (Table 12.1). With increase in free IAA levels, the conjugated IAA levels decrease in shoots prior to dormancy release (Nagar and Sood 2006). In addition, polyamines play a key role in winter dormancy and dormancy release of tea buds (Kakkar and Nagar 1997; Nagar and Kumar 2000). Characterization of novel microRNA (miRNA) in tea suggests that miRNA-mediated BR signaling might play an important role in regulating developmental and seasonal variations in tea (Mohanpuria and Yadav 2012). A recent study showed that a miRNA designated as cs-miR414 is profoundly expressed in dormant bud of tea compared with that in active bud, and cs-miR414 targets mRNAs that are involved in maintaining the endogenous concentration of BRs and its homeostasis, implying that BRs level is critical for bud dormancy in tea (Jeyaraj et al. 2014) (Table 12.1).

JAs act as positive regulators of cold and freezing tolerance. In *Arabidopsis*, cold treatment increases JA accumulation by inducing the expression of JA biosynthesis genes through promoting COI1-mediated degradation of JAZ repressors (Kazan 2015). *Arabidopsis* mutants deficient in JA biosynthesis or signaling display increased sensitivity to freezing stress. Exogenous JA treatment enhances freezing tolerance in *Arabidopsis* (Hu et al. 2013). However, the role of ET on freezing tolerance is species-specific. Strict control of ET homeostasis during cold adaptation seems to be crucial for freezing tolerance (Kazan 2015). Notably, cold stress increases ABA accumulation in a range of plant species and exogenous ABA enhances cold tolerance (reviewed by Rasool et al. 2018). Recently, we demonstrated that melatonin, a biostimulator, can also improve cold tolerance in tea plants (Li et al. 2018b). Exogenous melatonin ameliorates cold-induced reductions in photosynthetic capacity by reducing oxidative stress through enhanced antioxidant potential and redox homeostasis (Table 12.1).

C-repeat/dehydration-responsive element binding factor (CBF) plays a vital role in cold stress (Bali et al. 2017). CBFs can induce the expression of a suite of cold-responsive genes to increase plant cold tolerance. For example, low-temperature stress strongly and transiently induces the expression of the three *DREB1* and *CBF* genes, including *DREB1A/CBF3*, *DREB1B/CBF1*, and *DREB1C/CBF2*, which belong to the DREB1 subgroup in *Arabidopsis* (Lata and Prasad 2011; Mizoi et al. 2012). In *Arabidopsis*, expression of three cold-inducible *DREB1s/CBFs* is under the control of low-temperature signals (Mizoi et al. 2012). The cold-responsive CBF1 controls accumulation of DELLA proteins thus demonstrating regulatory function of DELLA proteins as a major component in the CBF regulon of plant cold adaptation (reviewed by Golldack et al. 2013). Furthermore, DELLA proteins contribute to the survival of *Arabidopsis* at low temperatures via targeting the CBF1 regulatory pathway of cold adaptation. In *Arabidopsis*, CBF1 activates GA2-oxidases, reduces the cellular GA content, and enhances accumulation of the growth repressing DELLA protein RGA. Excitingly, loss of function mutation of GAI and

RGA suppresses the freezing tolerance in *Arabidopsis*, confirming that DELLA proteins contribute to the survival of plants at low temperatures via targeting the CBF1 regulatory pathway of cold adaptation (Achard et al. 2008b). The DELLA-mediated growth restraint might allow the cellular reprogramming to activate stress adaptive mechanisms instead of cellular growth processes and thus support survival of plants (Golldack et al. 2013). DREB1B/CBF1 has been found to regulate expression of *AtGA2ox3* and *AtGA2ox6* in response to cold stress (Achard et al. 2008b; Colebrook et al. 2014). Moreover, SA, along with GA_3, elevate *CBF1* gene expression in tomato under cold stress (Bali et al. 2017). Nonetheless, CBF pathway is differentially regulated by JAs and ET during cold stress (Rasool et al. 2018).

Wang et al. (2012) showed that low temperature (4 °C) induces CsCBF1 expression immediately and significantly in tea leaves, suggesting that ICE1-CBF cold-response pathway is conserved in tea plants. In addition, tea *CsCBF3* gene in tea plant plays an important role in response to low temperature through ABA-dependent pathway. Transgenic *Arabidopsis* plants that overexpress *CsCBF3* show increased tolerance to cold stress (Yin et al. 2016) (Table 12.1). Moreover, two members of the DREB1/CBF family, a subfamily of the AP2/ERF group, have been found to regulate GA deactivation in response to cold stress in *Arabidopsis* (Achard et al. 2008b; Magome et al. 2008). Abiotic stresses, such as salt and cold, usually upregulate the transcription of *GA2ox* and *DELLA* gene, which in some cases are direct targets of stress-induced AP2/ERF-type transcription factors (Colebrook et al. 2014). Wu et al. (2015) reported that five *CsAP2/ERF* genes from each AP2/ERF subfamily (DREB, ERF, AP2, and RAV) are related to temperature stresses and two of them are downregulated in the cold-acclimated tea plants. Tea CsbZIP6 negatively regulates cold stress response through mediating ABA sensitivity. For example, overexpression of *CsbZIP6* decreases the tolerance of transgenic *Arabidopsis* to freezing stress and ABA exposure by negatively regulating cold- and drought-responsive genes under cold stresses (Wang et al. 2017) (Table 12.1).

Heat stress induces excessive production of ROS that cause damage to lipids, proteins, and nucleic acids in plants. To date, all major hormones such as ABA, auxin, GAs, CKs, SA, JA, ET, and BRs have been reported to play critical roles in response of plants to heat stress (Ahammed et al. 2016). Several lines of evidence suggest that JAs positively regulate heat stress tolerance in *Arabidopsis*. For example, exogenous JA application increases heat stress tolerance. Both JA biosynthesis and signaling are required for thermotolerance. ET also provides heat tolerance to plants (reviewed by Rasool et al. 2018). ET, despite being produced under heat stress, does not confer heat tolerance. However, the EIN2-mediated pathway negatively regulates thermotolerance as the *ein2* mutant shows increased thermotolerance (Khan et al. 2015). Khan et al. (2013) reported that SA treatment can alleviate heat stress in *T. aestivum* by increasing proline accumulation and restriction of the stress-induced ET formation. During heat stress, exogenous application of SA alleviates the damage to plants by limiting ET generation (reviewed by Bali et al. 2017).

Various regulate heat stress response by regulating the expression of heat shock proteins and activating the heat shock defense in tomato, soybean (*Glycine max*), and *Arabidopsis*. Wheat *HsfA6* is obviously upregulated during heat stress, and its overexpression improves thermotolerance in transgenic wheat. Temperature stress positively or negatively regulated *CsHsf* genes in tea plant, among which *CsHsfB1*, *CsHsfB2b*, and *CsHsfC1* significantly upregulate under heat treatment (Table 12.1) (Liu et al. 2016). MAPK-dependent responses to heat stress in alfalfa and tomato plants have previously been described (Smékalová et al. 2014). In *Arabidopsis*, MPK6 phosphorylates HsfA2, which is a major heat stress transcription factor, on its T249 residue during heat stress response (Smékalová et al. 2014). Aside from the specificity and crosstalk among transcription factors, the tea plant itself exhibits a particular defense mechanism and contains a large amount of secondary metabolites. Thus the transcriptional regulation mechanism in tea plant is considerably complex and requires further investigations (Liu et al. 2016).

3.4 Other Abiotic Stresses

A large number of studies showed that SA plays an important role in the minimization of metal-induced stress-impacts. Exogenous SA application improves growth and photosynthetic traits under lead, cadmium, and copper stress in several crop plant species such as *Oryza sativa*, *Zea mays*, and *Phaseolus vulgaris*, respectively (Khan et al. 2015). Previous studies reported that ET accumulation increases in response to various heavy metals in a wide range of plant species. For example, Cu induces the transcripts of *ACS* gene in potato (*Solanum tuberosum*) and tobacco (*Nicotiana tabacum*) (Thao et al. 2015). Transcriptome analysis of chromium-treated rice (*Oryza sativa*) roots also indicates enhanced expression of four ET biosynthetic genes (*AC51*, *AC52*, *ACO4*, and *ACO5*) (Steffens 2014; Trinh et al. 2014). Upon exposure to heavy metal stress, the levels of JA, SA, ABA, and ET increase, while the contents of GA_3 and auxin decrease in plants (Thao et al. 2015). Al-induced ET production may lead to auxin redistribution by affecting auxin polar transport systems through AUX1 and PIN2 (Sun et al. 2010). Cu treatment affects a greater number of signaling components including transcription factors *WRKY22*, *WRKY25*, and *WRKY29* as well as *MEKK1* and *MPK3* genes. Interestingly, heat shock transcription factor A4a (HsfA4a) improves the tolerance of rice to cadmium stress. Cd stress induces the activation of MPK3, MPK4, and MPK6 as well as WRKY25 (a downstream target of MPK4). Long-term exposure to Cd does not influence the transcription of ZAT12, a zinc finger transcription factor, in the roots, but short-term exposure to Cu alters *ZAT12* gene expression (Smékalová et al. 2014).

SA enhances the antioxidant response and maintains cellular redox state under ozone stress (Khan et al. 2015). In addition, SA is involved in signaling network integrating other phytohormones such as JA and/or ET in ozone-exposed plants. For example, ozone-mediated induced JAs biosynthesis can attenuate SA-dependent lesion initiation caused by ozone in *Arabidopsis* (Rao et al. 2000). Ozone-induced SA accumulation can be promoted by ET-mediated regulation of the expression of the *PAL* and *chorismate mutase* (*CM*) genes in *Nicotiana tabacum* (Ogawa et al.

2005). SA and ET both act to regulate ozone-induced cell death in *Arabidopsis* (Khan et al. 2015). Ozone-sensitivity in hybrid poplar is correlated with insensitivity to both SA and JA, where these phytohormones are associated with the PCD in lesion formation (Koch et al. 2000). Many MAPK cascades respond to hormones such as ABA, JA, SA, ET, auxins, and BRs (Smékalová et al. 2014). Ozone activates MAPK signaling pathways through triggering ROS production and accumulation of ET, JA, and SA, resulting in local PCD (Šamajová et al. 2013). The *MKK1/MPK6* module is an important component of the ABA-dependent signaling pathway that is responsible for H_2O_2 production and stress responses (Xing et al. 2008). Previous studies using inhibitor treatment, gene mutation, and transgenic technology all suggested that ABA-induced *CAT1* expression and H_2O_2 production were mediated by MKK1 (MAP2K) and MPK6 in *Arabidopsis* (Smékalová et al. 2014). The MAPK inhibitors are able to inhibit ABA-mediated stomatal closure in pea and *Vicia faba* (Zelicourt et al. 2016). In addition, MPK9 and MPK12 seem to be positive regulators of ABA signaling in stomatal guard cells and probably act upstream of ABA-dependent anion channel activation and downstream of ROS signaling (Jammes et al. 2009).

Transcriptomic analyses of *mekk1*, *mkk1/2*, and *mpk4* have confirmed the role of MAPK cascade as an integrating element in ROS- and SA-initiated stress pathways. MPK4 is a key regulator of plant defense mechanisms based primarily on negative regulation of SA signaling. Ozone activates MPK3 and MPK6 in *Arabidopsis*, whereas MPK3 activity is dependent on SA (Šamajová et al. 2013). The MKK3/MPK6 cascade mediates JA signaling and has negative effects on the gene expression patterns it induces in mechanical wounding caused by herbivore attacks (Smékalová et al. 2014).

A molecular link between oxidative stress and auxin signal transduction is well established (Smékalová et al. 2014). ERF1 might enhance oxidative stress tolerance through regulating ROS-responsive gene expression (Müller and Munné-Bosch 2015). SA-mediated activation of antioxidant enzymes is considered as a major factor for SA-mediated regulation of UV-induced oxidative stress in *Capsicum annuum* leaves (Khan et al. 2015). Wounding stress rapidly triggers the production of plant hormones (e.g., JA) (Smékalová et al. 2014).

During oxygen deficiency caused by flooding, ET levels increase, and several ET-responsive genes including a specific set of ERF/AP2 TFs are upregulated in *Arabidopsis* (Kazan 2015). Submergence induces ET accumulation and the expression of ET-responsive ERF TFs, which antagonize ABA and directly or indirectly increase the accumulation of bioactive GA. The above three hormone (ET, ABA, and GA) signal pathways trigger the elongation of internode and stem, which allows the shoot to outgrow the floodwaters and thus enhances resistance of the rice plant to submergence conditions (Kazan 2015; Colebrook et al. 2014). Increased GA biosynthesis and signaling promote growth in plant escape responses to shading and submergence (Colebrook et al. 2014).

4 Conclusions and Future Perspectives

Stabilization of agricultural yields and improvement of environmental tolerance in crops inevitably require the comprehensive identification of cellular key components and molecular links of plant adaptive mechanisms (Golldack et al. 2013). The long-term goal of genetic engineering is to develop plants that are more resistant to abiotic and biotic stresses (Smékalová et al. 2014). In this chapter, we tried to discuss roles and mechanisms of different hormones in abiotic stress response based on the knowledge of model plants and crop species. In addition to exogenous application, the use of inducible promoters for the conditional expression of hormone biosynthetic genes may have potential to control hormone levels without the negative effects on plant growth and development with large changes in hormone concentrations (Peleg and Blumwald 2011). Some important genes and transcription factors that are linked to phytohormone-mediated responses to abiotic stress are emphasized for their potential role in tea plants. We also provide references from tea plants based on the availability of the literature. However, it appears that research on hormone-mediated mechanisms of tea plants tolerance to environmental stress is still in its infancy. Recently, global transcriptome profiles of tea provide important clues to further dissect crucial mechanism of hormone-mediated response in tea plants to abiotic stress. Therefore, a major practical goal should be to clarify those "key" genes in tea plants that can enhance stress tolerance. Some molecular players such as CBFs, DREBs, WRKYs, and MAPKs as well as individual members of these modules can be considered as good targets to explore hormone-regulated mechanisms in tea plants (Šamajová et al. 2013). Moreover, elucidation of the role of miRNAs in regulating hormone homeostasis and subsequent stress response may shed new light on tea physiology and molecular biology under perturbed environmental conditions. The knowledge of crosslinks among multiple hormone signals and other modules offers opportunities to unravel new concepts for agronomic traits and will be a promising target for biotechnological applications to improve environmental tolerance in tea while maintaining tea yield and quality for future sustainability in tea production (Sah et al. 2016; Golldack et al. 2013).

Acknowledgments Research in the authors' laboratories is supported by grants from the Key Project of International Science & Technology Cooperation, National Key Research and Development Programme of China (2017YFE0107500), the Open Fund of State Key Laboratory of Tea Plant Biology and Utilization (SKLTOF20170106), the Innovation Project of the Chinese Academy of Agricultural Sciences (CAAS-ASTIP-2015-TRICAAS-08), Henan University of Science and Technology (HAUST) Research Start-up Fund for New Faculty (13480058), the National Natural Science Foundation of China (31600561), and the Central Public-interest Scientific Institution Basal Research Fund (1610212016013, 1610212018015).

References

Achard P, Renou JP, Berthomé R, Harberd NP, Genschik P (2008a) Plant DELLAs restrain growth and promote survival of adversity by reducing the levels of reactive oxygen species. Curr Biol 18:656–660

Achard P, Gong F, Cheminant S, Alioua M, Hedden P, Genschik P (2008b) The cold-inducible CBF1 factor-dependent signaling pathway modulates the accumulation of the growth-repressing DELLA proteins via its effect on gibberellin metabolism. Plant Cell 20:2117–2129

Ahammed GJ, Xia XJ, Li X, Shi K, Yu JQ, Zhou YH (2015) Role of brassinosteroid in plant adaptation to abiotic stresses and its interplay with other hormones. Curr Protein Pept Sci 16:462–473

Ahammed GJ, Li X, Zhou J, Zhou YH, Yu JQ (2016) Role of hormones in plant adaptation to heat stress. In: Plant hormones under challenging environmental factors. Springer, Dordrecht, pp 1–21

Ahmed S, Stepp JR, Orians C, Griffin T, Matyas C, Robbat A, Cash S, Xue D, Long C, Unachukwu U, Buckley S, Small D, Kennelly E (2014) Effects of extreme climate events on tea (*Camellia sinensis*) functional quality validate indigenous farmer knowledge and sensory preferences in tropical China. PLoS One 9(10):e109126

Akula A, Akula C, Bateson M (2000) Betaine a novel candidate for rapid induction of somatic embryogenesis in tea (*Camellia sinensis* (L.) O. Kuntze). Plant Growth Regul 30:241–246

Alvarez S, Marsh EL, Schroeder SG, Schachtman DP (2008) Metabolomic and proteomic changes in the xylem sap of maize under drought. Plant Cell Environ 31:325–340

Bali S, Poonam, Gautam V, Kaur P, Khanna K, Kaur R, Vig AP, Ohri P, Bhardwaj R (2017) Interaction of salicylic acid with plant hormones in plants under abiotic stress. In: Nazar R et al (eds) Salicylic acid: a multifaceted hormone. Springer, Singapore. https://doi.org/10.1007/978-981-10-6068-7_10

Bidarigh S, Hatamzadeh A, Azarpour E (2012) The study effect of IBA hormone levels on rooting in micro cuttings of tea (*Camellia sinensis* L.). World Appl Sci J 20:1051–1054

Chakravarthy S, Tuori RP, D'Ascenzo MD, Fobert PR, Després C, Martin GB (2003) The tomato transcription factor Pti4 regulates defense-related gene expression via GCC box and non-GCC box *cis* elements. Plant Cell 15:3033–3050

Chandra K, Pandey R (1998) Influence of some bioregulators on quality traits of pruned tea (*Camellia sinensis* (L) O Kuntze). J Sci Food Agric 77:429–434

Colebrook EH, Thomas SG, Phillips AL, Hedden P (2014) The role of gibberellin signaling in plant responses to abiotic stress. J Exp Biol 217:67–75

Fariduddin Q, Yusuf M, Ahmad I, Ahmad A (2014) Brassinosteroids and their role in response of plants to abiotic stresses. Biol Plantarum 58(1):9–17

Ghanati F, Ishkaa MR (2009) Investigation of the interaction between abscisic acid (ABA) and excess benzyladenine (BA) on the formation of shoot in tissue culture of tea (*Camellia sinensis* L.). Int J Plant Prod 3(4):7–14

Golldack D, Li C, Mohan H, Probst N (2013) Gibberellins and abscisic acid signal crosstalk: living and developing under unfavorable conditions. Plant Cell Rep 32:1007–1016

Han WY, Huang JG, Li X, Li ZX, Ahammed GJ, Yan P, Stepp JR (2016) Altitudinal effects on the quality of green tea in East China: a climate change perspective. Eur Food Res Technol 243(2):1–8

Han W, Li X, Yan P, Zhang L, Ahammed GJ (2018) Tea cultivation under changing climatic conditions. In: Global tea science current status and future needs. Burleigh Dodds Science Publishing, Cambridge, pp 455–472

Hasegawa S, Sogabe Y, Asano T, Nakagawa T, Nakamura H, Kodama H, Ohta H, Yamaguchi K, Mueller MJ, Nishiuchi T (2011) Gene expression analysis of wounding-induced root-to-shoot communication in *Arabidopsis thaliana*. Plant Cell Environ 34:705–716

Hu Y et al (2013) Jasmonate regulates the inducer of cbf expression-C-repeat binding factor/DRE binding factor1 cascade and freezing tolerance in *Arobidopsis*. Plant Cell 25:2907–2924

Jammes F, Song C, Shin D, Munemasa S, Takeda K, Gu D, Cho D, Lee S, Giordo R, Sritubtim S, Leonhardt N, Ellis BE, Murata Y, Kwak JM (2009) MAP kinases MPK9 and MPK12 are preferentially expressed in guard cells and positively regulate ROS-mediated ABA signaling. Proc Natl Acad Sci U S A 106(48):20520–20525

Jeyaraj A, Chandran V, Gajjeraman P (2014) Differential expression of microRNAs in dormant bud of tea [*Camellia sinensis* (L.) O. Kuntze]. Plant Cell Rep 33:1053–1069

Ju C, Yoon GM, Shemansky JM, Lin DY, Ying ZI, Chang J, Garrett WM, Kessenbrock M, Groth G, Tucker ML, Cooper B, Kieber JJ, Chang C (2012) CTR1 phosphorylates the central regulator EIN2 to control ethylene hormone signaling from the ER membrane to the nucleus in *Arabidopsis*. Proc Natl Acad Sci U S A 109:19486–19491

Kakkar RK, Nagar PK (1997) Distribution and changes in endogenous polyamines during winter dormancy in tea [*Camellia sinensis* L. (O) Kuntze]. J Plant Physiol 151:63–67

Kazan K (2015) Diverse roles of jasmonates and ethylene in abiotic stress tolerance. Trends Plant Sci 20(4):219–229

Khan MIR, Khan NA (2014) Ethylene reverses photosynthetic inhibition by nickel and zinc in mustard through changes in PSII activity, photosynthetic nitrogen use efficiency, and antioxidant metabolism. Protoplasma 251:1007–1019

Khan MIR, Iqbal N, Masood A, Per TS, Khan NA (2013) Salicylic acid alleviates adverse effects of heat stress on photosynthesis through changes in proline production and ethylene formation. Plant Signal Behav 8:e26374

Khan MIR, Fatma M, Per TS, Anjum NA, Khan NA (2015) Salicylic acid-induced abiotic stress tolerance and underlying mechanisms in plants. Front Plant Sci 6:462

Koch JR, Creelman RA, Eshita SM, Seskar M, Mullet JE, Davis KR et al (2000) Ozone sensitivity in hybrid poplar correlates with insensitivity to both salicylic acid and jasmonic acid. The role of programmed cell death in lesion formation. Plant Physiol 123:487–496

Koshioka M, Yamaguchi S, Nishijima T, Yamazaki H, Ferraren D (1993) Endogenous gibberellins in the developing liquid endosperm of tea. Biosci Biotechnol Biochem 57:1586–1588

Lata C, Prasad M (2011) Role of DREBs in regulation of abiotic stress responses in plants. J Exp Bot 62:4731–4748

Li X, Ahammed GJ, Li ZX, Zhang L, Wei JP, Shen C, Yan P, Zhang LP, Han WY (2016) Brassinosteroids improve quality of summer tea (*Camellia sinensis* L.) by balancing biosynthesis of polyphenols and amino acids. Front Plant Sci 7:1304

Li X, Zhang L, Ahammed GJ, Li ZX, Wei JP, Shen C, Yan P, Zhang LP, Han WY (2017) Nitric oxide mediates brassinosteroid-induced flavonoid biosynthesis in *Camellia sinensis* L. J Plant Physiol 214:145–151

Li X, Ahammed GJ, Li Z-X, Zhang L, Wei J-P, Yan P, Zhang L-P, Han W-Y (2018a) Freezing stress deteriorates tea quality of new flush by inducing photosynthetic inhibition and oxidative stress in mature leaves. Sci Hortic Amsterd 230:155–160

Li X, Wei JP, Scott ER, Liu JW, Guo S, Li Y, Zhang L, Han WY (2018b) Exogenous melatonin alleviates cold stress by promoting antioxidant defense and redox homeostasis in *Camellia sinensis* L. Molecules 23(1):165

Liang Y, Lu J, Shang S (1996) Effect of gibberellins on chemical composition and quality of tea (*Camellia sinensis* L). J Sci Food Agric 72:411–414

Liu ZW, Wu ZJ, Li XH, Huang Y, Li H, Wang YX, Zhuang J (2016) Identification, classification, and expression profiles of heat shock transcription factors in tea plant (*Camellia sinensis*) under temperature stress. Gene 576:52–59

Magome H, Yamaguchi S, Hanada A, Kamiya Y, Oda K (2008) The DDF1 transcriptional activator upregulates expression of a gibberellin-deactivating gene, GA2ox7, under high-salinity stress in *Arabidopsis*. Plant J 56:613–626

Miura K, Tada Y (2014) Regulation of water, salinity, and cold stress responses by salicylic acid. Front Plant Sci 5(2):4

Mizoi J, Shinozaki K, Yamaguchi-Shinozaki K (2012) AP2/ERF family transcription factors in plant abiotic stress responses. Biochim Biophys Acta 1819:86–96
Mohanpuria P, Yadav SK (2012) Characterization of novel small RNAs from tea (*Camellia sinensis* L.). Mol Biol Rep 39:3977–3986
Müller M, Munné-Bosch S (2015) Ethylene response factors: a key regulatory hub in hormone and stress signaling. Plant Physiol 169:32–41
Nagar PK, Kumar A (2000) Changes in endogenous gibberellin activity during winter dormancy in tea (*Camellia sinensis* (L.) O. Kuntze). Acta Physiol Plant 22(4):439–443
Nagar PK, Sood S (2006) Changes in endogenous auxins during winter dormancy in tea (*Camellia sinensis* L.) O. Kuntze. Acta Physiol Plant 28:165–169
Naser V, Shani E (2016) Auxin response under osmotic stress. Plant Mol Biol 91:661–672
Ogawa D, Nakajima N, Sano T, Tamaoki M, Aono M, Kubo A et al (2005) Salicylic acid accumulation under O_3 exposure is regulated by ethylene in tobacco plants. Plant Cell Physiol 46:1062–1072
Peleg Z, Blumwald E (2011) Hormone balance and abiotic stress tolerance in crop plants. Curr Opin Plant Biol 14:290–295
Popko J, Hänsch R, Mendel R-R, Polle A, Teichmann T (2010) The role of abscisic acid and auxin in the response of poplar to abiotic stress. Plant Biol 12:242–258
Rani A, Singh K, Ahuja PS, Kumar S (2012) Molecular regulation of catechins biosynthesis in tea [*Camellia sinensis* (L.) O. Kuntze]. Gene 495:205–210
Rao MV, Lee HI, Creelman RA, Mullet JE, Davis KR (2000) Jasmonate perception desensitizes O_3-induced salicylic acid biosynthesis and programmed cell death in *Arabidopsis*. Plant Cell 12:1633–1646
Rasool S, Urwat U, Nazir M, Zargar SM, Zargar MY (2018) Cross talk between phytohormone signaling pathways under abiotic stress conditions and their metabolic engineering for conferring abiotic stress tolerance. In: Zargar SM, Zargar MY (eds) Abiotic stress-mediated sensing and signaling. in plants: an Omics perspective. Springer, Singapore. https://doi.org/10.1007/978-981-10-7479-0_12
Riemann M, Dhakarey R, Hazman M, Miro B, Kohli A, Nick P (2015) Exploring jasmonates in the hormonal network of drought and salinity responses. Front Plant Sci 6:1077
Sah SK, Reddy KR, Li J (2016) Abscisic acid and abiotic stress tolerance in crop plants. Front Plant Sci 7:571
Šamajová O, Plíhal O, Al-Yousif M, Hirt H, Šamaj J (2013) Improvement of stress tolerance in plants by genetic manipulation of mitogen-activated protein kinases. Biotechnol Adv 31(1):118–128
Seo JS, Joo J, Kim MJ, Kim YK, Nahm BH, Song SI, Cheong JJ, Lee JS, Kim JK, Choi YD (2011) OsbHLH148, a basic helix-loop-helix protein, interacts with OsJAZ proteins in a jasmonate signaling pathway leading to drought tolerance in rice. Plant J 65:907–921
Shi J, Wang L, Ma CY, Lv HP, Chen ZM, Lin Z (2014) Aroma changes of black tea prepared from methyl jasmonate treated tea plants. J Zhejiang Univ Sci B 15:313–321. https://doi.org/10.1631/jzus.B1300238
Singh K, Kumar S, Rani A, Gulati A, Ahuja PS (2009) Phenylalanine ammonia-lyase (PAL) and cinnamate 4-hydroxylase (C4H) and catechins (flavan-3-ols) accumulation in tea. Funct Integr Genomics 9:125–134
Skubacz A, Daszkowska-Golec A, Szarejko I (2016) The role and regulation of ABI5 (ABA-insensitive 5) in plant development, abiotic stress responses and phytohormone crosstalk. Front Plant Sci 7:1884
Smékalová V, Doslcočilová A, Komis G, Samaj J (2014) Crosstalk between secondary messengers, hormones and MAPK modules during abiotic stress signalling in plants. Biotechnol Adv 32:2–11
Smigocki AC, Owens LD (1989) Cytokinin-to-auxin ratios and morphology of shoots and tissues transformed by a chimeric isopentenyl transferase gene. Plant Physiol 91:808–811

Steffens B (2014) The role of ethylene and ROS in salinity, heavy metal, and flooding responses in rice. Front Plant Sci 5:685

Sun P, Tian QY, Chen J, Zhang WH (2010) Aluminium-induced inhibition of root elongation in Arabidopsis is mediated by ethylene and auxin. J Exp Bot 61:347–356

Tao JJ, Chen HW, Ma B, Zhang WK, Chen SY, Zhang JS (2015) The role of ethylene in plants under salinity stress. Front Plant Sci 6:1059

Teige M, Scheikl E, Eulgem T, Dóczi R, Ichimura K, Shinozaki K, Dangl JL, Hirt H (2004) The MKK2 pathway mediates cold and salt stress signaling in *Arabidopsis*. Mol Cell 15:141–152

Thao NP, Khan MIR, Thu NBA, Hoang XLT, Asgher M, Khan NA, Tran L-SP (2015) Role of ethylene and its cross talk with other signaling molecules in plant responses to heavy metal stress. Plant Physiol 169:73–84

Trinh NN, Huang TL, Chi WC, Fu SF, Chen CC, Huang HJ (2014) Chromium stress response effect on signal transduction and expression of signaling genes in rice. Physiol Plant 150:205–224

Upadhyaya H, Dutta BK, Panda SK (2013) Zinc modulates drought induced biochemical damages in tea (*C. sinensis* (L) O Kuntze). J Agric Food Chem 61(27):6660–6670

Verslues PE (2016) ABA and cytokinins: challenge and opportunity for plant stress research. Plant Mol Biol 91:629–640

Wang Y, Jiang CJ, Li YY, Wei CL, Deng WW (2012) CsICE1 and CsCBF1: two transcription factors involved in cold responses in *Camellia sinensis*. Plant Cell Rep 31:27–34

Wang XC, Zhao QY, Ma CL, Zhang ZH, Cao HL, Kong YM, Yue C, Hao XY, Chen L, Ma JQ, Jin JQ, Li X, Yang YJ (2013) Global transcriptome profiles of *Camellia sinensis* during cold acclimation. BMC Genomics 14:415

Wang W, Xin H, Wang M, Ma Q, Wang L, Kaleri NA, Wang Y, Li X (2016) Transcriptomic analysis reveals the molecular mechanisms of drought-stress-induced decreases in *Camellia sinensis* leaf quality. Front Plant Sci 7:385

Wang L, Cao H, Qian W, Yao L, Hao X, Li N, Yang Y, Wang X (2017) Identification of a novel bZIP transcription factor in *Camellia sinensis* as a negative regulator of freezing tolerance in transgenic arabidopsis. Ann Bot London 119(7):1195–1209

Wei K, Wang LY, Wu LY, Zhang CC, Li HL, Tan LQ, Cao HL, Cheng H (2014) Transcriptome analysis of Indole-3-butyric acid-induced adventitious root formation in nodal cuttings of *Camellia sinensis* (L.). Plos One 9:e107201

Wu ZJ, Li XH, Liu ZW, Li H, Wang YX, Zhuang J (2015) Transcriptome-based discovery of AP2/ERF transcription factors related to temperature stress in tea plant (*Camellia sinensis*). Funct Integr Genomics 15:741–752

Xing Y, Jia W, Zhang J (2008) AtMKK1 mediates ABA-induced CAT1 expression and H_2O_2 production via AtMPK6-coupled signaling in *Arabidopsis*. Plant J 54:440–451

Yin Y, Ma Q, Zhu Z, Cui Q, Chen C, Chen X, Fang W, Li X (2016) Functional analysis of CsCBF3 transcription factor in tea plant (*Camellia sinensis*) under cold stress. Plant Growth Regul 80:335–343

Yoshida T, Mogami J, Yamaguchi-Shinozaki K (2014) ABA-dependent and ABA-independent signaling in response to osmotic stress in plants. Curr Opin Plant Biol 21:133–139

Zelicourt A, Colcombet J, Hirt H (2016) The role of MAPK modules and ABA during abiotic stress signaling. Trends Plant Sci 21(8):677–685

Zheng C, Wang Y, Ding Z, Zhao L (2016) Global transcriptional analysis reveals the complex relationship between tea quality, leaf senescence and the responses to cold-drought combined stress in *Camellia sinensis*. Front Plant Sci 7:1858

Zhou L, Xu H, Mischke S, Meinhardt LW, Zhang D, Zhu X, Li X, Fang W (2014) Exogenous abscisic acid significantly affects proteome in tea plant (*Camellia sinensis*) exposed to drought stress. Hortic Res 1:14029

Zwack PJ, Rashotte AM (2015) Interactions between cytokinin signalling and abiotic stress responses. J Exp Bot 66(16):4863–4871

Chapter 13
Genomics Approaches for Biotic and Abiotic Stress Improvement in Tea

Sangeeta Borchetia, Gautam Handique, Somnath Roy, and Shabir Hussain Wani

Abstract Tea (*Camellia sinensis*) plants are continuously exposed to a wide range of biotic and abiotic stresses. Being a perennial monoculture crop, tea relatively provides a favorable niche for more than 1000 arthropod pests and numerous disease-causing microorganisms. Abiotic stress factors such as drought, temperature, waterlogging, salinity, and nutrient stress considerably constrain the growth, productivity, and quality of tea plants. At present, global climatic changes have made notable impact in tea with a decline in its productivity. The main biotic and abiotic stresses that affect tea plants in the field are being extensively studied. Each stress elicits a complex molecular cascade starting from perception of stress to initiation of signal transduction and its manifestation in cellular and metabolic level. Construction of genetic map, by application of conventional molecular marker technologies, has been difficult in tea due to the problems of self-incompatibility, the absence of pure inbreed lines, high heterozygosity, 15 pairs of chromosomes, and large genome size. However, novel approaches such as next-generation sequencing (NGS) technologies have successfully helped in identification of several candidate genes which are responsible for biotic and abiotic stress regulation in different plants. It has accelerated the process of large-scale single-nucleotide polymorphism

S. Borchetia
Department of Biotechnology, Tocklai Tea Research Institute, Tea Research Association, Jorhat, Assam, India

G. Handique
Advisory Department, Tocklai Tea Research Institute, Tea Research Association, Jorhat, Assam, India

S. Roy
Department of Entomology, Tocklai Tea Research Institute, Tea Research Association, Jorhat, Assam, India

S. H. Wani (✉)
Mountain Research Centre for Field Crops, Sher-e-Kashmir University of Agricultural Sciences and Technology of Kashmir, Jammu, Jammu and Kashmir, India

Department of Plant, Soil and Microbial Sciences, Michigan State University, East Lansing, MI, USA

W.-Y. Han et al. (eds.), *Stress Physiology of Tea in the Face of Climate Change*,
https://doi.org/10.1007/978-981-13-2140-5_13

(SNP) discovery and genotyping in tea, thus, facilitating the construction of high-density genetic maps. The landmark tea whole genome will lead to identification and mapping of quantitative trait loci related to agronomical important traits underlying biotic and abiotic stress. It will also aid in comparative genomic analysis, marker-assisted selection, and map-based cloning for developing stress-tolerant tea plants.

Keywords Tea genomics · Biotic stress · Abiotic stress · Molecular response · Defense genes

1 Introduction

Tea, *Camellia sinensis* (L.) O. Kuntze, belonging to the family Theaceae, is an important commercial plantation crop which is popular as a nonalcoholic beverage. It also has nutritional and medicinal properties for which it is considered as a general health drink (Chacko et al. 2010). India is the second largest producer and consumer of tea except China. Cultivated tea belongs to genus *Camellia* and comprises of three main natural hybrids. They are *C. sinensis* (L.) O. Kuntze or China type, *C. assamica* (Masters) or Assam type, and *C. assamica* sub spp. *lasiocalyx* (planchon ex Watt.) or Cambod or Southern type. The two main types of tea, i.e., China and Assam tea, differ from one another in their morphological, physiological, and biochemical attributes. Since members of the genus *Camellia* are highly heterozygous and outcrossing, a lot of genetic variants are observed. However, these are still referred to as China or Assam depending on their morphological proximity with the main taxa (Wight 1962; Banerjee 1992).

Sustainable tea production has been a major challenge in today's scenario of climate change. Tea, being consumed worldwide, there is a growing concern among the planters regarding the huge losses in yield of the tea plants due to changing climatic conditions. Recent research has revealed of a more extreme and intense weather with longer dry spells, heavy downpours, and more hail and cyclonic storms (Marx et al. 2017). Tea productivity has undergone marked changes in recent years owing to changing climatic conditions such as erratic rainfall patterns and increase in average and maximum temperatures (Dutta 2014; Gunathilaka et al. 2016; Duncan et al. 2016; FAO 2016). Stresses trigger a wide response on tea plant from cellular metabolism to changes in growth and yield. As plants are sessile organisms, their survival depends on the way they respond to the different kinds of environmental stresses such as drought, cold, salinity, pathogens, insects, etc. (Gosal et al. 2009, 2010; Krishania et al. 2013; Pathak and Wani 2015; Hossain et al. 2016). Tea productivity is adversely affected by abiotic and biotic stresses. Abiotic stress affects the tea plants at the morphophysiological, biochemical, and molecular level with changes at the whole plant functions (Sourour et al. 2017). On the other hand, biotic stressors such as insects, mites, and pathogens also contribute significantly toward the enormous damage to the crops (Maron and Crone 2006; Hazarika et al. 2009).

In both cases, there is serious concern in case of disruption in plant metabolism as well as in terms of agricultural output. Adaptability of tea clones to various biotic and abiotic stresses along with inherent quality of the end product has become a deciding factor for their large-scale adoption. Although a variety of approaches have been employed in the past to overcome the problem of abiotic and biotic stresses, high-throughput omics technologies are facilitating the identification of new genes (Table 13.1) and also understanding their function (Telem et al. 2016). In addition, network reconstructions at the genome-scale will help to quantify and characterize the genotype to phenotype relationships. Next-generation sequencing (NGS) technologies have made large-scale genome-wide detection very easy in complex genomes. Complete genome of tea can be used for comparative mapping with transcriptomes, cytological maps, and even partial targeted genome resequencing. This information will further serve in molecular and conventional breeding programs for the development of new cultivars with desired traits to overcome biotic and abiotic stresses.

2 Advances in Tea Genomics

The tea genome is heterogeneous in nature due to a self-incompatible reproductive system (Zhang et al. 2016), making crop improvement via classical plant breeding difficult. It has a genome size of 4GB (2n = 30 chromosomes). Significant progress in tea genomic research has been achieved with construction of cDNA libraries, sequencing and annotation of expressed sequence tags (ESTs), cDNA microarray construction, gene expression profiling, identification of EST SSR marker and specific trait linked genes, etc. (Chen et al. 2009a, b). Over the years, EST sequencing strategies have filled the genomic pool of *Camellia sinensis* with partial, yet informative data. Starting from only five ESTs of tea in the GenBank dbEST database in 2004, it has increased to 49,760 EST sequences as on October 2016. The nucleotide database of National Center for Biotechnology Information (NCBI) till date contains a total of 370,886 sequences. The first report on functional genomics in tea was the isolation of chalcone synthase gene from "Yabukita," a Japanese tea cultivar (Takeuchi et al. 1994). Since then a considerable amount of genomic work has been carried out in tea targeting isolation, characterization, cloning, and expression of trait-specific gene (Liu and Han 2010; Borchetia et al. 2011; Paul et al. 2012a, b; Deng et al. 2013).

Molecular markers derived from functionally characterized sequences provide advantage over random DNA markers due to their linkage with specific trait loci as they may be associated with major agronomic traits of the tea plant. The presence of a mapped tea genome is more desirable for effective analysis of quantitative trait loci (QTL) and application of marker-assisted selection (MAS) in plant breeding. The first genetic map of tea was constructed using RAPD markers related to theanine content, date of bud sprouting, resistance to anthracnose, and tolerance to cold (Tanaka 1996). Genetic maps of *Camellia sinensis* have been created mainly

Table 13.1 List of genes cloned for abiotic and biotic stresses in *Camellia sinensis*

Genes identified	Gene name	Stress responsible	Function	References
14 CsGAox genes (GA20ox, GA3ox, and GA2ox)	Gibberellin-dioxygenases	Abiotic stress	Bioactive gibberellin (GA) levels are rapidly regulated by gibberellin-dioxygenases (GAox), which are involved in the biosynthesis and deactivation of gibberellin	Pan et al. (2017)
CsC4Ha, CsC4Hb, CsC4Hc	Cinnamate 4-hydroxylase	Biotic and abiotic stresses	A cytochrome P450-dependent monooxygenase participates in the synthesis of numerous polyphenolic compounds, such as flavonoids and lignins	Xia et al. (2017a)
Three complete *CsGolS* genes	Galactinol synthase (GolS)	Biotic and abiotic stress	A key biocatalyst for the synthesis of raffinose family oligosaccharides (RFOs). CsGolS1 gene was sensitive to water deficit, low temperature, and abscisic acid, while CsGolS2 and CsGolS3 genes were sensitive to pest attack and phytohormones	Zhou et al. (2017)
CsAOC	Allene oxide cyclase	Biotic stress	Coordinating plant defense responses to pests and pathogen attack through transcriptional and metabolic changes	Wang et al. (2016a)
CsNAC genes *CsNAC17* and *CsNAC30*	NAC family MTFs (membrane-associated transcription factors)	Abiotic stress	Identified as highly responsive to abiotic stress like drought, high temperature, low temperature, salinity	Wang et al. (2016b)
14 CsINV	Invertase	Growth and development, abiotic and biotic stresses	Participate in different signaling pathways that regulate the carbohydrate allocation and the ratio of hexose and sucrose for improving the resistance of the leaves and the roots of the tea plant to various abiotic stresses	Qian et al. (2016)
CsZfp	A20/AN1-zinc-finger domain	Abiotic stress	Expressed in response to winter dormancy (WD), abiotic stresses (polyethylene glycol, hydrogen peroxide, and sodium chloride), and plant growth regulators [abscisic acid (ABA) and gibberellic acid (GA 3)]	Paul and Kumar (2015)
CsPIPs, CsTIPs	Aquaporins	Abiotic stress	Induced by drought	Yue et al. (2014)
CsGPX2	Glutathione peroxidase	Abiotic stress	Strong responses to most abiotic stresses including salinity, heavy metal toxicity, drought, heat, and plant hormones, but could not be induced by biotic treatment	Fu (2014)
CsCAD	Cinnamyl alcohol dehydrogenase	Biotic stress	Cinnamyl alcohol dehydrogenase (CAD) is a key enzyme in lignin biosynthesis and induced by both biotic and abiotic stress	Deng et al. (2013)
CsDHN1 and *CsDHN2*	Dehydrin1, dehydrin2	Abiotic stress	Dehydrin1 in response to winter dormancy (WD) as well as abiotic stresses [low temperature, sodium chloride, polyethylene glycol, and hydrogen peroxide]. Dehydrin2 in response to abiotic stress	Paul and Kumar (2013)

CsTUA	alpha-tubulin	Abiotic stress	Expressed in response to winter dormancy, abiotic stresses (sodium chloride, polyethylene glycol, and hydrogen peroxide), and plant growth regulators [abscisic acid (ABA), gibberellic acid (GA 3), indole-3-butyric acid (IBA), and 6-benzylaminopurine (BA)]	Paul et al. (2012a)
CsICE1, *CsCBF1*	C-repeat/ dehydration-responsive element binding factors (CBFs)	Abiotic stress	Cold-responsive genes to increase plant cold tolerance	Wang et al. (2012)
CsNAM	NAC family transcription factors	Abiotic and biotic stress	Plant-specific transcription factors (TFs) implicated in diverse processes including development and biotic and abiotic stress responses	Paul et al. (2012b)
CsCOR1	CsCOR1	Abiotic stress	Cold-regulated gene expressed in low temperature and dehydration-stress	Li et al. (2010)
CsCBF	C-repeat binding factor	Abiotic stress	C-repeat/dehydration-responsive element binding transcription factors (CBF/DREB) are important regulators of abiotic stress response	Chen et al. (2009b)
CsCSH1	H1-histone	Abiotic stress	Linker (H1) histones expressed due to low temperature	Fang et al. (2009)
CsCHIi	Chitinase	Biotic stress	Plant pathogen defense	Roy and Chakraborty (2009)
CsGS	Glutamine synthetase	Abiotic and biotic stress	Stimulated in response to abscisic acid (ABA), salicylic acid (SA), and hydrogen peroxide (H_2O_2)	Rana et al. (2008)
CsSMT	Selenocysteine methyltransferase	Abiotic stress	Selenocysteine methyltransferase (SMT) specifically methylates selenocysteine (SeCys) to produce the nonprotein amino acid Se-methylselenocysteine (SeMSC) and removes selenium toxic effect	Zhu et al. (2008)
CsCYS	Cystatin	Biotic stress	Inhibitors of cysteine-proteases in plant defense	Wang et al. (2005)
CsVDE	Violaxanthin de-epoxidase	Abiotic stress	Induced by cold and drought stress	Wei et al. (2003)
CsPR	PR-1-like protein	Biotic stress	Plant pathogen defense	Tomimoto et al. (1999)

based on dominant markers such as RAPD, AFLP, and ISSR markers (Hackett et al. 2000; Huang et al. 2005; Huang et al. 2006). A number of polymorphic molecular markers such as SSR, CAPS, STS, AFLP, ISSR, and RAPD were utilized for construction of an integrated genetic map of tea based on their maternal and paternal lineages (Hu et al. 2013). Few researchers have used pseudo-testcross theory to create tea linkage maps (Hackett et al. 2000; Ota and Tanaka 1999; Tanaka et al. 1995) using dominant markers. As tea does not have pure inbred lines due to high level of heterozygosity, pseudo-testcross is the way out for linkage maps. However, dominant markers posed some limitation as they are not universal markers, and their utility depends on the particular material being tested. Taniguchi et al. (2012) on the other hand used codominant SSR markers as landmark markers to create high-density reference map of tea. QTLs were identified in tea for yield and drought tolerance using RAPD, AFLP, and SSR markers which showed segregation of the progeny in 1:1 ratio after backcrossing was used for construction of linkage map of tea (Kamunya et al. 2009; Kamunya et al. 2010). The main limitation of molecular markers like RAPD and AFLP is their repeatability across plants which does not permit easy comparisons of maps at interspecific or intergeneric levels. A moderately saturated genetic map was constructed by Ma et al. (2014) in *C. sinensis* varieties interbred with different catechin content to characterize the QTLs for catechin content in the tender shoots of the progeny population using SSR markers. Bali et al. (2015) constructed a linkage map of Indian teas using two-way pseudo-testcross approach for mapping drought tolerance trait. The map generated comprised of 18 linkage groups which is similar to previous report in Chinese tea (Hu et al. 2013). Tan et al. (2013) reported tea floral transcriptome sequencing for SSR marker development and linkage map construction and provided insights into the molecular mechanism of tea self-incompatibility. Chen et al. (2011) used cDNA-AFLP method to analyze differentially expressed genes in self-incompatible lines of tea and isolated 34 genes involved in energy metabolism, signal transduction, and defense. The application of DNA-based molecular marker in tea is mainly limited to genetic diversity, relationship and molecular identification analysis (Mondal 2002; Chen et al. 2005; Das et al. 2016), etc.; however, its use in mapping of QTL and gene introgression by MAS is still inadequate in tea (Kamunya et al. 2010). Though partial but preliminary work with tea genome started with the construction of bacterial artificial chromosome (BAC) library of tea by Lin et al. (2011) and led to generation of 4,01,280 clones with an average insert size of 135 kb. Thereafter with the advancement of new sequencing technology, large-scale RNA sequencing (RNA-seq) became a fast, cost-effective, and reliable approach in non-model species with un-sequenced genomes to generate large expression datasets for functional genomic analysis (Shi et al. 2011), and genotyping by sequencing (GBS) with generation of SNPs (single-nucleotide polymorphisms) became another tool for genomics-assisted breeding in tea.

Next-generation sequencing (NGS)-assisted marker selection has the potential to generate large dataset in shorter duration to select them as effective markers (Bhat et al. 2016) and facilitated the functional characterization of genes/genomes to identify the diversity and gene function responsible for many traits (Brady and Provart

2007). Considerable progress in *C. sinensis* transcriptome sequencing has been achieved in the last few years (Shi et al. 2011; Wu et al. 2013). cDNA-AFLP method has been widely used to study the differential gene expression in ESTs obtained from an SSH (suppression subtractive hybridization) library in tea (Mondal and Sutoh 2013; Das et al. 2013) in response to biotic and abiotic stress. Transcriptome generated enormous data to identify a substantial number of putative SSR sequences to develop new SSR markers and construct a moderately saturated SSR-based genetic linkage map for QTL mapping and MAS breeding in *C. sinensis* (Tan et al. 2013). Zhang et al. (2014) verified 253 putative SNPs from the EST database of tea in 17 tea varieties through sequencing, and 34 sequences containing SNP with functional roles have been identified. The transcriptome projects have led to the functional identification of many genes which are probably unique to *Camellia sinensis* and may serve good candidate for marker-assisted breeding. It also led to identification of many SNPs within tea which can be used for functional and genotypic identification aspects. The SNP identification has opened up a tremendous scope for utilizing SNP markers associated with desirable features like tea quality, pest resistance, etc. in molecular breeding and has become an essential requirement in molecular breeding and population genomics. SNPs have been identified by Huang et al. (2007) in the coding region of polyphenol oxidase from different genotypes of tea. Fang et al. (2014) developed SNP markers with a high-throughput genotyping protocol which could serve as a powerful tool for variety authentication and quality control for the tea industry.

NGS technologies have made the large-scale genome-wide SNP detection very easy even in complex genomes. Ma et al. (2015) undertook SLAF-seq approach to construct reference genetic map of *C. sinensis* using 6042 SNP markers identified from an F1 mapping population of tea cultivars. Yang et al. (2016) applied RAD-seq approach for genomic SNP discovery in 18 cultivated and wild tea accessions for studying phylogenetic relationships among the different accessions on a genome-wide scale.

Apart from the above approaches used to study partial targeted tea genome, two groups reported chloroplast genomes and phylogenetic relationships of a number of *Camellia* species and varieties (Yang et al. 2013; Huang et al. 2014). However, chloroplast genome with very limited resolving power is not of very much use in species undergoing rapid evolution (Parks et al. 2009; Moore et al. 2006). It is therefore imperative to develop more high-throughput genomic data, to facilitate genome-scale molecular genetics research in cultivated and wild teas.

Recently, Xia et al. (2017b) sequenced the genome of a diploid elite cultivar Yunkang 10 (*C. sinensis* var. *assamica)* via whole-genome shotgun sequencing. They generated ~29.7 Gb of RNA sequencing (RNA-seq) data obtained from a total of eight libraries representing major tissue types and developmental stages, including young leaf, tender shoot, flower bud, flower, stem, root, seed, and seedling. Noncoding RNA (ncRNA) genes represented 700 transfer RNA (tRNA) genes, 2860 ribosomal RNA (rRNA) genes, 454 small nucleolar RNA (snoRNA) genes, 223 small nuclear RNA (snRNA) genes, and 233 microRNA (miRNA) genes, while the majority of tea genome (~67.21%) contains repetitive DNA terminal repeat with

(LTR) retrotransposons. 36,951 protein-coding genes were annotated. They also annotated ~867,339 simple sequence repeats, which will provide valuable genetic markers to assist tea breeding programs.

The NGS-based genotyping has facilitated the functional characterization of genes and has provided a more comprehensive view of diversity and gene function in tea plants. The availability of the whole genome sequence will lead to tea genome map which will be an important resource for studies related to inheritance of agriculturally important traits in tea, for monitoring the transmission of specific genes or genomic regions from parents to progeny. But plant conventional breeding requires genetic variability to be selected in order to increase the frequencies of favorable alleles and genetic combinations (Pérez-de-Castro et al. 2012) for crop improvement. Therefore, proper evaluation and resequencing of available tea germplasm for agronomic traits, biotic and abiotic stress tolerance, and quality parameters will assist in molecular breeding for the development of new cultivars with desired traits in the face of climate change.

3 Genomic Advances in Tea Plants in Response to Biotic Stress

As in all agricultural crops, tea which is grown under diverse cultural conditions over a wide geographical range is subjected to several biotic stresses which are serious constrain for its production (Hazarika et al. 2009). Recently few transcriptomic studies on tea have been performed in response to biotic stresses. Tea is primarily a crop favoring a warm humid environment conducive to proliferation of arthropod pests, disease, and weeds. Moreover, being a perennial monoculture crop, it provides a reasonably stable microclimate and food supply for pests and diseases (Roy and Muraleedharan 2014). An assessment on tea arthropod pests of the world lists more than 1000 species of arthropods and 82 nematode pests which cause crop loss annually worth US$ 500 million to US$ 1 billion (Hazarika et al. 2009). Each geographical region has its own distinctive pest complex, while few of them are distributed globally (Chen and Chen 1989) with crop loss varying between 15% and 20% (Muraleedharan and Selvasundaram 2002). The magnitudes of losses are bound to be higher today in view of the increased production and productivity (Roy et al. 2016). The accumulation of arthropod pest on tea is influenced by various factors including the age of plants; the older tea plantations harbor more number of insect species (Banerjee 1983). On a global scale, some of the major tea pests that cause serious damage to tea plantations include the tea mosquito bug (*Helopeltis theivora*), mites (*Oligonychus coffeae, Brevipalpus californicus, B. phoenicis, B. obovatus, Acaphylla theae, Polyphagotarsonemus latus, Tetranychus kanzawai*), red slug caterpillar (*Eterusia magnifica*), thrips (*Scirtothrips dorsalis, S. bispinosus*), leafhoppers (*Empoasca flavescens, Empoasca vitis*), Geometridae moth (*Buzura suppressaria, Hyposidra talaca, H. infixaria, Ectropis obliqua*), termites (*Microtermes* sp., *Odontotermes* sp., *Postelectrotermes militaris, Coptotermes* sp.), shot hole borer (*Xyleborus fornicatus*, *Euwallacea fornicates*), tea tussock (*Euproctis pseudoconspersa*), tea weevil (*Myllocerinus aurolineatus*), and tea tortricids (*Homona coffearia*, *H. magnanima*, *Adoxophyes honmai*) (Hazarika et al. 2009; Ye et al. 2014). Some of the tea pests, viz., thrips, loopers, and leafhoppers, have gained considerable importance during the last few years owing to the severity of damage and magnitude of crop loss to tea. Besides many other minor pests like scale insects, mealy bugs, leaf roller, leaf miner, white grub, red borer, and aphids may also adversely affect the survival and production of tea bushes in some tea-producing regions.

Apart from arthropod pests, nearly 400 pathogens are known to attack tea all over the world (Tewari 2016). Majority of the diseases associated with tea are of fungal origin which are categorized into three classes on the basis of the plant part that is infected by the pathogen, viz., root, stem, and leaf diseases. The most common primary root diseases include charcoal stump rot (*Ustulina zonata*), red root rot (*Poria hypolateritia*, *Ganoderma philippii*), brown root rot (*Fomes noxius*, *Phellinus noxius*), black root rot (*Rosellinia arcuata*, *R. bunodes*), purple root rot (*Helicobasidium compactum*), and xylaria root rot (*Xylaria* sp.). Violet root rot (*Schaerostelde repens*, *Sphaerostilbe repens*) and diplodia (*Botryodiplodia theobromae*) are some of the secondary root diseases infecting tea. Some of the root disease such as root knot is, however, caused by nematodes such as *Meloidogyne incognita* and *Meloidogyne javanica*. Among the stem diseases, black rot (*Corticium* sp.), collar canker (*Phomopsis theae*), wood rot (*Hypoxylon serpens*), branch canker (*Macrophoma theicola*), Leptothyrium dieback (*Leptothyrium theae*), fusarium (*Fusarium solani*), dieback (*Leptothyrium theae*, *Nectria cinnabarina*), and stem canker (*Macrophoma theicola*) are some of the primary diseases, while the secondary stem diseases include red rust (*Cephaleuros parasiticus*), poria branch canker (*Poria hypobrunnea*), and thorny stem blight (*Tunstallia aculeata*). Blister blight (*Exobasidium vexans*), black rot (*Ceratobasidium* sp., *Corticium invisum*, *C. theae*), red rust (*Cephaleuros parasiticus*, *C. mycoides*), gray blight (*Pestalotiopsis theae*), brown blight (*Glomerella cingulata*, *Colletotrichum gloeosporioides*), and anthracnose (*Gloeosporium theae-sinensis*) are some of the important leaf diseases in tea. Apart from these, only three bacterial diseases occur in tea which are bacterial canker (*Xanthomonas campestris*, *Xanthomonas gorlencovianum*), bacterial shoot blight (*Pseudomonas avellanae*), and crown gall (*Agrobacterium tumefaciens*).

Among the weed flora, both monocotyledons and dicotyledons (mostly grasses and sedges) are known to occur in tea plantations and can cause an average yield loss of 12–18%. Economically important weeds include carpet grass (*Axonopus compressus*), crowfoot grass (*Eleusine aegyptia*), goose grass (*Eleusine indica*), thatch grass (*Imperata cylindrica*), wild cane (*Saccharum spontaneum*), nut grass (*Cyperus rotundus*), Indian mallow (*Abutilon indicum*), goat weed (Ageratum conyzoides), broadleaf buttonweed (*Borreria hispida*), *Cassia occidentalis*, *C. tora*, Indian pennywort (*Centella asiatica*), *Clerodendrum infortunatum*, etc. Apart from these, some ferns such as *Dicranopteris linearis*, *Gleichenia linearis*, *Drymoglossum piloselloides*, *Lygodium flexuosum*, etc. and water hyacinth (*Eichhornia crassipes*) also occur in tea plantations. Weeds are another biotic stress which reduce crop yields (Zimdahl 2004) and compete with tea plants for water, nutrients, and light.

Herbicides are generally used to manage weeds in tea plantations, but its application also causes stress. The hardiness and resilience of weed plants to abiotic and biotic stress factors contribute to their success (Horvath et al. 2015). The traits responsible for tolerance to stress factors can be exploited to develop tea plants tolerant to biotic and abiotic stress. Molecular mechanism(s) responsible for traits like competitiveness and invasiveness of weeds in tea plants is poorly understood till date.

3.1 Molecular Responses to Plant-Insect Interactions

Plants have evolved elaborate defense systems to resist insect herbivores. They respond to insect-induced stress via both direct and indirect defense mechanisms. The direct defense mechanism includes the plant characteristics that provide physical barriers such as thicker leaves, hairs, trichomes, thorns, and cuticles as well as production of qualitative defenses such as glucosinolates, cyanogenic glucosides, alkaloids, phenolics, and proteinase inhibitors (PIs) which function as toxins, repellents, or anti-digestives that can deter feeding, kill, or retard the development of herbivores (War et al. 2012). Many of the secondary metabolites in tea plants are powerful chemical weapons used in direct defense. Indirect defenses refer to the release of green leaf volatiles, volatile organic compounds, and extrafloral nectars that can attract natural predators of the herbivores by providing food and shelter to predators (Arimura et al. 2009). They act as elicitors triggering the activation of various induced, specific, and systemic defense responses. Both direct and indirect defense actions are mediated through various biochemical, morphological, and molecular mechanisms. Using a virus-induced gene silencing (VIGS) approach, Heinrich et al. (2011) revealed that both salicylic acid-induced protein kinase (SIPK) and wound-induced protein kinase (WIPK) play a central role in plant responses to herbivory. When an insect attacks the plant, its wound response is reconfigured at transcriptional, phytohormonal, and defensive levels due to the introduction of oral secretions (OS) into wounds during feeding. Herbivory highly elicits mitogen-activated protein kinase (MAPK) activity and OS-elicited jasmonic acid (JA), ethylene (ET), and the transcriptional regulation of many defense-related genes that are dependent on the MAPK pathway (Wu et al. 2007). JA is the most important hormone that controls plant defense against herbivores (Yan and Xie 2015). Ethylene (ET) regulates a wide range of physiological processes in plant development and in abiotic and biotic stress resistance. After plants were wounded or infested by insects, they rapidly activate the biosynthesis of ET (von Dahl et al. 2007). Even though many biochemical compounds, morphological changes, and gene regulations have been found to play roles in plant-insect defense responses, the molecular mechanism by which tea plants protect themselves from insects is being studied very recently. EST-based markers associated with resistances to insect (*Helopeltis theivora*) (Ahmed and Das 2009) were developed to aid the marker-based selection process in tea breeding. Recently few transcriptomic studies on tea have been performed to study its response to biotic stresses. Molecular research has been conducted in the tea geometrid (*Ectropis oblique*, a chewing insect) which feed on tea leaves and tender buds and cause considerable mechanical damage.

Wang et al. (2016c) classified the defense mechanisms induced by oral secretions of *Ectropis oblique* in tea plant into four main pathways: receptor recognition, signals perception, signal transduction, and transcription factors and defense genes. Various elicitors in oral secretions resulting from *Ectropis oblique* attack can enter into plants through wounds which can be recognized by plant multiple receptors, such as leucine-rich repeat receptor-like protein kinase (LRR-RLKs) and serine/threonine (Ser/Thr) protein kinase. These receptors in turn activate the signal transduction pathways which regulate distinct resistance-gene expression systems and induce the tea plant to produce diverse defense metabolites. Shukla et al. (2008) reported that small regulatory RNAs target multiple transcription factors, which in turn, regulate the expression of various downstream genes involved in the stress response. Jeyaraj et al. (2017) identified several miRNAs which may function by targeting LRR-RLKs and Ser/Thr-kinase receptors and trigger Ca2+ influxes and Ca2 + − binding protein activation through MAPK cascades, followed by a series of processes such as phytohormone and ROS biosynthesis pathways.

Genetic structure of a plant as well as insect infestation, particularly *Empoasca flavescens* and *Scirtothrips dorsalis*, was reported to trigger the aroma and flavor formation in the famous Darjeeling tea (Gohain et al. 2012). They identified several genes and transcription factors from the suppression subtractive hybridization (SSH) library of leaf that were heavily infested by *E. flavescens* and *S. dorsalis*. Some stress-responsive genes were upregulated in gene expression profile of Taiwan, "oriental beauty," a high-flavored tea from leafhopper-infested tea leaves (Choi et al. 2007).

The tea mosquito *Helopeltis theivora* is another major pest of tea in India, especially in northeast causing substantial (10–50%) loss in crop (Gurusubramanian and Bora 2007). In some cases, yield loss can be up to 100% (Ahmed et al. 2011). Bandyopadhyay et al. (2014) provided insights into the *Helopeltis theivora* induced changes at the transcriptomic levels in Assam tea in order to identify the potential tolerance mechanism. They identified 445 unigenes, out of which 35.5% did not show any homology to nonredundant database. They observed that the upregulated defense-related genes belonged to flavonoid biosynthesis, purine metabolism, plant cell wall metabolism, metabolite transport, salicylic acid signal transduction pathway, jasmonic acid biosynthesis, and signaling pathway in tea plants.

3.2 Plant-Pathogen Interaction and the Mechanism of Defense

Plant disease resistance involves complex exchange of signals and responses and is governed by the genotypic combination of host and pathogen, environmental conditions, and nature of infected tissue (Develey-Rivière and Galiana 2007). Some of these defense mechanisms either provide structural and chemical barriers to hinder pathogen infection or gets induced after pathogen attack. Inducible defense responses in plants include waxy cuticular layers; accumulation of callose and lignin; synthesis of various proteins and antimicrobial compounds such as phytoalexins, pathogenesis-related (PR) proteins, and proline-rich proteins; and synthesis of hydrolytic enzymes such as chitinase and proteinase inhibitors (Van Loon et al. 2006). The first level of pathogen recognition occurs via pattern recognition receptors (PRRs) leading to pathogen-associated molecular patterns (PAMPs) (Jones and Dangl 2006) or elicitors like microbial proteins, small peptides, oligosaccharides, etc. PAMPs trigger early responses (ion fluxes, oxidative burst), intermediate responses (MAPK/CDPK activation, ethylene production, stomatal closure, transcriptional reprogramming), and late responses (salicylic acid accumulation, callose deposition) (Zipfel and Robatzek 2010). Manickavelu et al. (2010) stated that primary pathogen elicitors as well as secondary endogenous signals activate a diverse array of plant defense genes like glutathione S-transferases (GST), peroxidases, cell wall proteins, proteinase inhibitors, hydrolytic enzymes, PR-proteins, and phytoalexin biosynthetic enzymes such as phenylalanine ammonia lyase (PAL) and chalcone synthase (CHS) in response to pathogen stress.

Being a perennial plant, tea comes across a wide range of biotic stresses during its life span. Among the biotic stresses, fungal pathogens causing leaf diseases are among the major bottlenecks as commercial tea production is mainly dependent on young succulent leaves (Muraleedharan and Chen 1997). Wang et al. (2005) cloned the cystatin gene with the help of degenerate primers from tea. The deduced amino acid sequence contained the motif QXVXG conserved among most members of cystatin superfamily. In another study, eight wound/pathogen inducible cDNAs showing high sequence homology with other wound/pathogen inducible genes of herbaceous plants were cloned in tea, and the structure was analyzed through in silico (Yoshida and Homma 2005). Gray blight disease, *Pestalotiopsis theae*, is one of the major diseases of tea. Senthilkumar et al. (2012) used SSH method to categorize the differential gene expression pattern due to *P. theae* infection in tea and observed upregulation of several genes related to hypersensitive response and reactive oxygen species production. Bhorali et al. (2012) identified a few genes regulating blister blight resistance through transcriptome study. Jayaswall et al. (2016) studied the molecular basis of the immune system against blister blight disease by performing a genome-wide transcriptome study in resistant and susceptible tea genotypes and identified putative pathways, genes and their interactions, and candidates involved in blister blight defense in tea. The study revealed that putative sites and sequence motifs, ubiquitously conserved in upstream regions of genes, are upregulated during systemic acquired resistance (SAR) or R-mediated basal defense. WRKY and NAM transcription factors play a significant role by binding to promoter elements of defense-related genes and regulating their expression during plant immunity. LTR retrotransposons which are present in tea genome in huge amounts are activated during disease transition which elicits defense responses and defense gene-related enzymes, resistance genes, multidrug-resistant transporters, transcription factors, retrotransposons, and other defense genes comprising metacaspases and chaperons. In total 25 resistance (R) genes that trigger the hypersensitive cell death resistance response during disease transition and provide tea plant immunity were identified.

4 Abiotic Stress in Tea Plants and Contribution of Genomics to Gene Discovery

Tea plants are sensitive to different types of abiotic stresses, namely, drought (Das et al. 2012), salinity (Li et al. 2010), waterlogging, temperature changes, heavy metals (Basak et al. 2001), hail, soil pH and nutrient, light intensity, etc., which negatively influence the growth and productivity of tea. Global warming and climate change are threats to crop production due to water and temperature stresses which may intensify. Effects of some individual environmental factors on the tea plant are well documented (Upadhyaya et al. 2012).

However, most of the published molecular work relates to drought and winter dormancy which are the two main abiotic stresses impacting tea productivity and survival. Other abiotic stresses are less obvious but can equally affect the tea plant. Addition of fertilizers and pesticides either in excess or deficit can also cause abiotic stress in tea plants either through nutrient imbalance or via toxicity. Few genes responsible for more uptake of nutrients were cloned in tea plants. Rana et al. (2010) reported a difference in expression of cytosolic glutamine synthetase in tea during different developmental periods and in light or dark situations on the utilization of nitrate as well as ammonia. The gene, namely, ammonium transporter which was responsible for nitrogen absorption of plant roots, was cloned from tea root (Taniguchi and Tanaka 2004). Two cDNAs encoding ATP sulfurylase useful for sulfate assimilation were cloned from tea (Zhu et al. 2008a). Full-length cDNA of selenocysteine methyltransferase important for toxicity removal in plants was also cloned from tea plants (Zhu et al. 2008b). High amounts of salt intake in tea plants can lead to salinity stress leading to cell desiccation, while uptake of heavy metals can hamper the physiological and biochemical activities such as photosynthesis in tea plants.

Suppression subtractive hybridization and transcript analyses were utilized by many researchers to understand the molecular mechanism underlying abiotic stress. Several drought-responsive genes in tea plant were identified using cDNA-amplified fragment length polymorphism (Das et al. 2012; Gupta et al. 2012; Gupta et al. 2013). It was also used to identify the cold-responsive genes (Wang et al. 2009), dormancy-associated genes (Wang et al. 2010; Thirugnanasambantham et al. 2013), and genes responsible for lignin biosynthesis and phenylpropanoid metabolism in light-induced calli (Wang et al. 2012). Xia et al. (2017a) cloned the three genes, *CsC4Ha*, *CsC4Hb*, and *CsC4Hc,* representing cinnamate 4-hydroxylase (C4H), a cytochrome P450-dependent monooxygenase which participates in the synthesis of numerous polyphenolic compounds, such as flavonoids and lignins playing a major role in abiotic stresses. Recently, several conserved and novel miRNAs were identified in response to drought (Liu et al. 2016) and cold stress (Zhang et al. 2014; Zheng et al. 2015) in the tea plant by high-throughput sequencing. However, the functions and mechanisms by which these miRNAs affect the stress response in *C. sinensis* are unclear.

The tea tree genome sequences published in 2017 have provided an opportunity to perform a genome-wide scan of gene families. Pan et al. (2017) performed a genome-wide survey of the Gibberellin-dioxygenase (GAox) gene family using the genomic and transcriptomic database and identified and confirmed 14 full-length GAox genes in tea plants (C. sinensis), using cloning and sequencing. These candidate genes strongly influence the growth as well as a few can be used as marker genes for abiotic stress resistance breeding in tea plants.

4.1 Temperature Stress and Winter Dormancy

Winter dormancy in tea plants is characterized by the diminished growth of apical buds during winter months, severely reducing the commercial yield of tea. Tea plants growing beyond 16° N or S latitudes undergo seasonal dormancy as light intensity as well as temperature changes (Barua 1989), which alone or in combination have the potential to induce oxidative stress. Transcriptomic library sequences of actively growing and the winter-dormant tissue of tea have revealed a distinct difference between the two libraries. Comparatively large number of transcripts related to cell rescue, defense, and virulence class was observed in winter-dormant tissue comprising sequences encoding late embryogenesis abundant protein family, dehydrin, and metallothionein-like protein. While the actively growing transcripts mainly comprised of cell cycle and DNA processing class with a large number of sequences for histone proteins. The number of genes involved in metabolic activity was six times higher in the actively growing library compared to that of the winter dormant one (Kumar et al. 2012) suggesting reduction in cell division and associated activities. Six novel small RNA candidates were isolated, cloned, and validated through expression analysis in young and old leaves, during nondormant and dormant growth phases of *C. sinensis* (*Mohanpuria and Yadav* 2012). Li et al. (2010) cloned a cold-responsive gene *CsCOR1* from tea leaves which is expressed in cold and dehydration stress. The effect of various abiotic stresses including low temperature is inevitably associated with formation of harmful reactive oxygen species (ROS) (Pospíšil 2016). Under moderate stress, when scavenging system maintains ROS level low, ROS serves as signaling molecules which in turn activate an acclimation response and programmed cell death in plants (Dietz et al. 2016). Superoxide dismutase (SOD) was found to play an important role in scavenging ROS (Vyas and Kumar 2005) in tea. To elucidate the molecular mechanisms of cold acclimation, Wang et al. (2013) elucidated the molecular mechanisms of cold acclimation in tea plants through RNA-seq and digital gene expression (DGE) technologies. They identified 1770 differentially expressed transcripts of which 1168 were upregulated and 602 downregulated. These included a group of cold sensor or signal transduction genes, cold-responsive transcription factor genes, plasma membrane stabilization-related genes, osmosensing-responsive genes, and detoxification enzyme genes with carbohydrate metabolism and calcium signaling pathway playing a major role in tea plants. Hao et al. (2017) threw insights into the molecular mechanisms of bud dormancy regulation in tea plants to survive the cold winters. They analyzed the global gene expression profiles of axillary buds at the paradormancy, endodormancy, ecodormancy, and bud flush stages by RNA-seq analysis and identified 16,125 differentially expressed genes associated with epigenetic mechanisms, phytohormone signaling pathways, and callose-related cellular communication regulation.

4.2 Drought Stress

Drought is another major abiotic stress which limits the crop productivity in general (Wani and Gosal 2011; Sah et al. 2016; Wani et al. 2016) and quality and productivity of tea in particular. Several tea clones were released to the tea industry for their tolerance to drought based on physiological traits (Barman et al. 2008) and extensive field trials (Singh 2006). Generally, tea tolerates drought, at the expense of photosynthates (Kumar et al. 2012), but if the moisture stress prolongs, defoliation and death of plants occur. At present, research on the effects of drought stress in *C. sinensis* plants is mainly focused on the morphological, physiological, and molecular mechanisms underlying the stress responses (Upadhyaya et al. 2013) and identifying the genes which are playing a major role in drought stress.

In response to drought stress, the tea plants alter the osmolyte level of the plant cell's cytosol and the vacuoles mainly proline, glycine betaine, late embryogenesis abundant proteins, carbohydrates (e.g., sucrose, trehalose, fructan), polyols (e.g., pinitol, mannitol), and polyamines to maintain an osmotic balance (Zivcak et al. 2016; Gupta et al. 2012). The genomic route to drought resistance is complex and involves signaling events associated with the sensing of stress and the activation of defense and acclimation pathways involving reactive oxygen species, calcium, calcium-dependent protein kinases (CDPKs), mitogen-activated protein kinase cascades (MAPKs), and cross talk between different transcription factors (Farooq et al. 2009). The transcription factors mainly dehydration-responsive element-binding (DREB) protein family, zinc-finger family, WRKY and MYB family, ethylene-responsive element-binding factor (ERF) family, basic helix-loop-helix family, basic domain-leucine zipper (bZIP) family, NAC family, and the homeodomain transcription factor family have been identified to play a major role in tea plants in response to drought stress (Das et al. 2012; Gupta et al. 2012). Few groups investigated the accumulation of main bioactive components of *C. sinensis* where a decrease in total polyphenols and free amino acid content and increase in flavonoids were observed in response to drought stress (Liang et al. 1990; Jeyaramraja et al. 2003; Cheruiyot et al. 2007; Chen et al. 2010; Wang et al. 2016b). Transcriptome analysis done by Wang et al. (2016b, d), indicated that almost all known genes related to flavonoid biosynthesis are differentially expressed under drought stress with upregulation of the FLS and FNS unigenes, which participate in flavonoid biosynthesis. And the levels of CHS, DFR, LAR, ANS, and ANR tended to decrease initially and subsequently increased in response to drought stress suggesting that the effects of drought stress on the catechins depend on the regulation of genes that participate in flavonoid biosynthesis pathway.

5 Conclusion

Overlapping as well as distinct transcriptional molecular patterns are observed for biotic and abiotic stress in tea. Overlapping set of genes that are expressed due to both these stresses could represent a generalized stress response, and these genes may be targets for improving multiple stress tolerance in tea plants. The distinct stress resistance-associated genes are needed to be validated to develop tea plants which are more resistant to biotic or abiotic stress. The development of such plants would require understanding of stress regulatory networks, their synergistic as well as antagonistic effects, and identification of master regulatory genes to combat stress without affecting the productivity. At present, different tea research groups are involved in genotyping and identification of candidate genes in response to quality and environmental stress in tea through large-scale SNP discovery and genotyping using NGS-based methods such as restriction site-associated DNA sequencing (RAD-seq), genotyping-by-sequencing (GBS), sequence-based genotyping (SBG), and specific-locus amplified fragment sequencing (SLAF-seq). The tea genome sequence as well as genotyping data in the future will lead to construction of high-density genetic map which is necessary for mapping of quantitative trait loci (QTL), marker-assisted selection, and map-based cloning. The ultimate goal is to create stress-tolerant tea plants, either through genetic transformation or through conventional breeding. However, tea, being a perennial crop, genetic modification research is still at laboratory level with low transformation ratio. Appropriate identification and characterization of tea plants will help in studying the variation in the species gene pool. Advances in genomic technologies and genetic map with known molecular markers to specific traits can lead to rapid breeding approaches such as genome-wide selection (GWS) to identify potential cultivars which may be utilized in conventional breeding and further reduce the generation time. The challenge for tea scientists for the twenty-first century will be to develop tea plants tolerant to multiple abiotic and biotic stresses with improved yield and quality.

References

Ahmed P, Das S (2009) A functional genomic approach to analyze herbivory related transcripts in tea. In: Proceedings of the Indo-US workshop on epigenetic regulation and genome control, Hyderabad, India, 16–18 December 2009

Ahmed M, Paul SK, Mamun MSA (2011) Field performance and economic analysis of some commonly used insecticides against tea mosquito bug, *Helopeltis theivora*W. Bangladesh J Agric Res 36(3):449–454

Arimura G, Matsui K, Takabayashi J (2009) Chemical and molecular ecology of herbivore-induced plant volatiles: proximate factors and their ultimate functions. Plant Cell Physiol 50(5): 911–923. https://doi.org/10.1093/pcp/pcp030

Bali S, Mamgain A, Raina SN et al (2015) Construction of a genetic linkage map and mapping of drought tolerance trait in Indian beveragial tea. Mol Breeding 35:112. https://doi.org/10.1007/s11032-015-0306-5

Bandyopadhyay T, Gohain B, Bharalee R et al (2014) Molecular landscape of *Helopeltis theivora* induced transcriptome and defense gene expression in tea. Plant Mol Biol Rep 33(4):1042–1057. https://doi.org/10.1007/s11105-014-0811-4
Banerjee B (1983) Arthropod accumulation on tea in young and old habitats. Ecol Entomol 18(2):339–342
Banerjee B (1992) Botanical classification of tea. In: Wilson KC, Clifford MN (eds) Tea cultivation to consumption. Chapman and Hall, London, pp 25–51
Barman TS, Baruah U, Saikia JK (2008) Seasonal changes in metabolic activities of drought tolerant and susceptible clones of tea (*Camelia sinensis* L.). J Plant Crop 36:259–264
Barua DN (1989) Science and practice in tea culture. Tea Research Association, Jorhat, pp 56–58
Basak M, Sharma M, Chakraborty U (2001) Biochemical responses of *C. sinensis* (L.) O. Kuntze to heavy metal stress. J Environ Biol 22(1):37–41
Bhat JA, Ali S, Salgotra RK et al (2016) Genomic selection in the *Era* of Next generation sequencing for complex traits in plant breeding. FrontGenet 7:221. https://doi.org/10.3389/fgene.2016.00221
Bhorali P, Gohain B, Gupta S et al (2012) Molecular analysis and expression profiling of blister blight defense related genes in tea. Indian J Gen Plant Breed 72:226–233
Borchetia S, Bora C, Gohain B et al (2011) Cloning and heterologous expression of a gene encoding lycopene-epsilon-cyclase, a precursor of lutein in tea (*Camellia sinensis var assamica*). Afr J Biotech 10(32):5934–5939
Brady SM, Provart NJ (2007) Extreme breeding: leveraging genomics for crop improvement. J Sci Food Agric 87(6):925–929. https://doi.org/10.1002/jsfa.2763
Chacko SM, Thambi PT, Kuttan R et al (2010) Beneficial effects of green tea: A literature review. Chinese Med 5:13. https://doi.org/10.1186/1749-8546-5-13
Chen Z, Chen X (1989) An analysis of the world tea fauna. J Tea Sci 9:13–22
Chen L, Gao QK, Chen DM et al (2005) The use of RAPD markers for detecting genetic diversity, relationship and molecular identification of Chinese elite tea genetic resources [*Camellia sinensis*(L.)O. Kuntze] preserved in tea germplasm repository. Biodivers Conserv 14(6):1433–1444
Chen L, Zhao LP, Ma CL et al (2009a) Recent progress in the molecular biology of tea (*Camellia sinensis*) based on the expressed sequence tag strategy: A review. J Hortic Sci Biotechnol 84(5):476–482
Chen X, Fang WP, Zou Z et al (2009b) Cloning and expression analysis of CBF gene in cold induced tea plant [*Camellia sinensis* (L.) O. Kuntze]. J Tea Sci 29:53–59
Chen XH, Zhuang CG, He YF et al (2010) Photosynthesis, yield, and chemical composition of Tieguanyin tea plants (*Camellia sinensis* (L.) O. Kuntze) in response to irrigation treatments. Agric Water Manag 97:419–425
Chen X, Liu T, Hao S et al (2011) Differential gene expression analysis of self-incompatible lines in tea by cDNA-AFLP. Afr J Biotechnol 10:10906–10913
Cheruiyot E, Mumera LM, Ng'etich W, Hassanali A, Wachira F (2007) Polyphenols as potential indicators for drought tolerance in tea (*Camellia sinensis* L.). Biosci Biotechnol Biochem 71:2190–2197
Choi JY, Mizutani M, Shimizu BI et al (2007) Chemical profiling and gene expression profiling during manufacturing process of Taiwan oolong tea, "Oriental Beauty". Biosci Biotech Biochem 71(6):1476–1486. https://doi.org/10.1271/bbb.60708
Das A, Das S, Mondal TK (2012) Identification of differentially expressed gene profiles in young roots of tea [*Camellia sinensis* (L.) O. Kuntze] subjected to drought stress using suppression subtractive hybridization. Plant Mol Biol Rep 30(5):1088–1101. https://doi.org/10.1007/s11105-012-0422-x
Das A, Saha D, Mondal TK (2013) An optimized method for extraction of RNA from tea roots for functional genomics analysis. Indian J Biotechnol 12:129–132
Das S, Zaman A, Borchetia S et al (2016) Genetic relationship in tea germplasms with drought contrasting traits. Plant Breed Biotech 4(4):484–494. https://doi.org/10.9787/PBB.2016.4.4.484

Deng WW, Zhang M, Wu JQ et al (2013) Molecular cloning, functional analysis of three cinnamyl alcohol dehydrogenase (CAD) genes in the leaves of tea plant, *Camellia sinensis*. J Plant Physiol 170(3):272–282. https://doi.org/10.1016/j.jplph.2012.10.010

Develey-Rivière MP, Galiana E (2007) Resistance to pathogens and host developmental stage: a multifaceted relationship within the plant kingdom. New Phytol 175(3):405–416. https://doi.org/10.1111/j.1469-8137.2007.02130.x

Dietz KJ, Turkan I, Krieger-Liszkay A (2016) Redox- and reactive oxygen species-dependent signaling into and out of the photosynthesizing chloroplast. Plant Physiol 171(3):1541–1550. https://doi.org/10.1104/pp.16.00375

Duncan JMA, Saikia SD, Gupta N et al (2016) Observing climate impacts on tea yield in Assam, India. Appl Geogr 77:64–71. https://doi.org/10.1016/j.apgeog.2016.10.004

Dutta R (2014) Climate change and its impact on tea in Northeast India. J Water Clim Change 5(4):625. https://doi.org/10.2166/wcc.2014.143

Fang W, Zou Z, Hou X et al (2009) Cloning and sequence analysis of cold-induced H1-histone gene from Camellia sinensis. Acta Bot Boreali-Occidentalia Sin 8:1514–1519

Fang WP, Meinhardt LW, Tan HW et al (2014) Varietal identification of tea (*Camellia sinensis*) using nanofluidic array of single nucleotide polymorphism (SNP) markers. Horti Res 35:1–8. https://doi.org/10.1038/hortres.2014.35

FAO (2016) Report of the Working Group on Climate Change of the FAO Intergovernmental Group on Tea. Food and Agriculture Organization of the United Nations, Rome, pp 3–25

Farooq M, Wahid A, Kobayashi N et al (2009) Plant drought stress: effects, mechanisms and management. In: Lichtfouse E, Navarrete M, Debaeke P, Véronique S, Alberola C (eds) Sustainable agriculture. Springer, Dordrecht, pp 153–188

Fu JY (2014) Cloning of a new glutathione peroxidase gene from tea plant (Camellia sinensis) and expression analysis under biotic and abiotic stresses. Bot Stud 55:7. https://doi.org/10.1186/1999-3110-55-7

Gohain B, Borchetia S, Bhorali P et al (2012) Understanding tea flavor on a molecular basis. Plant Mol Biol 78(6):577–597. https://doi.org/10.1007/s11103-012-9887-0

Gosal SS, Wani SH, Kang MS (2009) Biotechnology and drought tolerance. J Crop Improve 23(1):19–54

Gosal SS, Wani SH, Kang MS (2010) Biotechnology and drought tolerance. In: Kang MS (ed) Water and agricultural sustainability strategies. CRC Press, Bocca Raton/London/NewYork/Leiden, pp 229–259

Gunathilaka R, Smart J, Fleming C (2016) The impact of changing climate on perennial crops: the case of tea production in Sri Lanka. Clim Chang 140(3–4):577–592. https://doi.org/10.1007/s10584-016-1882-z

Gupta S, Bharalee R, Bhorali P et al (2012) Identification of drought tolerant progenies in tea by gene expression analysis. Funct Integ Genom 12(3):543–563. https://doi.org/10.1007/s10142-012-0277-0

Gupta S, Bharalee R, Bhorali P et al (2013) Molecular analysis of drought tolerance in tea by cDNA-AFLP based transcript profiling. Mol Biotechnol 53(3):237–248. https://doi.org/10.1007/s12033-012-9517-8

Gurusubramanian G, Bora S (2007) Relative toxicity of some commonly used insecticides against adults of *Helopeltis theivora* Waterhouse (Miridae: Hemiptera) collected from Jorhat area tea Plantations, South Assam India. Resist Pest Manag Newsl 17(1):8–12

Hackett CA, Wachira FN, Paul S et al (2000) Construction of a genetic linkage map for tea (*Camellia sinensis*). Heredity 85(4):346–355

Hao X, Yang Y, Yue C et al (2017) Comprehensive transcriptome analyses reveal differential gene expression profiles of *Camellia sinensis* axillary buds at para-, endo-, ecodormancy, and bud flush stages. Front Plant Sci 8:553. https://doi.org/10.3389/fpls.2017.00553

Hazarika LK, Bhuyan M, Hazarika BN (2009) Insect pests of tea and their management. Annu Rev Entomol 54:267–284

Heinrich M, Baldwin IT, Wu J (2011) Two mitogen-activated protein kinase kinases, MKK1 and MEK2, are involved in wounding and specialist lepidopteran herbivore *Manduca sexta*-induced responses in *Nicotiana attenuata*. J Exp Bot 62(12):4355–4365. https://doi.org/10.1093/jxb/err162

Horvath DP, Hansen SA, Moriles-Miller JP et al (2015) RNAseq reveals weed-induced PIF3-like as a candidate target to manipulate weed stress response in soybean. New Phytol 207(1):196–210. https://doi.org/10.1111/nph.13351

Hossain MA, Wani SH, Bhattachajee S, Burritt DJ, Tran LSP (2016) Drought stress tolerance in plants, Vol 1 Physiology and biochemistry. Springer, Cham. ISBN 978-3-319-28897-0

Hu CY, Lee TC, Tsai HT et al (2013) Construction of an integrated genetic map based on maternal and paternal lineages of tea (*Camellia sinensis*). Euphytica 191(1):141–152. https://doi.org/10.1007/s10681-013-0908-0

Huang JA, Li JX, Huang YH et al (2005) Construction of AFLP molecular markers linkage map in tea plant. J Tea Sci 25(1):7–15

Huang FP, Liang YR, Lu JL et al (2006) Genetic mapping of first generation of backcross in tea by RAPD and ISSR markers. J Tea Sci 26:171–176

Huang JA, Huang YH, Luo JW et al (2007) Identification of single nucleotide polymorphism in polyphenol oxidase gene in tea plant (*Camellia sinensis*). J Hunan Agric Univ 33:454–458

Huang H, Shi C, Liu Y et al (2014) Thirteen *Camellia* chloroplast genome sequences determined by high throughput sequencing: genome structure and phylogenetic relationships. BMC Evol Biol 14(1):151. https://doi.org/10.1186/1471-2148-14-151

Jayaswall K, Mahajan P, Singh G et al (2016) Transcriptome analysis reveals candidate genes involved in blister blight defense in tea (*Camellia sinensis* (L) Kuntze). Sci Rep 6:30412. https://doi.org/10.1038/srep30412

Jeyaraj J, Liu S, Zhang X et al (2017) Genome-wide identification of microRNAs responsive to *Ectropis oblique* feeding in tea plant (*Camellia sinensis* L.). Sci Rep 7:13634. https://doi.org/10.1038/s41598-017-13692-7

Jeyaramraja P, Pius P, Raj Kumar R et al (2003) Soil moisture stress-induced alterations in bioconstituents determining tea quality. J Sci Food Agric 83(12):1187–1191. https://doi.org/10.1002/jsfa.1440

Jones JDG, Dangl JL (2006) The plant immune system. Nat 444(7117):323–329. https://doi.org/10.1038/nature05286

Kamunya SM, Wachira FN, Pathak RS et al (2009) Genetic mapping and identification of quantitative trait loci for yield and drought tolerance in tea (*Camellia sinensis* (L.) O. Kuntze). Tea 30:19–41

Kamunya SM, Wachira FN, Pathak RS et al (2010) Genome mapping and testing for quantitative trait loci in tea (*Camellia sinensis* (L.) O. Kuntze). Tree Genet Genom 6(6):915–929

Krishania S, Dwivedi P, Agarwal K (2013) Strategies of adaptation and injury exhibited by plants under a variety of external conditions: a short review. Comunicata Scientiae 4(2):103–110

Kumar S, Paul A, Bhattacharya A (2012) Tea: present status and strategies to improve abiotic stress tolerance. In: Tuteja N, Gill SS, Tiburcio AF, Tuteja R (eds) Improving crop resistance to abiotic stress, 1st and 2nd edn. Wiley, New York, pp 1401–1424. https://doi.org/10.1002/9783527632930

Li XW, Feng ZG, Yang HM, Zhu XP, Liu J, Yuan HY (2010) A novel cold-regulated gene from *Camellia sinensis*CsCOR1, enhances salt- and dehydration tolerance in tobacco. Biochem Biophys Res Commun 394(2):354–359. https://doi.org/10.1016/j.bbrc.2010.03.011

Liang YR, Liu ZS, Xu YR et al (1990) A study on chemical composition of two special green teas (*Camellia sinensis*). J Sci Food Agric 53:541–548

Lin J, Kudrna D, Wing RA (2011) Construction, Characterization, and preliminary BAC-end sequence analysis of a bacterial artificial chromosome library of the tea plant (*Camellia sinensis*). J Biomed Biotechnol 2011:476723. https://doi.org/10.1155/2011/476723

Liu S, Han B (2010) Differential expression pattern of an acidic 9/13-lipoxygenase in flower opening and senescence and in leaf response to phloem feeders in the tea plant. BMC Plant Biol 10:228–238. https://doi.org/10.1186/1471-2229-10-228

Liu SC, Xu YX, Ma JQ et al (2016) Small RNA and degradome profiling reveals important roles for microRNAs and their targets in tea plant response to drought stress. Physiol Plant 158(4):435–451. https://doi.org/10.1111/ppl.12477

Ma JQ, Yao MZ, Ma CL et al (2014) Construction of a SSR-based genetic map and identification of QTLs for catechins content in tea plant (*Camellia sinensis*). PLoS One 9:e93131. https://doi.org/10.1371/journal.pone.0093131

Ma JQ, Huang L, Ma CL et al (2015) Large-scale SNP discovery and genotyping for constructing a high-density genetic map of tea plant using Specific-Locus Amplified Fragment Sequencing (SLAF-seq). PLoS One 10(6):e0128798. https://doi.org/10.1371/journal.pone.0128798. PMID: 26035838

Manickavelu A, Kawaura K, Oishi K et al (2010) Comparative gene expression analysis of susceptible and resistant near-isogenic lines in common wheat infected by *Puccinia triticina*. DNA Res 17(4):211–222. https://doi.org/10.1093/dnares/dsq009

Maron JL, Crone E (2006) Herbivory: effects on plant abundance, distribution and population growth. Proc R Soc Lond B: Biol Sci 273(1601):2575–2584. https://doi.org/10.1098/rspb.2006.3587

Marx W, Haunschild R, Bornmann L (2017) Global warming and tea production-the bibliometric view on a newly emerging research topic. Climacteric 5:46. https://doi.org/10.3390/cli5030046

Mohanpuria P, Yadav SK (2012) Characterization of novel small RNAs from tea (*Camellia sinensis* L.). Mol Biol Rep 39(4):3977–3986. https://doi.org/10.1007/s11033-011-1178-3

Mondal TK (2002) Detection of genetic diversity among the Indian tea (*Camellia sinensis*) germplasm by Inter-simple sequence repeats (ISSR). Euphytica 128(3):307–315

Mondal TK, Sutoh K (2013) Application of next-generation sequencing for abiotic stress tolerance. In: Barh D, Zambare V, Azevedo V (eds) The world of omics and applications in plant and agricultural sciences. CRC Press, Boca Raton, pp 383–402

Moore MJ, Dhingra A, Soltis PS et al (2006) Rapid and accurate pyrosequencing of angiosperm plastid genomes. BMC Plant Biol 6(1):17. https://doi.org/10.1186/1471-2229-6-17

Muraleedharan N, Chen ZM (1997) Pests and diseases of tea and their management. J Plant Crop 25:15–43

Muraleedharan N, Selvasundaram R (2002) An IPM package for tea in India. Planters Chron 98:107–124

Ota S, Tanaka J (1999) RAPD-based linkage mapping using F1 segregating populations derived from crossings between tea cultivar 'Sayamakaori' and strain 'Kana-Ck17'. Breed Res 1(1):16

Pan C, Tian K, Ban Q et al (2017) Genome-wide analysis of the biosynthesis and deactivation of gibberellin-dioxygenases gene family in *Camellia sinensis* (L.) O. Kuntze. Genes 8(9):E235. https://doi.org/10.3390/genes8090235

Parks M, Cronn R, Liston A (2009) Increasing phylogenetic resolution at low taxonomic levels using massively parallel sequencing of chloroplast genomes. BMC Biol 7(1):84. https://doi.org/10.1186/1741-7007-7-84

Pathak MR, Wani SH (2015) Salinity stress tolerance in relation to polyamine metabolism in plants. In: Managing salt tolerance in plants: molecular and genomic perspectives. CRC Press, pp 241–250

Paul A, Kumar S (2013) Dehydrin2 is a stress-inducible, whereas Dehydrin1 is constitutively expressed but up-regulated gene under varied cues in tea [*Camellia sinensis* (L.) O. Kuntze]. Mol Biol Rep 40:3859–3863

Paul A, Kumar S (2015) An A20/AN1-zinc-finger domain containing protein gene in tea is differentially expressed during winter dormancy and in response to abiotic stress and plant growth regulators. Plant Gene 1:1–7

Paul A, Lal L, Ahuja PS (2012a) Alpha-tubulin (CsTUA) up-regulated during winter dormancy is a low temperature inducible gene in tea [*Camellia sinensis* (L.) O. Kuntze]. Mol Biol Rep 39(4):3485–3490. https://doi.org/10.1007/s11033-011-1121-7

Paul A, Muoki RC, Singh K, Kumar S (2012b) CsNAMlike protein encodes a nuclear localized protein and responds to varied cues in tea [*Camellia sinensis* (L.) O. Kuntze]. Gene 502:69–74

Pérez-de-Castro AM, Vilanova S, Cañizares J et al (2012) Application of genomic tools in plant breeding. Curr Genomics 13(3):179–195. https://doi.org/10.2174/138920212800543084

Pospíšil P (2016) Production of reactive oxygen species by photosystem II as a response to light and temperature stress. Front Plant Sci 7:1950. https://doi.org/10.3389/fpls.2016.01950

Qian W, Yue C, Wang Y et al (2016) Identification of the invertase gene family (INVs) in tea plant and their expression analysis under abiotic stress. Plant Cell Rep 35:2269. https://doi.org/10.1007/s00299-016-2033-8

Rana NK, Mohanpuria P, Yadav SK (2008) Cloning and characterization of a cytosolic glutamine synthetase from Camellia sinensis (L.) O. Kuntze that is upregulated by ABA, SA, and H2O2. Mol Biotechnol 39:49–56

Rana NK, Mohanpuria P, Kumar V et al (2010) A CsGS is regulated at transcriptional level during developmental stages and nitrogen utilization in *Camellia sinensis* (L.) O. Kuntze. Mol Biol Rep 37(2):703–710. https://doi.org/10.1007/s11033-009-9559-6

Roy SC, Chakraborty BN (2009) Cloning and sequencing of chitinase gene specific PCR amplified DNA fragment from tea plant (Camellia sinensis) and analyzed the nucleotide sequence using bioinformatics algorithms. Can J Pure Appl Sci 3:795–800

Roy S, Muraleedharan N (2014) Microbial management of arthropod pests of tea: current state and prospects. Appl Microbiol Biotechnol 98(12):5375–5386. https://doi.org/10.1007/s00253-014-5749-9

Roy S, Handique G, Muraleedharan N et al (2016) Use of plant extracts for tea pest management in India. Appl Microbiol Biotechnol 100(11):4831–4844. https://doi.org/10.1007/s00253-016-7522-8

Sah SK, Kaur G, Wani SH (2016) Metabolic engineering of compatible solute trehalose for abiotic stress tolerance in plants. In: Osmolytes and Plants Acclimation to Changing Environment: Emerging Omics Technologies. Springer, New Delhi, pp 83–96

Senthilkumar P, Thirugnanasambantham K, Mandal AKA (2012) Suppressive subtractive hybridization approach revealed differential expression of hypersensitive response and reactive oxygen species production genes in tea (*Camellia sinensis* (L.) O.Kuntze) leaves during *Pestalotiopsis thea* infection. Appl Biochem Biotech 168(7):1917–1927. https://doi.org/10.1007/s12010-012-9907-1

Shi CY, Yang H, Wei CL et al (2011) Deep sequencing of the *Camellia sinensis* transcriptome revealed candidate genes for major metabolic pathways of tea-specific compounds. BMC Genomics 12:131. https://doi.org/10.1186/1471-2164-12-131

Shukla LI, Chinnusamy V, Sunkar R (2008) The role of microRNAs and other endogenous small RNAs in plant stress responses. Biochim Biophys Acta 1779(11):743–748. https://doi.org/10.1016/j.bbagrm.2008.04.004

Singh ID (2006) Selection and testing field performance of candidate tea clones in India. Int J Tea Sci 5:1–12

Sourour A, Afef O, Mounir R et al (2017) A review: morphological, physiological, biochemical and molecular plant responses to water deficit stress. Int J Eng Sci 6:1–4

Takeuchi A, Matsumoto S, Hayatsu M (1994) Chalcone synthase from *Camellia sinensis* isolation of the cDNAs and the organ-specific and sugar-responsive expression of the genes. Plant Cell Physiol 35(7):1011–1018

Tan LQ, Wang LY, Wei K et al (2013) Floral transcriptome sequencing for SSR marker development and linkage map construction in the tea plant (*Camellia sinensis*). PLoS One 8(11):e81611. https://doi.org/10.1371/journal.pone.0081611

Tanaka J (1996) RAPD linkage map of tea plant and the possibility of application in tea genetics and breeding. Tea Res J 84:44–45

Tanaka J, Sawai Y, Yamaguchi S (1995) Genetic analysis of RAPD markers in tea. Jpn J Breed 45(2):198

Taniguchi F, Tanaka J (2004) Characterization of genes for ammonium assimilation in *Camellia sinensis*. In: Proceedings of 2004 International Conference on O-Cha culture and Science, Shizuaka, Japan
Taniguchi F, Furukawa K, Ota-Metoku S et al (2012) Construction of a high-density reference linkage map of tea (*Camellia sinensis*). Breed Sci 62(3):263–273. https://doi.org/10.1270/jsbbs.62.263
Telem RS, Wani SH, Singh NB, Sadhukhan R, Mandal N (2016) Single Nucleotide Polymorphism (SNP) Marker for Abiotic Stress Tolerance in Crop Plants. In: Advances in plant breeding strategies: agronomic, abiotic and biotic stress traits. Springer, Cham, pp 327–343
Tewari A (2016) To check the bio-efficiency of deuce 2.8(deltamethrin) against *Helopeltis theivora* or more commonly known as the tea mosquito. Int J Innov Appl SciEng 2:31–35
Thirugnanasambantham K, Prabu G, Palanisamy S et al (2013) Analysis of dormant bud (Banjhi) specific transcriptome of tea (*Camellia sinensis* (L.) O. Kuntze) from cDNA library revealed dormancy-related genes. Appl Biochem Biotechnol 169(4):1405–1417. https://doi.org/10.1007/s12010-012-0070-5
Tomimoto Y, Ikehashi I, Kakeda K, Kowyama Y (1999) A pistal specific PR-1 like protein in Camellia, its expression, sequence and genealogical position. Breed Sci 49:97–104
Upadhyaya H, Dutta BK, Sahoo L et al (2012) Comparative effect of Ca, K, Mn and B on post-drought stress recovery in tea [*C. sinensis* (L.) O Kuntze]. Am J Plant Sci 3(4):443–460. https://doi.org/10.4236/ajps.2012.34054
Upadhyaya H, Dutta BK, Panda SK (2013) Zinc modulates drought induced biochemical damages in tea (*C. sinensis* (L) O Kuntze). J Agric Food Chem 61(27):6660–6670. https://doi.org/10.1021/jf304254z
van Loon LC, Rep M, Pieterse CMJ (2006) Significance of inducible defense-related proteins in infected plants. Annu Rev Phytopathol 44:135–162. https://doi.org/10.1146/annurev.phyto.44.070505.143425
von Dahl CC, Winz RA, Halitschke R, Kühnemann F, Gase K, Baldwin IT (2007) Tuning the herbivore-induced ethylene burst: the role of transcript accumulation and ethylene perception in *Nicotiana attenuata*. Plant J 51(2):293–307. https://doi.org/10.1111/j.1365-313X.2007.03142.x
Vyas D, Kumar S (2005) Purification and partial characterization of a low temperature responsive Mn-SOD from tea (*C. sinensis* (L.) O. Kuntze). Biochem Biophys Res Commun 329(3):831–838. https://doi.org/10.1016/j.bbrc.2005.02.051
Wang CX, Li YY, Jiang CJ, Yu YB (2005) Molecular cloning and sequence analysis on cDNA of cystatin gene from tea leaves. J Tea Sci 25:177–182
Wang L, Li X, Zhao Q et al (2009) Identification of genes induced in response to low temperature treatment in tea leaves. Plant Mol Biol Rep 27(3):257–265. https://doi.org/10.1007/s11105-008-0079-7
Wang X-C, Yang YJ, Chen L et al (2010) Construction and preliminary analysis of the suppression subtractive hybridization cDNA libraries between dormant and sprouting buds of tea plant (*Camellia sinensis*). J Tea Sci 30:129–135
Wang YS, Gao LP, Wang ZR et al (2012) Light induced expression of genes involved in phenylpropanoid biosynthetic pathways in callus of tea (*Camellia sinensis* (L.) O. Kuntze). Sci Hort 133(6):72–83. https://doi.org/10.1016/j.scienta.2011.10.017
Wang XC, Zhao QY, Ma CL et al (2013) Global transcriptome profiles of *Camellia sinensis* during cold acclimation. BMC Genomics 14:415. https://doi.org/10.1186/1471-2164-14-415
Wang M, Ma Q, Han B et al (2016a) Molecular cloning and expression of a jasmonate biosynthetic gene allene oxide cyclase from Camellia sinensis. Can J Plant Sci 96:109–116
Wang W, Xin H, Wang M et al (2016b) Transcriptomic analysis reveals the molecular mechanisms of drought-stress-induced decreases in *Camellia sinensis* leaf quality. Front Plant Sci 7:385. https://doi.org/10.3389/fpls.2016.00385
Wang YN, Tang L, Hou Y et al (2016c) Differential transcriptome analysis of leaves of tea plant (*Camellia sinensis*) provides comprehensive insights into the defense responses to Ectropis oblique attack using RNA-Seq. Funct Integr Genom 16(4):383–398. https://doi.org/10.1007/s10142-016-0491-2

Wang YX, Liu ZW, Wu ZJ et al (2016d) Transcriptome-wide identification and expression analysis of the NAC gene family in tea plant [*Camellia sinensis* (L.) O. Kuntze]. PLoS One 11(11):e0166727. https://doi.org/10.1371/journal.pone.0166727

Wani SH, Gosal SS (2011) Introduction of OsglyII gene into Oryza sativa for increasing salinity tolerance. Biol Plant 55(3):536–540

Wani SH, Sah SK, Hossain MA, Kumar V, Balachandran SM (2016) Transgenic approaches for abiotic stress tolerance in crop plants. In: Advances in plant breeding strategies: agronomic, abiotic and biotic stress traits. Springer, Cham, pp 345–396

War AR, Paulraj MG, Ahmad T et al (2012) Mechanisms of plant defense against insect herbivores. Plant Signal Behav 7(10):1306–1320. https://doi.org/10.4161/psb.21663

Wei CL, Jiang CJ, Tao HZ, Wan XC (2003) Cloning and bioinformatics analysis of sequence signature of violaxanthin de-epoxidase cDNA in tea plant (Camellia sinensis (L.) O. Kuntze). J Nanj Agric Univ 26:14–19

Wight W (1962) Tea classification revised. Curr Sci 31:298–299

Wu J, Hettenhausen C, Meldau S, Baldwin IT (2007) Herbivory rapidly activates MAPK signaling in attacked and unattacked leaf regions but not between leaves of *Nicotiana attenuata*. Plant Cell 19(3):1096–1122

Wu H, Chen D, Li J et al (2013) De novo characterization of leaf transcriptome using 454 sequencing and development of EST-SSR markers in tea (*Camellia sinensis*). Plant Mol Biol Rep 31(3):524–538. https://doi.org/10.1186/s12870-014-0347-7

Xia J, Liu Y, Yao S et al (2017a) Characterization and expression profiling of *Camellia sinensis* cinnamate 4-hydroxylase genes in phenylpropanoid pathways. Genes 8(8):193. https://doi.org/10.3390/genes8080193

Xia EH, Zhang HB, Sheng J et al (2017b) The tea tree genome provides insights into tea flavor and independentevolution of caffeine biosynthesis. Mol Plant 10(6):866–877. https://doi.org/10.1016/j.molp.2017.04.002

Yan C, Xie D (2015) Jasmonate in plant defence: sentinel or double agent? Plant Biotechnol J 13(9):1233–1240. https://doi.org/10.1111/pbi.12417

Yang JB, Yang SX, Li HT et al (2013) Comparative chloroplast genomes of *Camellia* species. PLoS One 8(8):e73053. https://doi.org/10.1371/journal.pone.0073053

Yang H, Wei C-L, Liu H-W et al (2016) Genetic divergence between *Camellia sinensis* and its wild relatives revealed via genome-wide SNPs from RAD sequencing. PLoS One 11(3):e0151424. https://doi.org/10.1371/journal.pone.0151424

Ye G-Y, Xiao Q, Chen M, Chen X-x, Yuan Z-j, Stanley D, Hu C (2014) Tea: Biological control of insect and mite pests in China. Biol Control 68:72–91

Yoshida K, Homma T (2005) Isolation of wound/pathogen inducible cDNA from tea by mRNA differential display. Inter Tea Symp-2005 62–65 (Hanzou, China)

Yue C, Cao H, Wang L et al (2014) Molecular cloning and expression analysis of tea plant aquaporin (AQP) gene family. Plant Physiol Biochem 83:65–76

Zhang Y, Zhu X, Chen X et al (2014) Identification and characterization of cold-responsive microRNAs in tea plant (*Camellia sinensis*) and their targets using high-throughput sequencing and degradome analysis. BMC Plant Biol 14:271. https://doi.org/10.1186/s12870-014-0271-x

Zhang CC, Wang LY, Wei K et al (2016) Transcriptome analysis reveals self-incompatibility in the tea plant (*Camellia sinensis*) might be under gametophytic control. BMC Genomics 17:359. https://doi.org/10.1186/s12864-016-2703-5

Zheng C, Zhao L, Wang Y et al (2015) Integrated RNA-seq and sRNA-seq analysis identifies chilling and freezing responsive key molecular players and pathways in tea plant (*Camellia sinensis*). PLoS One 10(4):e0125031. https://doi.org/10.1371/journal.pone.0125031

Zhou Y, Liu Y, Wang S et al (2017) Molecular cloning and characterization of galactinol synthases in *Camellia sinensis* with different responses to biotic and abiotic stressors. J Agric Food Chem 65(13):2751–2759. https://doi.org/10.1021/acs.jafc.7b00377

Zhu L, Jiang CJ, Deng WW et al (2008) Cloning and expression of selenocysteine methyltransferase cDNA from *Camellia sinensis*. Acta Physiol Plant 30:167. https://doi.org/10.1007/s11738-007-0105-7

Zhu L, Deng WW, Ye AH et al (2008a) Cloning of two cDNAs encoding a family of ATP sulfurylase from *Camellia sinensis* related to selenium or sulfur metabolism and functional expression in *Escherichia coli*. Plant Physiol Biochem 46(8–9):731–738. https://doi.org/10.1016/j.plaphy.2007.03.029

Zhu L, Jiang CJ, Deng WW et al (2008b) Cloning and expression of selenocysteine methyltransferase cDNA from *Camellia sinensis*. Acta Physiol Plnt 30(2):167–174

Zimdahl RL (ed) (2004) Weed-crop competition: a review. Blackwell, London

Zipfel C, Robatzek S (2010) Pathogen-associated molecular pattern-triggered immunity: Veni, Vidi…? Plant Physiol 154(2):551–554. https://doi.org/10.1104/pp.110.161547

Zivcak M, Brestic M, Sytar O (2016) Osmotic adjustment and plant adaptation to drought stress. In: Hossain M, Wani S, Bhattacharjee S, Burritt D, Tran LS (eds) Drought stress tolerance in plants. Springer, Cham, pp 105–143

Chapter 14
Tea Antioxidants As Affected by Environmental Factors

Lovedeep Kaur and Natthawuddhi Donlao

Abstract The polyphenolic compounds present in tea have been reported to vary considerably with the source of tea, the production location, the season of tea production and tea processing, ultimately affecting the antioxidant properties of tea. The studies done so far have indicated that both season and location need to be considered in order to produce teas with higher polyphenol contents and improved antioxidant quality. These factors are discussed in detail in this chapter.

Keywords *Camellia sinensis* · Tea · Polyphenol · Catechin · Altitude · Season · Antioxidant

1 Introduction

Tea, an important dietary source of natural phenolic antioxidants, is known for its beneficial health effects: reduced risk of cardiovascular disease and certain types of cancer; inflammatory bowel, liver and neurodegenerative diseases; and diabetes and weight loss (Dufresne and Farnworth 2001; Carloni et al. 2013; Kaur et al. 2014). Tea catechins, the main polyphenols present in tea, are the active components that are mainly responsible for these physiological effects. Catechins and other polyphenols act as antioxidants by sequestering metal ions and by scavenging reactive oxygen and nitrogen species (Wiseman et al. 1997). The concentrations of the phenolic compounds have long been used as a quality parameter of tea.

At present, there are over 2000 varieties of tea, and it is produced mainly in China, India, Kenya, Sri Lanka, Indonesia and Turkey. All of the varieties of tea are

---

L. Kaur (✉)
Riddet Institute and Massey Institute of Food Science and Technology, Massey University, Palmerston North, New Zealand
e-mail: L.Kaur@massey.ac.nz

N. Donlao
Graduate School of Horticulture, Chiba University, Matsudo, Japan

School of Agro-Industry, Mae Fah Luang University, Chiang Rai, Thailand

W.-Y. Han et al. (eds.), *Stress Physiology of Tea in the Face of Climate Change*,
https://doi.org/10.1007/978-981-13-2140-5_14

produced from young and tender leaves of *Camellia sinensis* (L.) (family Theaceae), an evergreen shrub. Freshly plucked leaves are processed to give black (fermented), oolong (semi-fermented), and green (unfermented) teas (Carloni et al. 2013), with black tea accounting for approximately 75% of global production. There are other tea products which are less commonly consumed or are limited to particular geographical locations or communities, such as pickled tea (made from tea leaves plucked in the rainy season and pickled in bamboo containers), brick tea (compressed black or green tea) and white tea (made from the delicate tea buds allowed to wither in natural sunlight before they are lightly processed to prevent further fermentation). Some specialty tea products have been developed which are more pharmacologically active due to high levels of biologically active molecules. Some examples of such teas include white tea, flavoured teas (ginger, lemongrass, lemon, vanilla, strawberry), scented tea, herbal teas and decaffeinated teas.

The content of phenolic compounds, antioxidant properties and hence the quality of tea are reportedly influenced by a variety of environmental (atmospheric temperature, rainfall and the amount of sunlight), including geographical location/season of tea picking (Carloni et al. 2013; Sharpe et al. 2016; Bancirova 2010; Astill et al. 2001; Lin et al. 2003; Chan et al. 2007), and processing factors (Astill et al. 2001; Sharpe et al. 2016). These factors are reviewed and discussed in this chapter.

Other factors that have been shown to have a pronounced impact on fresh tea leaf and tea component concentrations in the final brew are as follows:

- Cultivar type (Chan et al. 2007; Carloni et al. 2013; Astill et al. 2001; Lin et al. 2003).
- Age of the leaf, i.e. plucking position (Lin et al. 2003; Chan et al. 2007).
- Horticultural practices, e.g. mechanical- or hand-plucking and age of tea leaves (Carloni et al. 2013; Chan et al. 2007).
- The different technologies of the tea companies (Carloni et al. 2013).
- The type of tea, e.g. blended, decaffeinated and instant (Bancirova 2010).
- The preparation of tea infusion, e.g. amount of the product used, brewing time, water temperature (Bancirova 2010; Sharpe et al. 2016; Astill et al. 2001), reinfusions, the time between reinfusions (Sharpe et al. 2016) and amount of agitation (Astill et al. 2001).
- The grade (particle size) of the tea leaves, whether they are contained in a teabag, and, if so, the size of the bag and the material used in its construction (Astill et al. 2001).

2 Tea Polyphenols

Polyphenols are divided into several groups, and one of which is represented by flavonoids (Ananingsih et al. 2013). Flavonoids are plant secondary metabolites, containing an aromatic ring bearing one or more hydroxyl groups. The most common flavonoids in tea are the flavan-3-ol, which provide tea with its typical

Table 14.1 Chemical composition of young shoots of tea[a]

Component	g/100g (dry weight basis)
Flavanols:	
(–)-Epigallocatechin 3-gallate (EGCG)	9–13
(–)-Epigallocatechin (EGC)	3–6
(–)-Epicatechin 3-gallate (ECG)	3–6
(–)-Epicatechin (EC)	1–3
Gallocatectin (GC)	1–2
(+)-Catechin (C)	1–2
Flavonols and their glycosides	3–4
Leucoanthocyanins	2–3
Phenolic acids: theogallin and others	4
Total polyphenols	27–40[b]
Caffeine	3–4
Amino acids: theanine	2
Others	2
Carbohydrates	4
Organic acids	0.5
Volatile compounds	0.01
Polysaccharides: starch	2–5
Other	12
Protein	15
Ash	5
Cellulose	7
Lignin	6
Lipids	3

[a]Hilton (1973) © John Wiley and Sons (1973). This material is reproduced with permission of John Wiley & Sons, Inc
[b]Modder and Amarakoon (2002)

flavonoid pattern (Ananingsih et al. 2013; Labbé et al. 2006; Dreosti 2000; Green et al. 2007; Table 14.1). In some cases, the antioxidant activity of flavanols and flavonols derived from tea is up to five times more effective than vitamin C and vitamin E (Dreosti 2000). The major polyphenols (flavanols) in tea are catechins (Dreosti 2000; Green et al. 2007), and the compounds in this group show the most powerful antioxidant activity (Vinson et al. 1995; Stewart et al. 2005). The tea catechins are composed of a family of eight most abundant natural substances: (+)-catechin (C), (−)-epicatechin (EC), (−)-gallocatechin (GC), (−)-epigallocatechin (EGC), (−)-catechin gallate (CG), (−)-gallocatechin gallate (GCG), (−)-epicatechin gallate (ECG) and (−)-epigallocatechin gallate (EGCG) (Dalluge and Nelson 2000). The antioxidant activity and free radical-scavenging capacity of the tea catechins have been reported as one of the most important benefits of tea (Almajano et al. 2008). Among tea catechins, EGCG is considered as the most important due to its high content in tea and its antioxidant activity (Labbé et al. 2006; Bazinet et al. 2010). It has also been found that EGCG exhibits the carcinogenesis inhibition of many types of cancer in animal models (Higdon and Frei 2003). Tea and tea polyphenols have been reported to serve as antioxidants and induce the detoxification system, thereby inhibiting carcinogen metabolism and cancer initiation (Sur and Panda 2017). Apart from health benefits, tea catechins have been found to be correlated with quality of the tea as assessed by professional tea tasters (Liang et al. 2005).

Other polyphenols include flavonols (quercetin, kaempferol, myricetin), and their glycosides, and depsides such as chlorogenic acid. In addition, tea also contains amino acids (theanine, gamma-aminobutyric acid), carbohydrates, proteins, vitamins, minerals, trace elements, volatile compounds, carotenoids and alkaloids, namely, caffeine, theophylline and theobromine (Kerio et al. 2013). Caffeine is present at an average level of 3% along with very small amounts of methylxanthines, theobromine and theophylline. Anthocyanidins are also found in the leaf. The amino acid theanine is unique to tea. Vitamin C is lost during the processing of the fresh leaf, but carotenoids and vitamin K are present in brewed tea. Tea also accumulates aluminium, potassium, fluoride and manganese (Modder and Amarakoon 2002).

During black tea manufacture, the catechins are oxidised to theaflavins (TFs) and thearubigins (TRs) in the presence of polyphenol oxidase and oxygen. The TFs are catechin dimers (theaflavin, theaflavin 3-gallate, theaflavin 3′-gallate, theaflavin 3.3′-digallate), and TRs are catechin oligomers of unknown structure. These oxidised constituents are unique to black tea. TF formation requires the presence of both EC or ECG and EGC OR EGCG (Wickremasinghe, 1978). In green tea manufacture, the first operation is the heat destruction of polyphenol oxidase by steaming (in Japan) or pan firing (in China). The primary difference between black and green tea therefore is that the former contains a mixture of native unoxidised polyphenols together with oxidised products such as TF and TR, whereas green tea contains unoxidised native polyphenols only. Oolong tea is a semi-fermented tea that is allowed to oxidise only partially to lock in the rich flavour that is associated with the high quality of oolong tea. The polyphenol profile of oolong tea is similar to that of green or black tea, but it has a lower content of EGCG, EC and ECG than that found in green tea because the fresh oolong tea leaves are subjected to a partial fermentation step before drying (Zuo et al. 2002; Chen et al. 2010).

The concentrations of the phenolic compounds have long been used as a quality parameter of tea. The plain black teas are valued for their TF content, which is responsible for their taste and brightness and which also contributes to their colour, and for the TRs, which are responsible for thickness and colour of both the liquors and infusion (Biswas et al. 1973). Catechins and gallates are used as quality parameters for green tea. Bryce et al. (1970) and Coxon et al. (1970) found that flavonoids (C, EC, GC, ECG, GC, EGC, EGCG) and one phenolic acid (gallic acid) in the fresh leaf were responsible for the formation of TFs in the resultant black tea. The relative amounts of catechins and polyphenol oxidase in the tea leaf vary with the season, elevation, agricultural practices employed, maturity of the leaf, processing methods, manufacturing practices and storage and may affect tea quality. The strength of brewed tea and the infusion time may also affect the phenolic composition of brewed tea (Hilton 1973).

Catechins and other polyphenols act as antioxidants in vitro by sequestering metal ions and by scavenging reactive oxygen and nitrogen species (Wiseman et al. 1997). The antioxidant activity of teas generally decreases in the following order: green > oolong > black (Roginsky et al. 2003) although some studies have shown that black teas have greater activity than green teas (Venditti et al. 2010), whereas others have reported the absence of any significant differences (Carloni et al. 2013).

Ranking of total phenolic content (TPC), 2,2-diphenyl-1-picrylhydrazyl (DPPH), radical-scavenging activity (RSA), ferric-reducing power (FRAP) and ferrous-ion-chelating (FIC) ability has been reported to be as follows: green > oolong ≈ black, green ≈ oolong > black, green > oolong > black, and black > green ≈ oolong, respectively (Chan et al. 2010). Findings from the latter study deviated from that of Yokozawa et al. (1998), who reported that the ranking of RSA was green > black > oolong. Green teas have been reported to have significantly higher TPC, RSA and FRAP, but poorer FIC ability, than black teas (Chan et al. 2007). The FRAP values and the TPC values have been reported to increase with increasing brewing temperature from 70 °C to 100 °C and with stirring at both temperatures for both unfermented (green) and fermented (black) tea leaves (Jayasekera et al. 2011). For DPPH, values increased with increasing brewing temperature and stirring for unfermented leaves, but there was a statistically significant interaction effect for unfermented leaves, such that the effect of temperature was dependent upon stirring (Kaur et al. 2014).

3 Antioxidant Properties of Tea: In Vitro and In Vivo Studies

Tea polyphenols act as antioxidants by scavenging reactive oxygen and nitrogen species and chelating redox-active transition metal ions. They may also act as indirect antioxidants by (1) inhibiting redox-sensitive transcription factors, nuclear factor $_{\kappa}$B and activator protein-1; (2) inhibiting pro-oxidant enzymes such as inducible nitric oxide synthase, lipoxygenase, cyclooxygenases and xanthine oxidase; and (3) induction of phase II and antioxidant enzymes, such as glutathione S-transferases and superoxide dismutases (Frei and Higdon 2003).

Numerous in vitro studies have shown that tea catechin is a potent hydrophilic antioxidant that scavenges oxygen radicals (Rice-Evans et al. 1995; Salah et al. 1995) and chelates metal ions. Among catechins, ECG has shown the strongest antioxidant activity in vitro (Katiyar et al. 1994). TFs and TRs present in black tea have been reported to show antioxidant properties in the same way as green tea containing ECs (Yoshino et al. 1994).

Antioxidative activity and the order of effectiveness of green tea polyphenols as radical scavengers are as follows: ECG > EGCG > EGC > gallic acid > EC = catechins. The oxidation of low-density lipoprotein is inhibited by catechins, EC, ECG and EGCT to a similar degree, but not as much as in the presence of EGC or gallic acid (Salah et al. 1995). The relative antioxidative capacity of black tea polyphenols in both aqueous and lipophilic phases is: theaflavin digallate (TFDG) > theaflavin 3^1-gallate ($TF3^1G$) = theaflavin 3-gallate (TF3G) > TF. Furthermore, research has shown that the efficacy of antioxidant activity of a reconstituted green tea polyphenol mixture is lower than that of the extracted crude mixture itself which indicates that other components in the mixture possess their own antioxidant activity or act synergistically with the catechins (Amarowicz and Shahidi 1995).

Flavonoid compounds with similar chemical structures exhibit comparable trends in antioxidant activity (Fukumoto and Mazza 2000). This activity usually increases with an increase in the number of hydroxyl groups and a decrease in glycosylation. Relative antioxidant effectiveness of tea polyphenols in an aqueous system has been found to be different from that tested in a lipid system (Aruoma 1997), which indicates that the activity of flavonoids is influenced by environmental factors. Some flavonoids have shown synergism when they are tested in a mixture (Aruoma 1997).

Green tea is known to have a high antioxidant potential because of its greater concentration of hydroxyl-rich polyphenols. TFs and TRs in black tea have also been shown to be a mixture of heterogeneous polyphenols (Yoshino et al. 1994). Black tea has shown similar or higher antioxidative activity than green tea in the inhibition of lipid peroxidation in rat liver (Yoshino et al. 1994). While both green and black teas are good sources of antioxidant polyphenols, more research is necessary to establish the mechanisms associated with antioxidant activity of black teas.

Liu et al. (2000) reported that catechin derivates extracted from green tea were effective in preventing peroxidation of low-density lipoprotein (LDL), via trapping the initiating and/or propagating peroxyl radicals.

In contrast to in vitro studies, several in vivo studies have produced conflicting data in relation to the antioxidant activity of tea. Serafini et al. (1996) reported that in men a single dose equivalent to 300 ml (2 cups) of green tea increases resistance against radical initiator-induced peroxidation in plasma at 30–50 min after ingestion. Ishikawa et al. (1997) found that daily consumption of 750 ml (5 cups) of black tea for 4 weeks lowered the susceptibility of LDL to oxidative modification. However, recently, Van het Hof et al. (1997) reported that daily consumption of 900 ml (6 cups) of green tea for 4 weeks had no effect on the resistance of LDL to oxidation. In this study, human plasma antioxidant activity was measured after overnight fasting (i.e. at least 12 hours after they consumed tea). Therefore, it may be that the effects of the catechins were not observed due to the catechins having been metabolised by the time the measures were made.

Villaño et al. (2012) demonstrated, through FRAP and total radical-trapping antioxidant parameter (TRAP) assays, that acute ingestion of oolong tea increased plasma antioxidant status in healthy subjects. They reported that the increase obtained with FRAP assay was mainly due to changes in ascorbic acid and with flavonoids exerting a minor effect. However, changes in plasma TRAP were found to be related to the direct action of the absorbed phenolics and their metabolites.

In summary, the limited amount of research that has been undertaken has not generally revealed significant decreases in biomarkers for oxidative damage in humans. Lifestyle factors and genetic variability in the response of humans to oxidative stress may mask small changes in biomarkers induced by tea polyphenols.

Antioxidant capacity of the plasma after the consumption of tea remains as the main piece of indirect evidence for the absorption of tea polyphenols through the gut barrier (Serafini et al. 1996; Van het Hof et al. 1999) and their likely physiological effects.

The mechanisms related to TF and TR absorption however are far from clear.

4 Effects of Production and Processing

Normally, fresh tea shoots are very rich in polyphenols, especially catechin and its derivatives. However, recent findings indicated that the processing method (e.g. drying time, the mechanical force applied and enzymatic reaction during the processing) accounts for the marked difference in the chemical composition of tea (Ravichandran and Parthiban 2000; Astill et al. 2001). Moreover, plant variety and growth conditions, which influence the chemical compositions of fresh tea leaves, would be expected to produce quite wide variations in the chemical compositions of the final products (Astill et al. 2001). The fermentation stages of oolong tea and black tea can greatly affect oxidation of the tea catechins (Cabrera et al. 2003; Ananingsih et al. 2013).

In green tea, fresh green tea leaves, which are very rich in catechins, are not fermented; they are withered (Anesini et al. 2008; Bancirova 2010). The leaves are typically steamed or roasted to inactivate enzymes like polyphenol oxidase (Gulati et al. 2003; Chan et al. 2007; Anesini et al. 2008; Bancirova 2010; Carloni et al. 2013). The tea leaves retain their green colour and almost all of their original polyphenol content (Chan et al. 2007). However, some oxidation still occurs during the withering process, and consequently, about 20% to 30% of total flavonoids in green tea may be changed to oxidised catechin polymers (Higdon and Frei 2003). The operations in green tea processing, i.e. steaming, roasting and rolling also resulted in the reduction of total catechin in the final product. Moreover, several chemical changes usually take place during the drying process, and these changes are promoted by driving force of heat rather than enzymatic reaction (Temple et al. 2001). Heat treatment has been identified as the major cause for catechin degradation during manufacturing of green tea. The heat treatments alter catechin content due to thermal degradation, oxidation, epimerisation and polymerisation (Ananingsih et al. 2013).

During the production of oolong tea, young green shoots (usually the top three leaves of each branch) are freshly harvested in the early morning and allowed to wither under the sunlight for a few hours (Dou et al. 2007). The tea leaves are allowed to ferment to a limited extent and contain a mixture of catechins, TFs and TRs (Chan et al. 2007). After that, the tea leaves are pan fired at ~200 °C, rolled to form a ball shape and then dried in a specialised oven at various desired temperatures. The reaction time for the contact between phenolic compounds and oxidative enzymes is empirically controlled by experts during this semi-fermentation process, and the final fermentation degree of oolong tea ranges from 20% to 80%, depending on the desired final product. The amount of TFs in oolong tea has been reported to be very low and mostly transformed into TRs after drying (Dou et al. 2007). The polyphenol profile of oolong tea is similar to that of green or black tea, but it has a lower content of EGCG, EC and ECG than that found in green tea because the fresh oolong tea leaves are subjected to a partial fermentation step before drying (Chen et al. 2010).

In the manufacturing of black tea, tea leaves are subjected to crushing and a full fermenting process (Anesini et al. 2008; Bancirova 2010). During fermentation of fresh tea leaves, the enzymatic oxidation by endogenous polyphenol oxidases and peroxidases is accelerated by rupturing the withered tea leaves using orthodox rollers or machines (CTC: crush-tear-curl) (Carloni et al. 2013). In this step, some catechins are oxidised or condensed to larger polyphenolic molecules (dimer or polymer) such as TFs and TRs (Almajano et al. 2008; Lin et al. 2003; Subramanian et al. 1999; Carloni et al. 2013). These polymers are responsible for black tea's bitter taste and dark colour (Almajano et al. 2008; Chan et al. 2007; Subramanian et al. 1999).

In addition to the three main categories of tea, white tea is the least processed tea. In this special kind of tea, tea leaves are subjected to only sun-withering and drying. Furthermore, white teas are characterised by the fact that only tea buds still covered with fine white hair and one or two very young tea leaves are used (Carloni et al. 2013). As mentioned above, in black tea and oolong tea, the enzymatic oxidation changes some amounts of catechins into TFs and TRs (Damiani et al. 2014). Therefore, among the three main types of commercial teas, green tea provides a higher amount of flavanols than oolong and black teas, and green tea shows higher antioxidant activity than other semi-fermented and fermented teas (Wang et al. 2000). Chan et al. (2007) reported that the greatest amount of catechins was found in green tea (26.7%), followed by oolong tea (23.2%) and black tea (4.3%). Carloni et al. (2013) studied the effect of manufacturing on antioxidant properties of teas from a single cultivar but processed differently, and their results indicated that antioxidant profile was in the following order: green ≥ low-caffeine green ˃ white ≥ black Orthodox ˃ typical black CTC. Lin et al. (2003) compared catechins in the same tea but manufactured by different fermentation processes, and they found that the levels of EGCG and total catechins were in the following order: green tea ˃ oolong tea ˃ fresh tea leaf ˃ black tea.

Chan et al. (2007) reported that green and black teas produced from var. *assamica* had higher polyphenol contents (30%) than those from var. *sinensis* (20%). Damiani et al. (2014) investigated the antioxidant characteristics of white teas steeped in either hot water or water at room temperature, and their results showed that preparing tea infused in water at room temperature for approximately 2 h could constitute an alternative tea beverage potentially richer in healthful bioactive compounds compared to the more commonly consumed hot tea infusions. Sharpe et al. (2016) showed that water temperature and brewing time were directly related to antioxidant capacity, while the time between rebrews (up to 1 h) and the use of tap or de-ionised water had no effect.

5 Environmental Factors

The suitable conditions for tea cultivation include suitable temperature (15–25 °C), high relative humidity (80–90%) and high annual rainfall (around 1500–2000 mm) (Li et al. 2007). Generally, tea is cultivated in the highlands. Especially, in India and Sri Lanka, it is planted at elevations up to 2000 m above sea level (Chan et al. 2007). However, it is well known that tea can grow in varied ecological zones and especially in the monsoon climate of tropics, from almost sea level to the highlands. Although tea can grow in different areas, the quality of tea leaf is attributed to changes in environmental conditions during growth (Ryu et al. 2017). The concentration of total polyphenols and their constituent compounds in tea is also influenced by various environmental factors such as climate, season, soil fertility and altitude (Jayasekera et al. 2011).

Evidently, some biochemical parameters of green tea have been reported to vary with the soil properties. Li et al. (2007) compared green tea from Qian-zhang-yan tea plantations (Q) and Deng-cun tea plantations (D). These two plantations are located in the main tea production regions in Yichang, Hubei Province, southwestern China. At Q, the soil is a little acidic and yellow. The Q is part of the Karst region, where the bedrock is dolomite. On the other hand, at D, the soil is an acidic sandy loam. The D is part of the non-Karst region, where the bedrock is granite. The soil moisture content at Q was only 11–16%, whereas that at D is 17–24%. The biochemical constituents, such as total tea polyphenols, caffeine, amino acids and proline for green tea from two different plantations are presented in Table 14.2. The tea polyphenol content of green tea from D was higher than that from Q, but the contents of amino acids and caffeine exhibited no significant differences. The L-proline content of green tea from Q was clearly higher than that from D.

Previous studies have assumed that the health benefits of green tea and its products are primarily associated with the secondary metabolites in tea, and the metabolites also play an important role in the quality of green tea. Moreover, it has been thought that metabolomic studies provide a new approach for monitoring changes in nutraceutical quality during tea plant growth and cultivation. Ryu et al. (2017) investigated the distribution and alteration of bioactive secondary metabolites of tea during different growth stages. Metabolite profiling of tea leaves cultivated during different growth stages (April, April to early May, mid-May to late May, early June and after July) was evaluated. Their results indicated significant changes in the contents of metabolites during growth. Especially, the flavonols

Table 14.2 Biochemical parameters in green tea from Qian-zhang-yan and Deng-cun tea plantations in Yichang, Hubei Province, PR China

Site	Polyphenols (%)	Caffeine (%)	Amino acids (%)	Proline (μg/g)
Q	31.04 ± 0.50	4.08 ± 0.22	4.27 ± 0.20	234 ± 12.3
D	34.06 ± 0.50	4.06 ± 0.20	4.21 ± 0.19	180 ± 10.2

Reproduced from Li et al. (2007) © Elsevier (2007); with permission from Elsevier
The results are mean values, $n = 5$, $P < 0.05$

and theaflavins increased gradually, whereas gallotannins and flavan-3-ols decreased continuously depending on the growth stage.

Jayasekera et al. (2011) investigated TPC, FRAP and DPPH of water extracts from unfermented and fermented tea leaves grown in different regions of Sri Lanka. The scope of their study included four highland plantations and two lower altitude plantations. For high-grown teas, samples were also collected from the two harvesting seasons (dry and monsoon). Their results demonstrated that there was a variation in antioxidant activity of the tea leaves from different six plantations (Table 14.3). FRAP and DPPH for both the fermented and unfermented teas from different plantations showed a statistically significant interaction between season and plantation. Moreover, a similar interaction between season and plantation was also observed for TPC in unfermented teas from the four highland plantations.

Jayasekera et al. (2014) further investigated variation in the contents of the catechins, flavonols and theaflavins in fresh and fermented tea leaves grown in different regions of Sri Lanka. The scope of their study included high-, medium- and low-altitude regions. Tea samples were also collected in both dry and monsoon seasons in the high-altitude regions. In this study, they found considerable variation in the phenolic profiles of Sri Lankan teas. A significant interaction between plantation and season was also observed for phenolic constituents in both unfermented and fermented teas. Their results suggest that attention needs to be paid to both plantation and season in order to obtain teas with higher catechin contents.

Seasonal variations of phenolic compounds in fresh tea shoots grown in Australia have also been studied by Yao et al. (2005). Three main tea flavanols, i.e. EGCG, ECG and EGC, and four grouped phenolics, i.e. total catechins (Cs), total catechin gallates (CGs), total flavanols (Fla) and total polyphenols (PPs), in fresh tea shoots were analysed and compared during the commercial harvest seasons from April 2000 to May 2001 (Table 14.4). The levels of EGCG, ECG and CGs in the fresh tea shoots were higher in the warm months (April and May 2000) and lower during the cool months (July to September 2000). After that, the levels increased throughout the warmer months (October to December 2000) and remained high until May 2001. Their results indicate that seasonal variations of phenolic compounds in tea shoots may be due to one or all three of the following environmental conditions, i.e. day length, sunlight and/or temperature.

Studies in Central Africa have shown that the concentration of flavonoids in fresh apical shoots of tea was highest during the cold season (Hilton and Palmer-Jones 1973). In contrast, a lower level of catechin gallates was reported for Australian tea harvested in the cooler months compared to that in tea harvested in the warmer months (Caffin et al. 2004). In the Northern Hemisphere, total flavanol content is greatest during the height of the summer. Studies in Japan have shown that the EGC level is higher in spring, while ECG and EGCG levels are higher in summer (Nakagawa and Torri 1964).

The influence of growing altitude and season of tea picking on the contents of catechins and phenolic compound of 'Huang Zhi Xiang' oolong tea have been reported by Chen et al. (2010). According to their results in Table 14.5, the contents of EGCG, CG and total catechins (TC) in the oolong teas grown at a high altitude

Table 14.3 Effect of plantation, season and plantation x season on mean (n = 6) FRAP (mmol/l)[a], DPPH (% inhibition)[b] and TPC (mg GAE/g dry leaves)[c] values for unfermented and fermented teas[d]

	Plantation			Season			SEM	Significance		
	Pedro	Laxapana	Greatwestern	Sarnia	Dry	Monsoon		Plantation	Season	Interaction
Unfermented										
FRAP	10.2	11.04	10.8	11.2	10.8	10.7	0.096	<0.001	NS	<0.001
DPPH	46.3	43.3	38.4	41.4	42.3	42.3	1.03	<0.05	NS	<0.001
TPC	170.72	173.66	169.92	179.48	172.8	174.16	2.24	<0.0001	0.008	<0.001
Fermented										
FRAP	10.4	10.8	9.9	9.7	10.9	9.5	0.125	<0.001	<0.0001	<0.0001
DPPH	44.8	41.1	33.5	44.6	44.1	44.1	1	<0.001	0.004	<0.05
TPC	171.9^{a}	169.2^{b}	157.5^{c}	153.6^{c}	169.6^{a}	156.5^{b}	2.01	<0.0001	<0.0001	NS

Reproduced from Jayasekera et al. (2011) © Elsevier (2011); with permission from Elsevier

[a]FRAP, ferric reducing antioxidant power

[b]DPPH, 2,2-diphenyl-1-picrylhydrazyl radical-scavenging activity

[c]TPC, total phenolic content. Means sharing the same letter in a row within plantation and season effects were not significantly different at $P < 0.05$

[d]Means (±SE) of triplicate measurements for two separate runs (n = 6)

Table 14.4 Mean content of phenolic compounds in hand-plucked fresh tea shoots

Harvest date	Individual and grouped polyphenols[a] (mg/g, dry basis)						
	EGCG[b]	ECG[c]	EGC[d]	Cs[e]	CGs[f]	Fla[g]	PPs[h]
April 12, 2000	120.52gh	34.50abcd	47.65defg	75.25bcd	163.75efg	239.00bc	275.29ab
May 9, 2000	128.63h	44.26f	51.79g	80.28def	183.83h	264.11e	304.73c
July 19, 2000	91.39a	35.16abcd	50.65fg	91.96h	132.30ab	224.25ab	264.22a
Aug 23, 2000	91.31a	31.56a	50.58efg	89.86gh	128.64a	218.49a	258.23a
Sept 23, 2000	96.12ab	33.51ab	50.28efg	87.20fgh	136.90ab	224.10ab	260.75a
Oct 23, 2000	100.82bc	34.29abc	52.33g	91.42h	143.17bc	234.60abc	274.64ab
Nov 13, 2000	103.09bcd	33.37ab	47.47defg	82.67efg	144.65bcd	227.32abc	264.61a
Dec 6, 2000	110.97def	36.10bcd	45.83cdef	78.98cde	156.63cdef	235.61abc	274.95ab
Dec 27, 2000	112.93efg	37.89cde	41.99bc	75.83bcde	159.91efg	235.74bc	275.47ab
Jan 15, 2001	113.65efg	37.48cde	45.63cde	82.69efg	161.16efg	243.85cd	286.22bc
Feb 1, 2001	121.41gh	40.62ef	46.45cdef	83.05efg	173.51gh	256.56de	300.14c
Feb 22, 2001	112.14defg	37.33cde	41.79bc	72.34abc	160.19efg	232.53abc	268.69ab
March 19, 2001	108.63cde	37.55cde	35.49a	66.21a	157.03def	223.24ab	262.88a
April 9, 2001	118.38fg	38.04de	39.82ab	71.29ab	167.36fg	238.65bc	277.53ab
May 2, 2001	104.72bcde	37.32cde	43.94bcd	76.61bcde	152.39cde	228.99abc	268.28ab
LSD[i]	9.39	3.71	5.00	7.39	13.68	17.21	19.88

[a]Means in columns followed by a common letter are not significantly different ($P > 0.05$)
[b]EGCG, epigallocatechin gallate
[c]ECG, epicatechin gallate
[d]EGC, epigallocatechin
[e]Cs, total catechins
[f]CGs, total catechin gallates
[g]Fla, total flavanols
[h]PPs, total polyphenols
[i]LSD, Fisher's least significant difference

were significantly higher than those grown at a low altitude for both of the spring and autumn teas. However, the contents of GC, EGC, C and simple catechins (SC) from a low altitude were significantly higher than those from a high altitude for the spring oolong tea. In contrast, the content of C in the autumn oolong tea grown at a

Table 14.5 Comparative analyses of contents of catechins and polyphenols of the 'Huang Zhi xiang' Oolong tea (mg g^{-1} DM) and QS[a]

Quality parameters	Spring Oolong tea Altitude: 500 m	Spring Oolong tea Altitude: 350 m	Autumn Oolong tea Altitude: 500 m	Autumn Oolong tea Altitude: 350 m	LSD ($P < 0.05$)
QS[b]	79.5 ± 0.8a	68.8 ± 1.4c	80.7 ± 0.9a	75.5 ± 2.7b	3.0
GC[c]	8.2 ± 0.5b	10.1 ± 0.9a	8.0 ± 0.1b	8.0 ± 0.7b	1.2
EGC[d]	16.8 ± 2.0b	20.3 ± 2.3a	17.6 ± 1.9ab	16.0 ± 0.1b	3.4
C[e]	5.9 ± 0.1bc	7.9 ± 1.2a	7.0 ± 0.1ab	5.6 ± 0.9c	1.4
EC[f]	4.2 ± 0.6a	5.4 ± 1.2a	5.5 ± 0.6a	4.7 ± 0.6a	1.5
EGCG[g]	102.8 ± 0.7a	76.6 ± 5.6c	103.4 ± 3.1a	90.6 ± 5.5b	8.0
ECG[h]	20.8 ± 3.1b	20.9 ± 3.1b	25.2 ± 1.3ab	25.6 ± 1.0a	4.4
TC[i]	158.3 ± 2.1ab	142.0 ± 11.4c	167.2 ± 2.5a	151.5 ± 5.5bc	12.3
SC[j]	36.1 ± 3.2b	44.0 ± 5.0a	38.6 ± 3.1ab	34.4 ± 1.0b	5.7
CG[k]	124.4 ± 1.3ab	97.4 ± 5.5c	129.6 ± 0.7a	116.9 ± 6.9b	8.4
PC[m]	176.2 ± 3.7c	132.4 ± 4.9d	200.2 ± 9.9b	217.1 ± 7.8a	13.2

[a]The experimental results were analyzed in one-way analysis of variance with Duncan's Multiple Range Test (DMRT) ranking – means ± S.D. within a row followed by the same letter are not significantly different at $P < 0.05$ according to DMRT
[b]QS, quality scores
[c]GC, (−)-gallocatechin
[d]EGC, (−)-epigallocatechin
[e]C, (+)-catechins
[f]EC, (−)-epicatechin
[g]EGCG, (−)-epigallocatechin gallate
[h]ECG, (−)-epigallocatechin
[i]TC, total catechins
[j]SC, simple catechins
[k]CG, catechin gallate
[m]PC, phenolic compounds

low altitude was significantly lower than that grown at a high altitude. In addition, the contents of EC and ECG were not significantly changed in the oolong tea samples that were grown at different altitudes. Their study suggests that the production of the 'Huang Zhi Xiang' oolong tea grown at low altitudes should utilise the autumn tea leaves, while the production of oolong tea grown at high altitudes can use both the spring and autumn tea leaves.

Fang et al. (2017) investigated the abundance of 30 compounds from 21 cultivars of tea from China grown in a single plantation (under the same environmental conditions) and harvested in spring (March), summer (May) and late summer (September) in 2013. They found that harvest time impacted the abundance of EGC, theanine and afzelechin gallate. These differences in chemistry between harvesting times will influence the flavour of the tea. Moreover, the alteration in phenolic compounds among different cultivars grown in the same environment was observed. This behaviour indicates that the synthesis and accumulation of the compounds are influenced by the genetic variation.

Black tea plain quality parameters (TFs, TRs, colour, brightness and sensory characteristics) have been reported to vary widely with geographical area of production (Owuor et al. 1986). Owuor et al. (1990) showed that there was an increase in the quality of tea with an increase in altitude, which was explained by the decrease in growth rate with the increase in altitude. Zhang et al. (2018) have reported higher polyphenol (catechins and theaflavins) contents for black tea from low plantation elevation than those from high elevation. They also reported that small leaves contained 15% more polyphenols than larger leaves from similar elevation. Hilton and Palmer-Jones (1973) showed that in Central Africa, the quality of tea was inversely related to the growth rate of the tea bush. The impact of the plant's growth rate in Central Africa was also reflected in the tea's polyphenolic content. For the fast-growing tea, the polyphenolic content was lower and so was the TF content after fermentation. Similar observations have been made for tea produced in Himachal Pradesh in India (Gulati and Ravindranath 1996). Sud and Baru (2000) also reported a lower level of TF and TR in black tea produced in the rainy season in Himachal Pradesh. The decreased TF and TR contents, however, were partly attributed to a low degree of withering and high chlorophyll content in the rainy season. They concluded that when the more hot dry air was circulated through the withering troughs to increase the evaporation of leaf moisture, the quality of tea also increased.

Previous studies have assumed that large differences in climate are necessary for significant quality differences to be observed. It has been thought that a superior genotype selected in one location maintains its desirable attributes despite location. However, tea plants selected in one location and planted in other locations have usually not matched the performance at the site of selection (Wachira et al. 2002). One reason for such differences has been altitude, which affects rates of growth, even when other agronomic/cultural practices are similar. From the results obtained in Kenya, where trials were conducted within a 10-km radius, making climatic variations minimal or nearly constant (Owuor et al. 1990; Ng'etich et al. 2001), it was assumed that yields and quality will change with the altitude in a predictable manner, though not necessarily at the same rates for different cultivars. Consequently, a genotype with desirable attributes shall maintain these attributes relative to other genotypes, wherever they are grown within the same country. However, there are tea genotypes, which are more resistant to yield variations with the location of production (Wachira et al. 2002). Owuor et al. (2010) assessed 20 commercial genotypes under identical management in 3 locations within Kenya and revealed significant differences in plain black tea quality parameters. Their results suggested that it is not possible to produce tea of the same quality even from the same cultivars when the production location is varied.

The production of black tea from the same vegetatively propagated cultivars in Kenya and Malawi has shown variations in both chemical composition and quality (Owuor et al. 2008). Whereas it is possible to produce black teas with similar total TFs and individual TFs, brightness and total colour levels, black teas from Kenya generally have higher TRs, total volatile flavour compounds and flavour index. However, the pattern in the changes in the individual TFs or theaflavin digallate equivalent followed that of total (Flavognost) TFs, suggesting that the flavan-3-ol patterns in tea leaf might not have been affected by the geographical area of production. The variations in black tea quality between Malawi and Kenya were attributed to differences in environmental conditions, leading to different shoot growth rates and biochemical composition in the shoots (Owuor et al. 2008).

6 Concluding Remarks

Tea is now viewed as a product containing physiologically active molecules with antioxidant properties. The experimental evidence generated to date suggests that consumption of tea or tea polyphenols results in modest transient increases in plasma antioxidant capacity in humans. However, the polyphenolic compounds present in tea have been reported to vary considerably with the source of tea, the production location, the season of tea production and tea processing, ultimately affecting the antioxidant properties of tea. More studies are, however, needed to explore the interactions between various environmental/husbandry factors and how these interactions affect the tea polyphenols and overall antioxidant quality of different teas.

Acknowledgements Permission from Elsevier to reproduce parts of the chapter by Kaur et al. (2014) is gratefully acknowledged. The authors would also like to thank Shiromani Jayasekera and Distinguished Professor Paul Moughan from the Riddet Institute for their valuable contributions.

References

Almajano MP, Carbó R, Jiménez JAL, Gordon MH (2008) Antioxidant and antimicrobial activities of tea infusions. Food Chem 108(1):55–63. https://doi.org/10.1016/j.foodchem.2007.10.040

Amarowicz R, Shahidi F (1995) Antioxidant activity of green tea catechins in a β-carotene-linoleate model system. J Food Lip 2:47–56

Ananingsih VK, Sharma A, Zhou W (2013) Green tea catechins during food processing and storage: a review on stability and detection. Food Res Int 50(2):469–479. https://doi.org/10.1016/j.foodres.2011.03.004

Anesini C, Ferraro GE, Filip R (2008) Total polyphenol content and antioxidant capacity of commercially available tea (Camellia sinensis) in Argentina. J Agric Food Chem 56(19):9225–9229. https://doi.org/10.1021/jf8022782

Aruoma OI (1997) Antioxidant and pro-oxidant actions of plant extracts and plant-derived compounds. Proc Nutr Soc Aust 21:85–90

Astill C, Birch MR, Dacombe C, Humphrey PG, Martin PT (2001) Factors affecting the caffeine and polyphenol contents of black and green tea infusions. J Agric Food Chem 49(11):5340–5347. https://doi.org/10.1021/jf010759+

Bancirova M (2010) Comparison of the antioxidant capacity and the antimicrobial activity of black and green tea. Food Res Int 43(5):1379–1382. https://doi.org/10.1016/j.foodres.2010.04.020

Bazinet L, Araya-Farias M, Doyen A, Trudel D, Têtu B (2010) Effect of process unit operations and long-term storage on catechin contents in EGCG-enriched tea drink. Food Res Int 43(6):1692–1701. https://doi.org/10.1016/j.foodres.2010.05.015

Biswas AK, Sarkar AR, Biswas AK (1973) Biological and chemical factors affecting the valuations of North-East Indian plain teas. III. Statistical evaluation of the biochemical constituents and their effects on colour, brightness and strength of black teas. J Sci Food Agric 24:1457–1477

Bryce T, Collier PD, Fowlis I, Thomas PE, Frost D, Wilkins CK (1970) The structures of the theaflavins of black tea. Tetrahedron Lett 32:2789–2792

Cabrera C, Giménez R, López MC (2003) Determination of tea components with antioxidant activity. J Agric Food Chem 51(15):4427–4435. https://doi.org/10.1021/jf0300801

Caffin N, D'Arcy B, Yao L, Rintoul G (2004) Developing and index of quality for Australian Tea Rural Industries Research and Development Corporation Publication No:04/033

Carloni P, Tiano L, Padella L, Bacchetti T, Customu C, Kay A, Damiani E (2013) Antioxidant activity of white, green and black tea obtained from the same tea cultivar. Food Res Int 53(2):900–908. https://doi.org/10.1016/j.foodres.2012.07.057

Chan EWC, Lim YY, Chew YL (2007) Antioxidant activity of Camellia sinensis leaves and tea from a lowland plantation in Malaysia. Food Chem 102(4):1214–1222. https://doi.org/10.1016/j.foodchem.2006.07.009

Chan EWC, Lim YY, Chong KL, Tan JBL, Wong SK (2010) Antioxidant properties of tropical and temperate herbal teas. J Food Com Anal 23:185–189

Chen Y, Jiang Y, Duan J, Shi J, Xue S, Kakuda Y (2010) Variation in catechin contents in relation to quality of 'Huang Zhi Xiang' Oolong tea (Camellia sinensis) at various growing altitudes and seasons. Food Chem 119(2):648–652. https://doi.org/10.1016/j.foodchem.2009.07.014

Coxon DT, Holmes A, Ollis WD, Vora VC (1970) The constitution and configuration of the theaflavin pigments of black tea. Tetrahedron Lett 60:5237–5240

Dalluge JJ, Nelson BC (2000) Determination of tea catechins. J Chromatogr A 881(1):411–424. https://doi.org/10.1016/S0021-9673(00)00062-5

Damiani E, Bacchetti T, Padella L, Tiano L, Carloni P (2014) Antioxidant activity of different white teas: comparison of hot and cold tea infusions. J Food Composit Anal 33(1):59–66. https://doi.org/10.1016/j.jfca.2013.09.010

Dou J, Lee VSY, Tzen JTC, Lee M-R (2007) Identification and comparison of phenolic compounds in the preparation of oolong tea manufactured by semifermentation and drying processes. J Agric Food Chem 55(18):7462–7468. https://doi.org/10.1021/jf0718603

Dreosti IE (2000) Antioxidant polyphenols in tea, cocoa, and wine. Nutrition 16(7):692–694. https://doi.org/10.1016/S0899-9007(00)00304-X

Dufresne CJ, Farnworth ER (2001) A review of latest research findings on the health promotion properties of tea. J Nutr Biochem 12:404–421

Fang R, Redfern SP, Kirkup D, Porter EA, Kite GC, Terry LA, Berry MJ, Simmonds MSJ (2017) Variation of theanine, phenolic, and methylxanthine compounds in 21 cultivars of Camellia sinensis harvested in different seasons. Food Chem 220:517–526. https://doi.org/10.1016/j.foodchem.2016.09.047

Frei B, Higdon J (2003) Antioxidant activity of tea polyphenols *in vivo*: evidence from animal studies. J Nutr 133:3275–3284

Fukumoto LR, Mazza G (2000) Assessing antioxidant and pro-oxidant activities of phenolic compounds. J Agric Food Chem 48:3597–3604

Green RJ, Murphy AS, Schulz B, Watkins BA, Ferruzzi MG (2007) Common tea formulations modulate *in vitro* digestive recovery of green tea catechins. Mol Nutr Food Res 51(9):1152–1162. https://doi.org/10.1002/mnfr.200700086

Gulati A, Ravindranath SD (1996) Seasonal variations in quality of Kangra Tea (Camellia sinensis(L) O Kuntze) in Himachal Pradesh. J Sci Food Agric 71:231–236

Gulati A, Rawat R, Singh B, Ravindranath SD (2003) Application of microwave energy in the manufacture of enhanced-quality green tea. J Agric Food Chem 51(16):4764–4768. https://doi.org/10.1021/jf026227q

Higdon JV, Frei B (2003) Tea catechins and polyphenols: health effects, metabolism, and antioxidant functions. Crit Rev Food Sci Nutr 43(1):89–143. https://doi.org/10.1080/10408690390826464

Hilton PJ (1973) Tea. In: Snell FD, Ettre LS (eds) Encyclopedia of industrial chemical analysis, vol 18. Wiley, New York, pp 455–518
Hilton PJ, Palmer-Jones R (1973) Relationship between the flavanol composition of fresh tea shoots and the theaflavin content of manufactured tea. J Sci Food Agric 24:813–818
Ishikawa T, Szukawa M, Ito T, Yoshida H, Ayaori M, Nishiwaki M, Yonemura A, Hara Y, Nakamura H (1997) Flavonoid supplementation on the susceptibility of low-density lipoprotein to oxidative modification. Am J Clin Nutr 66:261–266
Jayasekera S, Molan AL, Garg M, Moughan PJ (2011) Variation in antioxidant potential and total polyphenol content of fresh and fully-fermented Sri Lankan tea. Food Chem 125(2):536–541. https://doi.org/10.1016/j.foodchem.2010.09.045
Jayasekera S, Kaur L, Molan A-L, Garg ML, Moughan PJ (2014) Effects of season and plantation on phenolic content of unfermented and fermented Sri Lankan tea. Food Chem 152:546–551. https://doi.org/10.1016/j.foodchem.2013.12.005
Katiyar SK, Agarwal R, Mukhtar H (1994) Inhibition of spontaneous and photoenhanced lipid peroxidation in mouse epidermal microsomes by epicatechin derivatives from green tea. Cancer Lett 79:61–66
Kaur L, Jayasekera S, Moughan PJ (2014) Chapter 13: Antioxidant quality of tea (Camellia sinensis) as affected by environmental factors A2 – Preedy, Victor. In: Processing and impact on antioxidants in beverages. Academic, San Diego, pp 121–129. https://doi.org/10.1016/B978-0-12-404738-9.00013-1
Kerio LC, Wachira FN, Wanyoko JK, Rotich MK (2013) Total polyphenols, catechin profiles and antioxidant activity of tea products from purple leaf coloured tea cultivars. Food Chem 136:1405–1413
Labbé D, Tremblay A, Bazinet L (2006) Effect of brewing temperature and duration on green tea catechin solubilization: basis for production of EGC and EGCG-enriched fractions. Sep Purifi Technol 49(1):1–9. https://doi.org/10.1016/j.seppur.2005.07.038
Li T, Yu L-J, Li M-T, Li W (2007) Comparative studies on the qualities of green teas in Karst and non-Karst areas of Yichang, Hubei Province, PR China. Food Chem 103(1):71–74. https://doi.org/10.1016/j.foodchem.2006.07.027
Liang Y, Lu J, Zhang L, Wu S, Wu Y (2005) Estimation of tea quality by infusion colour difference analysis. J Sci Food Agric 85(2):286–292. https://doi.org/10.1002/jsfa.1953
Lin Y-S, Tsai Y-J, Tsay J-S, Lin J-K (2003) Factors affecting the levels of tea polyphenols and caffeine in tea leaves. J Agric Food Chem 51(7):1864–1873. https://doi.org/10.1021/jf021066b
Liu Z, Ma LP, Zhou B, Yang L, Liu ZL (2000) Antioxidative effects of green tea polyphenols on free radical initiated and photosensitized peroxidation of human low density lipoprotein. Chem Phys Lipids 106:53–63
Modder WWD, Amarakoon AMT (2002) Tea and health. The Tea Research Institute, Colombo
Nakagawa M, Torri H (1964) Studies on the flavonols in tea. Part II. Variation in the flavonolic constituents during the development of tea leaves. Agric Biol Chem 28:497–504
Ng'etich WK, Stephens W, Othieno CO (2001) Responses of tea to environment in Kenya. 3. Yield and yield distribution. Exp Agric 37:361–372
Owuor PO, Tsushida T, Horita H, Murai T (1986) Comparison of the chemical composition of black teas from main tea producing parts of the world. Tea 7:71–78
Owuor PO, Obaga SO, Othieno CO (1990) Effects of altitude on the chemical composition of black tea. J Sci Food Agric 50:9–17
Owuor PO, Obanda M, Nyirenda HE, Mandala WL (2008) Influence of region of production on clonal black tea chemical characteristics. Food Chem 108:263–271
Owuor PO, Wachira FN, Ng'etich WK (2010) Influence of region of production on relative clonal plain tea quality parameters in Kenya. Food Chem 119:1168–1174
Ravichandran R, Parthiban R (2000) Lipid occurrence, distribution and degradation to flavour volatiles during tea processing. Food Chem 68(1):7–13. https://doi.org/10.1016/S0308-8146(99)00143-0

Rice-Evans CA, Miller NJ, Bolwell PG, Broamley PM, Pridham JB (1995) The relative antioxidant activities of plant-derived polyphenolic flavonoids. Free Rad Res 22:375–383
Roginsky V, Barsukova T, Hsu CF, Kilmartin PA (2003) Chain-breaking antioxidant activity and cyclic voltammetry characterization of polyphenols in a range of green, oolong, and black teas. J Agric Food Chem 51:5798–5802
Ryu HW, Yuk HJ, An JH, Kim D-Y, Song H-H, Oh S-R (2017) Comparison of secondary metabolite changes in Camellia sinensis leaves depending on the growth stage. Food Control 73:916–921. https://doi.org/10.1016/j.foodcont.2016.10.017
Salah N, Miller NJ, Paganga G, Tijburg L, Bolwell GP, Rice-Evans CA (1995) Polyphenolic flavanols as scavengers of aqueous phase radicals and as chain-breaking antioxidants. Arch Biochem Biophys 322:339–346
Serafini M, Ghiselli A, Ferro-Luzzi A (1996) *In vivo* antioxidant effect of green and black tea in man. Eur J Clin Nutr 50:28–32
Sharpe E, Hua F, Schuckers S, Andreescu S, Bradley R (2016) Effects of brewing conditions on the antioxidant capacity of twenty-four commercial green tea varieties. Food Chem 192:380–387. https://doi.org/10.1016/j.foodchem.2015.07.005
Stewart AJ, Mullen W, Crozier A (2005) On-line high-performance liquid chromatography analysis of the antioxidant activity of phenolic compounds in green and black tea. Mol Nutr Food Res 49:52–60
Subramanian N, Venkatesh P, Ganguli S, Sinkar VP (1999) Role of polyphenol oxidase and peroxidase in the generation of black tea theaflavins. J Agric Food Chem 47(7):2571–2578. https://doi.org/10.1021/jf981042y
Sud RG, Baru A (2000) Seasonal variations in theaflavins, thearubigins, total colour and brightness of Kangra orthodox tea (*Camellia sinensis* (L) O Kuntze) in Himachal Pradesh. J Sci Food Agric 80:1291–1299
Sur S, Panda CK (2017) Molecular aspects of cancer chemopreventive and therapeutic efficacies of tea and tea polyphenols. Nutrition 43–44:8–15
Temple SJ, Temple CM, Boxtel AJBV, Clifford MN (2001) The effect of drying on black tea quality. J Sci Food Agric 81(8):764–772. https://doi.org/10.1002/jsfa.881
Van het Hof KH, De Boer HSM, Wiseman SA, Lien N, Westsrtate JA, Tijburg LBM (1997) Consumption of green or black tea does not increase resistance of low density lipoprotein to oxidation in humans. Am J Clin Nutr 66:1125–1132
Van het Hof KH, Wiseman SA, Yang CS, Tijburg LB (1999) Plasma and lipoprotein levels of tea catechins following repeated tea consumption. Proc Soc Exp Biol Med 220:203–209
Venditti E, Bacchetti T, Carloni P, Greci L, Damiani E (2010) Hot vs. cold water steeping of different teas: do they affect antioxidant activity? Food Chem 119:1597–1604
Villaño D, Lettieri-Barbato D, Guadagni F, Schmid M, Serafini M (2012) Effect of acute consumption of oolong tea on antioxidant parameters in healthy individuals. Food Chem 132:2102–2106
Vinson JA, Dabbagh YA, Serry MM, Jang J (1995) Plant flavonoids, especially tea flavonols, are powerful antioxidants using an *in vitro* oxidation model for heart disease. J Agric Food Chem 43(11):2800–2802. https://doi.org/10.1021/jf00059a005
Wachira FN, Ng'etich W, Omolo J, Mamati G (2002) Genotype x environment interactions for tea yields. Euphtica 127:289–296
Wang L-F, Kim D-M, Lee CY (2000) Effects of heat processing and storage on flavanols and sensory qualities of green tea beverage. J Agric Food Chem 48(9):4227–4232. https://doi.org/10.1021/jf0003597
Wickremasinghe RL (1978) Monographs on tea production in Sri Lanka, vol 7. Facets of Tea Research in Practice, TRI, Colombo
Wiseman SA, Balentine DA, Frei B (1997) Antioxidants in tea. Crit Rev Food Sci Nutr 37:705–708
Yao L, Caffin N, D'Arcy B, Jiang Y, Shi J, Singanusong R, Liu X, Datta N, Kakuda Y, Xu Y (2005) Seasonal variations of phenolic compounds in Australia-grown tea (Camellia sinensis). J Agric Food Chem 53(16):6477–6483. https://doi.org/10.1021/jf050382y

Yokozawa T, Dong E, Nakagawa T, Kashiwagi H, Nakagawa H, Takeuchi S, Chung HY (1998) *In vitro* and *in vivo* studies on the radical scavenging activity of tea. J Agric Food Chem 46:2143–2150

Yoshino K, Hara Y, Sano M, Tomita I (1994) Antioxidative effects of black tea theaflavins and thearubigin on lipid peroxidation of rat liver homogenates induced by tert-butyl hydroperoxide. Biol Pharm Bull 17:146–149

Zhang C, Suen CL-C, Yang C, Quek SY (2018) Antioxidant capacity and major polyphenol composition of teas as affected by geographical location, plantation elevation and leaf grade. Food Chem 244:109–119

Zuo Y, Chen H, Deng Y (2002) Simultaneous determination of catechins, caffeine and gallic acids in green, oolong, black and pu-erh teas using HPLC with a photodiode array detector. Talanta 57:307–316

Chapter 15
Toward the Implementation of Climate-Resilient Tea Systems: Agroecological, Physiological, and Molecular Innovations

Selena Ahmed

Abstract Climate change is one of the most pressing issues for tea systems with major implications for farmer livelihoods. Changes in climate variables outside of the thresholds for tea production can impact both tea quality and productivity. It is thus critical to implement strategies for minimizing the forecasted trajectory of climate change and its impacts on tea systems, the environment, and society. Tea systems, as other agricultural systems, face two major challenges in the context of climate change. The first climate challenge is the mitigation challenge to reduce the negative impacts of tea systems throughout the entire value chain (from farm to cup to waste) that contribute to climate change. The second climate challenge for tea systems is the adaptation challenge which involves overcoming the effects of climate change through enhancing the resilience of tea systems. This chapter provides an overview of agricultural, physiological, and molecular innovations at the production level toward the development of climate-resilient tea systems. In addition, this chapter highlights priority areas for research and development including the cost-effectiveness, replicability, and adaptability of various good agricultural practices (GAPs) for climate mitigation and adaptation. Some climate adaptation innovations in the tea system also serve as climate mitigation strategies such as tea agroforestry and maintaining trees in tea farms; such practices should be further examined and promoted for replicability. Ultimately, multi-sectoral collaboration between governments, industry, farmers, and researchers is called for to tackle the issue of climate change in tea systems including removing barriers for the adoption of climate innovations. It is expected that such multi-sectoral collaboration will allow us to more effectively support farmer livelihoods and well-being while meeting consumer demand and maintaining stable tea systems.

Keywords Climate adaptation · Climate mitigation · Agroecology · Good agricultural practices (GAPs) · Agricultural diversification · Sustainability

---

S. Ahmed (✉)
Food and Health Lab, Sustainable Food and Bioenergy Systems Program, Montana State University, Bozeman, MT, USA
e-mail: selena.ahmed@montana.edu

W.-Y. Han et al. (eds.), *Stress Physiology of Tea in the Face of Climate Change*,
https://doi.org/10.1007/978-981-13-2140-5_15

1 Introduction

In a world of growing uncertainty and variability, the security of tea systems is threatened by multiple social, cultural, political, economic, health, and environmental factors associated with global change including climate change, urbanization, decreasing farm labor, market shifts, and political instability. Tea systems refer to the complex interconnected web of activities, resources, and people encompassing all aspects involved in cultivating, harvesting, processing, packaging, distributing, marketing, and preparing tea for human consumption while maintaining the ecological integrity of natural resources. Climate change is one of the most pressing issues for tea and other agricultural systems because it exacerbates many of the other sustainability challenges faced by these systems (Herforth et al. 2017; Medina et al. 2014; Springmann et al. 2016; Myers et al. 2017). Sustainability refers to the capacity of a system to provide nourishment for human needs in the long term while maintaining the ecological integrity of natural resources or "preserving the capacity of a system to function in the future" (Tendall et al. 2015).

Tea systems, from the production of tea to its consumption, are increasingly exposed to climate shocks ranging from sudden shocks such as extreme drought and rainfall to long-term stressors such as increased temperatures (Wijeratne 1996; Ahmed et al. 2014a, b; Boehm et al. 2016; Li et al. 2017). These climate factors impact tea systems in variable ways including increased soil erosion, decreased leaf quality, reduced resilience of tea crops, and increased vulnerability of tea farmers through decreased livelihoods (Ahmed et al. 2014a, b; Kowalsick et al. 2014; International Trade Centre 2014). Climate factors also exacerbate other socioeconomic shocks faced by tea systems including decreased farm labor, disruptions to value chains, political crises, shifts in consumer demand, and increased price volatility. Ultimately, all these stressors increase the vulnerability of tea systems to overall shocks from global change with major implications for farmer livelihoods and the quality and quantity of tea supply. Tea systems, as other agricultural systems, face two major challenges in the context of climate change: the mitigation challenge and the adaptation challenge (Pachauri et al. 2014).

1.1 Climate Mitigation

Human activities are responsible for the large increase in atmospheric greenhouse gasses over the last 150 years (Pachauri et al. 2014). The electricity, agricultural, industry, transportation, commercial, and residential sectors all notably contribute to greenhouse gas emissions. For example, it has been noted that the agricultural sector contributes around 19–29% of global anthropogenic greenhouse gas emissions (Vermeulen et al. 2012). Scholars and global policy makers have proposed that it is necessary to reduce greenhouse gas emissions notably by 2030 in order to achieve the global goal of limiting warming to 2° Celsius to mitigate detrimental impacts to socio-ecological systems (Wollenberg et al. 2016).

The climate mitigation challenge for tea systems involves reducing the negative impacts of tea systems throughout the entire value chain, from production through consumption and waste, in contributing to climate change. The main greenhouse gas emissions contributing to climate change are carbon dioxide and methane (Pachauri et al. 2014). Carbon dioxide is produced in the tea value chain during activities that involve, or rely on, the burning of fossil fuels and deforestation. Methane is produced from cattle as well as during the decomposition of organic waste. In addition, the application of fertilizers in agricultural systems also emits the greenhouse gas nitrous oxide. Sources of greenhouse gas emissions at the farm level in tea systems include application of fertilizer, land clearing, tree cutting, burning of biomass, reduced vegetation cover, and raising livestock for manure and food (International Trade Centre 2014). At the factory level where tea processing occurs, key sources of greenhouse gas emissions include using fossil fuels (including gas, coal, and oil) as well as fuel wood for processing and drying tea along with electricity for running tea processing equipment (International Trade Centre 2014). Other sources of greenhouse gas emissions in the tea system include utilization of fossil fuels for tea transportation and distribution, including over thousands of miles from tea-producing countries to countries that consume tea, but do not produce this botanical.

Climate mitigation activities are those that reduce greenhouse gas emissions in the earth's atmosphere that are responsible for climate change (Pachauri et al. 2014). In addition to activities that reduce greenhouse gas emissions through the tea value chain, climate mitigation strategies can involve carbon neutral activities that increase soil carbon storage during production. Tea production systems can serve as carbon sinks to remove greenhouse gasses from the atmosphere and store them in the soil as an emission sink through managing for rich soil organic matter along with the planting of trees and maintenance of dense vegetation (International Trade Centre 2014). Managing for rich soil organic matter is critical for mitigation on tea farms as soil is an emission sink that stores more carbon than all vegetation and the atmosphere combined (International Trade Centre 2014).

1.2 Climate Adaptation

The second challenge for tea systems is the adaptation challenge. Climate adaptation is variably defined including as "adjustment in natural or human systems in response to actual or expected climatic stimuli or their effects, which moderates harm or exploits beneficial opportunities" (Pachauri et al. 2014) and "activities that help to manage the social, environmental and economic impacts of climate change" (International Trade Centre 2014). Climate adaptation involves overcoming the effects of climate change through enhancing the resilience and adaptive capacity of tea systems using various agricultural, physiological, and molecular innovations. Adapting to climate change is critical for tea systems as the growth of tea is temperature-dependent, with a lack of tea growth when temperatures are either too low or too high outside of the thresholds of specific tea varieties, regardless of other climatic factors (Lemmesa 1996). Tea is also vulnerable to other climate variables including precipitation, water stress, solar radiation, and carbon dioxide; changes in the environmental variables outside of the thresholds for healthy tea production can impact both quality and productivity (see Chap. 1 by Ahmed et al. in this book). Some climate adaptation innovations in the tea system also serve as climate mitigation strategies. Additionally, many of these climate mitigation and adaptation strategies provide other environmental and societal benefits such as reduced cost of operations, improved health, reduced waste, enhanced working environment, and heightened marketability of products and profits (International Trade Centre 2014).

The concepts of resilience and adaptive capacity are increasingly used to strengthen the response of agriculture systems to climate change while reducing vulnerability (FAO 2013; Choptiany et al. 2014; Tendall et al. 2015). Resilience has been variably defined including the following definitions: "the capacity of a system to absorb disturbance and re-organize while undergoing change so as to maintain the same function, structure, identity and feedbacks" (Walker et al. 2004) and "the dynamic capacity to continue to achieve goals despite disturbances and shocks" (Tendall et al. 2015). Adaptive capacity has been defined as "the ability of ecological and social systems to adapt to environmental changes" (Gunderson and Holling 2002). Focusing on strengthening the resilience of agricultural systems emphasizes interactions of nonlinear dynamics, thresholds, and uncertainty during gradual and rapid change across temporal and spatial scales (Folke 2006). A loss of resilience and adaptive capacity of agricultural systems such as tea production systems indicates vulnerable systems (Folke 2006). The concept of resilience is often confused with the concept of sustainability. However, it is important to note that resilience is a complementary concept to sustainability. Resilience implies the capacity of a system to continue to provide a function over time despite shocks and change (Tendall et al. 2015). Thus, strengthening the resilience of a system implies advancing its sustainability through enhancing adaptive capacity to respond to environmental changes (Rees 2010).

Building the resilience of tea systems involves a holistic and long-term perspective that integrates multiple scales, levels, and approaches (Tendall et al. 2015). Since climate change is a long-term and unpredictable process, it is important to approach climate mitigation and adaptation as a continuous, flexible, and long-term process with different timeframes for implementation that plan for potential future risks (Dinesh et al. 2017). This involves the need to continuously design and implement strategies that are suitable for different periods of time in response to a continually changing environment (Dinesh et al. 2017). For example, sectors within the tea industry can adapt to their new realities through improved management practices including agroecological production methods that facilitate farmers' ability to respond to climate risk while preserving local biological diversity. At the same time, improved technologies including water harvesting and precision irrigation can be implemented in response to drought shocks while conserving water resources. Molecular technologies can also enable the development of new tea varieties with enhanced resistance to pests, disease, and extreme climate events. This chapter provides an overview of agricultural, physiological, and molecular innovations toward the development of climate-resilient tea systems that support farmer livelihoods and consumer demand for high-quality tea.

2 Agroecological Strategies for Climate Resilience

There are many ways in which humans manage agricultural systems for the cultivation of crops that vary along a continuum of intensification (Ahmed et al. 2015). Climate change is calling for increased management attention in agricultural systems with heightened pressure resulting from variability in water supply, soil fertility, weeds, pests, and disease. Management factors in agricultural systems comprise of cultural practices, perceptions, values, resource rights, and regulations (Ahmed et al. 2015). An agroecological approach has been promoted as a way to adapt to and mitigate climate change by implementing agricultural production strategies that mimic or augment natural processes through the incorporation of plants, animals, insects, people, and natural resources as well as their dynamic interactions (Altieri et al. 2015; Jiggins 2014).

Unlike modern industrial agriculture that relies on agricultural chemical input and intensification, an agroecological approach seeks to more closely resemble naturally functioning ecosystems through emphasizing complex ecological interactions capable of providing ecosystem services for the management of soil fertility, productivity, pollination, pest and disease control, and microclimate regulation (Altieri 1995). Agroecological management practices focus on recycling nutrients and energy on a farm, diversifying species and genetic resources spatially and temporally, and focusing on interactions and productivity across the agricultural system rather than on individual species (De Schutter 2012). An agroecological approach for building resilience to climate risk emphasizes interactions among ecological, economic, and social dimensions of agricultural systems (Altieri 1995) and the entire food system (Francis et al. 2008; Gliessman 2015).

However, in most modern-day agricultural systems, energy and nutrients are directed out of the system at harvest with external inputs being required to keep these systems functioning (Ahmed et al. 2015). Managers and laborers working in agricultural systems must input energy and nutrients back into the system through management practices including the application of fertilizer in order to achieve productivity. Historically, agricultural inputs included human and animal labor until the advent of modern agriculture in the 1930s when inputs shifted to an inclusion of fertilizer, pesticides, growth hormones, and, indirectly, fossil fuel to increase productivity. Many of the agricultural management practices that are being implemented in response to the pressures of global change such as the addition of chemical fertilizers, pesticides, and herbicides contribute to environmental externalities including air and water pollution, soil depletion, loss of biodiversity, pressures on limited fossil fuels, and emissions of greenhouse gasses (Francis et al. 2008; Gliessman 2015). For example, extensive use of chemical agricultural inputs can degrade soil organic matter on tea farms which results in a cascade of effects including decreasing water-holding capacity and fertility of soils that ultimately degrade the integrity of tea production systems as well as surrounding landscapes. At the same time, tea enterprises, particularly those of smallholder tea farmers, may lack the purchasing power for adaptation practices that involve high-cost input materials including fertilizers, pesticides, and herbicides.

2.1 Good Agricultural Practices (GAPs) for Climate Mitigation and Adaptation

Investing in agroecological strategies is recognized as an effective way of building climate resilience in tea production systems as they support both environmental and human well-being (Ahmed and Stepp 2016) without high-cost input materials. Multiple agroecological management practices have been identified in the literature for climate resilience while overcoming the environmental externalities associated with chemical agricultural inputs. These practices may also be known as good agricultural practices (GAPs) for climate mitigation and adaptation and include the following (as summarized in Ahmed and Stepp 2016):

1. *Agricultural diversification* including agroforestry, crop rotations, mixed cropping, landscape mosaics, polycultures, cover cropping, and maintenance of diverse landraces (Altieri et al. 2015; Schwendenmann et al. 2010; Lin 2011; Howden et al. 2007; Fraser 2007; Di Falco and Perrings 2003)
2. *Tree planting and maintaining vegetative cover* including planting hedges as windbreaks, planting shade trees, using cover crops, and preventing deforestation (Bhattarai et al. 2015; Kotecký 2015)
3. *Management of soil organic matter and carbon sequestration* through no-tillage (Fuhrer and Chervet 2015; Lal 2004), mixed crop-livestock systems (Thornton and Herrero 2014), organic agriculture (Lotter 2003), manure management, mulching, cover crops, good agricultural practices, and integrated nutrient management (INM)
4. *Water management* through improvement of water allocation or irrigation efficiency (Chartzoulakis and Bertaki 2015), precision agricultural management (Lal et al. 2011), rainwater harvesting (Pandey et al. 2003), cover cropping with nurse plants or other plants (Delgado et al. 2007), and conservation (Delgado et al. 2011)
5. *Controlling pests and disease* through integrated pest management (IPM) and "pull-push strategies" (Midega et al. 2015)
6. *Migration and relocating agroecosystems* to more suitable locations (Bardsley and Hugo 2010; Farauta et al. 2012)

Agroecological strategies that serve as both climate adaptation and climate mitigation strategies include those that reduce soil erosion, reduce leaching of nitrogen and phosphorus, conserve soil moisture, modify the microclimate to reduce temperature extremes, avoid cultivation of new land, and sequester soil carbon. Agricultural practices that increase the amount of carbon sequestered in the soil include agroforestry, tree planting, preventing deforestation, reduced or zero tillage, cover cropping, mulching, sustainable forest management, and various soil and water conservation structures (Ahmed and Stepp 2016).

2.2 Agricultural Diversification

Agricultural diversification (Schwendenmann et al. 2010; Vandermeer et al. 1998; Lin 2011; Yachi and Loreau 1999; Altieri 1999) has been identified as a key strategy for building resilience (Altieri et al. 2015; Ahmed and Stepp 2016). Central to agricultural diversification is enhancing biodiversity at the landscape, species, and genetic levels in order to maintain ecosystem services including nutrient cycling, soil protection, pollination, flood control, and genetic resources (Magurran 2003) while strengthening resilience in responding to natural and anthropogenic change (Gunderson and Holling 2002). Biodiversity refers to the variation among living organisms and their environment (Magurran 2003); it is often operationalized as species richness (the total number of species in a defined space at a given time) and species abundance (the proportion of individuals of each species in relation to the total population size) (Ahmed et al. 2015). Agricultural biodiversity is defined by the Convention of Biodiversity (2000) as "all components of biological diversity of relevance to food and agriculture, and all components of biological diversity that constitute the agricultural ecosystems: the variety and variability of animals, plants, and microorganisms, at the genetic, species, and ecosystem levels, which are necessary to sustain key functions of the agro-ecosystem, its structure and processes."

Surveys with tea farmers in China's southern Yunnan indicate that multiple agricultural diversification practices that enhance or maintain agricultural biodiversity also serve to reduce the vulnerability of tea systems to climate risk (Fig. 15.1; Ahmed et al. 2014a, b). These agricultural diversification practices include growing tea as part of agroforests and diversified cropping systems, cultivating tea from diverse seed, maintaining trees within tea farms, and managing diverse forest buffers around tea farms (Ahmed et al. 2014a, b). Tea agroforests are agroecosystems that have been created by either thinning forests for the production of tea or where fields such as old swidden plots have undergone succession either unintentionally or by intentional human management (Ahmed et al. 2010, 2013). Farmers often prune tea plants in agroforests to spread out their branch formation and reduce their height for ease of harvest (Ahmed et al. 2010). Agroforests resemble forest systems in their multistoried vegetative structure with a high canopy layer, mid-level tree layer of tea plants, and herbaceous ground layer (Long and Wang 1996). The multistoried vegetative structure and biodiverse species composition of tea agroforests provide multiple ecosystem services of a forest system including pest and disease control, modifying microclimate, windbreak, soil fertility, prevention of landslides, and provision of other useful plants such as for food, medicine, ritual, and construction (Ahmed et al. 2013). Tea agroforests do not require agrochemical input for production.

Fig. 15.1 Diversified enrichment planting of trees within tea farms is a key good agricultural practice (GAP) that serves for both climate mitigation and climate adaptation. Planting trees in tea farms helps mitigate climate change by building soil organic matter and increasing the amount of carbon sequestered and stored in the soil and biomass vegetation. Trees in tea farms help with climate adaptation by buffering climate variability and extreme events, diversifying farming systems both ecologically and economically, reducing soil and water erosion, improving water management, enhancing biodiversity, reducing variability of crop output, reducing heat and excessive light radiation, facilitating water cycles through water infiltration and groundwater recharge, and preventing high evapotranspiration and water stress. Research and development is needed for this GAP to be replicable for effective implementation in diverse contexts. In addition, policy and donor support is needed to help tea producers overcome barriers to adopting this GAP for climate mitigation and adaptation as well as to popularize these innovations

Good agricultural practices (GAPs) for climate mitigation and adaptation within tea agroforests include the following (Ahmed et al. 2010, 2013, 2015):

1. *Careful site selection of tea farms* based on soil type, slope, microclimate, and elevation
2. *Diversified enrichment planting of trees* (Fig. 15.1) with different functional classes, ages, and sizes that contribute to diversified vegetative structure, composition, and seasonal availability of plant products with various cultural uses (such as fruit trees, medicinal plants, sacred trees, etc.) that support livelihoods and well-being
3. *Creation and preservation of forested edges around tea farms* between tea fields to build soils, serve as windbreaks and buffers, as well as prevent pests from spreading

4. *Maintenance of an herbaceous understory in tea farms* including plants that serve as weed suppressors, soil enrichers, disease suppressors, and pollinator attractors
5. *Experimentation and conservation of genetically diverse tea germplasm in tea farms* with varied morphological and phytochemical traits including seeds and seedlings procured from wild forest populations and seed exchange networks
6. *Planned spacing between tea plants and other plants* to ensure individuals are meeting nutrient needs without the need for fertilizers while having healthy competition
7. *Tolerance of specific pests and epiphytes* on tea plants with selective manual removal
8. *Manual fertilizing through mulching* by turning over the weedy understory at the base of tea trees and burying of broken branches to promote decomposition and soil fertility
9. *Shade management* of canopy to reduce bare soil and associated evaporation of soil moisture while improving crop water use efficiency and reducing heat and excessive light

Agroforestry has been identified as beneficial for climate adaptation through buffering climate variability and extreme events, diversifying farming systems both ecologically and economically, reducing soil and water erosion, improving water management, enhancing biodiversity, capturing soil carbon, and reducing variability of crop output (Ajayi et al. 2003; Franzel and Scherr 2002). Shade is recognized as essential on tea farms for reducing heat and excessive light radiation (Mukhopadhyay and Mondal 2017). Tree cover plays an important role in water cycles through water infiltration and groundwater recharge. Windbreaks in and around tea farms are essential to prevent high evapotranspiration and water stress that can inflict unprotected tea plants (Lemmesa 1996). Shade trees in tea farms can be classified as two types: permanent and temporary. Permanent shade trees are planted for the long term, while temporary shade trees are generally fast-growing trees that are planted in order to support the establishment of tea plants and protect the tea plants from direct light. Once the permanent shade trees become established, the temporary shade trees are removed (Mukhopadhyay and Mondal 2017). Popular trees for shade and windbreaks include leguminous trees which also serve a nitrogen-fixing role in enhancing nitrogen levels and fertility of soils (Mukhopadhyay and Mondal 2017). However, it is important to use multiple shade tree species on tea farms in order to prevent damage from epidemic pests and diseases (Mukhopadhyay and Mondal 2017). Mukhopadhyay and Mondal (2017) suggest that using four species of shade trees in combination is an ideal mixture.

Furthermore, trees in and around agroforestry systems increase soil organic matter (Mukhopadhyay and Mondal 2017) and carbon storage in biomass vegetation and in soils (Dinesh et al. 2017). Agroforestry systems have been shown to accumulate between 1.1 and 28.2 t CO2 ha-1 yr-1 in biomass and between 3.7 and 27.3 t CO2 ha-1 yr-1 in soils (Dinesh et al. 2017). In addition, the diversity of trees and other plant species in agroforests may contribute to household livelihoods and food security through the provisioning of diverse fruits, fodder, fuel, building materials, and firewood (Ajayi et al. 2003; Franzel et al. 2004; Ahmed et al. 2013). Concurrently, trees within agroforest also contribute to mitigating greenhouse gas emissions through carbon sequestration both above and below ground (Verchot et al. 2007).

While agroforestry has been demonstrated to be successful in certain contexts, further research and development is called for regarding the suitability and replicability of its implementation and effectiveness in varied contexts, including those where agroforestry has not previously been practiced. This includes determining which specific trees to plant within agroforests in different locations as well as modifying agroforestry schemes based on location. Such research and development efforts should focus on determining the cost-effectiveness of various agroforestry schemes as well as barriers and opportunities that producers face for adopting these strategies. In addition, policy support is needed to help popularize this management practice and to facilitate building capacity of smallholder farmers in adopting this innovation.

2.3 *Soil Management*

Since tea is a perennial crop, management of soil on tea farms is of great importance for productivity and quality. Managing for rich soil organic matter is central to creating healthy soils with active biological activity as well as suitable physical and chemical characteristics that can support agriculture (Altieri et al. 2015). Rich soil organic matter helps to improve the soil's water retention capacity and thereby improves infiltration and enhances drought tolerance of crops. This prevents runoff of soil particles under intense rains. In addition, soil organic matters improve surface soil aggregation by tightly holding soil particles during rain or windstorms (Altieri et al. 2015). Soils with rich soil organic matter typically contain symbiotic mycorrhizal fungi, such as arbuscular mycorrhizal fungi, that form a key component of the microbial populations that influence plant growth and soil productivity (Altieri et al. 2015). For example, arbuscular mycorrhizal fungi are known to improve plant-water relations and thereby increase the drought resistance of host plants (Garg and Chandel 2010).

Key agroecological management practices for building rich soil organic matter include cover cropping, mulching, no-tillage (Fuhrer and Chervet 2015; Lal 2004), mixed crop-livestock systems (Thornton and Herrero 2014), organic agriculture (Lotter 2003), manure management, and integrated nutrient management (INM). Several of these practices are recognized to conserve soil, improve soil ecology, stabilize and enhance crop yield, and conserve water (Altieri et al. 2015). For example, mulching disrupts the soil drying process by protecting the soil surface with residues that reduce wind speed by up to 99 % and thereby reduce losses due to evaporation (Altieri et al. 2015). Likewise, cover cropping in agricultural systems can improve water penetration and decrease water runoff losses by two- to sixfold (Altieri et al. 2015).

Integrated nutrient management is a practice that focuses on building soil health and reducing nutrient loss through application of organic matter, inorganic, and biofertilizers to soils (Gruhn et al. 2000). Organic matter added to the soil includes compost, vermicompost, and farm waste. Mixed cropped systems with the inclusion of legume crops also support the building of soil organic matter. Biofertilizers that can be applied for nutrient management of tea systems include azolla, blue green algae, and rhizobium (Adesemoye et al. 2008). Previous research has demonstrated that biofertilizers and microbial inoculants that increase plant growth and yield can also enhance nutrient uptake of plants as part of an integrated nutrient management system (Adesemoye et al. 2008).

2.4 Water Management

With shifts in the hydrological cycle with climate change that are resulting in increased rainfall intensities, rainfall variability, changes in the timing of precipitation, and heightened extreme events including drought and flooding, managing for water in tea farms is crucial. These climate strategies are exacerbating other sustainability challenges associated with water in tea systems including the detrimental impacts of high-intensity industrial agriculture. High-intensity industrial agriculture can deplete freshwater resources through intense irrigation that results in surface- and groundwater runoff as well as water wastage. At the same time, rainfall is generally concentrated within a few months of the year with large volumes of water lost through surface runoff, soil evaporation, and deep percolation (Altieri et al. 2015). This results in a water management challenge regarding how to effectively capture water during rainfall periods and store it for crop use during times of water scarcity such as during droughts or dry spells (Altieri et al. 2015).

Water management of tea is especially needed where the distribution of rainfall is uneven and scant (Mukhopadhyay and Mondal 2017) or where rainfall is increasingly extreme. Sprinklers and drip irrigation systems are widely used in many tea production areas with scant rainfall (Mukhopadhyay and Mondal 2017). Tea production areas that are vulnerable to shifts in precipitation can benefit from multiple rainwater harvesting and floodwater harvesting techniques including the creation of structures for water and nutrient catchment (Reji et al. 2013; Barrow 1999). Other on-farm water management practices for enhancing the resilience of tea farms to climate change include precision agricultural techniques for monitoring of crop and soil moisture status and implementing context-specific schedules and strategies for natural irrigation such as planting of nurse plant guilds that help conserve water resources (Ahmed et al. 2015).

Tea production areas that are experiencing increased rainfall can benefit from drainage strategies as tea plants are vulnerable to stagnant water and cannot thrive in areas with waterlogging (Mukhopadhyay and Mondal 2017). Sound drainage plans on tea farms include removal of excess moisture near the root zone of tea plants within a precise interval without causing soil erosion and with having an efficient drainage outlet (Mukhopadhyay and Mondal 2017). Previous research has documented that proper drainage systems can increase tea yield between 30% and 35% (Deka et al. 2006).

2.5 Pest and Disease Management

Tea farms attract a host of pests and disease that can be detrimental for crop loss. The prevalence of pests and disease is being impacted by changes in temperature and precipitation that are becoming more variable with climate change. The major pests for tea can be classified as chewing pests, sucking pests, and mite pests (Mukhopadhyay and Mondal 2017). Chewing pests include bunch caterpillar (*Andraca bipunctata*), looper caterpillar (*Buzura suppressaria*), red slug caterpillar (*Eterusia magnifica*), and flush worm (*Laspeyresia leucostoma*). Sucking pests include tea mosquito bug (*Helopeltis theivora*), jassid (*Empoasca flavescens*), and thrips (*Scirtothrips dorsalis*). Mite pests include purple mite (*Calacarus carinatus*), pink mite (*Acaphylla theae*), scarlet mite (*Brevipalpus phoenicis*), and red spider mite (*Oligonychus coffeae*). In addition, major diseases of tea include blister blight (caused by the *Exobasidium vexans* pathogen), red rust (caused by the *Cephaleuros parasiticus* algae), and black rot (caused by two fungi, *Corticium theae* and *Corticium invisum*; Mukhopadhyay and Mondal 2017).

Agroecological strategies for pest and disease management are increasingly used due to concerns of pesticide residues that may render tea inappropriate for consumption, environmental externalities of the use of pesticides, as well as the rising costs of the pesticides (Mukhopadhyay and Mondal 2017). Agroecological strategies for pest and disease management include on-farm diversification, intercropping, biological control through enhancing beneficial insects on farms, managing soils for minimizing crop pests, and integrated pest management (IPM). Numerous research highlights that increasing on-farm biodiversity can beneficially contribute to the control of pests and disease (Altieri and Nicholls 2005). For example, multiple successful on-farm trials have shown the benefits of intercropping to enhance biodiversity while serving to reduce pests (Altieri and Nicholls 2005). IPM includes strategies for monitoring pests for early detection, introduction of predators, manual control, use of biological pesticides, and discretion on the choice of pesticides to be used (Mukhopadhyay and Mondal 2017).

2.6 Agroecological Strategies for Enhancing Insect Pollinators

Some insects play a critical role for pollination in tea production systems. Pollinators including bees, butterflies, beetles, moths, wasps, and flies contribute to the productivity of hundreds of food crops globally and allow the majority of flowering plants to reproduce (Goulson 2003). Unfortunately, many modern agricultural practices result in farms being poor habitats for wild bees and other pollinators with the use of pesticides and fungicides that are decimating bee populations worldwide (Kevan 1999). Strategies for enhancing pollination services include maintaining uncultivated land along farm margins, managing wildflower growth within farms and farm borders, cultivating specific plants to maximize pollinator diversity, and intercropping a diversity of pollinator species (Altieri and Nicholls 2005).

3 Physiological and Molecular Strategies for Climate Resilience

New physiological and molecular technologies are being developed in response to climate change that improve crop yield and reduce pest and disease damage while reducing the need for land, fertilizer, and water input. Physiological approaches seek to understand and modify the response of tea plants to various abiotic and biotic stressors. Molecular approaches involve utilization of high-throughput sequencing, bioinformatics, DNA-based markers, linkage maps, differential gene expression, and molecular cloning to improve the response of tea plants to various stressors. Key physiological and molecular innovations for overcoming the impacts of climate change on agricultural systems include:

1. *Cultivation of tea from seed* versus clonal propagules and grafting.
2. *Micropropagation and somatic embryogenesis.*
3. *Traditional breeding methods of climate-smart cultivars* including those with enhanced resistance to climate variability such as extreme drought, early-maturing varieties, and biofortified crops that accumulate more minerals and vitamins.
4. *Genetically modified crops* including transgenic biofortified crops.
5. *Sustainable intensification* includes the coupling of conventional plant breeding with plant biotechnology (Timmer 2003; Christou and Twyman 2004).

3.1 Cultivation of Tea from Seed

The selection of planting tea from seed versus from clonal propagules via grafting impacts the drought and overall stress tolerance of tea plants. While tea plants cultivated from seeds generally give lower yields, they also require less fertilization and produce a heterogeneous stand and product compared to plants propagated by vegetative cuttings (Ahmed and Stepp 2013). At the same time, seed propagation also results in tea plants with an extensive taproot that helps protect soils, prevent erosion, and enhance fertility. This propagation method has been adapted as an alternative method to enhance fertility, along with other management practices such as intercropping, alley cropping, and organic agriculture (Ahmed and Stepp 2013). In contrast, clonal propagules generally require greater agrochemical inputs that are associated with greenhouse gas emissions and have narrower genetic diversity compared to tea plants grown from seed (Ahmed and Stepp 2013). Some agroclimatic conditions may favor clonal propagules, while other environmental conditions result in greater vulnerability for clonal propagules compared to seed plants (Williges 2004).

3.2 *Traditional Breeding Methods with Modern Science*

Varietal selection, germplasm diversification, and breeding of climate-smart cultivars are key physiological strategies for enhancing on-farm climate resilience (Kurukulasuriya and Mendelsohn 2008; Moniruzzaman 2015; Bradshaw et al. 2004; Seo and Mendelsohn 2008). Tea varieties are required in different agroclimatic zones which are capable of surviving in a variety of rapidly changing and extreme environmental conditions including heat, drought, and other environmental stresses. For example, in tea-producing areas where droughts are becoming more prevalent and/or extreme, the planting of drought-tolerant cultivar is a key adaptation strategy. In order to breed climate-smart cultivars, basic and applied science is called for on tea ecophysiology in response to climate change across all tea-producing countries. These efforts should take a long-term collaborative approach research between countries to document how climate change is impacting various components on tea agro-ecosystems including productivity, quality, soil, pests and disease, and socioeconomic implications. In addition, these efforts should take an evidence-based approach to document how climate-smart tea cultivars and other climate innovations are responding to climate change. Implementation of a shared "big data" bank approach will facilitate these efforts.

Traditional breeding methods coupled with modern plant science are being used to develop tea cultivars that are more pest and disease resistant, have specific flavor and phytochemical profiles, and are better able to respond to climate events including extreme droughts, high temperatures, and high-salinity environments (Hefferon 2011). Crop varieties that have traits of pest and disease resistance can reduce carbon emissions by decreasing demand for pesticides and the number of in-field applications (Lybbert and Sumner 2012). Breeding practices typically focus on the selection of specific genotypes and phenotypes as well as hybridization of desired traits (Ahmed et al. 2015) such as early-maturing varieties that allow cropping calendars to be adjusted to cope with seasonally unfavorable conditions (Wassmann et al. 2009).

The development of new tea varieties would widen the genetic diversity base of tea; it would allow tea producers to enhance the genetic diversity of tea plants in their tea fields toward reducing the vulnerability of tea plants to various shocks.

Modern plant science has enabled researchers to develop new strategies to search for and identify traits which could help crops withstand extreme environmental conditions. For example, molecular science methods can help understand the transcriptional changes induced by environmental stressors in the identification of signaling proteins and transcription factors which regulate the stress-induced gene expression (Bhardwaj and Yadav 2001). Transcription factors are regulatory proteins that modulate gene expression through interactions and are active in plant responses to environmental stress. The examination of the genetic material from crop wild relatives of crops using molecular biology techniques has enabled the identification of genes which have been lost over the course of crop evolution (Hefferon 2011). In addition, molecular biology techniques can also enable identification of genes linked to chemical metabolites in plants that support defense and resistance (Bhardwaj and Yadav 2001). Retrieval and reintegration of this crop genetic material from wild relatives can strengthen the genetic base of tea systems including genes that are better able to respond to climate change. A key component of enhancing the genetic base of crops is efforts that encourage more open international exchange of germplasm (Atlin et al. 2017; Grüneberg et al. 2015).

A major recent molecular advancement supporting tea systems is the sequencing of the complete tea genome by researchers at the Kunming Institute of Botany at the Chinese Academy of Sciences (Xia et al. 2017). The researchers sequenced the tea tree genome (cultivar Yunkang 10) from Yunnan Province, China, and found a total of 714 gene clusters containing 2170 genes unique to the tea tree, potentially related to environmental adaptation and phytochemical properties within the tea lineage (Xia et al. 2017). Findings from this genome sequencing effort revealed that the tea plant has a larger genome compared to coffee, pepper, potato, and tomato. In addition, findings revealed that the tea plant harbors the highest content of repetitive DNA compared to these other plants (Xia et al. 2017). Furthermore, the research team discovered lineage-specific expansions of genes associated with flavonoid metabolic biosynthesis that are recognized to enhance catechin production, terpene enzyme activation, and stress tolerance (Xia et al. 2017). The aforementioned characteristics are important features for tea flavor and adaptation to climate change. This molecular advancement is enabling the development of new tea varieties that are resistant to environmental pressures including climate variability and shifts in pests and disease while enabling the development of a more diversified set of tea flavors.

The generation of biofortified crops using either traditional breeding or genetically modified methods can result in plants that are more nutrient-rich through the accumulation of a greater quantity of vitamins and minerals. Plants which are nutrient-rich are beneficial both for the integrity of the plant and for consumers on the context of climate change. Nutrient-rich plants are better able to endure extreme environmental conditions imposed by climate change and exhibit more vigorous growth, higher yield, and greater disease resistance (Welch and Graham 2004; Bouis 2003). At the same time, biofortified crops support the diets and health of farmers and consumers. It has been argued that people with access to biofortified foods may be better prepared to withstand deleterious effects on their livelihoods due to climate change (Hotz and McClafferty 2007).

3.3 Genetically Modified Crops

Along with traditional breeding techniques, more advanced biotechnology techniques such as genetic modification are being applied to leverage existing varieties that are well suited to specific environments in the context of climate change (Lybbert and Sumner 2012) while designing new varieties with desired traits including those that make them resistant to climate change (Hefferon 2011). Genetically modified crops may involve the introduction of functional genes from related or unrelated sources (Passioura 2006). Scientists are designing transgenic crops that can thrive in poor soils, tolerate drought and heat along with other extreme conditions, accumulate minerals and vitamins, and extract heavy metals and pollutants from contaminated soils. The cultivation of transgenic crops is recognized by some scientists to provide more arable land and support crop value chains and food security in the context of climate change (Hefferon 2011). Genetically modified crops have been shown to produce dramatic improvements in yield and reductions in production costs (Lybbert and Sumner 2012). However, it is important to note that such agricultural biotechnology remains controversial.

4 Technological, Monetary, and Certification Strategies for Climate Resilience

In addition to agroecological, physiological, and molecular innovations for climate resilience, new technologies are being developed that are advancing the ability of agricultural systems to respond to climate change. Technological and monetary innovations for strengthening climate resilience include (Dinesh et al. 2017):

1. *Precision irrigation/micro-irrigation powered by solar* that expands access to affordable irrigation
2. *Digital agriculture* from tailored advice to shared value with millions of farmers
3. *Climate-informed advisories* that reduce uncertainty in agricultural systems by generating, translating, and communicating locally relevant climate information to support climate-informed decision-making, policy, and planning toward enhancing production and resilience
4. *Weather index-based agricultural insurance* that provides countries and farmers with risk reduction in response to erratic weather and extreme climate events

5. *Technology for tea production that reduces carbon emissions* including inputs and machinery
6. *Climate-smart and low-carbon tea certification guidelines and schemes* that educate consumers on the benefits of reducing the carbon footprint of tea throughout the value chain (from farm to cup to waste)

5 Overcoming Costs and Barriers to Adopting Climate-Resilient Tea Practices

While various good agricultural practices (GAPs) for climate mitigation and adaptation have been identified through science and practice, there remain major barriers for the adoption of these practices. The literature identifies five broad categories of costs and barriers associated with the adoption climate mitigation and adaptation practices including the following: (1) investment costs, (2) maintenance costs, (3) opportunity costs, (4) transaction costs, and (5) risk costs. For example, while many climate resilience strategies such as agricultural diversification are relatively affordable because they rely on mimicking natural systems for supporting ecosystem services rather than expensive external inputs (Lin 2011), smallholder tea farmers may still face severe constraints to investing in good agricultural practices (GAPs) for climate mitigation and adaptation. These constraints include the inability to tolerate additional risk of changing management strategies or waiting to receive economic and ecological benefits from more long-term strategies such as agroforestry, which may range from 3 to 10 years (depending on site conditions, how the system was created, and the specific species within the systems).

Further research and development is needed at the farm level on the cost-effectiveness, replicability, and adaptability of various GAPs for climate mitigation and adaptation practices in different contexts as one approach to overcome barriers for the adoption of climate innovations. In addition, policy and donor support is needed to build capacity of tea producers in overcoming barriers to adopting climate-resilient tea systems as well as for popularizing tried and tested GAPs. This calls for a multi-sectoral approach with collaboration between governments, industry, smallholder farmers, and researchers. There are several examples of policy, donor, and NGO support working together to implement climate resilience programs in tea systems. The following is an example of a case study from Kenya that demonstrates how multiple sectors have worked together to tackle the issue of climate change in the tea system.

5.1 Multi-sectoral Case Study in Kenya

The collaborative climate change adaptation and climate mitigation program of the Kenya Tea Development Agency Management Services (KTDA-MS), the Ethical Tea Partnership, the German Development Agency GIZ, the Kenyan Institute of Organic Farming, and the Kenyan Forestry Research Institute provide a promising example of multi-sectoral collaboration of how coupled policy, donor, and NGO support can help smallholder farmers overcome costs and barriers for adopting climate-resilient tea practices. These organizations launched a program that provided training on sustainable agricultural practices for climate change adaption to over hundreds of thousands of tea growers in Kenya. The sustainable agricultural practices that were promoted seek to hedge against fluctuations in the supply of tea and associated tea prices and farmer livelihoods. KTDA disseminated the climate change adaption training to the smallholder communities through lead farmers, demonstration plots, manuals, and poster. As part of this effort, a comprehensive manual on climate change adaptation techniques was developed as well as another manual on climate change mitigation techniques with the goal to help tea producers overcome knowledge and skill gaps to build resilience of tea systems in response to climate change.

The training that the Kenya collaborative of agencies provided smallholder farmers includes the following topics:

1. *Soil, water, and bush management* including creating bench terraces, water harvesting and retention through troughs, and drip irrigation.
2. *Composting* kitchen and livestock waste and applying it as a soil fertility amendment.
3. *Mulching* through the application of tea prunings as a mulch.
4. *The use of shade trees*. A key component of the training was on the use of shade trees and which varieties are most suitable to use in specific tea-growing areas in different agroclimatic zones. Tree planting within and around tea fields both helps with sequestering carbon and serves as a fuel source for tea processing in a closed-loop production system.
5. *Conservation and crop diversification* including planting trees to reduce water stress on tea plants.
6. *Pest management* through partitioning farms into plots with napier grass as a pull-push strategy to attract insect pests away from tea crops.
7. *Access to drought- and frost-resistant tea clones.*
8. *Fuel wood conservation.*
9. *Access to energy-efficient stoves.*

The climate change adaptation manual has been adjusted for Malawi by the Tea Research Foundation of Central Africa in order to fit the local tea production context. The climate mitigation manual provides a set of tools and resources to support the tea sector on how to reduce energy use and carbon emissions in growing and processing tea. As part of this effort, the KTDA-MS helped provide affordable finance to smallholder tea farmers in order to increase their ability to adopt climate-resilient practices. Furthermore, the KTDA-MS led an agreement with the Tea Research Foundation of Kenya and other organizations to increase access for farmers to tea germplasm that is more tolerant to extreme weather including drought-resistant and frost-resistant varieties.

6 Conclusion

Climate change is threatening tea security by direct and indirect impacts on tea quality and productivity that ultimately impact farmers and consumers. It is thus critical to implement climate mitigation and adaptation strategies for minimizing the forecasted effects of climate change on tea systems while enhancing the resilience of tea systems. Multiple agricultural, physiological, and molecular innovations for tea at the production level have been identified toward the development of climate-resilient tea systems. The implementation of climate mitigation and adaptation actions should be coupled with monitoring strategies to evaluate their effectiveness (Dinesh et al. 2017). This will enable these strategies to be modified as needed across diverse contexts and remain relevant to changing contexts. In addition, further research and development is needed at the farm level on the cost-effectiveness, replicability, and adaptability of tea agroforestry and various GAPs for climate mitigation and adaptation practices in different contexts. These efforts should take a long-term collaborative approach research between tea-producing countries to document how climate change is impacting various components on tea agro-ecosystems including productivity, quality, soil, pests and disease, and socioeconomic implications. In addition, an evidence-based approach with a shared "big data" bank is called for to document how various climate innovations in the tea system are responding to climate change. These efforts should be coupled with policy and donor support to build capacity of tea producers to overcome barriers for adopting climate-resilient tea systems as well as to popularize tried and tested GAPs for climate mitigation and adaptation that are cost-effective and replicable. Ultimately, multi-sectoral collaboration between governments, industry, farmers, and researchers is called for to tackle the issue of climate change in the tea system toward more effectively supporting farmer livelihoods and well-being while meeting consumer demand and maintaining stable tea markets.

Acknowledgments Funding support for this research was supported by the US National Science Foundation Grants: NSF CNH BCS-1313775 and NSF RII Track-2 FEC OIA 1632810.

References

Adesemoye A, Torbert H, Kloepper J (2008) Enhanced plant nutrient use efficiency with PGPR and AMF in an integrated nutrient management system. Can J Microbiol 54(10):876–886. https://doi.org/10.1139/W08-081

Ahmed S, Stepp JR, Toleno R, Peters CM (2010) Increased market integration, value, and ecological knowledge of tea agro-forests in the akha highlands of Southwest China. Ecol Soc 15(4):27

Ahmed S, Stepp J (2016) Beyond yields: climate change effects on specialty crop quality and agroecological management. Elem Sci Anthro 4:1–16

Ahmed S, Peters CM, Chunlin L, Meyer R, Unachukwu U, Litt A et al (2013) Biodiversity and phytochemical quality in indigenous and state-supported tea management systems of Yunnan, China. Conservation Lett 5(6):28–23

Ahmed S, Orians C, Griffin T, Buckley S, Unachukwu U, Stratton A et al (2014a) Effects of water availability and pest pressures on tea (*Camellia sinensis*) growth and functional quality. AoB Plants 6

Ahmed S, Stepp J, Orians C, Griffin T, Matyas C, Robbat A et al (2014b) Effects of extreme climate events on tea (*Camellia sinensis*) functional quality validate indigenous farmer knowledge and sensory preferences in tropical China. PLoS One 9(10):E109126

Ahmed S, Stepp JR, Xue D (2015) Cultivating botanicals for sensory quality: from good agricultural practices (GAPs) to taste discernment by smallholder tea farmers. In: Reynertson K (ed) Botanicals: methods for quality and authenticity. CRC Press, Boca Raton, pp 16–29

Ajayi O, Franzel S, Kuntashula E, Kwesiga F (2003) Adoption of improved fallow technology for soil fertility management in Zambia: empirical studies and emerging issues. Agrofor Syst 59(3):317–326

Altieri MA, Nicholls CI (2005) Biodiversity and pest management in agroecosystems. Haworth Press, New York

Altieri MA (1995) Agroecology: the science of sustainable agriculture, 2nd edn. Westview Press, Boulder, CO

Altieri M (1999) The ecological role of biodiversity in agroecosystems. Agric Ecosyst Environ 74(1):19–31. https://doi.org/10.1016/S0167-8809(99)00028-6

Altieri M, Nicholls A, Henao C, Lana I (2015) Agroecology and the design of climate change-resilient farming systems. Agron Sustain Dev 35(3):869–890. https://doi.org/10.1007/s13593-015-0285-2

Atlin GN, Cairns JE, Das B (2017) Rapid breeding and varietal replacement are critical to adaptation of cropping systems in the developing world to climate change. Glob Food Sec 12:31–37

Bardsley D, Hugo K (2010) Migration and climate change: examining thresholds of change to guide effective adaptation decision-making. Popul Environ 32(2):238–262

Barrow C (1999) Alternative irrigation: the promise of runoff agriculture. Earthscan, London

Bhardwaj J, Yadav SK (2001) Genetic mechanisms of drought stress tolerance, implications of transgenic crops for agriculture. In: Lichtfouse E (ed) Agroecology and strategies for climate change, sustainable agriculture reviews. Springer, New York, pp 213–235

Bhattarai B, Beilin R, Ford R (2015) Gender, agrobiodiversity, and climate change: a study of adaptation practices in the Nepal Himalayas. World Dev 70:122–132

Boehm R, Cash S, Anderson B, Ahmed S, Griffin T, Robbat A et al (2016) Association between empirically estimated monsoon dynamics and other weather factors and historical tea yields in China: results from a yield response model. Climate 4(2):20

Bouis H (2003) Micronutrient fortification of plants through plant breeding: can it improve nutrition in man at low cost? Proc Nutr Soc 62(2):403–411

Bradshaw B, Dolan H, Smit B (2004) Farm-level adaptation to climatic variability and change: crop diversification in the Canadian prairies. Clim Chang 67(1):119–141

Convention on Biological Diversity (2000) Programme of work on agricultural biodiversity. Decision V/5 of the conference of the parties to the convention. www.fao.org/fileadmin/templates/soilbiodiversity/Downloadable_files/cs-agr-outline.pdf. Accessed 17 May 2018.
Chartzoulakis K, Bertaki M (2015) Sustainable water management in agriculture under climate change. Agri Agricultural Sci Procedia 4:88–98
Choptiany J, Graub B, Dixon J, Phillips S (2014) Self-evaluation and holistic assessment of climate resilience of farmers and pastoralists (SHARP). FAO, Rome, IT
Christou P, Twyman R (2004) The potential of genetically enhanced plants to address food insecurity. Nutr Res Rev 17(1):23–42
De Schutter O (2012) Agroecology, a tool for the realization of the right to food. In: Lichtfouse E (ed) Agroecology and strategies for climate change, sustainable agriculture reviews. Springer, Dordrecht, pp 1–6
Deka A, Deka PC, Mondal TK (2006) Tea. In: Parthasarathy VA, Chattopadhyay PK, Bose TK (eds) Plantation crops-I. Calcutta, Naya Udyog, pp 1–148
Delgado JA, Dillon MA, Sparks RT, Essah SYC (2007) A decade of advances in cover crops: cover crops with limited irrigation can increase yields, crop quality, and nutrient and water use efficiencies while protecting the environment. J Soil Water Conserv 62(5):110A–117A
Delgado JA, Groffman PM, Nearing MA, Goddard T, Reicosky D, Lal R et al (2011) Conservation practices to mitigate and adapt to climate change. J Soil Water Conserv 66(4):118A–129A
Di Falco S, Perrings C (2003) Crop genetic diversity, productivity and stability of agroecosystems. A theoretical and empirical investigation. Scottish J Pol Eco 50(2):207–216
Dinesh D, Campbell B, Bonilla-Findji O, Richards M (2017) 10 best bet innovations for adaptation in agriculture: a supplement to the UNFCCC NAP technical guidelines. CCAFS Working Paper no. 215. Wageningen, NL: CGIAR Research Program on Climate Change, Agriculture and Food Security.
Farauta BK, Egbule CL, Agwu AE, Idrisa YL, Onyekuru NA (2012) Farmers' adaptation initiatives to the impact of climate change on agriculture in northern Nigeria. J Agri Ext 16(1):132–144
Food and Agriculture Organization (2013) Resilient livelihoods- disaster risk reduction for food and nutrition security framework programme. FAO, Rome
Folke C (2006) Resilience: the emergence of a perspective for social–ecological systems analyses. Glob Environ Chang 16(3):253–267
Francis C, Lieblein G, Breland T, Salomonsson L, Geber U, Sriskandarajah N, Langer V (2008) Transdisciplinary research for a sustainable agriculture and food sector. Agron J 100(3):771–776
Franzel S, Scherr SJ (2002) Introduction. In: Franzel S, Scherr SJ (eds) Trees on the farm: assessing the adoption potential of agroforestry practices in Africa. CABI International, Wallingford, UK
Franzel S, Denning G, Lillesø J, Mercado A (2004) Scaling up the impact of agroforestry: lessons from three sites in Africa and Asia. Agrofor Syst 61(1):329–344
Fraser E (2007) Travelling in antique lands: using past famines to develop an adaptability/resilience framework to identify food systems vulnerable to climate change. Clim Chang 83(4):495–514
Fuhrer J, Chervet A (2015) No-tillage: long-term benefits for yield stability in a more variable climate? Procedia Environ Sci 29:194–195
Garg N, Chandel S (2010) Arbuscular mycorrhizal networks: process and functions. A review. Agron Sustain Dev 30(3):581–599. https://doi.org/10.1007/978-94-007-0394-0_40
Gliessman S (2015) Agroecology: the ecology of sustainable food systems. CRC Press/Taylor & Francis, Boca Raton, FL
Goulson D (2003) Conserving wild bees for crop pollination. J Food Agri Environ 1(1):142–144
Gruhn P, Goletti F, Yudelman M (2000) Integrated nutrient management, soil fertility, and sustainable agriculture: current issues and future challenges. IDEAS Working Paper Series from RePEc. https://www.econbiz.de/Record/integrated-nutrient-management-soil-fertility-and-sustainable-agriculture-current-issues-and-future-challenges-gruhn-peter/10005038272. Accessed 17 May 2018.

Grüneberg WJ, Ma D, Mwanga ROM, Carey EE, Huamani K, Diaz F et al (2015) Advances in sweet potato breeding from 1993 to 2012 potato and sweet potato in Africa: transforming the value chains for food and nutrition security. CABI International, Wallingford, UK

Gunderson, L., & Holling, C. (2002). Panarchy: Understanding transformations in human and natural systems. Washington, DC: Island Press

Hefferon KL (2011) Innovations in agricultural biotechnology in response to climate change. In: Blanco J, Kheradmand H (eds) Climate change – research and technology for adaptation and mitigation. https://doi.org/10.5772/23024

Herforth A, Ahmed S, Remans R, Fanzo J, DeClerck F (2017) Creating sustainable, resilient food Systems for healthy diets. UN Stand Comm Nut News 42:15–22

Hotz C, McClafferty B (2007) From harvest to health: challenges for developing biofortified staple foods and determining their impact on micronutrient status. Food Nutr Bull 28(2):S271–S279

Howden S, Soussana J, Tubiello F, Chhetri N, Dunlop M, Meinke H (2007) Adapting agriculture to climate change. Proc Natl Acad Sci 104(50):19691–19696

International Trade Centre (ITC) (2014) Mitigating climate change in tea sector. ITC, Geneva xvi, 102 pages (Technical paper) Doc. No. SC-14-245.E

Pachauri RK, Allen MR, Barros VR, Broome J, Cramer W, Christ R et al (2014) Contribution of working groups I, II and III to the fifth assessment report of the intergovernmental panel on climate change. In: Climate change synthesis report. IPCC, Geneva, p 151

Jiggins J (2014) Agroecology: adaptation and mitigation potential and policies for climate change. In: Freedman B (ed) Global environmental change. Springer, Dordrecht, pp 733–743

Kevan P (1999) Pollinators as bioindicators of the state of the environment: species, activity and diversity. Agric Ecosyst Environ 74(1):373–393

Kotecký V (2015) Contribution of afforestation subsidies policy to climate change adaptation in the Czech Republic. Land Use Policy 47:112–120

Kowalsick A, Kfoury N, Robbat A, Ahmed S, Orians C, Griffin T et al (2014) Metabolite profiling of *Camellia sinensis* by automated sequential, multidimensional gas chromatography/mass spectrometry reveals strong monsoon effects on tea constituents. J Chromatogr A 1370:230–239

Kurukulasuriya P, Mendelsohn R (2008) Crop switching as a strategy for adapting to climate change. African J Agri Res Econ 2(1):105–125

Lal R (2004) Soil carbon sequestration to mitigate climate change. Geoderma 123(1):1–22

Lal R, Delgado JA, Groffman PM, Millar N, Dell C (2011) Management to mitigate and adapt to climate change. J Soil Water Conserv 66(4):276–285

Lemmesa F (1996) Tea production and management. In: Lemmesa F (ed) Teaching handout. Jimma College of Agriculture, Ethiopia, pp 1–14

Li X, Zhang L, Ahammed GJ, Li ZX, Wei JP, Shen C et al (2017) Stimulation in primary and secondary metabolism by elevated carbon dioxide alters green tea quality in *Camellia sinensis L.* Sci Rep 7(1):7937. https://doi.org/10.1038/s41598-017-08465-1

Lin BB (2011) Resilience in agriculture through crop diversification: adaptive management for environmental change. Bioscience 61(3):183–193

Long C, Wang J (1996) Studies of traditional tea-garden of jinuo nationality, China. In: Jain SK (ed) Ethnobiology in human welfare. Deep Publications, New Delhi, pp 339–344

Lotter D (2003) Organic agriculture. J Sustain Agric 21(4):59–128

Lybbert T, Sumner D (2012) Agricultural technologies for climate change in developing countries: policy options for innovation and technology diffusion. Food Policy 37(1):114–123

Magurran AE (2003) Measuring biological diversity. Blackwell Publishing, Malden, MA

Medina A, Rodriguez A, Magan N (2014) Effect of climate change on *Aspergillus flavus* and aflatoxin B1 production. Front Microbiol 5:348

Midega CAO, Bruce TJA, Pickett JA, Pittchar JO, Murage A, Khan ZR (2015) Climate-adapted companion cropping increases agricultural productivity in East Africa. Field Crop Res 180:118–125

Moniruzzaman S (2015) Crop choice as climate change adaptation: evidence from Bangladesh. Ecol Econ 118:90–98

Mukhopadhyay M, Mondal TK (2017) Cultivation, improvement, and environmental impacts of tea subject. Agri Environ Online Publ. https://doi.org/10.1093/acrefore/9780199389414.013.373
Myers S, Smith M, Guth S, Golden C, Vaitla B, Mueller N et al (2017) Climate change and global food systems: potential impacts on food security and undernutrition. Annu Rev Public Health 38:259–277
Pandey DN, Gupta AK, Anderson DM (2003) Rainwater harvesting as an adaptation to climate change. Curr Sci 85(1):46–59
Passioura J (2006) Increasing crop productivity when water is scarce—from breeding to field management. Agric Water Manag 80(1):176–196
Reji C, Scoones I, Toulmin C (2013) Sustaining the soil: indigenous soil and water conservation in Africa. Routledge, London
Rees WE (2010) Thinking resilience. The post carbon reader series. Post-carbon Institute, Santa Rosa, CA
Schwendenmann L, Veldkamp E, Moser G, Hölscher D, Köhler M, Clough Y et al (2010) Effects of an experimental drought on the functioning of a cacao agroforestry system, Sulawesi, Indonesia. Glob Chang Biol 16(5):1515–1530
Seo SN, Mendelsohn R (2008) An analysis of crop choice: adapting to climate change in South American farms. Ecol Econ 67(1):109–116
Springmann M, Godfray H, Rayner M, Scarborough P (2016) Analysis and valuation of the health and climate change cobenefits of dietary change. Proc Natl Acad Sci U S A 113(15):4146–4151
Tendall DM, Joerin J, Kopainsky B, Edwards P, Shreck A, Le QB et al (2015) Food system resilience: defining the concept. Glob Food Sec 6:17–23
Thornton PK, Herrero M (2014) Climate change adaptation in mixed crop–livestock systems in developing countries. Glob Food Sec 3(2):999–107
Timmer CP (2003) Biotechnology and food systems in developing countries. Am Soc Nutrit Sci 133:3319–3322
Vandermeer J, Van Noordwijk M, Anderson J, Ong C, Perfecto I (1998) Global change and multispecies agroecosystems: concepts and issues. Agric Ecosyst Environ 67(1):1–22
Verchot L, Noordwijk V, Kandji M, Tomich S, Ong T, Albrecht C et al (2007) Climate change: linking adaptation and mitigation through agroforestry. Mitig Adapt Strateg Glob Chang 12(5):901–918
Vermeulen S, Campbell B, Ingram J (2012) Climate change and food systems. Annu Rev Environ Resour 37:195–222
Walker B, Holling C, Carpenter S, Kinzig A (2004) Resilience, adaptability and transformability in social–ecological systems. Ecol Soc 9(2):5
Wassmann R, Jagadish SVK, Heuer S, Ismail A, Redona E, Serraj R et al (2009) Climate change affecting rice production: the physiological and agronomic basis for adaptation strategies. Adv Agron 101:59–122
Welch R, Graham R (2004) Breeding for micronutrients in staple food crops from a human nutrition perspective. J Exp Bot 55(396):353–364
Wijeratne M (1996) Vulnerability of Sri Lanka tea production to global climate change. Water Air Soil Pollut 92(1):87–94
Williges U (2004) Status of organic agriculture in Sri Lanka with special emphasis on tea production systems (*Camellia sinensis L.*) O. Kuntze) (Doctoral dissertation). Retrieved from https://www.scribd.com/document/243439359/Status-of-Organic-Agriculture-in-Sri-Lanka-with-Special-Emphasis-on-Tea-Production-Systems
Wollenberg E, Richards M, Smith P, Havlík P, Obersteiner M, Tubiello F et al (2016) Reducing emissions from agriculture to meet the 2 °C target. Glob Chang Biol 22(12):3859–3864
Xia EH, Zhang HB, Sheng J, Kui L, Zhang QJ, Kim C et al (2017) The tea tree genome provides insights into tea flavor and independent evolution of caffeine biosynthesis. Mol Plant 10(6):866–877
Yachi S, Loreau M (1999) Biodiversity and ecosystem productivity in a fluctuating environment: the insurance hypothesis. Proc Natl Acad Sci U S A 96(4):1463–1468

Index

W.-Y. Han et al. (eds.), *Stress Physiology of Tea in the Face of Climate Change*,
https://doi.org/10.1007/978-981-13-2140-5

Printed by Printforce, the Netherlands